Plant Genetics: Principles and Practices

Plant Genetics: Principles and Practices

Edited by Herbert McCoy

SYRAWOOD
PUBLISHING HOUSE

New York

Published by Syrawood Publishing House,
750 Third Avenue, 9th Floor,
New York, NY 10017, USA
www.syrawoodpublishinghouse.com

Plant Genetics: Principles and Practices
Edited by Herbert McCoy

International Standard Book Number: 978-1-68286-786-0 (Hardback)

Cataloging-in-Publication Data

Plant genetics : principles and practices / edited by Herbert McCoy.
    p. cm.
Includes bibliographical references and index.
ISBN 978-1-68286-786-0
1. Plant genetics. 2. Plant genetics--Technique. 3. Genetics. I. McCoy, Herbert.
QK981 .P53 2019
572.86--dc23

TABLE OF CONTENTS

PREFACE

Plant genetics is the science of studying genes, variation and heredity in plants. Plants display distinct characteristics such as their ability to self-fertilize and adaptability to polyploidy. The study of these aspects of plants has an economic significance. Staple crops can be engineered to develop traits of increased yields, better nutritional value, disease resistance, etc. The genetic engineering of plants is possible with the use of techniques such as, electroporation, microinjection, gene guns and agrobacterium. Introducing a new gene into a plant involves the presence of a promoter in the area where the new gene is required to be introduced. Such genetic manipulation can be transgenic, cisgenic or subgenic. This book contains some path-breaking studies in the area of plant genetics. It outlines the principles and practices of genetic engineering of plants in detail. This book is a complete source of knowledge on the present status of this important field.

After months of intensive research and writing, this book is the end result of all who devoted their time and efforts in the initiation and progress of this book. It will surely be a source of reference in enhancing the required knowledge of the new developments in the area. During the course of developing this book, certain measures such as accuracy, authenticity and research focused analytical studies were given preference in order to produce a comprehensive book in the area of study.

This book would not have been possible without the efforts of the authors and the publisher. I extend my sincere thanks to them. Secondly, I express my gratitude to my family and well-wishers. And most importantly, I thank my students for constantly expressing their willingness and curiosity in enhancing their knowledge in the field, which encourages me to take up further research projects for the advancement of the area.

Editor

Trait stacking via targeted genome editing

William M. Ainley[1,†], Lakshmi Sastry-Dent[1,†], Mary E. Welter[1], Michael G. Murray[1], Bryan Zeitler[2], Rainier Amora[2], David R. Corbin[1], Rebecca R. Miles[1], Nicole L. Arnold[1], Tonya L. Strange[1], Matthew A. Simpson[1], Zehui Cao[1], Carley Carroll[1], Katherine S. Pawelczak[1], Ryan Blue[1], Kim West[1], Lynn M. Rowland[1], Douglas Perkins[1], Pon Samuel[1], Cristie M. Dewes[1], Liu Shen[1], Shreedharan Sriram[1], Steven L. Evans[1], Edward J. Rebar[2], Lei Zhang[2], Phillip D. Gregory[2], Fyodor D. Urnov[2], Steven R. Webb[1] and Joseph F. Petolino[1,*]

[1]Dow AgroSciences LLC, Indianapolis, IN, USA
[2]Sangamo BioSciences, Inc., Richmond, CA, USA

*Correspondence

email jfpetolino@dow.com
†Both authors contributed equally to this work.

Keywords: gene targeting, designed zinc finger nucleases, transgene stacking.

Summary

Modern agriculture demands crops carrying multiple traits. The current paradigm of randomly integrating and sorting independently segregating transgenes creates severe downstream breeding challenges. A versatile, generally applicable solution is hereby provided: the combination of high-efficiency targeted genome editing driven by engineered zinc finger nucleases (ZFNs) with modular 'trait landing pads' (TLPs) that allow 'mix-and-match', on-demand transgene integration and trait stacking in crop plants. We illustrate the utility of nuclease-driven TLP technology by applying it to the stacking of herbicide resistance traits. We first integrated into the maize genome an herbicide resistance gene, *pat*, flanked with a TLP (ZFN target sites and sequences homologous to incoming DNA) using WHISKERS™-mediated transformation of embryogenic suspension cultures. We established a method for targeted transgene integration based on microparticle bombardment of immature embryos and used it to deliver a second trait precisely into the TLP via cotransformation with a donor DNA containing a second herbicide resistance gene, *aad1*, flanked by sequences homologous to the integrated TLP along with a corresponding ZFN expression construct. Remarkably, up to 5% of the embryo-derived transgenic events integrated the *aad1* transgene precisely at the TLP, that is, directly adjacent to the *pat* transgene. Importantly and consistent with the juxtaposition achieved via nuclease-driven TLP technology, both herbicide resistance traits cosegregated in subsequent generations, thereby demonstrating linkage of the two independently transformed transgenes. Because ZFN-mediated targeted transgene integration is becoming applicable across an increasing number of crop species, this work exemplifies a simple, facile and rapid approach to trait stacking.

Introduction

The stacking of traits into a single product for enhanced and sustainable crop performance requires the introduction and expression of multiple transgenes. Different approaches for transgene stacking have been used including crossing of independently generated transgenic plants (Bizily et al., 2000; Cao et al., 2002; Datta et al., 2002) and sequential transformation of previously transformed material (Ntui et al., 2011; Qi et al., 2004; Ramana Rao et al., 2011). The main issue with these approaches is the independent segregation of randomly integrated transgenes and the concomitant breeding complexity involved with their subsequent introgression. Transformation with multigene constructs, that is, molecular stacks, results in transgenes at a single locus (Lin et al., 2003; Naqvi et al., 2010), but challenges associated with building and effectively delivering such large DNA sequences may limit this approach (Dafny-Yelin and Tzfira, 2007; Halpin, 2005).

The problem of random integration of independent transgenes in sequentially transformed plant material could be addressed with site-specific DNA integration technology. Site-specific recombinases have been used to mediate integration of transgenes into pre-integrated loci (Day et al., 2000; Fladung and Becker, 2010; Li et al., 2010; Srivastava and Ow, 2002; Vergunst

et al., 1998). Similarly, double-strand break repair via homologous recombination at pre-integrated meganuclease cleavage sites also allows for targeted DNA integration for subsequent transgene stacking (D'Halluin et al., 2008, 2013; Puchta et al., 1996). These technologies can enable transgenes to be targeted to pre-integrated sequences, which then segregate as single loci making downstream breeding less cumbersome. However, the number of sequences available to use as targeted integration sites, that is, for recombination or cleavage, is somewhat limited with these systems. Nonetheless, targeting a native, endogenous genomic loci for trait stacking has recently been achieved in cotton using re-engineered meganucleases (D'Halluin et al., 2013).

One of the most exciting recent developments in genome engineering is the ability to design nucleases with investigator-determined recognition sequence specificity (Urnov et al., 2010). Modular Cys2-His2 zinc finger protein domains can be engineered and combined to recognize and bind any given stretch of DNA sequence (Pabo et al., 2001). When fused to sequence-independent nuclease domain monomers from the bacterial type IIS restriction endonuclease FokI (Kim et al., 1996), zinc finger nucleases (ZFNs) can be designed to bind and cleave the targeted DNA sequence upon Fok1 dimerization at the recognition site (Urnov et al., 2010). ZFNs have been shown to create double-strand DNA breaks at endogenous plant loci of maize (Shukla *et al.*, 2009) and tobacco (Townsend *et al.*, 2009) as well as pre-integrated sequences in tobacco (Cai *et al.*, 2009; Wright *et al.*, 2005) and Arabidopsis (de Pater *et al.*, 2009), thereby facilitating homology-directed repair and concomitant targeted transgene integration. Thus, the use of designed nucleases for site-specific DNA integration could allow for sequential transgene stacking at specified endogenous genomic loci or pre-integrated sites (Figure 1).

The present study was motivated by a desire to practice in maize a general method of transgene stacking into genomic loci using homology-directed repair of pre-integrated 'trait landing pads' (TLPs). Biochemical, bioinformatic, functional and proxy system screening criteria were used to identify a suite of ZFNs for use in the context of the TLPs. Each TLP was comprised of ZFN recognition sites flanked by unique sequences otherwise absent in the maize genome and capable of providing homology to incoming DNA (Figures 1 and 2a). Two herbicide resistance genes were chosen for sequential stacking: *pat* encoding phosphinothricin acetyltransferase (Wohlleben *et al.*, 1988) and *aad1* encoding aryloxyalkanoate dioxygenase (Wright *et al.*, 2010), which impart resistance to the herbicides Ignite® and Assure® II, respectively. Random integration of the first transgene, *pat*, flanked by a TLP followed by cotransformation of the second transgene, *aad1*, flanked by sequences homologous to the integrated TLP along with a corresponding ZFN resulted in stacking of the two herbicide resistance traits via site-specific integration into the TLP.

Results

Design and validation of a TLP platform

To develop a robust transgene stacking platform, a suite of ZFNs and corresponding recognition sequences was required. Four pre-existing ZFNs were used to create a candidate panel of novel ZFN recognition sequence combinations for screening. The selected ZFNs targeted CCR5 (Perez *et al.*, 2008), AAVS1 (DeKelver *et al.*, 2010), Rosa26 (Orlando *et al.*, 2010) and Prmt1 (not previously published), thus offering the potential for up to 32 novel ZFN recognition sequence combinations, that is, targetable sites. These ZFNs were chosen based on high activity at their intended endogenous target sites in the human genome and on the absence of their target sequence in the maize genome. Each of these ZFNs, when transfected into human (CCR5, AAVS1 and Prmt1) or mouse (Rosa26) cells, reproducibly yielded >40% single-step targeted genome editing.

A budding yeast system was used to determine whether the initial ZFN panel was robust with respect to double-strand break induction. In this study, a highly active ZFN engineered to cut the human *CCR5* locus was used as a reference (Perez *et al.*, 2008), because this particular ZFN pair had previously been successfully expressed in plants to drive transgene excision (Petolino *et al.*, 2010). Novel combinations of monomers from the four ZFNs were generated, resulting in a matrix of 'mix-and-match' ZFN proteins (Table S1). Each ZFN pair was evaluated for *in vivo* compatibility with respect to cleavage activity, and four ZFN pairs were chosen

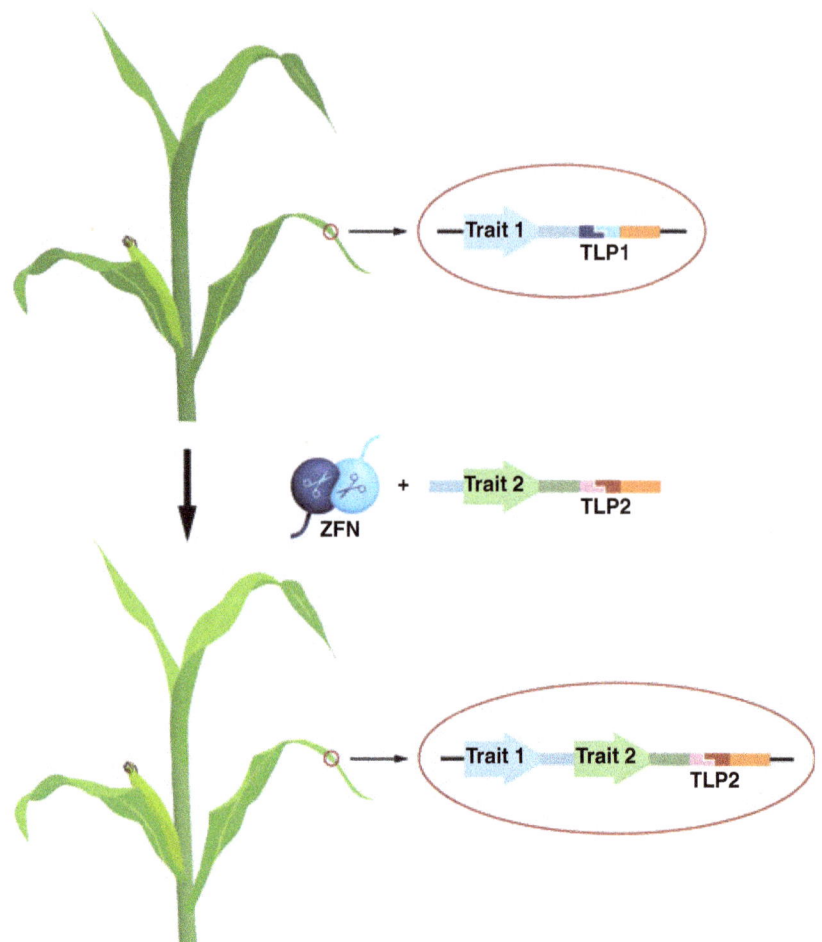

Figure 1 Sequential transgene stacking using trait landing pads (TLPs). Trait 1 (indicated as a blue arrow) is flanked by TLP1 comprising a ZFN target site (blue stepped box) and sequences homologous to an incoming donor construct containing trait 2 (shown in blue and pink). Cotransformation of trait 2 donor DNA and ZFN-1 results in targeted addition of Trait 2 into TLP1— thus stacking traits 1 and 2. Trait 2 donor DNA is engineered to contain a novel TLP2 site for additional stacking using a distinct (ZFN-2) and distinct homologous donor DNA.

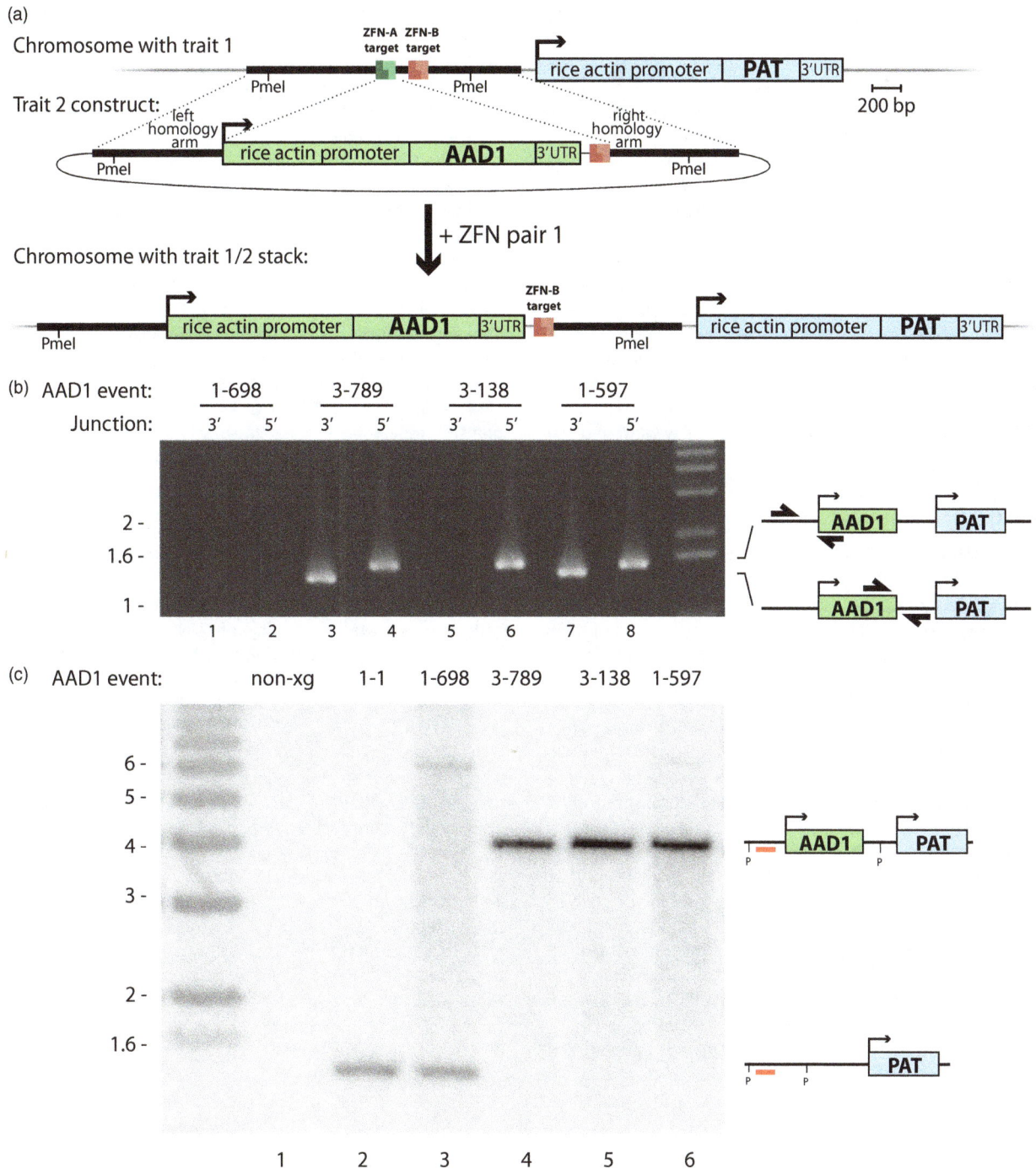

Figure 2 Transgene targeting strategy and analysis. (a) (top) Integrated TLP construct with trait 1 (*PAT*). Zinc finger pairs are comprised of ZFN-A and ZFN-B: TLP-1 = ZFN-1 and ZFN-3; and TLP-2 = ZFN-2 and ZFN-5. *Pme1* = restriction enzyme recognition sequence for Southern blot analysis liberates a 1.5-kb digestion product. (Middle) Donor DNA construct with trait 2 (*AAD1*). (Bottom) Predicted targeted sequence following donor DNA integration into TLP via homology-directed repair. *Pme1* = restriction enzyme recognition sequence for Southern blot analysis liberates a 4.2-kb digestion product. *PAT* = phosphinothricin acetyltransferase and *AAD-1* = aryloxyalkanoate dioxygenase-1. (b) PCR analysis. 3′ and 5′ 'in–out' PCR results. Events AAD-1/TLP-1-1-597 and AAD-1/TLP-1-3-789 amplified both the 1.3-kb and 1.5-kb product indicative of 3′ and 5′ donor integration, respectively. Event AAD-1/TLP-1-3-138 amplified only the 5′ product. Event AAD-1/TLP-1-1-698 did not amplify either product. (c) Southern blot analysis. Genomic DNA digested with *Pme1* and probed with a TLP sequence (red line) outside donor DNA. Events AAD-1/TLP-1-1-597, AAD-1/TLP-1-3-138 and AAD-1/TLP-1-3-789 display the 4.2-kb hybridization product. Event AAD/TLP-1-1-698 is nontargeted. p*AAD-1*/TLP1-1 is a positive plasmid control. Nontransgenic (non-xg) = 'Hi-II'.

for further study based upon on-target cleavage activity comparable to that displayed by the reference. None of these artificial hybrid target sequences (recognized by the selected ZFN pairs) were found to occur in the maize genome. The demonstration that ZFN monomers and corresponding recognition sequences can be 'mixed and matched' to create novel ZFN cleavage site combinations allows for the generation of large numbers of unique sequences for subsequent targeting.

Four selected ZFN recognition sequences were used to generate two different target TLP constructs (pPAT/TLP-1 and pPAT/TLP-2). Each TLP construct consists of the trait of interest (in this case, the herbicide resistance gene *pat)* immediately adjacent to a TLP for subsequent targeted gene addition (Figure 2a). Specifically, each TLP employs two ZFN target sequences flanked by 1 kb of random sequence that provides a unique region of sequence homology between the target TLPs and incoming donor DNA for subsequent homology-directed gene targeting (Figure 2a). pPAT/TLP-1 and pPAT/TLP-2 comprised binding sites for ZFN-1/ZFN-3 and ZFN-2/ZFN-5, respectively, and were integrated into the maize genome via WHISKERS™ transformation. Southern blot analysis of genomic DNA isolated from leaves of regenerated T_0 plants revealed a single hybridization product when probed with *pat* and TLP sequences (Figure S1). The intactness of the integrated TLP was confirmed using standard sequencing of PCR amplification products. Two events from pPAT/TLP-1 (TLP-1-004 and TLP-1-007) and one event from pPAT/TLP-2 (TLP-2-011) were selected for retargeting based on confirmed intact, single-copy TLP integration.

Targeted trait transgene stacking to a TLP

Having established transgenic events comprising a single copy of the TLP-flanked herbicide resistance gene, *pat*, we next attempted to target a second herbicide resistance gene, *aad1*, into this same locus directly at the TLP site. From a total of 8558 TLP-containing immature embryos cobombarded with donor DNA and ZFN constructs, 1738 haloxyfop-resistant transgenic events were generated (Table 1). Transgenic events were generated from bombarded embryos derived from all three target events with transformation frequencies ranging from 16.9 to 28.8%. To interrogate the target site for ZFN-mediated cleavage and/or targeted transgene integration, a high-throughput qPCR assay was designed and used to monitor the intactness of the ZFN recognition sites using primers anchored to sequences proximal to the ZFN cleavage site in the homology arms. In an intact TLP-1 or TLP-2, qPCR is expected to result in a 193-bp or 198-bp amplification product, respectively, comprising the intact ZFN recognition sequences (Figure S2). Donor integration or insertions/deletions at the ZFN cleavage site in the TLP (Figure 2a) prevent amplification resulting in a reduced qPCR signal. Overall, 158 of the events (9.1%) failed to amplify the 193-bp or 198-bp product (Table 1), consistent with TLPs carrying disrupted ZFN

recognition sequences. This high frequency of editing is comparable to that observed in transformed human cells (Urnov *et al.*, 2005), highlighting the potency of these TLP ZFN pairs when delivered by microparticle bombardment into immature embryos. Differences were observed among the target TLP events with TLP-1-007 cobombarded with pZFN-1 and pAAD1/TLP-1-1, displaying the highest TLP disruption frequency (15.8%).

To further characterize the transgenic events for bona fide targeted integration of the second herbicide resistance transgene, we performed junctional 'in–out' PCR to specifically score for the presence of donor sequences at the 5′ and 3′ ends of the target TLP loci. These PCRs were designed such that one primer was located within the pre-integrated target TLP, outside the donor homology region, and a second primer was located on the incoming donor DNA whereby only targeted events produce amplification products (Figure 2a). Based on the ability to amplify either a 1.5-kb or a 1.3-kb product using 5′ and 3′ 'in–out' primers, respectively (Figures 2a, b), the estimated donor DNA integration frequency on an event basis ranged from 0 to 7.5% in the various TLP event x ZFN combinations (Table 1). Events were obtained that amplified either 5′ or 3′ products (Figure 2b, event AAD-1/ELP-1-3-138), that is, imperfect integration, or both (Figure 2b, events AAD-1/ELP1-1-597 and AAD-1/ELP-1-3-789). Sequence analysis of these amplification products confirmed donor DNA integration into the pre-integrated ELP (Figure S3). As with the estimated frequency of ZFN recognition sequence disruption, embryos from event TLP-1-007 bombarded with pZFN-1 and pAAD-1/TLP-1-1 had the highest integration frequency (7.5%) based on junctional 'in–out' PCR analysis (Table 1).

To further confirm the integrity of ZFN-driven TLP transgene integration, we performed Southern blot analysis on genomic DNA isolated from these events. Following digestion of genomic DNA with the restriction enzyme *Pme*1, 54 events (3.1%) displayed the expected 4.2-kb hybridization product indicative of targeted integration of the intact donor into the TLP (Figure 2c). As with the frequencies of TLP sequence disruption and donor DNA integration estimated by PCR, intact donor integration, estimated by Southern blot analysis, was highest in embryos of target event TLP-1-007 cobombarded with pZFN-1 and pAAD-1/TLP-1-1 (5.3%). Interestingly, of the two TLP-1 target events, TLP-1-007 displayed a targeting frequency roughly 2X that observed for TLP-1-004 based on PCR amplification of ZFN

Table 1 Targeted integration of *AAD-1*-containing constructs into pre-integrated TLPs following cobombardment with ZFN expression constructs

| | TLP-1 | | | | TLP-2 | | |
| | TLP-1-004 | | TLP-1-007 | | TLP-2-011 | | |
	ZFN-1	ZFN-3	ZFN-1	ZFN-3	ZFN-2	ZFN-5	Total/Ave.
Total immature embryos bombarded	1896	1864	1943	1951	406	498	8558
Haloxyfop-resistant transgenic events generated	321	383	360	453	117	104	1738
Transformation frequency (% of bombarded embryos)	16.9	20.5	18.5	23.2	28.8	20.9	20.3
Events with disrupted TLP	33	26	57	30	10	2	158
Disruption frequency (% of total events)	10.3	6.8	15.8	6.6	8.5	1.9	9.1
Events with 3′ and/or 5′ PCR amplification products	12	6	27	19	5	0	69
Donor integration frequency (% of total events)	3.7	1.8	7.5	4.2	4.3	0	4.0
Events with targeted digestion product on Southern blot	4	12	19	16	3	0	54
Intact donor integration frequency (% of total events)	1.2	3.1	5.3	3.5	2.6	0	3.1

recognition sequences, TLP donor junction sequences or Southern blot analysis (Table 1). Given that these two events differ only in integration site, these data suggest some position effect on targeting efficiency. Nonetheless, targeted transgene integration was observed in all three TLP target events.

Functional cosegregation of the stacked traits

Demonstrating the genetic linkage of the two independently transformed herbicide resistance genes involved analysing segregating progenies (Figure 3a). Plants were regenerated from transgenic event AAD-1/TLP-1-3-792 with confirmed integration of a single copy of the *aad1* construct into the PAT/TLP locus, and crosses were made to nontransgenic plants to generate progenies. Because the targeted AAD1/TLP-1-3::PAT/TLP locus was hemizygous in the regenerated plants, it was anticipated that the targeted locus would segregate 1:1 in the progenies. Moreover, because the *aad1* construct was integrated into the PAT/TLP locus, the two linked genes would be expected to cosegregate. PCR amplification of *aad1* and *pat* gene sequences in 42 progeny is shown in Figure 3b. A total of 19 of the 42 progeny plants were positive for both *aad1* and *pat* ($\chi^2 = 0.38$; $P < 0.01$). Progenies positive for the 1.5-kb amplification product indicative of *aad1* were always positive for the 1.3-kb *pat* gene product. All PCR-positive progenies were also positive for both AAD1 and PAT proteins using an ELISA strip test and were resistant to Ignite® and Assure® II herbicide spray treatments (Figure 3c). These results clearly demonstrate functional, cosegregation of the stacked transgenes.

Discussion

Given the complexity and number of traits required for modern agriculture, the need for facile and efficient methods for sequential transgene stacking is paramount. An example that underscores the trend towards transgenic products with multiple traits is SmartStax™, an 8-trait stack, codeveloped by Dow AgroSciences and Monsanto (Marra *et al.*, 2010). In the present study, an herbicide resistance transgene, *aad1*, was precisely targeted to the genomic locus of a previously integrated herbicide resistance transgene, *pat*, resulting in a sequential stack. The ability to integrate multiple trait genes into a single locus to enable simple inheritance addresses a significant agricultural challenge.

The large suite of novel ZFN recognition sequence combinations, made possible by 'mix-and-match', facilitates the engineering of transgenic loci for subsequent editing. These attributes provide considerable flexibility for commercial trait deployment including targeted transgene integration (Shukla *et al.*, 2009) and excision (Petolino *et al.*, 2010) of both selectable marker and trait genes.

A current limitation of the technology is the need to couple transgene addition, that is, transformation, with the use of a selectable marker. As in the current study where the second trait transgene involves a selectable phenotype, for example herbicide resistance, this is not an issue. However, when a trait does not lend itself to *in vitro* selection, the use of selectable marker genes will be required. The number of selectable marker genes appropriate for commercial product development is limited. In such cases, the selectable marker gene sequence could be flanked with ZFN cleavage sites for subsequent excision (Petolino *et al.*, 2010) and future use for additional stacking.

An important consideration for trait stacking is the inheritance and expression of multiple transgenes at a particular genomic

locus. Transformation of a large construct, that is, ~34 kb, containing 10 transgenes was successfully transformed into rice, and all but one gene was expressed in the resulting transgenic plants (Lin *et al.*, 2003). However, obtaining intact integration of such large constructs and stable gene expression over multiple generations may be a challenge. In a study more aligned with the current approach (Day *et al.*, 2000), reporter gene expression was found to be similar in multiple transgenic events targeted to certain chromosomal locations, although differences due to transgene silencing were observed. Similarly, four genes targeted to a single locus in cotton using re-engineered meganucleases were transmitted to the next generation as a single genetic unit and accumulated the expected protein products (D'Halluin *et al.*, 2013). These results suggest that functional co-expression of multiple transgenes stacked at a single locus is possible.

The results of the present study point towards a novel means of transgenic product development. The ~5% intact transgene integration frequency, isolated from a background of random integration highlights the practical utility of this approach for transgene stacking. Moreover, the use of microparticle bombardment of immature embryos represents a commercially viable path to deployment of this technology across a broad spectrum of commercial maize germplasms. Indeed, transgene stacking via ZFN-mediated TLP targeting has been demonstrated in a commercial maize germ plasm (data not shown). The use of TLPs with unique ZFN cleavage sites, enabling multiple cycles of transgene addition and deletion, represents a significant advance in the ability to deliver genetic solutions to agricultural challenges.

Experimental procedures

Yeast-based ZFN testing

The following four ZFNs and corresponding recognition sequences were used to generate novel nuclease/target sequence combinations: CCR5 (gtcatcctcatcctgat aaactgcaaaagg), AAVS1 (caccc cacagtggggccactagggacaggatt), Rosa26 (ctgcaactccagtcttctag aagatgggcgggagtct) and Prmt1 (gtccatgctcaacaccgtgctctatgcccg ggacaagt). ZFNs resulting from combinations of these four protein pairs (Table S1) were cloned into budding yeast expression vectors and screened for *in vivo* double-strand DNA break induction as previously described (Doyon *et al.*, 2008).

Vector assembly

Selected ZFNs (CL:AR, RL:PR, AL:PR and PL:AR) were cloned into plant expression vectors resulting in pZFN-1, pZFN-2, pZFN-3 and pZFN-5, respectively, whereby the nuclease was under the control of the Zmubi-1 promoter (Christensen and Quail, 1996) and terminated by the Zmlip 3' untranslated region (Cowen *et al.*, 2007). Two target TLPs, each consisting of two tandem ZFN recognition sequences flanked by 1 kb of random sequence, were synthesized. TLP-1 and TLP-2 comprised recognition sequences for ZFN-1/ZFN-3 and ZFN-2/ZFN-5, respectively (Figure 2a). The two TLP sequences were cloned into pPAT, that is, the *OsAct1* promoter (McElroy *et al.*, 1990), the phosphinothricin acetyltransferase (*pat*) selectable marker gene (Wohlleben *et al.*, 1988) and the *ZmLip* 3' untranslated region (Cowen *et al.*, 2007), using the GATEWAY® LR clonase reaction (Invitrogen, Carlsbad, CA) resulting in plasmids, pPAT/TLP-1 and pPAT/TLP-2.

Donor DNA constructs were assembled by inserting a *Sphingobium herbicidovorans* aryloxyalkanoate dioxygenase (*aad1*) gene

(a)   Identify plant carrying AAD1-PAT cassette at TLP

↓

Cross to wild-type plant

↓

Genotype and phenotype progenies for herbicide resistance

(b)

(c)

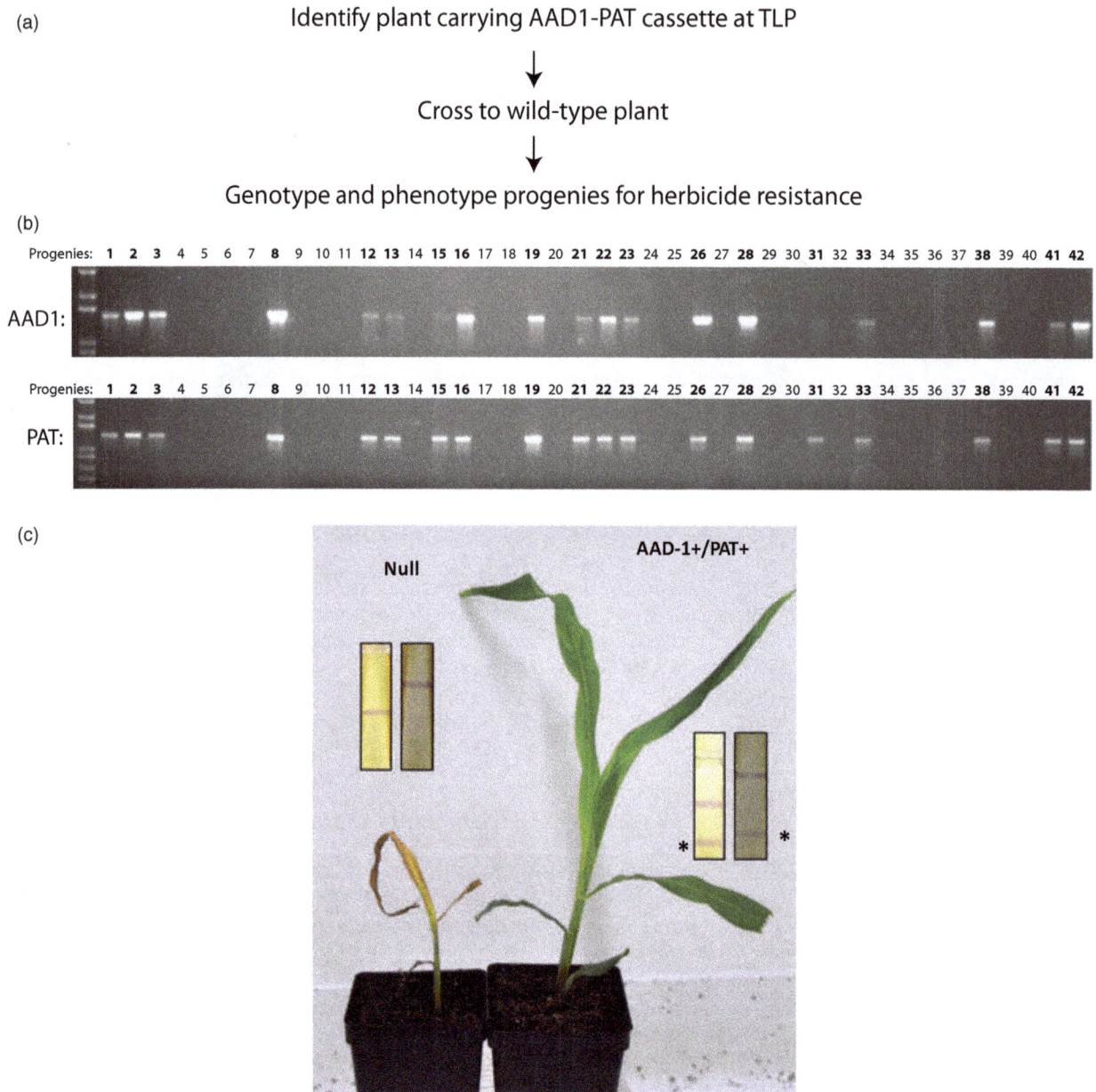

Figure 3 Functional cosegregation of stacked transgenes. (a) Experimental outline. (b) Cosegregation of *AAD-1* and *PAT*. 5′ (upper) and 3′ (lower) PCR amplification of TLP sequences representative of AAD-1 and PAT expression cassettes in AAD-1/TLP-1-3-792 crossed to nontransgenic 'Hi-II'. Of 42 progenies, 19 were positive and 23 were negative for both products. (c) A dual-transgene-positive plant and a null sib segregant were phenotyped for herbicide resistance, and ELISA was used to detect transgene-encoded protein (inset) as described in Materials and Methods. *denotes positive reaction for protein detection.

(Wright *et al.*, 2010) expression cassette between the appropriate homologous sequences that were truncated at their proximal ends to facilitate subsequent analysis of targeted events. The *aad-1* expression cassette consisted of the *OsAct-1* promoter, the *aad-1* gene encoding an enzyme with an α-ketoglutarate-dependent dioxygenase activity that confers resistance to aryloxyphenoxy-propionate herbicides, for example Assure® II, and the *ZmLip* 3′ untranslated region. Two donor plasmids were created for targeting each TLP, which differed based on the ZFN half-sites flanking the insert (*i.e.* either ZFN-1 or ZFN-3 recognition sequences for targeting to events produced using pPAT/ELP-1 or ZFN-2 or ZFN-5 recognition sequences for targeting events produced using pPAT/ELP-2) The resulting donor plasmids are

designated according to the TLP and ZFN recognition sequence targeted: pAAD-*1*/ELP1-1, pAAD-1/ELP1-3, pAAD-1/ELP2-2 and pAAD-1/ELP2-5 (Figure 2a).

Transgenic event production

WHISKERS™ transformation for TLP integration

Embryogenic maize suspension cultures of the genotype 'Hi-II' were generated and transgenic plants produced as previously described (Petolino and Arnold, 2009) using WHISKERS™-mediated DNA delivery and selection on 1 mg/L Herbiace® (Meija Seika, Tokyo, Japan). Regenerated plants were self-pollinated, and T$_1$ seeds were harvested and dried. The seed was planted and leaf tissue sampled for zygosity determination using qPCR described below. Plants homozygous for PAT/ELP-1 (PAT/ELP-1-004 and PAT/ELP-1-007) or PAT/ELP-2 (PAT/ELP-2-011) were selected and used as pollen parents to cross onto nontransgenic 'Hi-II' ears to generate immature embryos hemizygous for PAT/ELP-1 or PAT/ELP-2 for use in subsequent targeting.

Targeting ELPs via microparticle bombardment

Immature embryos from 'Hi-II' crosses with events PAT/ELP-1-004 or PAT/ELP-1-007 were cotransformed with donor DNA (pAAD-1/ELP-1-1 or pAAD-1-1-3) and ZFN expression constructs (pZFN-1 and pZFN-3), and crosses with PAT/ELP-2-011 were cotransformed with donor DNA (pAAD-1/ELP-1-1 or pAAD-1-1-3) and ZFN expression constructs (pZFN-1 and pZFN-3) using microparticle bombardment (Wang and Frame, 2009). All cotransformations involved coprecipitation of 9 parts of donor DNA to one part ZFN expression construct on a weight basis. R-haloxyfop acid (Dow Chemical, Midland, MI) at 0.181 mg/L was used to select transgenic events, and plants were regenerated as previously described (Wright *et al.*, 2010). Following molecular analysis, plant regenerated from confirmed targeted events was crossed to nontransgenic 'Hi-II' for segregation analysis.

Molecular analysis

Genomic DNA isolation

Tissue samples were collected in 96-well collection plates and lyophilized for 48 h. Tissue disruption was performed with a Kleco™ tissue pulverizer (Garcia Manufacturing, Visalia, CA) in Biosprint® 96 RLT lysis buffer (Qiagen, Gathersburg, MD) with one stainless steel bead. Following tissue maceration, genomic DNA was isolated in high-throughput format using the Biosprint® 96 Plant Kit and Extraction Robot (Qiagen). Genomic DNA was then diluted to 2 ng/μL prior to qPCR.

Quantitative PCR

Transgene detection by hydrolysis probe assay was performed by real-time PCR using the LightCycler® 480 System (Roche Applied Science, Indianapolis, IN). The assays were multiplexed with an internal reference assay, invertase, to ensure gDNA was present in each assay. For amplification, LightCycler® 480 Probes Master Mix (Roche Applied Science) was prepared at $1\times$ final concentration in a 10-μL volume multiplex reaction containing 0.4 μM of each primer and 0.2 μM of each probe. A two-step amplification reaction was performed with an extension at 60°C for 40 s (for the *aad1* gene) and 60°C for 30 s (for the ZFN recognition sequence) and with fluorescence acquisition. Cp scores, the point at which the fluorescence signal crosses the background threshold using the fit points algorithm (Light Cycler® Software Release 1.5) and the Relative Quant Module (based on the $\Delta\Delta C_t$ method), were used to perform the analysis.

Locus-specific 'In–Out' PCR

'In–out' PCR was conducted using a Takara Ex Taq HS Kit (Clontech Laboratories, Inc., Mountain View, CA). Each PCR was carried out in 15 μL final volumes, which contained $1\times$ Ex Taq buffer, 200 nM of forward and reverse primers, 10–20 ng of genomic DNA template and final concentration of 0.05 unit/μL Ex Taq HS polymerase (Clontech Laboratories). For real-time 'in–out' PCR, a SYTO-13 dye from Invitrogen (Grand Island, NY) was included in the PCR mix at a final concentration of 4 μM. Real-time 'in–out' PCR was performed on an ABI ViiA7 PCR system

(Life Technologies Corporation, Carlsbad, CA). After initial denaturing, the amplification programme contained 35 cycles of 98 °C for 10 s, 66 °C for 30 s and 68 °C for 2 min with fluorescence acquisition followed by a melting temperature analysis programme. The reaction mixture was then placed at 65 °C for 30 s and 72 °C for 10 min and finally held at 4 °C. Samples with positive signals were run on 1% agarose gels. PCR products were sequenced by standard Sanger sequencing methods (Eurofins MGW Operon, Huntsville, AL).

Southern blot analysis

Genomic DNA (5 μg) was digested in $1\times$ Buffer 3 with 50 units of *Pme*I (New England BioLabs, Ipswich, MA) in a final volume of 125 μL. *Pme*1 was selected as it cuts at the target locus outside the homology arms and allows interrogation of the locus. Samples were incubated at 37°C overnight. The digested DNA was concentrated by reprecipitation with quick precipitation solution (Edge Biosystems, Inc., Gaithersburg, MD). Recovered digests were resuspended in 30 μL of $1\times$ loading buffer and incubated at 65°C for 30 min. Resuspended samples were loaded onto a 0.8% agarose gel prepared in $1\times$ TAE [0.8 M Tris–acetate (pH8.0)/0.04 mM EDTA] and electrophoresed in $1\times$ TAE buffer. The gel was sequentially subjected to denaturation (0.2 M NaOH/0.6 M NaCl) for 30 min and neutralization [0.5 M Tris–HCl (pH7.5)/1.5 M NaCl] for 30 min. Transfer of DNA fragments was performed by passively wicking 20XSSC solution overnight through the gel onto treated Immobilon NY+ (Millipore, Billerica, MA) using a chromatography paper wick and paper towels. Following transfer, the membrane was briefly washed with 2XSSC, cross-linked with the Stratalinker 1800 (Stratagene, La Jolla, CA) and baked at 80°C for 1 h. Blots were incubated with Perfect Hyb™ Plus prehybridization solution (Sigma, St. Louis, MO) for 1 h at 65°C in glass roller bottles using a model 400 hybridization incubator (Robbins Scientific, Sunnyvale, CA). For probe preparation, sequence outside the donor homology region was PCR-amplified and purified on agarose gels using a QIAquick gel extraction kit (Qiagen). The fragment was labelled with 3000 Ci/mmol α^{32}P-dCTP (Perkin/Elmer, Waltham, MA) using Prime-it® II Random Primer Labeling Kit (Stratagene). Blots were hybridized overnight at 65°C with denatured probe at approximately 2×10^6 counts per mL/hybridization buffer. Following hybridization, blots were sequentially washed at 65°C with 0.1XSSC/0.1%SDS for 40 min. Blots were exposed to phosphor screen and imaged using a Storm 860 Imaging Systems (Molecular Dynamics, Sunnyvale, CA).

Functional analysis of *pat* and *aad-1*

The presence of PAT and AAD-1 protein in leaves of 5-day-old seedling was detected using lateral flow device tests from Romer Labs (Union, MO) and American Bionostica, Inc. (Swedesboro, NJ), respectively. Foliar applications of Ignite® (Bayer Crop Science, Kansas City, MO) and Assure® II (Du Pont Crop Protection, Wilmington, DE) were applied to 5-day-old seedlings using a track sprayer calibrated to deliver 187 L/ha at a rate of 560 g ae/ha and 70 g ae/ha, respectively. Phenotypic ratings were taken 5 days after spraying.

Acknowledgements

We thank Gardenia Gonzales Gil and Erica Moehle for helping with the figures.

References

Bizily, S.P., Rugh, C.L. and Meagher, R.B. (2000) Phytodetoxification of hazardous organomercurials by genetically engineered plants. *Nat. Biotechnol.* **18**, 213–217.

Cai, C.Q., Doyon, Y., Ainley, W.M., Miller, J.C., Dekelver, R.C., Moehle, E.A., Rock, J.M., Lee, Y.L., Garrison, R., Schulenberg, L., Blue, R., Worden, A., Baker, L., Faraji, F., Zhang, L., Holmes, M.C., Rebar, E.J., Collingwood, T.N., Rubin-Wilson, B., Gregory, P.D., Urnov, F.D. and Petolino, J.F. (2009) Targeted transgene integration in plant cells using designed zinc finger nucleases. *Plant Mol. Biol.* **69**, 699–709.

Cao, J., Zhao, J.Z., Tang, D., Shelton, M. and Earle, D. (2002) Broccoli plants with pyramided cry1Ac and cry1C Bt genes control diamondback moths resistant to Cry1A and Cry1C proteins. *Theor. Appl. Genet.* **105**, 258–264.

Christensen, A.H. and Quail, P.H. (1996) Ubiquitin promoter-based vectors for high-level expression of selectable and/or screenable marker genes in monocotyledonous plants. *Transgenic Res.* **5**, 213–218.

Cowen, N.M., Armstrong, K. and Smith, K.A. (2007) *Use of regulatory sequences in transgenic plants.* (USPTO ed).

Dafny-Yelin, M. and Tzfira, T. (2007) Delivery of multiple transgenes to plant cells. *Plant Physiol.* **145**, 1118–1128.

Datta, K., Baisakh, N., Thet, K.M., Tu, J. and Datta, S.K. (2002) Pyramiding transgenes for multiple resistance in rice against bacterial blight, yellow stem borer and sheath blight. *Theor. Appl. Genet.* **106**, 1–8.

Day, C.D., Lee, E., Kobayashi, J., Holappa, L.D., Albert, H. and Ow, D.W. (2000) Transgene integration into the same chromosome location can produce alleles that express at a predictable level, or alleles that are differentially silenced. *Genes Dev.* **14**, 2869–2880.

DeKelver, R.C., Choi, V.M., Moehle, E.A., Paschon, D.E., Hockemeyer, D., Meijsing, S.H., Sancak, Y., Cui, X., Steine, E.J., Miller, J.C., Tam, P., Bartsevich, V.V., Meng, X., Rupniewski, I., Gopalan, S.M., Sun, H.C., Pitz, K.J., Rock, J.M., Zhang, L., Davis, G.D., Rebar, E.J., Cheeseman, I.M., Yamamoto, K.R., Sabatini, D.M., Jaenisch, R., Gregory, P.D. and Urnov, F.D. (2010) Functional genomics, proteomics, and regulatory DNA analysis in isogenic settings using zinc finger nuclease-driven transgenesis into a safe harbor locus in the human genome. *Genome Res.* **20**, 1133–1142.

D'Halluin, K., Vanderstraeten, C., Stals, E., Cornelissen, M. and Ruiter, R. (2008) Homologous recombination: a basis for targeted genome optimization in crop species such as maize. *Plant Biotechnol. J.* **6**, 93–102.

D'Halluin, K., Vanderstraeten, C., Van Hulle, J., Rosolowska, J., Van Den Brande, I., Pennewaert, A., D'Hont, K., Bossut, M., Jantz, D., Ruiter, R. and Broadhvest, J. (2013) Targeted molecular trait stacking in cotton through targeted double-strand break induction. *Plant Biotechnol. J.* doi: 10.1111/pbi.12085.

Doyon, Y., McCammon, J.M., Miller, J.C., Faraji, F., Ngo, C., Katibah, G.E., Amora, R., Hocking, T.D., Zhang, L., Rebar, E.J., Gregory, P.D., Urnov, F.D. and Amacher, S.L. (2008) Heritable targeted gene disruption in zebrafish using designed zinc-finger nucleases. *Nat. Biotechnol.* **26**, 702–708.

Fladung, M. and Becker, D. (2010) Targeted integration and removal of transgenes in hybrid aspen (Populus tremula L. × P. tremuloides Michx.) using site-specific recombination systems. *Plant Biol.* **12**, 334–340.

Halpin, C. (2005) Gene stacking in transgenic plants–the challenge for 21st century plant biotechnology. *Plant Biotechnol. J.* **3**, 141–155.

Kim, Y.G., Cha, J. and Chandrasegaran, S. (1996) Hybrid restriction enzymes: zinc finger fusions to Fok I cleavage domain. *Proc. Natl Acad. Sci. USA*, **93**, 1156–1160.

Li, Z., Moon, B.P., Xing, A., Liu, Z.B., McCardell, R.P., Damude, H.G. and Falco, S.C. (2010) Stacking multiple transgenes at a selected genomic site via repeated recombinase-mediated DNA cassette exchanges. *Plant Physiol.* **154**, 622–631.

Lin, L., Liu, Y.-G., Xu, X. and Li, B. (2003) Efficient linking and transfer of multiple genes by a multigene assembly and transformation vector system. *Proc. Natl Acad. Sci. USA*, **100**, 5962–5967.

Marra, M.C., Piggott, N.E. and Goodwin, B.K. (2010) The anticipated value of SmartStax™ for US corn growers. *AgBioForum*, **13**, 1–12.

McElroy, D., Zhang, W., Cao, J. and Wu, R. (1990) Isolation of an efficient actin promoter for use in rice transformation. *Plant Cell*, **2**, 163–171.

Naqvi, S., Farré, G., Sanahuja, G., Capell, T., Zhu, C. and Christou, P. (2010) When more is better: multigene engineering in plants. *Trends Plant Sci.* **15**, 48–56.

Ntui, V.O., Azadi, P., Thirukkumaran, G., Khan, R.S., Chin, D.P., Nakamura, I. and Mii, M. (2011) Increased resistance to fusarium wilt in transgenic tobacco lines co-expressing chitinase and wasabi defensin genes. *Plant. Pathol.* **60**, 221–231.

Orlando, S.J., Santiago, Y., DeKelver, R.C., Freyvert, Y., Boydston, E.A., Moehle, E.A., Choi, V.M., Gopalan, S.M., Lou, J.F., Li, J., Miller, J.C., Holmes, M.C., Gregory, P.D., Urnov, F.D. and Cost, G.J. (2010) Zinc-finger nuclease-driven targeted integration into mammalian genomes using donors with limited chromosomal homology. *Nucleic Acids Res.* **38**, e152.

Pabo, C.O., Peisach, E. and Grant, R.A. (2001) Design and selection of novel Cys2His2 zinc finger proteins. *Annu. Rev. Biochem.* **70**, 313–340.

de Pater, S., Neuteboom, L.W., Pinas, J.E., Hooykaas, P.J. and van der Zaal, B.J. (2009) ZFN-induced mutagenesis and gene-targeting in Arabidopsis through Agrobacterium-mediated floral dip transformation. *Plant Biotechnol. J.* **7**, 821–835.

Perez, E.E., Wang, J., Miller, J.C., Jouvenot, Y., Kim, K.A., Liu, O., Wang, N., Lee, G., Bartsevich, V.V., Lee, Y.L., Guschin, D.Y., Rupniewski, I., Waite, A.J., Carpenito, C., Carroll, R.G., Orange, J.S., Urnov, F.D., Rebar, E.J., Ando, D., Gregory, P.D., Riley, J.L., Holmes, M.C. and June, C.H. (2008) Establishment of HIV-1 resistance in CD4+ T cells by genome editing using zinc-finger nucleases. *Nat. Biotechnol.* **26**, 808–816.

Petolino, J.F. and Arnold, N.L. (2009) Whiskers-mediated maize transformation. *Methods Mol. Biol.* **526**, 59–67.

Petolino, J.F., Worden, A., Curlee, K., Connell, J., Strange Moynahan, T.L., Larsen, C. and Russell, S. (2010) Zinc finger nuclease-mediated transgene deletion. *Plant Mol. Biol.* **73**, 617–628.

Puchta, H., Dujon, B. and Hohn, B. (1996) Two different but related mechanisms are used in plants for the repair of genomic double-strand breaks by homologous recombination. *Proc. Natl Acad. Sci. USA*, **93**, 5055–5060.

Qi, B., Fraser, T., Mugford, S., Dobson, G., Sayanova, O., Butler, J., Napier, J.A., Stobart, A.K. and Lazarus, C.M. (2004) Production of very long chain polyunsaturated omega-3 and omega-6 fatty acids in plants. *Nat. Biotechnol.* **22**, 739–745.

Ramana Rao, M., Parameswari, C., Sripriya, R. and Veluthambi, K. (2011) Transgene stacking and marker elimination in transgenic rice by sequential Agrobacterium-mediated co-transformation with the same selectable marker gene. *Plant Cell Rep.* **30**, 1241–1252.

Shukla, V.K., Doyon, Y., Miller, J.C., DeKelver, R.C., Moehle, E.A., Worden, S.E., Mitchell, J.C., Arnold, N.L., Gopalan, S., Meng, X., Choi, V.M., Rock, J.M., Wu, Y.Y., Katibah, G.E., Zhifang, G., McCaskill, D., Simpson, M.A., Blakeslee, B., Greenwalt, S.A., Butler, H.J., Hinkley, S.J., Zhang, L., Rebar, E.J., Gregory, P.D. and Urnov, F.D. (2009) Precise genome modification in the crop species Zea mays using zinc-finger nucleases. *Nature*, **459**, 437–441.

Srivastava, V. and Ow, D. (2002) Biolistic mediated site-specific integration in rice. *Mol. Breed.* **8**, 345–349.

Townsend, J.A., Wright, D.A., Winfrey, R.J., Fu, F., Maeder, M.L., Joung, J.K. and Voytas, D.F. (2009) High-frequency modification of plant genes using engineered zinc-finger nucleases. *Nature*, **459**, 442–445.

Urnov, F.D., Miller, J.C., Lee, Y.L., Beausejour, C.M., Rock, J.M., Augustus, S., Jamieson, A.C., Porteus, M.H., Gregory, P.D. and Holmes, M.C. (2005) Highly efficient endogenous human gene correction using designed zinc-finger nucleases. *Nature*, **435**, 646–651.

Urnov, F.D., Rebar, E.J., Holmes, M.C., Zhang, H.S. and Gregory, P.D. (2010) Genome editing with engineered zinc finger nucleases. *Nat. Rev. Genet.* **11**, 636–646.

Vergunst, A.C., Jansen, L.E. and Hooykaas, P.J. (1998) Site-specific integration of Agrobacterium T-DNA in Arabidopsis thaliana mediated by Cre recombinase. *Nucleic Acids Res.* **26**, 2729–2734.

Wang, K. and Frame, B. (2009) Biolistic gun-mediated maize genetic transformation. *Methods Mol. Biol.* **526**, 29–45.

Wohlleben, W., Arnold, W., Broer, I., Hillemann, D., Strauch, E. and Puhler, A. (1988) Nucleotide sequence of the phosphinothricin N-acetyltransferase gene from Streptomyces viridochromogenes Tu494 and its expression in Nicotiana tabacum. *Gene*, **70**, 25–37.

Wright, D.A., Townsend, J.A., Winfrey Jr, R.J., Irwin, P.A., Rajagopal, J., Lonosky, P.M., Hall, B.D., Jondle, M.D. and Voytas, D.F. (2005) High-frequency homologous recombination in plants mediated by zinc-finger nucleases. *Plant J.* **44**, 693–705.

Regulation of the alkaloid biosynthesis by miRNA in opium poppy

Hatice Boke[1], Esma Ozhuner[1], Mine Turktas[1], Iskender Parmaksiz[2], Sebahattin Ozcan[3] and Turgay Unver[1],*

[1]Department of Biology, Faculty of Science, Cankiri Karatekin University, Cankiri, Turkey
[2]Department of Molecular Biology and Genetics, Faculty of Science, Gaziosmanpasa University, Tokat, Turkey
[3]Department of Field Crops, Faculty of Agriculture, Ankara University, Ankara, Turkey

*Correspondence

email turgayunver@gmail.com

Summary

Opium poppy (*Papaver somniferum*) is an important medicinal plant producing benzylisoquinoline alkaloids (BIA). MicroRNAs (miRNAs) are endogenous small RNAs (sRNAs) of approximately 21 nucleotides. They are noncoding, but regulate gene expression in eukaryotes. Although many studies have been conducted on the identification and functions of plant miRNA, scarce researches on miRNA regulation of alkaloid biosynthesis have been reported. In this study, a total of 316 conserved and 11 novel miRNAs were identified in opium poppy using second-generation sequencing and direct cloning. Tissue-specific regulation of miRNA expression was comparatively analysed by miRNA microarray assays. A total of 232 miRNAs were found to be differentially expressed among four tissues. Likewise, 1469 target transcripts were detected using *in silico* and experimental approaches. The Kyoto Encyclopedia of Genes and Genomes pathway analyses indicated that miRNA putatively regulates carbohydrate metabolism and genetic-information processing. Additionally, miRNA target transcripts were mostly involved in response to stress against various factors and secondary-metabolite biosynthesis processes. Target transcript identification analyses revealed that some of the miRNAs might be involved in BIA biosynthesis, such as pso-miR13, pso-miR2161 and pso-miR408. Additionally, three putatively mature miRNA sequences were predicted to be targeting BIA-biosynthesis genes.

Keywords: alkaloid biosynthesis, deep sequencing, gene ontology, microRNA, microarray, *Papaver somniferum*.

Introduction

Opium poppy (*Papaver somniferum*) is an agronomically and economically important medicinal plant used for main morphinan alkaloid production (Ziegler and Facchini, 2008). The opium poppy synthesizes large amount of benzylisoquinoline alkaloids (BIA) as secondary metabolites. The latex of the poppy capsule mainly includes morphine and relatively low levels of other BIA alkaloids, such as codeine, papaverine, thebaine and noscapine (Page, 2005).

The BIA biosynthesis is a multistep pathway in which many genes and enzymes are involved. The biosynthesis starts with decarboxylations, deaminations and/or hydroxylations of tyrosine to form dopamine and 4-hydroxyphenylacetaldehyde (4-HPAA) by tyrosine decarboxylase (TYDC; Ziegler and Facchini, 2008). Then, dopamine or 4-HPAA is converted into (S)-norcoclaurine by norcoclaurine synthase (Samanani *et al.*, 2004). To form (S)-methylcoclaurine, (S)-norcoclaurine is methylated by the norcoclaurine 6-O-methyltransferase (6-OMT) and coclaurine *N*-methyltransferase (Desgagné-Penix *et al.*, 2010). The (S)-*N*-methylcoclaurine-3'-hydroxylase (NMCH or CYP80B3) P450-dependent monooxygenase and 3'-hydroxy-*N*-methylcoclaurine 4-O-methyltransferase (4-OMT) generate (S)-reticuline. Such molecule is a central intermediate of the isoquinoline alkaloid biosynthesis (Weid *et al.*, 2004). After that, different types of modifications of the (S)-reticuline produce many BIA, including the morphinan alkaloids, by at least eight enzymatic steps (Wijekoon and Facchini, 2012).

On the other hand, it has been found that the gene expression can be inhibited by small ribonucleic acids (sRNAs) via RNA interference (RNAi). This can be accomplished by two types of sRNAs, namely microRNA (miRNA) and small interfering RNA (siRNA). The miRNAs are endogenous small RNAs (sRNAs) of approximately 21 nucleotides (nt). They are noncoding, but regulate gene expression in eukaryotes (Khraiwesh *et al.*, 2010). The miRNAs are transcribed as long precursors from their genes in the nucleus, being further processed into their mature forms (Unver and Budak 2009). Different plant miRNAs have been identified and characterized in several species (Eldem *et al.*, 2013). Thus, a total of 7057 plant miRNAs were reported from 73 species, which are publicly available in the miRBase (version 21) database (http://www.mirbase.org). Of those, 427 and 713 mature sequences belong to *Arabidopsis thaliana* and *Oryza sativa*, respectively. A limited number of opium poppy miRNAs are available in the literature. Thus, 20 opium poppy miRNAs were *in silico* identified and further experimentally validated (Unver *et al.*, 2010b). The miRNAs repress specific messenger RNA (mRNA) targets by cleavage and post-transcriptional inhibition (Jones-Rhoades *et al.*, 2006). Indeed, the miRNAs play important regulatory roles in different biological processes in plants, including growth and development (Aukerman and Sakai, 2004; Chen, 2004), physiologic response to biotic and abiotic stresses such as bacterial pathogen attacks (Navarro *et al.*, 2006; Zhang *et al.*, 2007), drought (Kantar *et al.*, 2010; Zhou *et al.*, 2010) and cold (Zhang *et al.*, 2009), as well as metabolism, signal transduction and protein degradation (Achard *et al.*, 2004; Guo *et al.*, 2005; Zhang and Wang, 2015; Zhang *et al.*, 2006) by regulating their specific target mRNAs.

One of the biological functions of miRNA is to regulate secondary-metabolite syntheses in plants. For instance, it has been found that miRNA393 redirects the secondary-metabolite productions via perturbing the auxin signallings. Thus, the overexpression of miRNA393 changed the level of biosyntheses of glucosinolate and camalexin (Robert-Seilaniantz *et al.*, 2011). Another study showed that the overexpression and loss of miRNA163 in *Arabidopsis* species alter the production of secondary-metabolite profiles (Ng *et al.*, 2011). Therefore, miRNA might also act on the regulation of the morphine-biosynthesis pathway. Thus, direct cloning and deep-sequencing strategies have been used in this work to identify and characterize opium poppy miRNA. Moreover, tissue-specific expression levels of miRNA were measured using miRNA microarray assays. The link between miRNA regulation and BIA biosynthesis was studied by target identification and ontology analyses.

Results

miRNA identification and characterization

High-throughput sRNA-sequencing overview

A total of 11 999 328 raw reads were generated by the high-throughput Illumina platform. After processing of the primary reads, 10 510 914 (87.6%) total clean reads, including 2 999 700 unique reads, were counted for the mixed tissue sRNA library (Table 1). The sequencing output was uploaded to NCBI SRA archive (accession no. SRR1731699). Due to the lack of the opium poppy reference genome or related-organism genome data, such unique sequences were mapped to the *Arabidopsis thaliana* genome, using SOAP2 (Li *et al.*, 2009). A total of 150 795 tRNAs, 289 634 rRNAs, 2721 snRNAs, 1345 snoRNAs, 1 839 895 siRNAs, 31 158 exon-sense RNAs, 1257 exon-antisense RNAs, 464 intron-sense RNAs and 495 intron-antisense RNAs were identified. Also, 12 550 (0.42%) unique reads were considered as miRNAs. The size distribution of the reads in the data set was between 19 and 24 nt, and most of the sRNAs were found as 24 nt in length (Figure S1).

Table 1 Classification of small RNA sequences belonging to opium poppy

Category	Unique sRNA	Percentage	Total sRNA	Percentage
Total	2 999 700	100	10 510 914	100
Exon antisense	360	0.01	1257	0.01
Exon sense	1395	0.05	31 158	0.30
Intron antisense	220	0.01	495	0.00
Intron sense	116	0.00	464	0.00
miRNA	12 550	0.42	488 373	4.65
rRNA	20 210	0.67	289 634	2.76
Repeat	5280	0.18	93 094	0.89
siRNA	77 732	2.59	1 839 895	17.50
snRNA	796	0.03	2721	0.03
snoRNA	426	0.01	1345	0.01
tRNA	7158	0.24	150 793	1.43
Unannotated	2 873 457	95.79	7 611 685	72.42

Conserved miRNA

A total of 316 miRNAs were identified, belonging to 111 conserved miRNA families (Table 2). These miRNA families were also found to be conserved in various plant species (Eldem *et al.*, 2012; Ozhuner *et al.*, 2013; Unver and Budak, 2009; Vahap Eldem *et al.*, 2013; Yanik *et al.*, 2013). The expression level of known miRNAs showed a broad range. Thus, many of the detected miRNAs had moderate expression levels, while 17 miRNAs were counted more than 1000 times (Table 2). Among the detected miRNAs, the highest expression was measured in pso-miR535 as 112 835, and almost half of identified poppy miRNAs (48%) were counted <10 times.

In consistency with previously reported results, the uracil nucleotide was found to be dominant in the first position of the 5'-end for the majority of these putatively novel miRNAs, and the higher proportion of the sequences was identified at the 5'-end of the hairpins, when compared to the 3'-end. We also calculated the minimum folding-free energies of putative opium poppy miRNA precursors for the sRNA library, ranging from −34.09 to −75.80 kcal/mol.

To discover additional miRNAs not captured by deep sequencing, a directional cloning strategy was applied. The obtained clones were sequenced via the Sanger technique. A total of 12 conserved miRNAs were identified this way (Table 2b).

Novel miRNA

Two new miRNAs (pso-miR1 and pso-miR13) were discovered by the deep-sequencing analyses. Additionally, nine novel miRNAs were identified by directional cloning. The list of the novel opium poppy miRNAs is shown in Table 3. Pre-miRNA sequences of the directly cloned miRNA are listed in Table S1. Similarly to the conserved miRNA, the majority of the nucleotides in the mature sequences of the novel miRNA were uracil.

Tissue-specific miRNA expression

In this study, miRNA microarrays were used to investigate miRNA expression patterns in opium poppy tissues. Root, stem, leaf and young capsule tissues with three biological replicates were compared for miRNA expression levels. A total of 12 miRNA chips were used for hybridization and data analyses.

Several miRNAs showed differential expression among different tissues (Figure 1; Table S2). A total of 232 miRNAs were differentially expressed in four different tissues. Among them, miR396a, miR396b and miR535 were mostly expressed in root; miR165a, miR166g and miR172e in stem; miR530, miR535 and miR2911 in leaf; and miR156c, miR164a and miR167 were mostly expressed in the young capsule tissue.

Some miRNAs were differentially regulated in only one tissue (Table 4). As an example, the miR167 family members showed higher expression values in the capsule than in other tissues. The stem tissue had the lowest expression level of miR156 family members when compared to the other three tested tissues. Additionally, the significant up-regulation of miR156 was detected in the capsule tissue. The data also indicated that a higher number of miRNAs were differentially regulated in the capsule tissue than in the others.

Some miRNAs (38) were differentially regulated between the capsule and leaf tissues. A broad range of expression differences were found in the capsule, including sevenfold suppression for some miRNAs (Table S3). A similar number of miRNAs were up-regulated and down-regulated. On the other hand, 35

Table 2 Conserved miRNA families. (a) miRNAs and read numbers identified by deep sequencing; and (b) miRNAs identified by directional cloning

(a)

miRNA family	Read number	miRNA family	Read number
pso-miR535	112 835	pso-miR3631	12
pso-miR156	92 129	pso-miR3633	12
pso-miR166	82 486	pso-miR3699	11
pso-miR167	34 560	pso-miR3701	11
pso-miR168	29 380	pso-miR3705	11
pso-miR157	17 899	pso-miR5181	11
pso-miR172	10 767	pso-miR855	11
pso-miR164	6218	pso-miR3707	10
pso-miR2911	3800	pso-miR3933	10
pso-miR159	3757	pso-miR4249	10
pso-miR408	2849	pso-miR2092	10
pso-miR780	2572	pso-miR390a-3p	10
pso-miR2916	1654	pso-miR2411	10
pso-miR2108	1592	pso-miR4352	9
pso-miR5225	1473	pso-miR4376	9
pso-miR6027	1113	pso-miR4403	9
pso-miR4351	1011	pso-miR4408	9
pso-miR319	972	pso-miR4415	9
pso-miR396	942	pso-miR4995	9
pso-miR1439	940	pso-miR4996	9
pso-miR1037	854	pso-miR5012	9
pso-miR390	792	pso-miR5015	9
pso-miR2673	775	pso-miR2615	9
pso-miR1851	712	pso-miR1850	9
pso-miR1089	706	pso-miR5021	8
pso-miR1310	596	pso-miR5032	8
pso-miR168-3p	547	pso-miR5039	8
pso-miR5558	469	pso-miR5054	8
pso-miR171	431	pso-miR5059	8
pso-miR530	398	pso-miR5064	8
pso-miR2628	354	pso-miR5072	8
pso-miR827	314	pso-miR5077	8
pso-miR401	287	pso-miR5079	8
pso-miR5205	281	pso-miR5083	8
pso-miR160	277	pso-miR1040	8
pso-miR1514	265	pso-miR1508	8
pso-miR894	258	pso-miR5144	7
pso-miR405	225	pso-miR5151	7
pso-miR414	213	pso-miR5164	7
pso-miR1511	208	pso-miR5174	7
pso-miR1507	203	pso-miR5175	7
pso-miR4233	176	pso-miR5180	7
pso-miR395	176	pso-miR5300	7
pso-miR1850	157	pso-miR5301	7
pso-miR472	154	pso-miR5337	7
pso-miR1023	152	pso-miR5368	7
pso-miR437	141	pso-miR5385	7
pso-miR165	141	pso-miR5490	7
pso-miR1863	139	pso-miR5493	7
pso-miR443	130	pso-miR5508	7
pso-miR845	119	pso-miR2091	7
pso-miR393	115	pso-miR5509	6
pso-miR1432	114	pso-miR5523	6
pso-miR2948	113	pso-miR5535	6

Table 2 Continued

(a)

miRNA family	Read number	miRNA family	Read number
pso-miR1044	111	pso-miR5539	6
pso-miR479	98	pso-miR5543	6
pso-miR1312	98	pso-miR5544	6
pso-miR823	93	pso-miR5565	6
pso-miR2867	90	pso-miR5568	6
pso-miR1114	87	pso-miR5635	6
pso-miR481	83	pso-miR5639	6
pso-miR1051	82	pso-miR5641	6
pso-miR397	82	pso-miR5645	6
pso-miR482	81	pso-miR5647	6
pso-miR2099	79	pso-miR5648	6
pso-miR1852	79	pso-miR5655	6
pso-miR2111	78	pso-miR5665	6
pso-miR2643	76	pso-miR5677	6
pso-miR529	75	pso-miR5679	6
pso- miR166a-5p	73	pso-miR5715	6
pso-miR774	72	pso-miR5720	6
pso-miR854	69	pso-miR5761	6
pso-miR398	68	pso-miR2636	6
pso-miR950	67	pso-miR5774	5
pso-miR1847	65	pso-miR5783	5
pso-miR1534	60	pso-miR5788	5
pso-miR810	57	pso-miR5792	5
pso-miR1106	57	pso-miR5794	5
pso-miR901	56	pso-miR5815	5
pso-miR812	53	pso-miR5816	5
pso-miR814	53	pso-miR5817	5
Pso-miR4393	53	pso-miR5825	5
pso-miR6466	53	pso-miR5837	5
pso-miR815	51	pso-miR6024	5
pso-miR818	51	pso-miR6035	5
pso-miR844	50	pso-miR6137	5
pso-miR2119	48	pso-miR6150	5
pso-miR835	48	pso-miR6158	5
pso-miR1533	47	pso-miR6164	5
pso-miR845	47	pso-miR6224	5
pso-miR821	45	pso-miR6231	5
pso-miR2055	44	pso-miR6232	5
pso-miR5565	44	pso-miR2634	5
pso-miR829	43	pso-miR6254	4
pso-miR824	42	pso-miR6266	4
pso-miR837	40	pso-miR6273	4
pso-miR169	39	pso-miR6283	4
pso-miR1153	39	pso-miR6300	4
pso-miR394	37	pso-miR6423	4
pso-miR1061	37	pso-miR6426	4
pso-miR838	36	pso-miR6437	4
pso-miR842	34	pso-miR6455	4
pso-miR1862	34	pso-miR6457	4
pso-miR859	33	pso-miR6462	4
pso-miR528	33	pso-miR6472	4
pso-miR867	32	pso-miR6475	4
pso-miR950	31	pso-miR6478	4
pso-miR1223	31	pso-miR7487	4
pso-miR952	29	pso-miR7489	4
pso-miR5658	29	pso-miR7495	4

Table 2 Continued

(a)

miRNA family	Read number	miRNA family	Read number
pso- miR2950	29	pso-miR7496	4
pso-miR1146	29	pso-miR7692	4
pso-miR1117	28	pso-miR7713	4
pso-miR1118	27	pso-miR7716	4
pso-miR1120	27	pso-miR7723	4
pso-miR156e-3p	26	pso-miR7724	4
pso-miR1122	25	pso-miR7730	4
pso-miR1127	24	pso-miR7743	4
pso-miR1128	24	pso-miR7747	4
pso- miR1028	24	pso-miR7748	4
pso-miR1134	23	pso-miR7757	4
pso-miR1135	20	pso-miR7758	4
pso-miR1136	20	pso-miR7760	4
pso-miR1137	20	pso-miR7761	4
pso-miR1510	20	pso-miR7765	4
pso-miR1320	19	pso-miR7770	4
pso-miR1425	19	pso-miR7783	4
pso-miR1152	19	pso-miR7786	4
pso-miR2931	19	pso-miR7839	4
pso-miR1436	18	pso-miR7840	4
pso-miR1440	18	pso-miR161	4
pso-miR1441	18	pso-miR1314	4
pso-miR1517	18	pso-miR1867	4
pso-miR1520	18	pso-miR1433	4
pso-miR1527	18	pso-miR2097	4
pso-miR1514	18	pso-miR473	4
pso-miR1528	17	pso-miR1080	4
pso-miR1846	17	pso-miR8000	3
pso-miR1873	17	pso-miR8005	3
pso-miR1879	16	pso-miR8010	3
pso-miR1886	15	pso-miR8013	3
pso-miR399	15	pso-miR8019	3
pso-miR1881	15	pso-miR8021	3
pso-miR2118	14	pso-miR8024	3
pso-miR2919	14	pso-miR8029	3
pso-miR2922	14	pso-miR8036	3
pso-miR2930	14	pso-miR8048	3
pso-miR2932	14	pso-miR912	3
pso-miR158	14	pso-miR1217	3
pso-miR860	14	pso-miR2275	2
pso-miR1150	14	pso-miR2936	2
pso-miR2947	13	pso-miR6434	2
pso-miR2950	13	pso-miR415	2
pso-miR3434	13	pso-miR1070	2
pso-miR3435	12	pso-miR162	2
pso-miR3444	12	pso-miR426	2
pso-miR3628	12	pso-miR1154	1
pso-miR3630	12	pso-miR820	1

(b)

miRNA name	miRNA sequence (5′→3′)	Mature miRNA length (nt)	Pre-miRNA length (nt)
pso-miR3640	UACAGGAUUGAUGGUGCCUAC	21	74
pso-miR480b	GCCUACAGGAUUGAUGGUGCC	22	131

Table 2 Continued

(b)

miRNA name	miRNA sequence (5′→3′)	Mature miRNA length (nt)	Pre-miRNA length (nt)
pso-miR2161	UUUUUUUUUUUUUACAUAUAAA	22	69
pso-miR855	AAAGGAUAAAGAAAAGGAGUU	21	61
pso-miR807a	CGCCUGUGGGAUGACGCGCUG	21	54
pso-miR771	CUGUGGGGGCCCUCAAUGCUG	21	132
pso-miR2118p	UGCCCGAGGCUGCCCGUGCCUA	22	54
pso-miR480a	CUACAGGAUUGAUGUGGAAUU	21	62
pso-miR2644b	CGGGCACCAUCGAUCUGUAGG	21	89
pso-miR480b	CUACAGGAUUGAUGGUGCCUA	21	57
pso-miR3640	GAUUGAUGGUGCCUACAGGAU	21	86
pso-miR2644a	CACCAUCAAUCUGUAGGCACC	21	96

Table 3 Potential novel miRNAs found in poppy

miRNA name	miRNA sequence (5′→3′)	Mature miRNA length (nt)	Pre-miRNA length (nt)
pso-miRc3	ACAGGAUUGGGGGGGGGCCAGG	21	114
pso-miRc4	UGGAUGGUUCAUCCUGGAAGGA	22	51
pso-miRc5	GUGGUUUGUUUUUUUUUUUU	21	105
pso-miRc6	GGGGGGGGGUUUUUUUUUUUU	21	51
pso-miRc8	AAAAAAACACCCACACCAAAA	21	70
pso-miRc12	UGCUGUGGGGGCGCUCAAUGC	21	133
pso-miRc15	GGCCGCCGGCAGGUCGACCA	21	132
pso-miRc18	GGGCGAUUGGGCCCGACGUCGCA	23	91
pso-miRc21	GCCGCGGGAUUGAUUGAUGGU	21	70
pso-miR1*	GAAAACUGUUGUAAUUGGCA	20	252
pso-miR13*	UGAUGGAGGAGAAGAGAAGA	20	165

*Identified by deep sequencing.

miRNAs were differentially regulated in the capsule when compared to stem tissue, including 23 up-regulated for the former (Table S4).

A higher number of differentially regulated miRNAs were observed in the leaf vs. root than in other tissues (Table S5). Thus, the expression level of 56 miRNAs was differentially regulated between leaf and root tissues. Namely, miR159, miR319, miR396 and miR2916 were highly expressed in root relative to leaf tissue, while the opposite pattern was observed for miR157, miR164, miR167, miR168 and miR530. In addition, pso-miR535 was one of the most abundant miRNAs in both deep sequencing and microarray approaches (Tables 2a and S2).

Target gene identification and functional analyses

Target transcript prediction and annotation

A total of 1469 mRNA transcripts putatively targeted by 105 individual opium poppy miRNAs were predicted (Table S6). For a comprehensive annotation, both the Gene Ontology (GO) and the Kyoto Encyclopedia of Genes and Genomes (KEGG) pathway

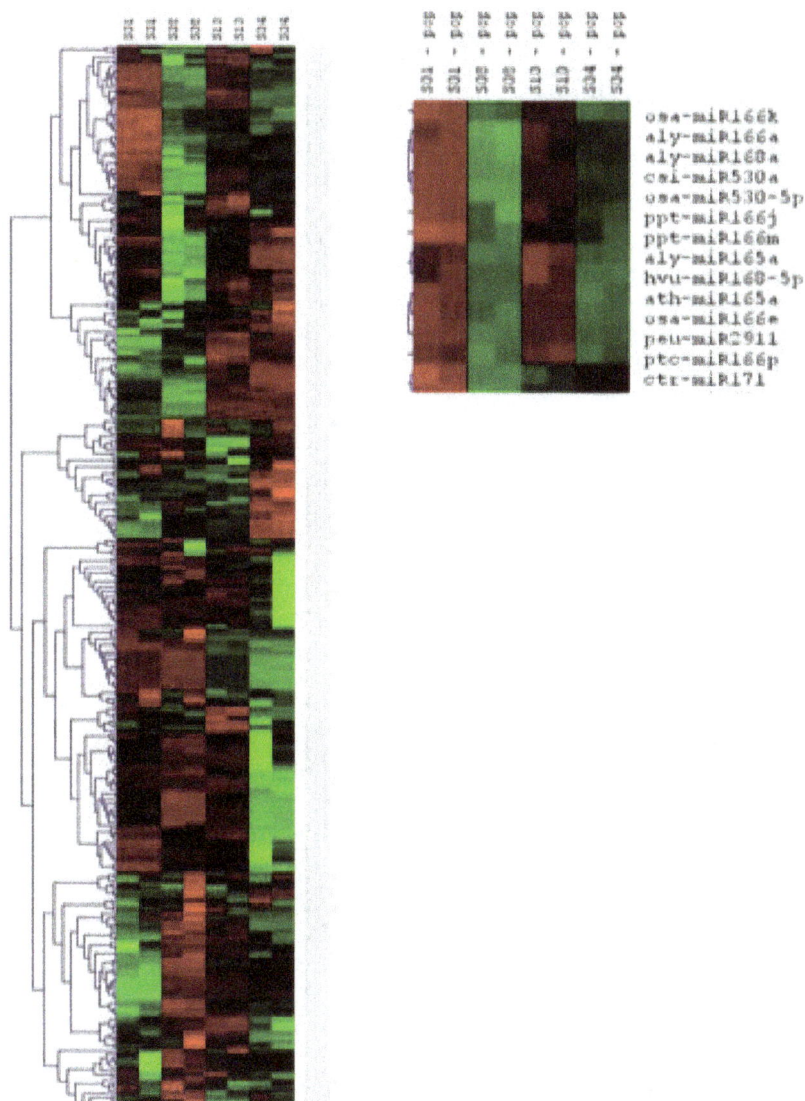

Figure 1 Microarray analyses of miRNA belonging to four different opium poppy tissues. The cluster analysis was performed for the miRNA with signal strength ≥32.

Table 4 Some of the differentially expressed miRNAs found by the microarray analyses

Tissue	Up-regulated	Down-regulated
Capsule	pso-miR156, pso-miR164, pso-miR167, pso-miR1436, pso-miR2919, pso-miR319 and pso-miR159	pso-miR535, pso-miR530-5p, pso-miR172, pso-miR166, 109mm pso-miR2911 and pso-miR529
Stem	pso-miR168	pso-miR156, pso-miR2916 and pso-miR1450
Leaf	pso-miR166, pso-miR171, pso-miR395, pso-miR2911 and pso-miR529	pso-miR1436, pso-miR319, pso-miR2919, pso-miR390 and pso-miR159
Root	pso-miR172, pso-miR396, pso-miR390, pso-miR2916 and pso-miR1450	pso-miR164, pso-miR167, pso-miR171, pso-miR168 and pso-miR395

analyses were performed. The results are summarized in Table S7. Multiple genes were found to be targeted by a single miRNA. For instance, miR156 targets 13 transcripts which are involved in diverse biological processes. A total of 286 ontology terms were observed, with most of the target genes being involved in response to biotic and abiotic stresses (214 target genes), secondary-metabolite biosyntheses (15 target genes) and several transcription events (16 genes) (Table S7).

The target identification analyses also showed that pso-miR2161 and pso-miR13 cleaved BIA-synthesis transcripts of S-adenosyl-L-methionine:3'-hydroxy-N-methylcoclaurine 4'-O-

methyltransferase 2 (4-OMT; GenBank accession AY217334) and 7-O-methyltransferase (7-OMT; GenBank accession FJ156103), respectively. Using computational analyses, although pre-miRNA sequences were absent, some of the small RNA sequences were predicted as putative miRNAs. The sequences were subjected to target transcript analyses. This way, three putative miRNAs were identified as targeting BIA-biosynthesis pathway genes. Thus, pso-miR t0047847 (read number: 18), pso-miR t0013376 (read number: 68) and pso-miR t0000199 (read number: 2611) target codeinone reductase (COR; GenBank accession FJ624147), (7S)-salutaridinol-7-O-acetyltransferase (SAT; GenBank accession

FJ200354) and tyrosine/dihydroxyphenylalanine decarboxylase (*TYDC*; GenBank accession AF025435), respectively.

A total of 43 pathways were identified by the KEGG analyses. Most of the miRNA target genes were involved in carbohydrate metabolism (14), followed by translation (11). Some KEGG pathways (15) were represented only once (Figure 2).

Experimental identification of miRNA targets

To identify the target transcript sequences for selected miRNA, modified 5′-RNA ligase-mediated (RLM) and rapid amplification of cDNA ends (RACE) (5′-RLM-RACE) experiments were performed. Four of the experimentally confirmed transcripts included miRNA cleavage sites (Table S8). One of the target transcripts was the *Lox* gene (EST stem_S098_C07.SEQ; GenBank accession FG613414), which was cleaved by pso-miR535. The 5′-end

sequencing identified the cleavage site, located at the 11th nucleotide upstream of the miR535-binding site.

qRT-PCR measurements

miRNA and target mRNA expression quantification

Quantitative real-time polymerase chain reaction (qRT-PCR) assays were performed to experimentally verify the expression of eight known miRNAs (miR156c, miR159c, miR160b-g, miR164a, miR166g, miR390a, miR395b and miR2950), one novel miRNA (miR13) and their target transcripts. As expected, qRT-PCR results of miRNA expression in different tissues were negatively correlated with the ones of target transcripts (Figure 3). Thus, the pso-miR159 expression was higher in stem compared to capsule, while an opposite pattern was detected for its target transcript

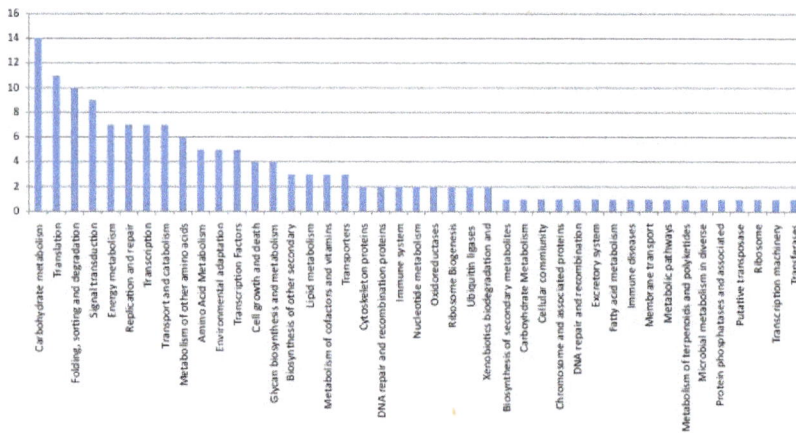

Figure 2 Kyoto Encyclopedia of Genes and Genomes (KEGG) pathway results. The *y*-axis represents the abundance of the pathways.

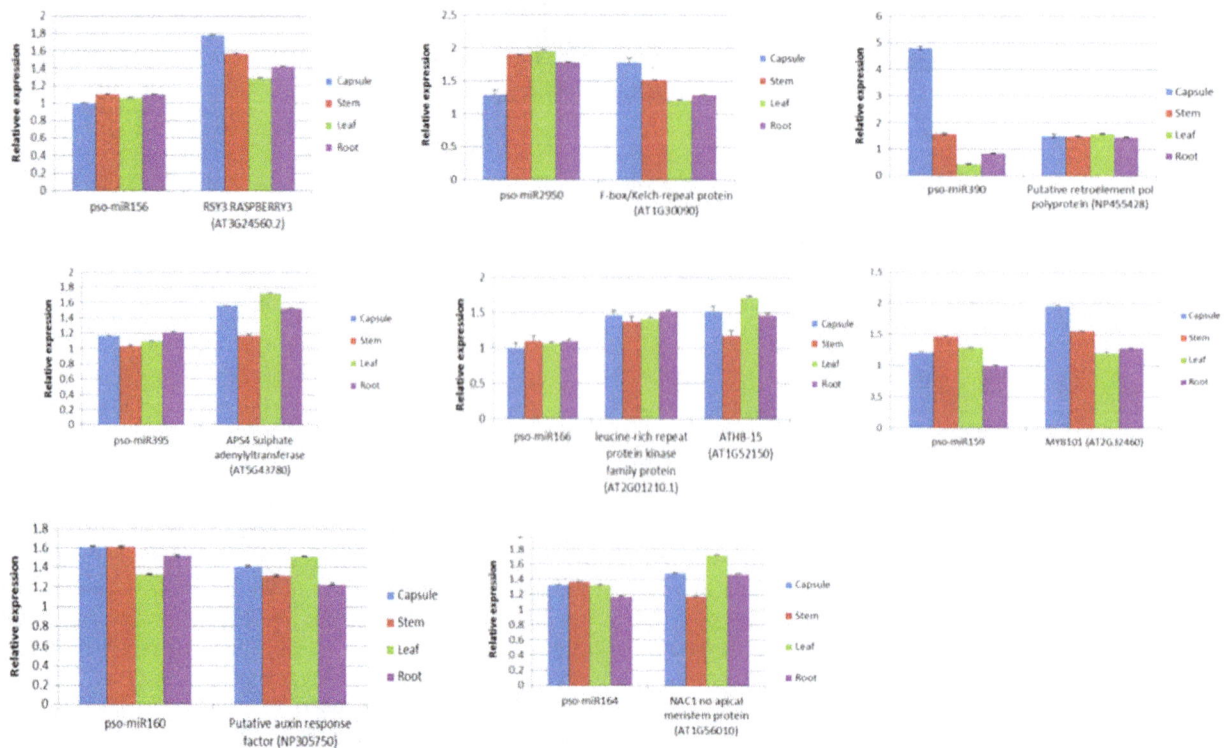

Figure 3 qRT-PCR expression levels of selected miRNAs and target transcripts in four different opium poppy tissues.

(*MYB101*). Additionally, the data obtained from comparative microarray analyses were also validated via qRT-PCR measurements. In both analyses, the expression of pso-miR160 and pso-miR166 was lower in leaf and capsule tissues compared to root and stem tissues. The pso-miR2950 expression was lower in capsule compared to other tissues, as confirmed by both qRT-PCR and miRNA microarray analyses, as well. Furthermore, the target transcript expression was found to be highest in capsule for pso-miR2950.

miRNA-regulating genes involved in the alkaloid-biosynthesis pathway

The expression of the pso-miR13 novel opium poppy miRNA was also measured, and its presence in all tissues was further validated with qRT-PCR. Its expression level in capsule was higher than in root tissue. Expectedly, the expression of the *7-O-methyltransferase* pso-miR13 target gene was repressed more in capsule than in root tissue (Figure 4). The expression of pso-miR2161 targeting the *4-OMT* gene was higher in capsule than in stem. Reversely, as expected, the target gene was expressed more in the latter tissue.

Moreover, the three putative miRNAs and their possible target transcripts were also analysed. It was discovered that the expression levels of the three putative miRNAs were slightly different between each other. Similarly, minor differences were observed for the target gene expressions (Figure 4).

Discussion

The miRNAs play an important role in the regulation of many biological processes, such as plant growth and development, biotic- and abiotic-stress responses and metabolic pathways. Some plant miRNAs have been discovered and characterized, revealing their regulatory roles (Vahap Eldem, 2013). Their functions in plant development and various stresses have been described (Eldem *et al.*, 2012; Inal *et al.*, 2014; Seefried *et al.*, 2014; Yanik *et al.*, 2013). Moreover, it has been shown that in Arabidopsis, the secondary-metabolite profiles are altered by the expression of miR163 and miR393 (Ng *et al.*, 2011; Robert-Seilaniantz *et al.*, 2011).

However, the possible relation between alkaloid biosynthesis and miRNA regulation is still unclarified. Indeed, although different miRNAs have been extensively investigated in plant species in recent years, there are no previous studies on their profiles in opium poppy (Eldem *et al.*, 2012; Gupta *et al.*, 2014; He *et al.*, 2014; Li *et al.*, 2013; Yu *et al.*, 2012; Zeng *et al.*, 2014). Indeed, only a limited number of computationally identified opium poppy

miRNAs have been reported (Unver *et al.*, 2010b). Thus, to characterize opium poppy miRNAs, comprehensive experimental approaches have been performed in this work. In addition to high-throughput sRNA sequencing, microarray analyses were utilized to determine tissue-specific miRNA expression profiles. Furthermore, target transcript analyses were performed using both bioinformatics and experimental methods.

By sequencing of sRNA, 316 conserved and 11 novel opium poppy miRNAs were identified. A broad variation in miRNA expression was observed. Some of the conserved miRNAs, such as miR156, miR166, miR167 and miR172, were found to be the most abundant, which is in agreement with previous findings (Lukasik *et al.*, 2013; Yanik *et al.*, 2013). In our study, miR535 has been identified as one of the most up-regulated miRNAs (Table 2), similarly as reported by others (Xia *et al.*, 2012). On the other hand, the miR535 was shown as moderately expressed in Californian poppy (Barakat *et al.*, 2007). Our results confirmed that the plant miRNA expression level varies across the species. Moreover, the miRNA expression may also be different between the plant tissues (Breakfield *et al.*, 2012). To comparatively analyse miRNA profiles, we performed miRNA microarray assays in four tissues. This way, 232 differentially expressed miRNAs were identified in opium poppy. Some miRNAs such as miR156 and miR167 showed a higher expression in capsule than in the other three tissues. The microarray results revealed that root had the highest pso-miR535 expression, being followed by leaf tissue. A similar pattern has been reported in grapevine (Mica *et al.*, 2009). The modified 5′-RLM-RACE experiments revealed that pso-miR535 targets the *Lox* gene. The lipoxygenase enzyme (LOX) encoded by *Lox* gene is involved in the production of jasmonic and linoleic acids, thus contributing to the alkaloid biosynthesis (Holkova *et al.*, 2010). The induction of the *Lox* gene caused the accumulation of sanguinarine in opium poppy suspension cultures. Because the microarray data presented the lowest expression of miR535 in capsule, this might lead to enhance *Lox* induction, generating more alkaloids (Holkova *et al.*, 2010).

Consistent with previous reports (Eldem *et al.*, 2012; Yanik *et al.*, 2013), more than one target mRNA was found to be regulated by a single opium poppy miRNA (Table S6). Carbohydrate metabolism, transcription and protein folding–sorting–degradation were primary pathways regulated by opium poppy miRNA. In our previous study, involvement of carbohydrate metabolism in BIA biosynthesis was also reported (Gurkok *et al.*, 2014). Due to their post-transcriptional regulatory roles, it is highly expected that target genes of the miRNA were involved in several genetic-information processing mechanisms. GO analyses

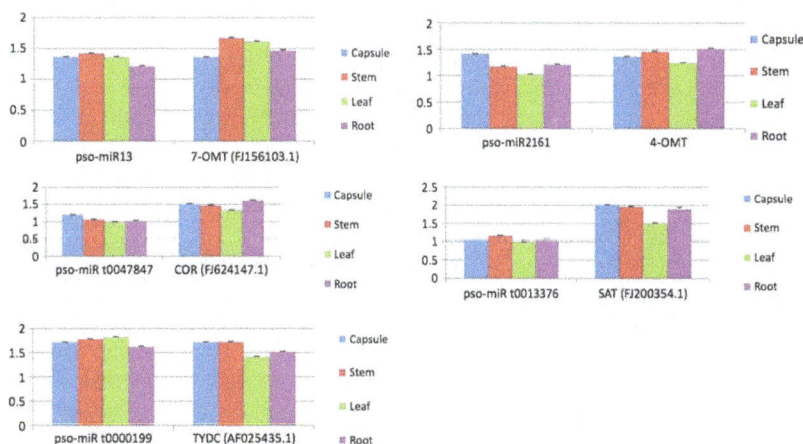

Figure 4 qRT-PCR quantifications of miRNAs and their target transcripts involved in alkaloid biosynthesis.

revealed that opium poppy miRNA target transcripts mostly involved in response mechanisms, which is in agreement with our previous transcriptome-wide conclusions (Gurkok et al., 2014).

The S-reticuline is an intermediate molecule in BIA biosynthesis, which is generated from S-norcoclaurine by 4-OMT. Using bioinformatics tools, 4-OMT was predicted to be the target transcript of pso-miR2161 (Table S6). The total amount of alkaloid content increased upon suppression of the 4-OMT mRNA (Desgagne-Penix and Facchini, 2012). Comparing the tissue-specific miRNA accumulation, the stem has higher level of pso-miR2161 than that of capsule tissue (Figure 4). A similar pattern for 4-OMT mRNA has been also shown by others (Samanani et al., 2006). Due to the central role of phloem sieve elements for alkaloid accumulation, up-regulation of 4-OMT transcript is expected. Therefore, the 4-OMT mRNA accumulation in stem is supposed to be caused by suppression of pso-miR2161.

The pso-miR13 is a novel opium poppy miRNA discovered in this work. Target prediction analysis suggested that pso-miR13 might cleave 7-OMT transcript. The 7-OMT is involved in conversion of S-reticuline to morphinan alkaloids (Desgagne-Penix et al., 2012; Gurkok et al., 2014; Weid et al., 2004). Indeed, the silencing of 7-OMT gene caused the accumulation of BIA in latex (Desgagne-Penix and Facchini, 2012). Additionally, the qRT-PCR data indicated the lowest accumulation of 7-OMT mRNA in capsule (Figure 4). Thus, the cleavage of 7-OMT transcript by pso-miR13 might be one of the regulatory mechanisms for BIA biosynthesis. On the other hand, the pso-miR408 possibly targets mRNA from a gene encoding FAD-binding and BBE domain-containing protein, also known as reticuline oxidase-like protein. The role of reticuline oxidase is the conversion of S-reticuline to (S)-scoulerine in the BIA pathway. Therefore, it is possible that miR408 might be one of the BIA-regulatory miRNAs (Figure 5). Furthermore, deep-sequencing outputs showed putative miRNA candidates targeting BIA-biosynthesis genes. In silico analyses revealed that SAT, TYDC and COR transcripts were possibly cleaved by three putative opium poppy miRNAs.

Thus, the miRNAs were involved in the regulation of BIA biosynthesis in this study, in addition to their functions in diverse biological processes.

Experimental procedures

Plant growth and sample preparation

The opium poppy (Papaver somniferum cv Ofis 95, TMO, Turkey) plants were grown in a growth cabin with 16 h/light at 23 °C and 8 h/dark at 20 °C with 70% humidity. The seeds were sown in pods (10 cm diameter) including perlite, compost and soil (1 : 1 : 1). Four-month-old poppy plants were used as tissue sources. The root, stem, leaf and young capsule prior to flowering tissues were collected and directly transferred into liquid nitrogen. The samples were stored at −80 °C until used. Total RNAs were isolated using TRIzol reagent from Invitrogen/Life Technologies/ Thermo Fisher Scientific (Waltham, MA) according to the man-ufacturer's instructions. The samples were collected from three different plants as biological replicates. The total RNA amount and quality were determined using a NanoDrop 2000c from NanoDrop Technologies/Thermo Fisher Scientific (Waltham, MA).

Small RNA isolation

All opium poppy tissues were separately used for small RNA isolation to identify and analyse expression profiles. The sRNAs

were isolated using the mirVana kit from Ambion/Life Technol-ogies/Thermo Fisher Scientific (Waltham, MA), according to the manufacturer's instructions. The RNAs were subjected to 15% (w/v) denaturing polyacrylamide gel electrophoresis (PAGE), after which the 18- to 30-nt-long RNAs were excised and isolated from the gel.

Library preparation and sequencing

To detect and characterize the expressed miRNA, equal amounts of sRNA from all tissues were mixed and used for library construction. A total of 8 µg pooled sRNA was used for sequencing by the Genome Analyzer GA-I from Illumina (San Diego, CA), following the manufacturer's protocols. Briefly, purified sRNAs were ligated with 5′- and 3′-linkers. These sRNA libraries were reversely transcribed into cDNA, amplified and sequenced by BGI (Hong Kong, China).

Deep-sequencing data analyses

Firstly, the raw sequences were processed to excise the linker sequences and to filter out low-copy and low-quality reads by Illumina's Genome Analyzer Pipeline software. Extracted small RNA sequences longer than 15 nt were filtered to identify mRNA using the Rfam database (http://www.sanger.ac.uk/resources/ databases/rfam.html; Griffiths-Jones et al., 2005) and noncoding RNA (e.g. tRNA, rRNA, snRNA and snoRNA) using the noncoding RNA database (ncRNAdb; http://biobases.ibch.poznan.pl/ncRNA). Additionally, the unique sequences were matched to the known miRNA from miRBase 20.0 (Griffiths-Jones et al., 2008; http:// www.mirbase.org) using the basic local alignment search tool for translated nucleotide database using a protein query (BLASTn; http://blast.ncbi.nlm.nih.gov/Blast.cgi? CMD=Web&PAGE_TYPE=BlastHome). Only three mismatches were allowed between the identified opium poppy miRNA and the known plant miRNA (Unver, 2009). All the sRNA and matched known miRNA sequences from miRBase were searched against expressed sequence tags (ESTs) using the SOAP 2.0 program (Li et al., 2008).

To identify novel opium poppy miRNA, the unique small RNA sequences were aligned with known EST sequences of opium poppy deposited in GenBank (http://www.ncbi.nlm.nih.gov/ genbank). By alignment of small RNA and EST, miRNA precursor (pre-miRNA) sequences were obtained. The potential miRNA candidates were predicted from their precursor-sequence folding. Previously published criteria (Allen et al., 2005) were applied using Mireap developed in BGI, to analyse structural features of precursor miRNA sequences. The filtered pre-miRNA sequences were additionally checked by Mfold (Zuker, 2003) in relation to their highest negative minimal folding energy and minimal free-folding energy indexes (Eldem et al., 2013). To predict putative miRNA target gene candidates, miRNA sequences were aligned with (i) plant EST using psRNATarget software with default parameters (Dai and Zhao, 2011), and (ii) opium poppy EST deposited in GenBank, using the parameters mentioned in Unver and Budak (2009a,b).

miRNA directional cloning

The pooled sRNAs were used as templates for RNA cloning experiments, according to the miRCat Small RNA Cloning Kit (https://www.idtdna.com/pages/products/mirna/mirna-cloning-pr oducts) from Integrated DNA Technologies (IDT; Coralville, IA) instructions. Briefly, 5′- and 3′-adapters were ligated to the isolated sRNA for cDNA synthesis and amplification. The serial analysis of

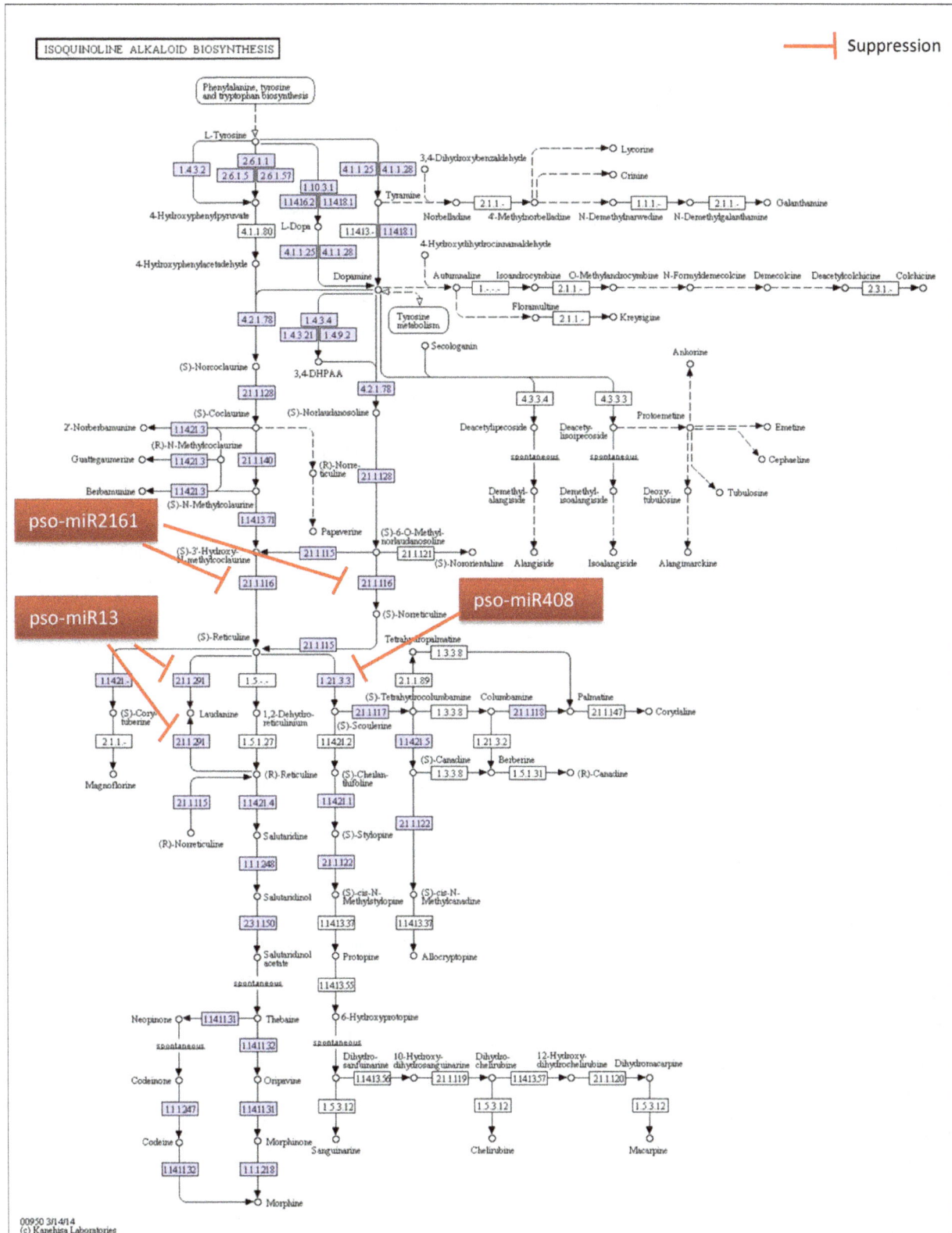

Figure 5 Benzylisoquinoline alkaloids (BIA)-synthesis pathway and involvement of opium poppy miRNA (KEGG pathway). The miRNAs suppressing the transcripts involved in the pathway are shown in red boxes. KEGG, Kyoto Encyclopedia of Genes and Genomes.

gene expression (SAGE)-like serial miRNA-linker (concatamerized SAGE tags) cloning method was applied to clone more than one miRNA into the pGEM-T Easy Vector System I (https://www.pro mega.com/products/pcr/pcr-cloning/pgem_t-easy-vector-systems)

from Promega (Fitchburg, WI). Obtained clones were sequenced via ABI PRISM 3700 Genetic Analyzer (Applied Biosystems/Life Technologies/Thermo Fisher Scientific, Waltham, MA), using the Sanger terminator methodology.

miRNA microarray analyses

The miRNA microarray experiment was performed by service provider LC Sciences (http://www.lcsciences.com; Houston, TX). The miRNA microarray chip involved triplicates of 4025 unique mature miRNA probes corresponding to miRNA transcripts listed in miRBase release 20.0 belonging to 72 species including wheat (40 probes), rice (553 probes) and barley (65 probes), with multiple control probes. The chip also covered 5690 unique mature miRNAs from the plant microRNA database (http://bioinformatics.cau.edu.cn/PMRD) with multiple control probes (74 × 1–16). The experiment was performed with two technical and three biological replicates of each sample. Four different tissue samples were hybridized with miRNA chips for comparative expression analyses. The small RNAs of root, stem, leaf and young capsule tissues were isolated with the mirVana kit previously described and hybridized with the miRNA chip. The labelling was achieved with Cy3 and Cy5 fluorescent dyes, and hybridization was performed overnight on a μParaflo microfluidics chip using a microcirculation pump from Atactic Technologies (Houston, TX).

After hybridization, the images were captured using a GenePix 4000B Microarray Scanner from Molecular Devices (Sunnyvale, CA) and processed by the Array-Pro Analyzer software from Media Cybernetics (Rockville, MD). The data were analysed by subtraction of background and normalized by obtained signals with the locally weighted scatterplot smoothing method to eliminate system-related variations. For the two-colour experiments, P values of the t-tests and the ratio of the two sets of detected signals (log_2-transformed and balanced) were calculated. Differentially expressed probes with P-values ≤ 0.01 were designated as significant (Unver et al., 2010a). The microarray data were deposited in the Gene Expression Omnibus (http://www.ncbi.nlm.nih.gov/geo; accession GSE62716).

Experimental target identification with modified 5'-RLM-RACE

To experimentally discover the potential target transcripts of identified opium poppy miRNA, a modified version of the 5'-RLM-RACE assay was performed (Arenas-Huertero et al., 2009; Kantar et al., 2010; Wei et al., 2009). Target ESTs of pso-miR156, pso-miR157, pso-miR160, pso-miR166, pso-miR167, pso-miR390, pso-miR535 and pso-miR2950 were predicted using psRNATarget (plant small RNA target analysis server) (http://plantgrn.noble.org/psRNATarget; Dai and Zhao, 2011) and BLASTn, as noted above. The RACE PCRs were performed using computationally proposed miRNA target ESTs: miRNA156 (GenBank accession TC368029), miRNA157 (GenBank accession NP1659796), miRNA160 (GenBank accession NP305750), miRNA166 (GenBank accession TC368138), miRNA167 (GenBank accession TC364720), miRNA390 (GenBank accession NP455428), miRNA535 (GenBank accession NR1121) and miRNA2950 target (GenBank accession TC392432).

To select the target transcripts of each miRNA for RLM-RACE analyses, firstly the corresponding putative miRNA target transcripts were in silico translated. Then, these were blasted against protein databases of Viridiplantae using BLASTp for standard protein searches (http://blast.ncbi.nlm.nih.gov/Blast.cgi/). The ESTs with the highest matches were selected to design specific primers for the assay. For each specific EST, including specific miRNA cleavage site, three specific primers (Table S9) were designed using Primer3 software (http://frodo.wi.mit.edu). Then,

1 μg of the mixed total RNA was ligated to adapters and used as template for cDNA synthesis using the FirstChoice RLM-RACE Kit from Ambion/Life Technologies/Thermo Fisher Scientific, following the manufacturer's procedures. The obtained sequences were searched against GenBank and miRBase v.20.

miRNA target transcript annotations

The putative mature miRNA sequences were compared with the National Center for Biotechnology Information (NCBI) Expressed Sequence Tags database (dbEST; http://www.ncbi.nlm.nih.gov/dbEST). Alignments between each miRNA and its putative miRNA target(s) should meet certain criteria in the literature (Allen et al., 2005; Schwab et al., 2005). To functionally annotate and categorize the putative miRNA targets, Blast2GO software suite v.2.3.1 (http://www.blast2go.com/b2ghome) with default parameters was applied (Gotz et al., 2008). Putative miRNA target sequences were used as queries against the KEGG database (http://www.genome.jp/kegg) for pathway analyses (Moriya et al., 2007).

miRNA validation and measurement via qRT-PCR

To confirm the presence and to measure the expression of the identified opium poppy miRNA, nine previously described miRNAs (miR156, miR159c, miR160, miR164, miR166, miR390, miR395b, miR2161 and miR2950) and one novel miRNA(miR13) were selected for qRT-PCR. Additionally, three putative miRNAs (pso-miR t0047847, pso-miR t0013376 and pso-miR t0000199) were subjected to qRT-PCR analyses. Total RNA was isolated from roots, stems, leaves and young capsules using TRIzol reagent, as previously described. The total RNAs were treated with DNase I from Fermentas/Thermo Fisher Scientific (Waltham, MA) to remove the genomic DNA. Specific miRNA stem-loop RT-PCR primers were designed for reverse transcription. In short, 1 μg of DNase-treated RNA from each tissue sample was used to generate single-stranded miRNA cDNA by Superscript III First-Strand Synthesis System from Invitrogen/Life Technologies/Thermo Fisher Scientific, according to Unver et al. (2010a). Specific forward primers for the selected miRNAs were also designed (Table S10).

The differential-expression levels of 11 opium poppy miRNAs were quantified in all four tissues with a SYBR Green I qRT-PCR assay, using the FastStart Universal SYBR Green Master Mix from Roche Diagnostics (Basel, Switzerland) and miRNA-specific primers in a LightCycler 480 Instrument II Roche (Mannheim, Germany). To normalize the miRNA expression, serial dilutions (1/2, 1/4, 1/8 and 1/16) of each sample were used as triplicate. Moreover, to remove the false and background signals, cDNAs of NoRT (without stem-loop RT primer), NoRNA (without RNA) and No template (without template) were synthesized for each miRNA qRT-PCR (Yanik et al., 2013).

miRNA target transcript validation and measurement

To further measure and validate the expression levels of the predicted miRNA target transcripts of the poppy miRNA, qRT-PCR assays were performed with 13 miRNA target transcript pairs. The target transcripts were identified using the psRNATarget previously described, selecting the 'User-submitted small RNA/user-submitted transcripts' tab and the BLASTn algorithm noted above. Specific PCR primers were designed using the online Primer3Plus software (http://www.bioinformatics.nl/cgi-bin/primer3plus/primer3plus.cgi; Untergasser et al., 2007) (Table S11). Firstly, 1.5 μg of total RNA was cleaned up with the RNeasy Plant

Mini Kit from Qiagen (Hilden, Germany) and used to synthesize cDNA by the Superscript III First-Strand Synthesis System from Invitrogen/Life Technologies/Thermo Fisher Scientific, according to the manufacturer's instructions.

In brief, the qRT-PCR was performed in a 96-well plate of the LightCycler 480 Instrument II with 20 μL reactions that contained 1–2 μL of the cDNA, 10 nm of each specific forward and reverse primers and FastStart SYBR Green Master Mix (Roche) indicated above. Each experiment was run in triplicate for each target transcript, and their relative quantities were calculated based on the *18S rRNA* as a normalizer gene (GenBank accession DQ912880), using 5'-TAGCGGGCCTCTTCTCTTTC-3' (forward primer) and 5'-CGCATTTCGCTACGTTCTTC-3' (reverse primer). The qRT-PCR profile included preheating for 10 min at 95 °C and 35 cycles (95 °C for 30 s, 57 °C for 30 s and 72 °C for 30 s). The melting curves of the real-time PCR products were analysed for each run to filter out the false-positive peaks. The data of the fluorescence signals were obtained from 57 to 95 °C, as the temperature increased at 0.5 °C per second (Eldem *et al.*, 2012; Vahap Eldem, 2013; Unver, 2009; Yanik *et al.*, 2013).

Acknowledgements

The study was kindly supported by TUBITAK with grant numbers 109O661 and 111O036. The manuscript was critically revised by Prof Gabriel Dorado of Cordoba University, Spain.

References

Achard, P., Herr, A., Baulcombe, D.C. and Harberd, N.P. (2004) Modulation of floral development by a gibberellin-regulated microRNA. *Development*, **131**, 3357–3365.

Allen, E., Xie, Z.X., Gustafson, A.M. and Carrington, J.C. (2005) microRNA-directed phasing during trans-acting siRNA biogenesis in plants. *Cell*, **121**, 207–221.

Arenas-Huertero, C., Pérez, B., Rabanal, F., Blanco-Melo, D., De la Rosa, C., Estrada-Navarrete, G., Sanchez, F., Covarrubias, A.A. and Reyes, J.L. (2009) Conserved and novel miRNAs in the legume Phaseolus vulgaris in response to stress. *Plant Mol. Biol.* **70**, 385–401.

Aukerman, M.J. and Sakai, H. (2004) Regulation of flowering time and floral organ identity by a microRNA and its APETALA2-like target genes (vol 15, pg 2730, 2003). *Plant Cell*, **15**, 2730–2741.

Barakat, A., Wall, K., Leebens-Mack, J., Wang, Y.J., Carlson, J.E. and Depamphilis, C.W. (2007) Large-scale identification of microRNAs from a basal eudicot (*Eschscholzia californica*) and conservation in flowering plants. *Plant J.* **51**, 991–1003.

Breakfield, N.W., Corcoran, D.L., Petricka, J.J., Shen, J., Sae-Seaw, J., Rubio-Somoza, I., Weigel, D., Ohler, U. and Benfey, P.N. (2012) High-resolution experimental and computational profiling of tissue-specific known and novel miRNAs in Arabidopsis. *Genome Res.* **22**, 163–176.

Chen, X. (2004) A microRNA as a translational repressor of APETALA2 in Arabidopsis flower development. *Science*, **303**, 2022–2025.

Dai, X.B. and Zhao, P.X. (2011) psRNATarget: a plant small RNA target analysis server. *Nucleic Acids Res.* **39**, W155–W159.

Desgagne-Penix, I. and Facchini, P.J. (2012) Systematic silencing of benzylisoquinoline alkaloid biosynthetic genes reveals the major route to papaverine in opium poppy. *Plant J.* **72**, 331–344.

Desgagne-Penix, I., Farrow, S.C., Cram, D., Nowak, J. and Facchini, P.J. (2012) Integration of deep transcript and targeted metabolite profiles for eight cultivars of opium poppy. *Plant Mol. Biol.* **79**, 295–313.

Desgagné-Penix, I., Khan, M.F., Schriemer, D.C., Cram, D., Nowak, J. and Facchini, P.J. (2010) Integration of deep transcriptome and proteome analyses reveals the components of alkaloid metabolism in opium poppy cell cultures. *BMC Plant Biol.* **10**, 252.

Eldem, V., Akcay, U.C., Ozhuner, E., Bakir, Y., Uranbey, S. and Unver, T. (2012) Genome-wide identification of miRNAs responsive to drought in peach (*Prunus persica*) by high-throughput deep sequencing. *PLoS ONE*, **7**, e50298.

Eldem, V., Okay, S. and Unver, T. (2013) Plant microRNAs: new players in functional genomics. *Turk. J. Agric. For.* **37**, 1–21.

Gotz, S., Garcia-Gomez, J.M., Terol, J., Williams, T.D., Nagaraj, S.H., Nueda, M.J., Robles, M., Talon, M., Dopazo, J. and Conesa, A. (2008) High-throughput functional annotation and data mining with the Blast2GO suite. *Nucleic Acids Res.* **36**, 3420–3435.

Griffiths-Jones, S., Moxon, S., Marshall, M., Khanna, A., Eddy, S.R. and Bateman, A. (2005) Rfam: annotating non-coding RNAs in complete genomes. *Nucleic Acids Res.* **33**, D121–D124.

Griffiths-Jones, S., Saini, H.K., van Dongen, S. and Enright, A.J. (2008) miRBase: tools for microRNA genomics. *Nucleic Acids Res.* **36**, D154–D158.

Guo, H., Xie, Q., Fei, J.F. and Chua, N.H. (2005) MicroRNA directs mRNA cleavage of the transcription factor NAC1 to downregulate auxin signals for Arabidopsis lateral root development. *Plant Cell*, **17**, 1376–1386.

Gupta, O., Sharma, P., Gupta, R. and Sharma, I. (2014) MicroRNA mediated regulation of metal toxicity in plants: present status and future perspectives. *Plant Mol. Biol.* **84**, 1–18.

Gurkok, T., Turktas, M., Parmaksiz, I. and Unver, T. (2014) Transcriptome profiling of alkaloid biosynthesis in elicitor induced opium poppy. *Plant Mol. Biol. Rep.* doi:10.1007/s11105-014-0772-7.

He, H., He, L. and Gu, M. (2014) Role of microRNAs in aluminum stress in plants. *Plant Cell Rep.* **33**, 831–836.

Holkova, I., Bezakova, L., Bilka, F., Balazova, A., Vanko, M. and Blanarikova, V. (2010) Involvement of lipoxygenase in elicitor-stimulated sanguinarine accumulation in *Papaver somniferum* suspension cultures. *Plant Physiol. Biochem.* **48**, 887–892.

Inal, B., Turktas, M., Eren, H., Ilhan, E., Okay, S., Atak, M., Erayman, M. and Unver, T. (2014) Genome-wide fungal stress responsive miRNA expression in wheat. *Planta*, **240**, 1287–1298.

Jones-Rhoades, M.W., Bartel, D.P. and Bartel, B. (2006) MicroRNAs and their regulatory roles in plants. In *Annual Review of Plant Biology*. **57**, pp. 19–53. Palo Alto, CA: Annual Reviews.

Kantar, M., Unver, T. and Budak, H. (2010) Regulation of barley miRNAs upon dehydration stress correlated with target gene expression. *Funct. Integr. Genomics*, **10**, 493–507.

Khraiwesh, B., Arif, M.A., Seumel, G.I., Ossowski, S., Weigel, D., Reski, R. and Frank, W. (2010) Transcriptional control of gene expression by microRNAs. *Cell*, **140**, 111–122.

Li, R.Q., Li, Y.R., Kristiansen, K. and Wang, J. (2008) SOAP: short oligonucleotide alignment program. *Bioinformatics*, **24**, 713–714.

Li, R.Q., Yu, C., Li, Y.R., Lam, T.W., Yiu, S.M., Kristiansen, K. and Wang, J. (2009) SOAP2: an improved ultrafast tool for short read alignment. *Bioinformatics*, **25**, 1966–1967.

Li, B., Duan, H., Li, J., Deng, X., Yin, W. and Xia, X. (2013) Global identification of miRNAs and targets in *Populus euphratica* under salt stress. *Plant Mol. Biol.* **81**, 525–539.

Lukasik, A., Pietrykowska, H., Paczek, L., Szweykowska-Kulinska, Z. and Zielenkiewicz, P. (2013) High-throughput sequencing identification of novel and conserved miRNAs in the *Brassica oleracea* leaves. *BMC Genomics*, **14**, 801.

Mica, E., Piccolo, V., Delledonne, M., Ferrarini, A., Pezzotti, M., Casati, C., Del Fabbro, C., Valle, G., Policriti, A., Morgante, M., Pesole, G., Pe, M.E. and Horner, D.S. (2009) High throughput approaches reveal splicing of primary microRNA transcripts and tissue specific expression of mature microRNAs in *Vitis vinifera*. *BMC Genomics*, **10**, 558.

Moriya, Y., Itoh, M., Okuda, S., Yoshizawa, A.C. and Kanehisa, M. (2007) KAAS: an automatic genome annotation and pathway reconstruction server. *Nucleic Acids Res.* **35**, W182–W185.

Navarro, L., Dunoyer, P., Jay, F., Arnold, B., Dharmasiri, N., Estelle, M., Voinnet, O. and Jones, J.D.G. (2006) A plant miRNA contributes to antibacterial resistance by repressing auxin signaling. *Science*, **312**, 436–439.

Ng, D.W.K., Zhang, C.Q., Miller, M., Palmer, G., Whiteley, M., Tholl, D. and Chen, Z.J. (2011) cis- and trans-Regulation of miR163 and target genes confers natural variation of secondary metabolites in two Arabidopsis species and their allopolyploids. *Plant Cell*, **23**, 1729–1740.

Ozhuner, E., Eldem, V., Ipek, A., Okay, S., Sakcali, S., Zhang, B.H., Boke, H. and Unver, T. (2013) Boron stress responsive microRNAs and their targets in barley. *PLoS ONE*, **8**, e59543.

Page, J.E. (2005) Silencing nature's narcotics: metabolic engineering of the opium poppy. *Trends Biotechnol.* **23**, 331–333.

Robert-Seilaniantz, A., MacLean, D., Jikumaru, Y., Hill, L., Yamaguchi, S., Kamiya, Y. and Jones, J.D.G. (2011) The microRNA miR393 re-directs secondary metabolite biosynthesis away from camalexin and towards glucosinolates. *Plant J.* **67**, 218–231.

Samanani, N., Liscombe, D.K. and Facchini, P.J. (2004) Molecular cloning and characterization of norcoclaurine synthase, an enzyme catalyzing the first committed step in benzylisoquinoline alkaloid biosynthesis. *Plant J.* **40**, 302–313.

Samanani, N., Alcantara, J., Bourgault, R., Zulak, K.G. and Facchini, P.J. (2006) The role of phloem sieve elements and laticifers in the biosynthesis and accumulation of alkaloids in opium poppy. *Plant J.* **47**, 547–563.

Schwab, R., Palatnik, J.F., Riester, M., Schommer, C., Schmid, M. and Weigel, D. (2005) Specific effects of microRNAs on the plant transcriptome. *Dev. Cell*, **8**, 517–527.

Seefried, W.F., Willmann, M.R., Clausen, R.L. and Jenik, P.D. (2014) Global regulation of embryonic patterning in Arabidopsis by microRNAs. *Plant Physiol.* **165**, 670–687.

Untergasser, A., Nijveen, H., Rao, X., Bisseling, T., Geurts, R. and Leunissen, J.A.M. (2007) Primer3Plus, an enhanced web interface to Primer3. *Nucleic Acids Res.* **35**, W71–W74.

Unver, T. and Budak, H. (2009) Conserved microRNAs and their targets in model grass species *Brachypodium distachyon*. *Planta*, **230**, 659–669.

Unver, T., Bakar, M., Shearman, R.C. and Budak, H. (2010a) Genome-wide profiling and analysis of *Festuca arundinacea* miRNAs and transcriptomes in response to foliar glyphosate application. *Mol. Genet. Genomics*, **283**, 397–413.

Unver, T., Parmaksiz, I. and Dundar, E. (2010b) Identification of conserved micro-RNAs and their target transcripts in opium poppy (*Papaver somniferum* L.). *Plant Cell Rep.* **29**, 757–769.

Wei, B., Cai, T., Zhang, R., Li, A., Huo, N., Li, S., Gu, Q.Y., Vogel, J., Jia, J., Qi, Y. and Mao, L. (2009) Novel microRNAs uncovered by deep sequencing of small RNA transcriptomes in bread wheat (*Triticum aestivum* L.) and *Brachypodium distachyon* (L.) Beauv. *Funct. Integr. Genomics*, **9**, 499–511.

Weid, M., Ziegler, J. and Kutchan, T.M. (2004) The roles of latex and the vascular bundle in morphine biosynthesis in the opium poppy, *Papaver somniferum*. *Proc. Natl Acad. Sci. USA*, **101**, 13957–13962.

Wijekoon, C.P. and Facchini, P.J. (2012) Systematic knockdown of morphine pathway enzymes in opium poppy using virus-induced gene silencing. *Plant J.* **69**, 1052–1063.

Xia, R., Zhu, H., An, Y.Q., Beers, E.P. and Liu, Z. (2012) Apple miRNAs and tasiRNAs with novel regulatory networks. *Genome Biol.* **13**, R47.

Yanik, H., Turktas, M., Dundar, E., Hernandez, P., Dorado, G. and Unver, T. (2013) Genome-wide identification of alternate bearing-associated microRNAs (miRNAs) in olive (*Olea europaea* L.). *BMC Plant Biol.* **13**, 10–31.

Yu, X., Wang, H., Lu, Y.Z., de Ruiter, M., Cariaso, M., Prins, M., van Tunen, A. and He, Y.K. (2012) Identification of conserved and novel microRNAs

that are responsive to heat stress in *Brassica rapa*. *J. Exp. Bot.* **63**, 1025–1038.

Zeng, H., Wang, G., Hu, X., Wang, H., Du, L. and Zhu, Y. (2014) Role of microRNAs in plant responses to nutrient stress. *Plant Soil*, **374**, 1005–1021.

Zhang, B. and Wang, Q. (2015) MicroRNA-based biotechnology for plant improvement. *J. Cell. Physiol.* **230**, 1–15.

Zhang, B., Pan, X.P. and Anderson, T.A. (2006) Identification of 188 conserved maize microRNAs and their targets. *FEBS Lett.* **580**, 3753–3762.

Zhang, B., Wang, Q.L. and Pan, X.P. (2007) MicroRNAs and their regulatory roles in animals and plants. *J. Cell. Physiol.* **210**, 279–289.

Zhang, J.Y., Xu, Y.Y., Huan, Q. and Chong, K. (2009) Deep sequencing of *Brachypodium* small RNAs at the global genome level identifies microRNAs involved in cold stress response. *BMC Genomics*, **10**, 449–465.

Zhou, L.G., Liu, Y.H., Liu, Z.C., Kong, D.Y., Duan, M. and Luo, L.J. (2010) Genome-wide identification and analysis of drought-responsive microRNAs in *Oryza sativa*. *J. Exp. Bot.* **61**, 4157–4168.

Ziegler, J. and Facchini, P.J. (2008) Alkaloid biosynthesis: metabolism and trafficking. *Annu. Rev. Plant Biol.* **59**, 735–769.

Zuker, M. (2003) Mfold web server for nucleic acid folding and hybridization prediction. *Nucleic Acids Res.* **31**, 3406–3415.

3

Small RNA and degradome sequencing reveals important microRNA function in *Astragalus chrysochlorus* response to selenium stimuli

Ozgur Cakir[1,3,]*, Bilgin Candar-Cakir[2,3] and Baohong Zhang[3,]*

[1]*Department of Molecular Biology and Genetics, Faculty of Science, Istanbul University, Istanbul, Turkey*

[2]*Program of Molecular Biology and Genetics, Institute of Science, Istanbul University, Istanbul, Turkey*

[3]*Department of Biology, East Carolina University, Greenville, NC, USA*

*Correspondence
emails zhangb@ecu.
edu (BZ) and ozgurckr@istanbul.edu.tr (OC)

Keywords: microRNA, selenium, *Astragalus chrysochlorus*, degradome analysis, high-throughput deep sequencing.

Summary

Selenium (Se), an essential element, plays important roles in human health as well as environmental sustainability. Se hyperaccumulating plants are thought as an alternative selenium resource, recently. *Astragalus* species are known as hyperaccumulator of Se by converting it to nonaminoacid compounds. However, Se-metabolism-related hyperaccumulation is not elucidated in plants yet. MicroRNAs (miRNAs) are key molecules in many biological and metabolic processes via targeting mRNAs, which may also play an important role in Se accumulation in plants. In this study, we identified 418 known miRNAs, belonging to 380 families, and 151 novel miRNAs induced by Se exposure in *Astragalus chyrsochlorus* callus. Among known miRNAs, the expression of 287 families was common in both libraries, besides 71 families were expressed only in Se-treated sample, whereas 60 conserved families were expressed in control tissue. miR1507a, miR1869 and miR2867-3p were mostly up-regulated, whereas miR1507-5p and miR8781b were significantly down-regulated by Se exposure. Computational analysis shows that the targets of miRNAs are involved in different types of biological mechanisms including 47 types of cellular component, 103 types of molecular function and 144 types of biological process. Degradome analysis shows that 1256 mRNAs were targeted by 499 miRNAs. We conclude that some known and novel miRNAs such as miR167a, miR319, miR1507a, miR4346, miR7767-3p, miR7800, miR9748 and miR-n93 target transcription factors, disease resistance proteins and some specific genes like cysteine synthase and might be related to plant hormone signal transduction, plant–pathogen interaction and sulphur metabolism pathways.

Introduction

MicroRNAs (miRNAs) are a class of small noncoding RNA molecules with the length of 21–24 nucleotides (Bartel, 2004). They regulate gene expression post-transcriptionally via targeting mRNAs for degradation and/or translational inhibition in all eukaryotic organisms (Saini *et al.*, 2012). miRNAs play an essential role in biological and metabolic processes in plants such as growth, development, maturation, cell differentiation and response to various abiotic and biotic stresses (Barrera-Figueroa *et al.*, 2012; Dalmay, 2006; Jones-Rhoades *et al.*, 2006; Zhang, 2015; Zhang and Wang, 2015). Thus, determination of functional aspects of miRNAs and their targets are important for breeding strategies and plant biotechnology.

Selenium is one of the nonmetallic elements, and it is a component of selenocysteine (Birringer *et al.*, 2002; Whanger, 2002). Se shares the same assimilation pathway with sulphur (S) as they have a similar chemical structure, so Se can be assimilated in plants as well (Sors *et al.*, 2005). Humans and animals need this element as a micronutrient in low concentrations, but it could be very toxic in higher concentrations (Hung *et al.*, 2012). In some cases, plants that can accumulate high level of Se are desired for bioremediation and biofortification studies as well as human health, and recent findings revealed that Se transportation could be closely associated with phytoremediation and

biofortification (Hung *et al.*, 2012; Schiavon *et al.*, 2015). The genus *Astragalus* L. belongs to Leguminosae, the largest flowering family and known as accumulator of high level of Se (Freeman *et al.*, 2006; Shrift and Virupaksha, 1965). *Astragalus bisulcatus* is the best known plant of *Astragalus* species that accumulates selenium in high concentrations with the accumulation level of 0.65% in its tissues (Pickering *et al.*, 2003). *Astragalus* plants contain many active secondary compounds such as saponins, phenolics and polysaccharides. Species of *Astragalus* are known for their immunostimulant, hepatoprotective, antiperspirant, diuretic, antiviral and tonic effects (Benchadi *et al.*, 2013). Although *Astragalus* species are hyperaccumulator of Se, their adverse characteristics such as slow growth, low biomass and nonedibility restrict the use of these plants directly for human dietary and biological applications such as bioremediation (Hung *et al.*, 2012). Therefore, biotechnology approaches are needed to improve undesirable properties of high level Se accumulated plants. miRNAs can be the key molecules for promoting Se accumulation in plants.

The effects of metal stress on miRNAs have been studied with boron in barley (Ozhuner *et al.*, 2013), aluminium in *Medicago truncatula* (Chen *et al.*, 2012) and cadmium in rice (Ding *et al.*, 2011). In these studies, it was found that plants respond to metal stress by altering miRNA expressions. All these stressors are affecting plants growth and development. It is known that they

(a)

(b)

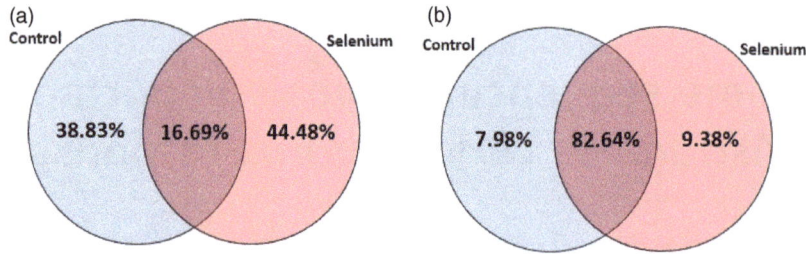

Figure 1 Common and unique sequences between Se-treated and control libraries (a) unique sRNAs (b) total sRNAs

can be toxic when their concentration is above a certain limit. The aim of our study was to identify Se-related miRNAs and their putative targets in *Astragalus chrysochlorus* Boiss. & Kotschy (2n = 16). This plant is shown to accumulate selenium as a secondary accumulator (Arı et al., 2010), and although several studies have been reported on identification and characterization of Se-related genes in *Astragalus* species (Arı et al., 2010; Çakır and Arı, 2013; Neuhierl and Bock, 1996; Sors et al., 2009), there is no study about Se-induced miRNA discovery and expression of miRNAs and their targets in *Astragalus*. For this purpose, we carried out high-throughput sequencing analysis of small RNAs in *A. chrysochlorus*. We also performed degradome sequencing for miRNA target identification.

Results

Small RNA library sequencing

A total of 23 646 078 and 20 850 840 small RNA reads (Table S1) were obtained for Se-treated and untreated samples, respectively. After removed the unnecessary sequences (adaptor, RNAs shorter than 18 nt, polyA), a total of 22 646 781 and 20 044 787 sequences were remained in Se-treated and control libraries, respectively. These total reads contained miRNA, rRNA, snRNA, tRNA, snoRNA, tRNA and unannotated sequences (Table S2). The small RNA sequences were ranged from 16 to 27 nt in length with the majority were 20–24 nt in length (Figure S1). In both Se-treated and control libraries, 21–22 nt long small RNAs were the most abundant. While the 21 nt long small RNAs were 26.68% and 28.39%, the 22 nt long small RNAs were 27.28% and 27.14% in libraries constituted from Se-treated and control callus tissues, respectively. In total small RNAs, 9.38% (unique, 44.48%) were specifically found in Se-treated sample, whereas 7.98% (unique, 38.83%) were specific in control sample; there are 82.64% (unique, 16.68%) small RNAs commonly existed in both samples (Figure 1).

Identification and expression patterns of known miRNAs

To identify known miRNAs in *A. chrysochlorus*, clean reads generated from two libraries aligned against miRNA database (Release 20) (Kozomara and Griffiths-Jones, 2011). A total of 418 miRNAs, belonging to 380 families, were detected in both Se-treated and control samples. Of these miRNAs, 71 were expressed only in Se-treated samples; 60 were expressed only in control sample. A total of 287 miRNA families were expressed in both treated and untreated samples (Figure 2). For example, miR1869 and miR6195 were only detected in Se-treated samples, whereas miRNAs, such as miR156, miR157 and miR159, were detected in both libraries. Among the 380 miRNA families, 160 miRNAs were differentially expressed in both libraries after normalization of miRNA reads to 'reads per million' (RPM) (Table 1, Figures 3 and 4). miR2867-3p was the miRNA with most fold change with 17.8-

fold up-regulated by Se exposure, followed by miR1869 and miR1507a. Their fold changes were 17.25 and 16.65, respectively. miR319b, miR535a, miR846-5p, miR3633a-3p, miR3711, miR3946, miR4414b, miR5232, miR5241a, miR5369, miR9662a-3p and miR9741 were also found to be significantly up-regulated in Se-treated tissues. On the other hand, miR165a-5p miR397a, miR399i, miR419, miR848-3p, miR1507-5p, miR2920, miR5077, miR5225-5p, miR5239, miR5721, miR6266a, miR7503, miR7539 and miR8675c were significantly down-regulated by Se treatment (Figure 4). The most significantly down-regulated one was miR1507-5p.

Identification of potentially novel miRNAs and their expression pattern

Among a total of 151 novel miRNAs, 55 and 57 miRNAs were identified in selenium treated and control samples, respectively, whereas 39 miRNAs found in both libraries (Figure 2). Of them, 30 miRNAs were differentially expressed after Se treatment (Table 2). Among these 30 miRNAs, 14 were only expressed in Se-treated sample, whereas 12 were only expressed in control.

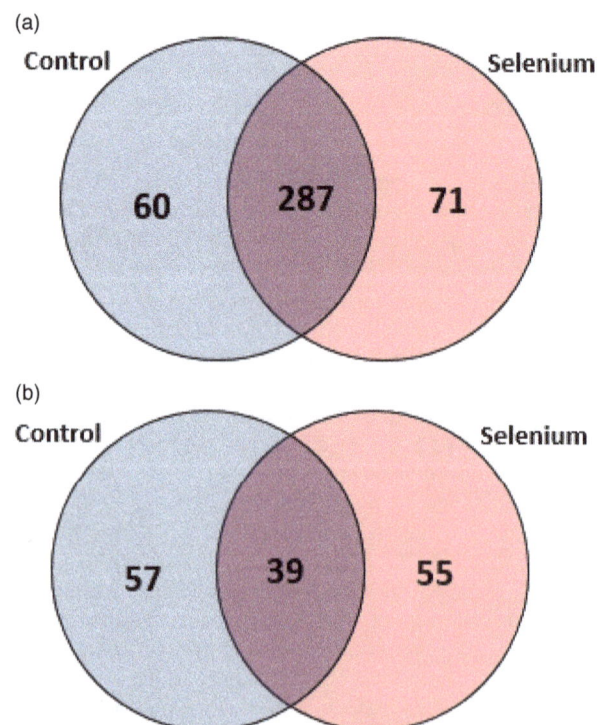

(a)

(b)

Figure 2 Distribution of miRNAs between Control and Selenium Treatment. (a) Conserved miRNAs; (b) Novel miRNAs.

Table 1 Differentially expressed known miRNAs after Se exposure in *Astragalus chrysochlorus*

miRNA	Normalized expression level*		Fold change (log$_2$ Se treatment/control)
	Se treatment	Control	
miR1044-3p	12.49	0	10.28
miR1081	1.45	0	7.18
miR1085-3p	30.11	0	11.55
miR1112-3p	3.79	0	8.56
miR1114	0	2.64	−8.04
miR1120b-3p	0	1.94	−7.60
miR1147.2	5.03	0	8.97
miR1440b	0.01	33.97	−11.73
miR1507-5p	0.01	1884.97	−17.52
miR1507a	1030.47	0	16.65
miR1531-3p	0	15.56	−10.60
miR158a-3p	1.28	0	7.00
miR161-5p.1	13.73	0	10.42
miR165a-5p	0	24.59	−11.26
miR166c	0	3.94	−8.62
miR166g-5p	4.28	0	8.74
miR167a	3721.19	8271.42	−1.15
miR167f-3p	32.14	105.71	−1.71
miR169n-3p	0	18.00	−10.81
miR171b-3p	33.07	11.62	1.50
miR171m	5.60	0	9.13
miR171n	0	4.14	−8.69
miR1861c	0	26.24	−11.35
miR1869	1566.09	0	17.25
miR1873	0	4.53	−8.82
miR2084	0	12.62	−10.30
miR2097-5p	3.62	0	8.50
miR2108b	1.28	0	7.00
miR2628	1.85	0	7.53
miR2670e	91.13	0	13.15
miR2867-3p	2285.31	0	17.80
miR2920	0	8.43	−9.71
miR2937	0	3.59	−8.48
miR319a-3p	4.54	0	8.82
miR319b	32.27	3.14	3.36
miR3437-3p	0	4.68	−8.87
miR3446-5p	14.88	0	10.53
miR3476	1.89	0	7.56
miR3633a-3p	8.30	1.74	2.24
miR3633a-5p	374.84	995.72	−1.40
miR3637-5p	2.11	0	7.72
miR3711	1.05	0	6.72
miR393a	12.05	4.88	1.30
miR3946	23.44	9.87	1.24
miR395	0	1.64	−7.36
miR395n	41.06	0	12.00
miR397a	246.92	811.38	−1.71
miR399b-3p	0	3.84	−8.58
miR399i	3.62	23.79	−2.71
miR415	6.97	1008.59	−7.17
miR419	3.92	9.42	−1.26
miR4240	51.13	0	12.32
miR4244	75.86	31.57	1.26
miR4346	67.29	30.03	1.16
miR4348a	8.87	0	9.79
miR4386	26.97	0	11.39
miR4388	0	6.23	−9.28

Table 1 Continued

miRNA	Normalized expression level*		Fold change (log$_2$ Se treatment/control)
	Se treatment	Control	
miR4414a-5p	0	33.32	−11.70
miR4414b	278.80	0	14.76
miR4415a-3p	0.52	1.14	−1.11
miR4415b-5p	64.90	0	12.66
miR447a.2-3p	15.63	6.88	1.18
miR477a	8.52	0	9.73
miR477d	0	5.48	−9.10
miR5029	0	19.70	−10.94
miR5049-3p	34.53	0	11.75
miR5070-3p	4.32	0	8.75
miR5077	31.26	63.30	−1.01
miR5083	0	1.19	−6.90
miR5176-3p	10.64	0	10.05
miR5208a	0	8.93	−9.80
miR5224b	22.56	0	11.13
miR5225-5p	0	44.65	−12.12
miR5232	19.38	6.48	1.57
miR5239	59.16	136.94	−1.21
miR5241a	70.34	13.32	2.40
miR5258	0	11.67	−10.18
miR5264	0	2.09	−7.71
miR5265	1.72	0	7.42
miR5270a	1.54	0.69	1.14
miR5273	6.18	0	9.27
miR5287a	0	13.76	−10.42
miR5287b	5.47	0	9.09
miR530-3p	0	12.02	−10.23
miR5337a	2.69	0	8.07
miR535a	188.85	0	14.20
miR535d	0	1.64	−7.36
miR5368	42.65	8.78	2.28
miR5369	68.66	0	12.74
miR5485	10.59	0	10.04
miR5503	22.16	9.82	1.17
miR5519	584.01	272.53	1.09
miR5528	4.81	0	8.91
miR5536	81.33	40.35	1.01
miR5575	0	4.83	−8.91
miR5634	0	1.496	−7.22
miR5664	0	11.17	−10.12
miR5671a	425.35	205.48	1.04
miR5721	5.12	147.22	−4.84
miR5763a	0	23.39	−11.19
miR5776	45.17	0	12.14
miR5792	20.48	0	11.00
miR5837.2	93.30	0	13.18
miR6021	33.60	11.72	1.51
miR6029	0	6.53	−9.35
miR6116-5p	0	33.47	−11.70
miR6151f	7.94	0	9.63
miR6172	4.54	0	8.82
miR6195	193.75	0	14.24
miR6196	18.85	45.24	−1.26
miR6209	137.06	53.87	1.34
miR6266a	0	9.778	−9.93
miR6267c-3p	1.67	0	7.39
miR6290	25.47	0	11.31
miR6449	1.41	0	7.14

Table 1 Continued

miRNA	Normalized expression level*		Fold change (log$_2$ Se treatment/control)
	Se treatment	Control	
miR6462c-5p	8.25	0	9.68
miR6470	0	1.945	−7.60
miR7503	0	108.90	−13.41
miR7533a	101.42	46.29	1.13
miR7539	3.57	79.27	−4.47
miR7700-5p	41.28	0	12.01
miR7752-3p	0	5.487	−9.10
miR7767-3p	11.43	0	10.15
miR7779-5p	2.51	0	7.97
miR7797a	0	29.23	−11.51
miR7800	1.45	0	7.18
miR7819	7.68	3.392	1.17
miR7825	0	10.37	−10.01
miR812g	1.54	0	7.27
miR8141	296.68	109.95	1.43
miR8149	60.18	0	12.55
miR8155	1.05	2.24	−1.08
miR8182	1.72	0	7.42
miR845a	7.33	0	9.51
miR846-5p	4.28	0	8.74
miR848-3p	0	1.496	−7.22
miR857	1.32	0	7.04
miR8633	84.47	0	13.04
miR8638	0	10.17	−9.99
miR8658	0	5.23	−9.03
miR866-5p	78.37	157.59	−1.00
miR8675c	0	1.097	−6.77
miR8691	1.89	0	7.56
miR869.2	2.82	0	8.14
miR8704	0	7.98	−9.64
miR8717	0	7.08	−9.46
miR8743a	0	55.87	−12.44
miR8757a	3.97	1.7960	1.14
miR8781b	0	120.28	−13.55
miR908.2	6.27	1.74	1.84
miR9471a-5p	3.92	0	8.61
miR9478-3p	2.78	7.38	−1.40
miR9497	0	8.73	−9.76
miR9568-3p	2.69	1.19	1.16
miR9657b-5p	8.96	0	9.80
miR9662a-3p	1.81	0	7.50
miR9741	37.93	0	11.88
miR9742	0	84.26	−13.04
miR9748	112.81	55.67	1.01
miR9766	0	6.78	−9.40

*All miRNA expressions were normalized to read per million (RPM). If miRNA expression measured as zero, normalized expression valued as 0.01 according to Murakami et al. (2006). Fold change was calculated using the formula, fold change = log$_2$(treatment/control) (Marsit et al., 2006). Significance was calculated as fold change log$_2$ > 1 or log$_2$ < −1 and P-value < 0.01. The miRNAs given in the table above are significantly expressed.

Only 4 of them were expressed in both treated and untreated samples. Among these novel miRNAs, 16 were up-regulated and 14 were down-regulated (Table 2). The most up-regulated miRNAs are miR-n1 (11.05), miR-n53 (11.06), miR-n61 (10.86),

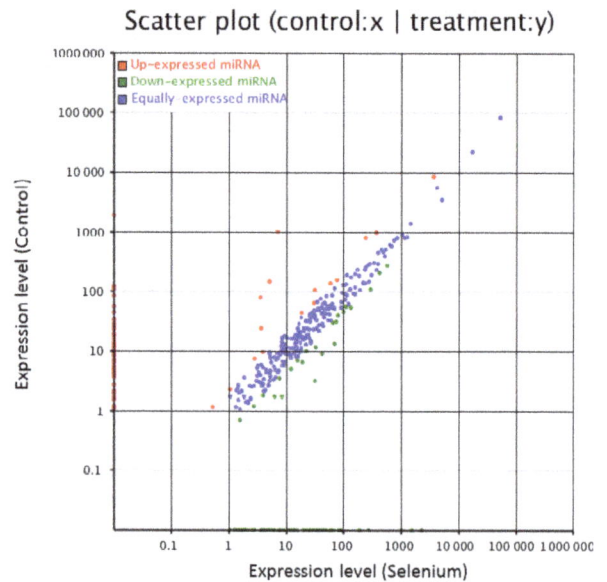

Figure 3 Small RNA expression profiles of control and Se-treated callus of *Astragalus chrysochlorus*.

miR-n74 (10.44), miR-n87 (10.46), miR-n90 (17.65) and miR-n93 (12.48). Of them, miR-n90 was up-regulated most with 17.65-fold change. The most significantly down-regulated one was miR-n146 with 9.49-fold change and miR-n122 was also down-regulated by 9.24-fold.

Degradome sequencing analysis

To validate the cleavage sites of miRNAs, we performed high-throughput degradome sequencing. A total of 29 371 471 reads were obtained by degradome sequencing. After removing the adaptors and other RNAs, a total of 4 955 825 unique reads were obtained. Figure 5 shows the target plots of identified targets of randomly selected genes. In total, 1339 predicted sites were identified. The predicted sites were determined to be cleaved by 499 miRNAs. The total predicted sites were in 1256 genes with 2027 cleavage events. The target genes were annotated and classified as transcription factors and their subunits (WRKY, trihelix transcription factor, bHLH143, RF2b, MYC2, GTE4, TCP8, Myb family transcription factor APL, ethylene-responsive transcription factors, heat stress transcription factors), enzyme coding genes such as kinases and transferases (probable leucine-rich repeat receptor-like protein kinase, probable LRR receptor-like serine/threonine protein kinase, calcium-dependent protein kinase 3, uracil phosphoribosyltransferase, mRNA cap guanine-N7 methyltransferase 1, chloroplastic homogentisate phytyltransferase 2, mitochondrial aminomethyltransferase), resistance proteins (TMV resistance protein, pleiotropic drug resistance protein), leucine-rich repeat, leucine zipper and zinc finger proteins and other structural and functional proteins.

Target identification and GO and KEGG pathway analyses

The targets of identified miRNAs were subjected to Gene Ontology (GO) and Kyoto Encyclopedia of Genes and Genomes (KEGG) to perceive their roles biologically. It was determined that the target genes are involved in 47 types of cellular component, 103 types of molecular function and 144 types of

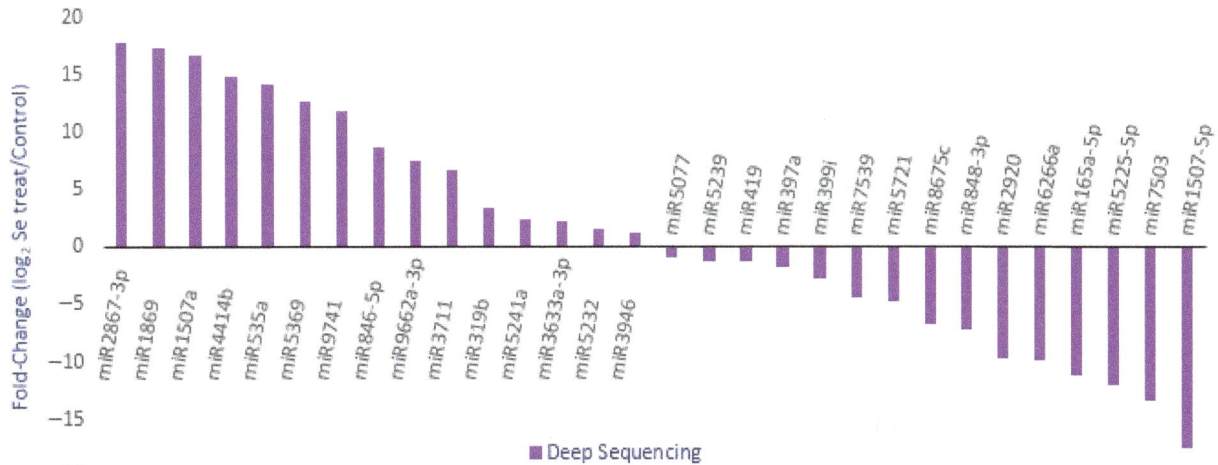

Figure 4 Expression profiles of randomly selected miRNAs with different abundance in Se-treated *Astragalus chrysochlorus* calli.

Table 2 Differentially expressed novel miRNAs after Se exposure in *Astragalus chrysochlorus*

| miRNA | Normalized expression level* | | Fold change (log2 Se treatment/control) |
	Se treatment	Control	
miR-n1	21.23	0	11.05
miR-n101	0	3.14	−8.29
miR-n109	0	1.09	−6.77
miR-n113	0	1.69	−7.40
miR-n117	0	3.29	−8.36
miR-n122	0	6.08	−9.24
miR-n135	0	1.19	−6.90
miR-n142	0	1.696	−7.40
miR-n144	0	2.99	−8.22
miR-n146	0	7.23	−9.49
miR-n147	0	1.59	−7.31
miR-n148	0	5.88	−9.20
miR-n15	2.11	0	7.72
miR-n21	5.87	0	9.19
miR-n26	2.47	6.43	−1.37
miR-n28	1.45	0	7.18
miR-n35	2.82	9.32	−1.72
miR-n42	1.19	0	6.89
miR-n5	2.34	0	7.87
miR-n53	21.37	0	11.06
miR-n61	18.63	0	10.86
miR-n70	14.26	5.28	1.43
miR-n71	562.11	214.96	1.38
miR-n74	13.90	0	10.44
miR-n79	1.32	0	7.04
miR-n8	1.63	0	7.35
miR-n87	14.08	0	10.46
miR-n90	2062.14	0	17.65
miR-n93	57.31	0	12.48
miR-n95	0	1.49	−7.22

*All miRNA expressions were normalized to read per million (RPM). If miRNA expression measured as zero, normalized expression valued as 0.01 according to Murakami *et al.* (2006). Fold change was calculated using the formula, fold change = log₂(treatment/control) (Marsit *et al.*, 2006). Significance was calculated as fold change log₂ > 1 or log₂ < −1 and *P*-value < 0.01. The miRNAs given in the table above are significantly expressed.

biological process. The detailed summary of GO classification was given in Figure 6. In Se-treated tissues, there were 3017 targets genes classified into different groups, but the biological regulations (206), cellular processes (637), metabolic processes (588), regulation of biological processes (201), single-organism processes (351) and response to stimulus (245) were the most abundant ones in biological process categories. About cellular component category, there were 2183 target genes involved in different groups. The most abundants were cell (538), cell part (538), macromolecular complexes (153), membrane (186), organelle (422) and organelle parts (162). The last category was molecular function and 1397 targets were determined. The most abundant groups were binding (689) and catalytic activity (572), respectively. The most abundant categories obtained by known and novel miRNA analysis were summarized in Tables 3 and 4, respectively. According to KEGG analysis, 968 target genes were annotated to 239 pathways. Consistent with GO analysis, we determined the putative targets of down-regulated and up-regulated miRNAs (Tables 5 and 6). The KEGG pathway related to plant–pathogen interaction pathway and the related miRNAs were shown in Figure 7. According to these analysis, calcium-dependent protein kinase 1 (CPDK) which is related to this pathway was affected by miR5049-3p expression. Cyclic nucleotide gated channel 10 (CGNGs) is an ion channel and affected by miR5485 expression. Calcium-binding calmodulin-like protein 7 was affected by miR4244 expression. Disease resistance protein (RPM1) was affected by miR1507a and miR1507c-5p. Disease resistance protein (RPS2) was affected by miR1510a, miR5652 and miR3633a-3p. Disease resistance protein (RSP5) was affected by miR5255, miR1510a, miR2118, miR2118a-3p and miR3633a-3p. Chloroplast heat-shock protein 90 (HSP90), which has a role in protein processing in endoplasmic reticulum, was affected by miR9722 and miR9748. Transcription factor MYC2 is involved in environmental information processing and plant hormone signal transduction and affected by miR9748. All these proteins are involved in environmental adaptation. Also leucin rich repeats receptor-like serine/threonine protein kinase (FLS2) and somatic embryogenesis receptor kinase 4 were also nonspecific serine/threonine protein kinase and affected by miR414 and miR5205b, respectively. WRKY transcription factor 25 is affected by miR5766 and miR831-5p.

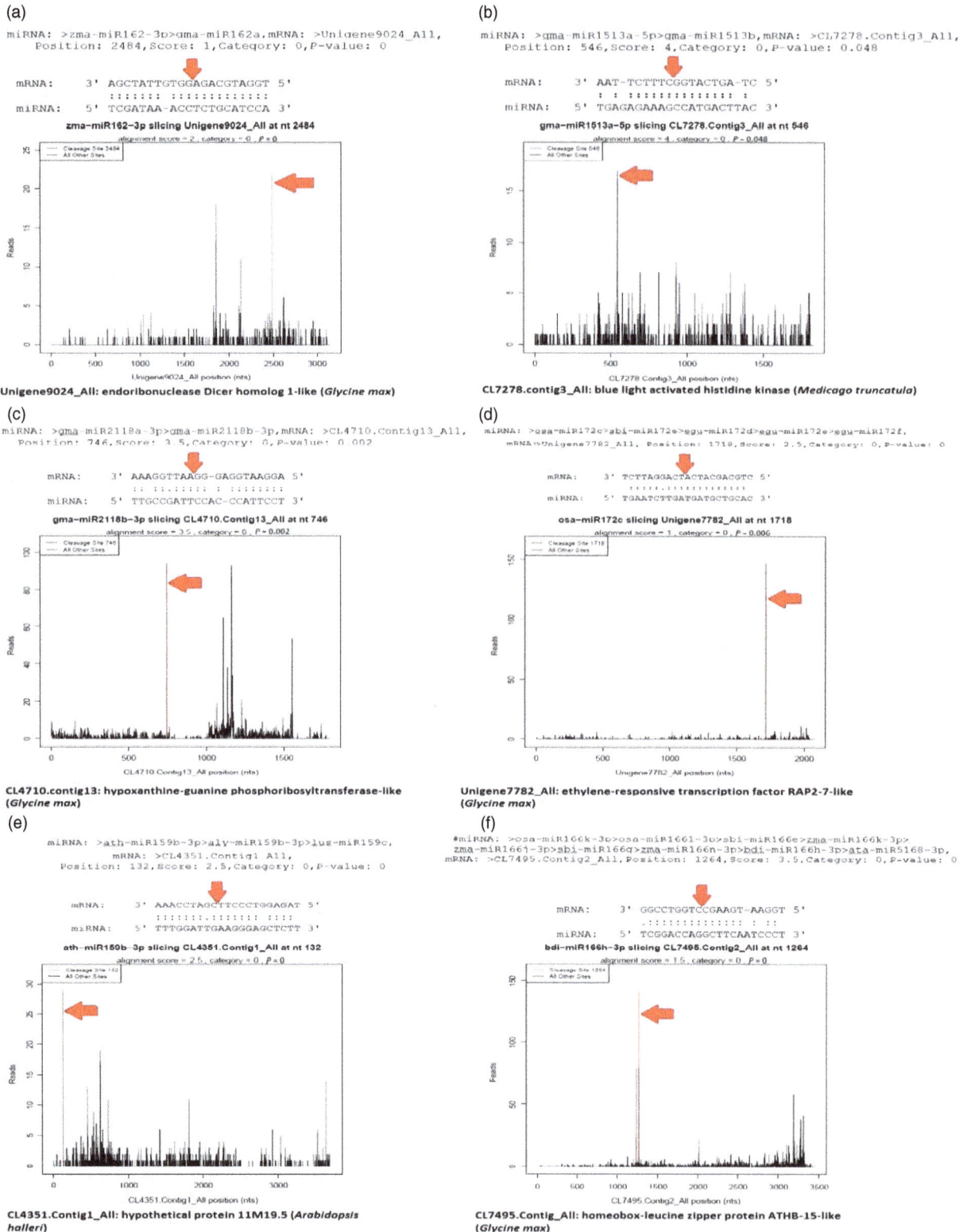

Figure 5 Target plots (t-plots) of miRNAs and their targets. The red arrows indicate the most abundant peaks and cleavage sites. (a) miR162-3p targeting endonuclease Dicer homologue-1-like protein, (b) miR1513a targeting blue light-activated histidine kinase, (c) miR2118b targeting hypoxanthine-guanine phosphoribosyltransferase-like protein, (d) miR172c targeting putative ethylene-responsive transcription factor RAP-2-7-like protein, (e) miR159b-3p targeting hypothetical protein 11M19.5, (f) miR166 h-3p targeting homeobox leucin zipper protein ATHB-15-like protein.

Serine/threonine protein kinase (PBS1) is affected by miR6196, miR6180, miR5658 and miR419. Jasmonate ZIM domain-containing protein (JAZ5) is involved in plant hormone signal transduction and affected by miR7127a and miR5248. Chitin elicitor receptor kinase 1 (CERK1) is a kind of serine/threonine protein kinases, affected by miR1850.1.

Table 3 The most abundant GO categories in Se-treated tissues obtained by known miRNA analysis

Go term (Se treatment)	Category	Enrichment factor	P-value
GO:0046499 S-adenosylmethioninamine metabolic process	Process	30.65813	0.00119
GO:0008215 spermine metabolic process	Process	30.65813	0.00119
GO:0006597 spermine biosynthetic process	Process	30.65813	0.00119
GO:0006557 S-adenosylmethioninamine biosynthetic process	Process	30.65813	0.00119
GO:0008295 spermidine biosynthetic process	Process	20.43875	0.01696
GO:0008216 spermidine metabolic process	Process	20.43875	0.01696
GO:2001251 negative regulation of chromosome organization	Process	20.43875	0.01696
GO:0042149 cellular response to glucose starvation	Process	14.52227	2.91e-06
GO:0000819 sister chromatid segregation	Process	10.90067	4.00e-10
GO:0009567 double fertilization forming a zygote and endosperm	Process	9.81060	0.00086
GO:0006476 protein deacetylation	Process	9.58067	4.59e-05
GO:0035601 protein deacylation	Process	9.58067	4.59e-05
GO:0004014 adenosylmethionine decarboxylase activity	Function	29.16322	0.00044
GO:0010385 double-stranded methylated DNA binding	Function	29.16322	6.04e-10
GO:0042301 phosphate ion binding	Function	29.16322	0.00044
GO:0070403 NAD+ binding	Function	21.87242	1.28e-12
GO:0043130 ubiquitin binding	Function	19.44215	0.00635
GO:0032266 phosphatidylinositol-3-phosphate binding	Function	19.44215	0.00635
GO:0032182 small conjugating protein binding	Function	19.44215	0.00635
GO:0030295 protein kinase activator activity	Function	13.12345	2.45e-06
GO:0019209 kinase activator activity	Function	11.41170	1.08e-05

Table 3 Continued

Go term (Se treatment)	Category	Enrichment factor	P-value
GO:0004564 beta-fructofuranosidase activity	Function	10.74434	0.00062
GO:0004575 sucrose alpha-glucosidase activity	Function	10.74434	0.00062
GO:0051765 inositol tetrakisphosphate kinase activity	Function	9.20944	0.00964
GO:0047325 inositol tetrakisphosphate 1-kinase activity	Function	9.20944	0.00964
GO:0052726 inositol-1,3,4-trisphosphate 5-kinase activity	Function	9.20944	0.00964
GO:0052725 inositol-1,3,4-trisphosphate 6-kinase activity	Function	9.20944	0.00964
GO:0005677 chromatin silencing complex	Cellular Component	23.44956	5.63e-11
GO:0000808 origin recognition complex	Cellular Component	17.14749	9.78e-35
GO:0000795 synaptonemal complex	Cellular Component	9.94057	1.47e-09
GO:0000794 condensed nuclear chromosome	Cellular Component	8.79359	1.07e-08

Discussion

In plants, small RNAs regulate the gene expression post-transcriptionally. Deep sequencing strategy is a powerful technology to discover miRNAs in plant species. It has been employed to identify miRNAs in many plant species. *Astragalus* species are known to accumulate Se in their tissues by converting it to nonamino acid compounds. However, there are still mysterious parts of selenium tolerance mechanism in plants. The aim of this study was to identify Se-responsive miRNAs and their putative targets by deep sequencing. To achieve this, small RNA libraries were constructed from both control and Se-treated callus tissues of *Astragalus chrysochlorus*. Se treatment significantly affected the expression of miRNAs. In total, 418 known and 151 novel miRNAs were identified. When the expression of Se-treated and control tissues were compared, average normalized reads for known miRNAs were 83.8 and 251.54, respectively, but unknown miRNAs were 93.54 and 65.59, respectively (Tables 1 and 2). The most significant expression difference was occured for miR2867-3p by 17.8-fold up-regulation. miR1869 and miR1507a were up-regulated by 17.25 and 16.65, respectively. Among the down-regulated miRNAs, miR1507-5p was the most down-regulated one with 17.52-fold change. Although it was the most up-regulated miRNA, the potential targets of miR2867-3p

Table 4 The most abundant GO categories in Se-treated tissues obtained by novel miRNA analysis

Go term (Se Treatment)	Category	Enrichment Factor	P-value
GO:0043174 nucleoside salvage	Process	27.33333	4.46e-17
GO:0006166 purine ribonucleoside salvage	Process	27.33333	4.46e-17
GO:0043101 purine-containing compound salvage	Process	23.76812	1.30e-15
GO:0042547 cell wall modification involved in multidimensional cell growth	Process	20.82540	0.01871
GO:0009831 plant-type cell wall modification involved in multidimensional cell growth	Process	20.82540	0.01871
GO:0018298 protein-chromophore linkage	Process	15.84541	1.28e-16
GO:0006465 signal peptide processing	Process	11.09179	0.00192
GO:0052657 guanine phosphoribosyltransferase activity	Function	34.53737	2.53e-21
GO:0004422 hypoxanthine phosphoribosyltransferase activity	Function	34.53737	2.53e-21
GO:0008442 3-hydroxyisobutyrate dehydrogenase activity	Function	13.81495	0.04382
GO:0016168 chlorophyll binding	Function	13.12420	2.14e-14
GO:0051539 4 iron, 4 sulphur cluster binding	Function	9.18082	2.00e-12
GO:0005852 eukaryotic translation initiation factor 3 complex	Cellular Component	15.08777	0.00016
GO:0009522 photosystem I	Cellular Component	9.22030	1.29e-13

Table 5 Targets of miRNAs up-regulated by Se exposure

miRNA	Putative target(s)
miR1085-3p	Cyclin A-like protein [Medicago truncatula]
	Rhomboid protease gluP [M. truncatula]
miR1147.2	Putative retrotransposon protein, identical [Solanum demissum]
miR1507a	NB-LRR type disease resistance protein Rps1-k-2 [M. truncatula]
	NBS-containing resistance-like protein [M. truncatula]
	NBS resistance protein [M. truncatula]
	Disease resistance protein RGA2 [M. truncatula]
miR2108b	Methyltransferase-like protein [M. truncatula]
	Ribonuclease H [M. truncatula]
miR2867-3p	hypothetical protein MTR_5g051130 [M. truncatula]
miR319a-3p	Transcription factor PCF5 [M. truncatula]
miR3443-5p	hypothetical protein MTR_056s0017 [M. truncatula]
miR3633a-3p	PREDICTED: TMV resistance protein N-like [Glycine max]
	Disease resistance protein RPS2 [M. truncatula]
	NBS-containing resistance-like protein [M. truncatula]
	TIR-NBS-LRR type disease resistance protein [M. truncatula]
	Kinase-like protein [M. truncatula]
	Disease resistance protein [M. truncatula]
	TMV resistance protein N [M. truncatula]
	Disease resistance protein [M. truncatula]
	TIR-NBS-LRR type disease resistance protein [M. truncatula]
	Disease resistance protein [M. truncatula]
miR395n	putative polyprotein [Cicer arietinum]
	Polynucleotidyl transferase, Ribonuclease H fold [M. truncatula]
miR4244	retrotransposon gag protein [Arachis hypogaea]
	PREDICTED: pentatricopeptide repeat-containing protein At2g01860-like [Glycine max]
	hypothetical protein VITISV_037041 [Vitis vinifera]
miR4346	Auxin influx protein [M. truncatula]
miR4348a	Pentatricopeptide repeat-containing protein [M. truncatula]
miR447a.2-3p	PREDICTED: WD and tetratricopeptide repeats protein 1-like [G. max]
miR5049-3p	calcium-dependent protein kinase [Swainsona canescens]
miR5070-3p	Mitochondrial protein, putative [M. truncatula]
	ATPase subunit 1 (mitochondrion) [Lotus japonicus]
miR5077	Pentatricopeptide repeat-containing protein [M. truncatula]
miR5167a-5p	S-adenosylmethionine decarboxylase [Medicago falcata]
miR5224b	Pol polyprotein [M. truncatula]
miR5287b	PREDICTED: LOW-QUALITY PROTEIN: cytokinin hydroxylase-like [G. max]
	isoflavonoid glucosyltransferase [Glycyrrhiza echinata]
	PREDICTED: macrophage migration inhibitory factor homologue isoform 2 [Fragaria vesca subsp. vesca]
miR5337a	PREDICTED: DNA topoisomerase 2-like [Solanum lycopersicum]
miR535a	PREDICTED: transcription factor GTE10-like [G. max]
	PREDICTED: Niemann-Pick C1 protein-like [G. max]
	ATP-dependent DNA helicase Q1 [M. truncatula]
miR5368	Cell wall-associated hydrolase, partial [M. truncatula]
miR5369	hypothetical protein MTR_8g103420 [M. truncatula]
	PREDICTED: phosphatidylserine synthase 2-like [G. max]
	PREDICTED: tubby-like F-box protein 7-like [G. max]
miR5485	hypothetical protein 11M19.5 [Arabidopsis halleri]
	PREDICTED: nuclear-pore anchor-like [G. max]

were not annotated. In wheat, miR2867-3p related to fungal stress was found to target disease resistance protein rga3-like and categorized in the group of response to stimulus (Inal et al., 2014). miR1507a was found to be related in nitrogen fixation in soya bean (Wang et al., 2009). Despite the fact that it was reported to be down-regulated by the phosphorus deficiency in Glycine max roots (Zeng et al., 2010), our results showed that miR1507a was up-regulated significantly in Se-treated tissues in A. chrysochlorus. The targets of this miRNA were determined as disease resistance protein RGA2 and NBS resistance protein. In another study, Chen et al. (2012) found that miR1507 expression was also decreased after Al treatment. High level of selenium accumulation in hyperaccumulator plant species enables to protection against herbivore and fungal pathogens (Freeman et al., 2010). Consistent with the KEGG pathway analysis, we

Table 5 Continued

miRNA	Putative target(s)
	PREDICTED: GPN-loop GTPase 2-like [*G. max*]
	hypothetical protein 11M19.5 [*Arabidopsis halleri*]
miR5528	PREDICTED: ALBINO3-like protein 2, chloroplastic-like [*G. max*]
miR5671a	hypothetical protein MTR_5g051130 [*M. truncatula*]
miR5837.2	Retrotransposon gag protein, putative [*M. truncatula*]
miR6021	retrotransposon gag protein [*Arachis hypogaea*]
miR6151f	DnaJ protein-like protein [*M. truncatula*]
miR6195	Alanyl-tRNA synthetase [*M. truncatula*]
miR6462c-5p	Solute carrier family 25 member [*M. truncatula*]
	PREDICTED: omega-hydroxypalmitate O-feruloyl transferase-like [*G. max*]
miR7533a	Vacuolar protein sorting protein [*M. truncatula*]
	Cell wall-associated hydrolase, partial [*M. truncatula*]
	SAM domain family protein [*M. truncatula*]
	PREDICTED: putative pentatricopeptide repeat-containing protein At5g59900-like [*G. max*]
miR7767-3p	Disease resistance protein [*M. truncatula*]
	TIR-NBS-LRR RCT1-like resistance protein [*Medicago sativa*]
	TIR-NBS-LRR RCT1-like resistance protein [*M. truncatula*]
	PREDICTED: TMV resistance protein N-like [*G. max*]
	TIR-NBS-LRR RCT1-like resistance protein [*M. truncatula*]
	TIR-NBS-LRR RCT1-like resistance protein [*M. sativa*]
miR7779-5p	Disease resistance protein RGA2 [*M. truncatula*]
miR7800	Cysteine synthase [*M. truncatula*]
miR8141	Ycf68 [*M. truncatula*]
miR8182	PREDICTED: flavonol synthase/flavanone 3-hydroxylase-like [*G. max*]
miR857	PREDICTED: tRNA (adenine-N(1)-)-methyltransferase noncatalytic subunit trm6-like [*G. max*]
	Chitin-inducible gibberellin-responsive protein [*M. truncatula*]
	PREDICTED: probable histone-arginine methyltransferase 1.4-like [*G. max*]
	PREDICTED: putative kinase-like protein TMKL1-like [*G. max*]
	Rhomboid protease gluP [*M. truncatula*]
	Acetylglutamate kinase-like protein [*M. truncatula*]
miR8633	Chloroplast inner envelope protein (IEP110) [*M. truncatula*]
miR9568-3p	PREDICTED: photosystem II CP47 chlorophyll apoprotein-like [*Solanum lycopersicum*]
miR9748	E3 ubiquitin-protein ligase HUWE1 [*M. truncatula*]
	Serine/threonine protein phosphatase 4 regulatory subunit [*M. truncatula*]
	PREDICTED: putative receptor-like protein kinase At1g80870-like [*G. max*]
	Lin-9-like protein [*M. truncatula*]
	Mitochondrial chaperone BCS1 [*M. truncatula*]
	PREDICTED: transcription initiation factor IIA large subunit-like [*G. max*]
	Zinc finger protein [*M. truncatula*]
	Zinc finger CCCH domain-containing protein [*M. truncatula*]
	PREDICTED: 97 kDa heat-shock protein-like [*G. max*]
	PREDICTED: putative receptor-like protein kinase At1g80870-like [*G. max*]
	Nucleosome assembly protein 1-like protein [*M. truncatula*]
	PREDICTED: BTB/POZ domain-containing protein At1g03010-like [*G. max*]

Table 5 Continued

miRNA	Putative target(s)
	PREDICTED: FACT complex subunit SPT16-like [*G. max*]
	PREDICTED: protein CWC15 homologue [*G. max*]
	PREDICTED: heparan-alpha-glucosaminide N-acetyltransferase-like [*G. max*]
	BCCIP-like protein [*M. truncatula*]
	PREDICTED: coiled-coil domain-containing protein 75-like [*G. max*]
	PREDICTED: ATP-dependent helicase BRM-like [*G. max*]
	Knotted-1 homeobox protein [*M. truncatula*]
	PREDICTED: pre-mRNA-processing factor 6-like [*G. max*]
	PREDICTED: mRNA-capping enzyme-like [*G. max*]
	PREDICTED: anaphase-promoting complex subunit 4 [*G. max*]
	Ascorbate peroxidase [*M. truncatula*]
	Ycf68 [*M. truncatula*]
	Cytochrome c oxidase subunit 5B [*M. truncatula*]
	TdcA1-ORF2 protein [*M. truncatula*]
	PREDICTED: protein DA1-related 2-like [*G. max*]
	4-hydroxy-3-methylbut-2-en-1-yl diphosphate synthase [*M. truncatula*]
	BZIP transcription factor ATB2 [*M. truncatula*]
	putative basic helix-loop-helix protein BHLH2 [*L. japonicus*]
	Zinc finger-like protein [*M. truncatula*]
	PREDICTED: zinc finger CCCH domain-containing protein 24-like [*G. max*]
	PREDICTED: transformation/transcription domain-associated protein-like [*G. max*]
	GT-2 factor [*M. truncatula*]
	alpha 1,4-fucosyltransferase [*M. truncatula*]
	PREDICTED: probable exocyst complex component 6-like [*G. max*]
	PREDICTED: probable WRKY transcription factor 40-like isoform 2 [*G. max*]
	Manganese-dependent ADP-ribose/CDP-alcohol diphosphatase [*M. truncatula*]
	cell division control protein 2 homologue 2 [*Saccharum* hybrid cultivar R570]

found that plant–pathogen interaction metabolism is one of the most affected pathway by Se exposure in *A. chyrsochlorus*. known as secondary accumulator of selenium, in our high-throughput results. miR535a was also found to be up-regulated significantly in Se-related tissues, and its potential targets, transcription factor GTE10-like, Niemann-Pick C1 protein-like, and ATP-dependent DNA helicase Q1, play role in plant hormone signal transduction pathway.

Van Hoewyk *et al.* (2008) investigated 40 μM Se stress in *Arabidopsis* shoots and roots by microarray analysis. The Se-responsive genes are involved in calcium signalling, ethylene and jasmonic acid synthesis and ethylene-responsive transcription factor family; stress-induced and disease-induced proteins were up-regulated significantly by selenate treatment as a defence response (Van Hoewyk *et al.*, 2008). In our study, miRNAs, such as miR1507a, miR3633a-3p, miR7767-3p and miR395, were found to target stress-induced and disease-induced proteins according to GO analysis. While miR1507a, miR3633a-3p and miR7767-3p were expressed in Se-treated

Table 6 Target of miRNAs down-regulated by Se exposure

miRNA	Putative target(s)
miR1531-3p	DAG protein, chloroplast precursor, putative [*Ricinus communis*]
miR165a-5p	Pentatricopeptide repeat-containing protein [*Medicago truncatula*]
miR167a	PREDICTED: auxin response factor 8-like [*G. max*]
miR1873	PREDICTED: protein GPR107-like [*G. max*]
miR3437-3p	gag polyprotein [*Cicer arietinum*]
miR395	Disease resistance protein [*M. truncatula*]
	NBS-LRR type disease resistance protein [*C. arietinum*]
	putative NBS-LRR type disease resistance protein [*Pisum sativum*]
	RGA-D protein [*C. arietinum*]
miR399i	PREDICTED: probable ubiquitin-conjugating enzyme E2 24-like [*G. max*]
	phosphate transporter 5 [*G. max*]
miR415	PREDICTED: LOW-QUALITY PROTEIN: pre-mRNA-splicing factor cwc22-like [*Glycine max*]
	PREDICTED: pre-mRNA-splicing factor CWC22 homologue [*G. max*]
miR419	PREDICTED: cysteine-rich receptor-like protein kinase 10-like [*G. max*]
	PREDICTED: condensin-2 complex subunit H2-like [*G. max*]
	dehydration responsive element binding protein [*Halimodendron halodendron*]
miR477d	Nuclear transcription factor Y subunit A-7, partial [*M. truncatula*]
	Replication protein A 70 kDa DNA-binding subunit [*M. truncatula*]
	hypothetical protein MTR_8g086620 [*M. truncatula*]
miR5029	PREDICTED: histidine decarboxylase-like [*G. max*]
miR5287a	PREDICTED: pre-mRNA-processing factor 40 homologue B-like [*G. max*]
	Mitochondrial protein, putative [*Medicago truncatula*]
miR530-3p	PREDICTED: fimbrin-like protein 2-like [*G. max*]
miR5575	hypothetical protein VITISV_032489 [*Vitis vinifera*]
miR5634	PREDICTED: kinesin-4-like [*G. max*]
	PREDICTED: HIPL1 protein-like [*G. max*]
	putative non-LTR retroelement reverse transcriptase [*Arabidopsis thaliana*]
	PREDICTED: peptide methionine sulfoxide reductase B2, chloroplastic-like isoform 1 [*G. max*]
miR5636	Polynucleotidyl transferase, Ribonuclease H fold [*M. truncatula*]
miR6116-5p	PREDICTED: arginyl-tRNA synthetase, cytoplasmic-like [*G. max*]
miR7503	hypothetical protein MTR_5g051130 [*M. truncatula*]
	Mitochondrial protein, putative [*M. truncatula*]
	ATPase subunit 1 (mitochondrion) [*Lotus japonicus*]
miR7752-3p	Ubiquitin [*M. truncatula*]
miR7797a	Putative retrotransposon protein, identical [*Solanum demissum*]
	putative polyprotein [*C. arietinum*]
	Thioredoxin fold [*M. truncatula*]
	LIM and UIM domain-containing [*M. truncatula*]
	Cc-nbs-lrr resistance protein [*M. truncatula*]
miR8658	PREDICTED: putative ribonuclease H protein At1g65750-like [*G. max*]

Table 6 Continued

miRNA	Putative target(s)
miR8717	non-ltr retroelement reverse transcriptase [*Rosa rugosa*]
	PREDICTED: putative ribonuclease H protein At1g65750-like [*Fragaria vesca* subsp. *vesca*]
miR9742	delta-pyrroline-5-carboxylate synthetase [*G. max*]
	hypothetical protein MTR_3g030260 [*M. truncatula*]
	PREDICTED: probable leucine-rich repeat receptor-like serine/threonine protein kinase At5g15730-like [*G. max*]
miR9766	Pyruvate kinase [*M. truncatula*]
	PREDICTED: transaldolase-like [*G. max*]

tissues, only miR395 was expressed in control tissues. These findings suggest that Se tolerance mechanism may differ by other stress mechanisms. Another miRNA affected by Se treatment in our study is miR393. It is reported that expression of miR393 is regulated by Cd, Hg and Al (Xie *et al.*, 2007). miR393 down-regulates the F-box auxin receptors TIR1/AFBs and bHLH transcription factors (Jones-Rhoades and Bartel, 2004; Navarro *et al.*, 2006). Zhou *et al.* (2008) reported that miR393, miR171, miR319 and miR529 were up-regulated in the leaves, but miR166 and miR398 were down-regulated when *Medicago truncatula* was treated with Cd, Hg and Al. Decrease in miR166 expression after Cd and Al exposure was also shown by Chen *et al.* (2012) and Ding *et al.* (2011). In callus tissues of *A. chrysochlorus*, miR319 (miR319a-3p and miR319b) and miR393 (miR393a) family miRNAs were up-regulated by Se treatment. The fold change of these miRNAs was 8.82, 3.36 and 1.3, respectively. miR319 targets TCP (TeosinteBranched/Cycloidea/PCF) transcription factor (Zhou *et al.*, 2008) and affects the plant growth and development. miR171 family miRNAs (miR171b-3p, miR171 m, miR171n) were also affected by Se treatment. miR171b-3p and miR171 m were up-regulated, and miR171n was significantly down-regulated by Se exposure. Zhou *et al.* (2012) reported that miR169 and miR395 were up-regulated in response to Hg-toxicity and miR171 was down-regulated in *Medicago truncatula* (Zhou *et al.*, 2012). In our study, miR169n-3p and miR395 were down-regulated by Se treatment by 10.81- and 7.36-fold, respectively.

Gene ontology analysis was applied for understanding the function of miRNAs' target genes. Among the known miRNAs, miR7800 was found to target the gene encoding cysteine synthase gene of sulphur/selenocompound metabolism that functions in selenocysteine and cysteine synthesis. In this study, miR7800 was up-regulated by Se exposure, which suggests that cysteine synthase may be down-regulated. We also found that miR6196 targets γ-glutamylcysteine synthetase that combines glutamate and methyl-selenocysteine to generate gamma-glutamyl-methylselenocysteine. This compound is thought to be the storage form of Se in hyperaccumulators (Freeman *et al.*, 2007; Kubachka *et al.*, 2007). In our study, miR6196 was down-regulated by Se treatment. It is thought that Se could be accumulated in different forms. The opposite expression of these miRNAs may be due to plant species, different Se concentration or maybe different exposure time. Certain novel miRNAs target the genes functioning in sulphur metabolism and selenocompound metabolism. miR-n9 was

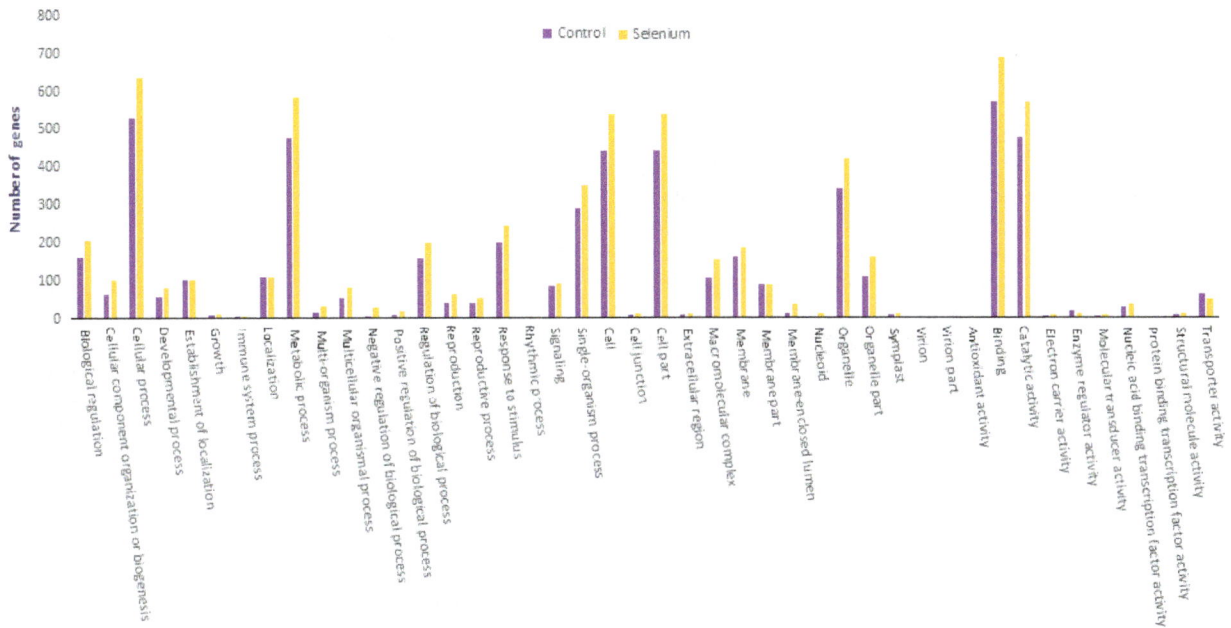

Figure 6 Summary of GO classifications of miRNA targets in *Astragalus chrysochlorus*.

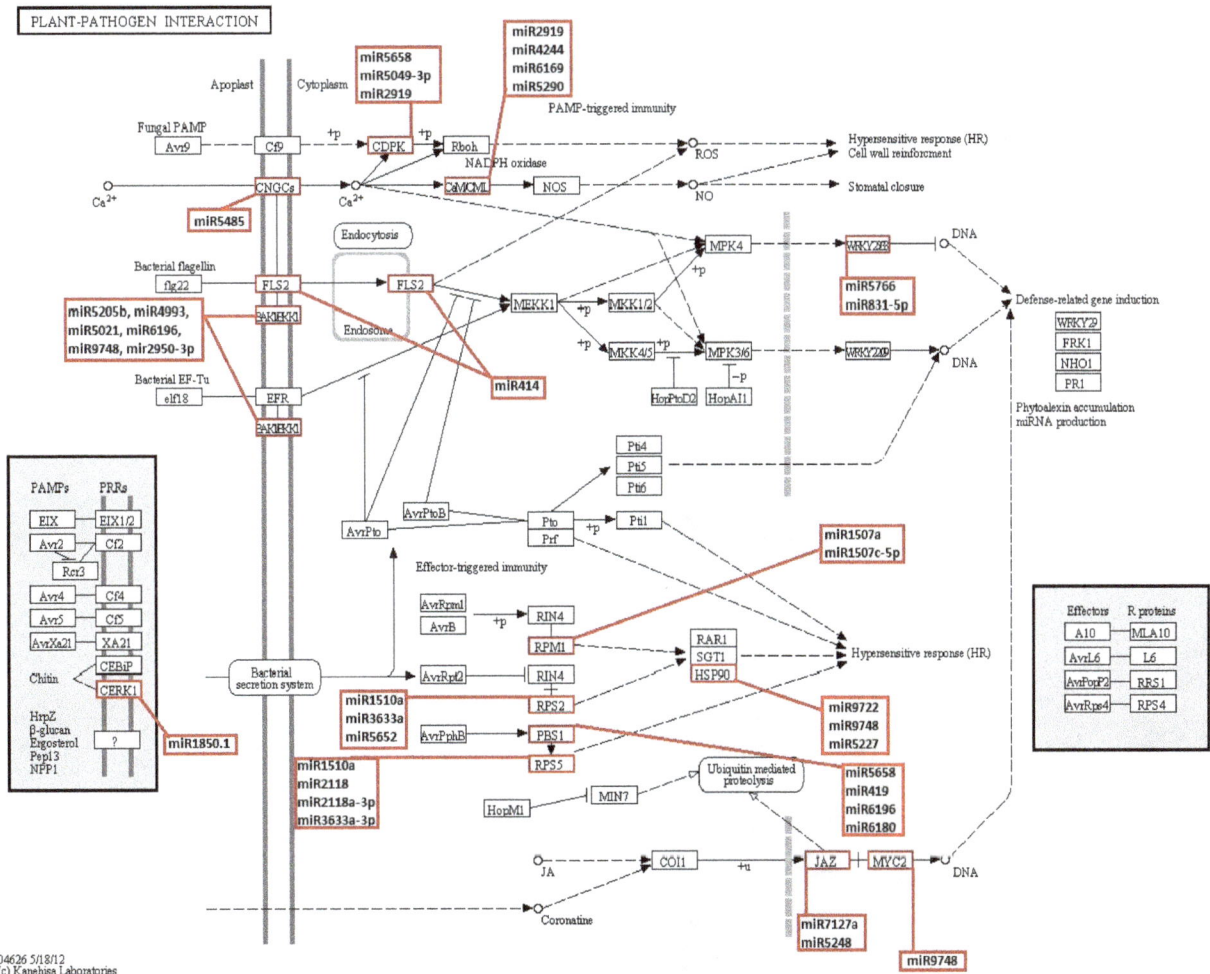

Figure 7 The KEGG Plant–pathogen interaction pathway and novel and known miRNAs which obtained in this study possibly targeting the genes involved in this pathway.

found to target NADPH-dependent thioredoxin reductase 3-like protein. miR-n69 targets cystathionine beta lyase. This enzyme also plays a role both in sulphur metabolism and selenocompound metabolism. miR-n116, miR-n131 and miR-n147 target serine acetyltransferase in sulphur metabolism. It is thought that these miRNAs should be investigated to understand whether they play a role in Se tolerance and accumulation. KEGG pathway analysis showed that plant–pathogen interaction pathway could be affected in high level by Se treatment in *A. chrysochlorus* calli. It is found that many genes that encode important proteins and enzymes in this pathway may be repressed by miRNAs with Se treatment. In this pathway, CDPK (calcium-dependent protein kinase) is thought to be down-regulated by miR5049-3p, while MYC2 (transcription factor MYC2) and HSP90 (heat-shock protein 90 kDa) expressions were decreased by miR9748. Disease resistance proteins RPS2 and RPS5 were down-regulated by miR3633a-3p, whereas RPM1 protein was down-regulated by miR1507a. All these proteins can change the hypersensitive response of the plant, and these results are consistent with the results in Van Hoewyk *et al.* (2008).

In conclusion, we identified a large number of known and novel miRNAs responsive to Se exposure. We detected 151 novel and 418 known miRNAs, some of them are expected to be regulated by Se treatment. *In silico* analysis showed that some miRNAs are found to be involved in targeting genes leading to Se metabolism. Se-responsive miRNAs are mostly involved in plant–pathogen interactions, plant hormone signal transduction, calcium signalling and sulphur metabolism; all these might be related to selenium tolerance and accumulation directly/indirectly (Figure 8). miR1507a, miR3633a-3p, miR4244, miR5049-3p, miR5485, miR7767-3p and miR9748 were expressed significantly by Se exposure. The results we obtained from our study show that miRNAs involved in plant–pathogen interaction pathway may

decrease the plant's hypersensitive response and increase the tolerance.

Materials and methods

Plant material and culturing conditions

The seeds of *Astragalus chrysochlorus*, collected in 2004 from Sertavul, Karaman, Turkey, were surface-sterilized with 70% alcohol for 1 min followed by 15 min in 5% bleach and then were rinsed with sterile distilled water three times. The sterilized seeds were germinated on growth-regulator-free MS (Murashige and Skoog, 1962) medium (pH 5.8) supplemented with 3% (w/v) sucrose and 0.8% agar (w/v). Callus tissues were proliferated on MS medium supplemented with 0.5 mg/L 2,4-dichlorophenoxy-acetic acid (2,4-D) and were subcultured every 3 weeks. All tissue culture experiments were carried out in a growth chamber illuminated with fluorescent light (ca. 1400 mol^{-2} ms^{-1}) over a 16/8 day and night at 25 ± 2 °C. The selenium treatment of calli was carried out in MS medium supplemented with 0 and 5 mg/L sodium selenate for 21 days. Each treatment was replicated three times and each time treatment was replicated in five individual culture dishes, and each dish contained nine callus tissues. Twenty-1-day-old callus tissues belong to control, and Se treatment were collected and frozen in liquid nitrogen and then stored at −80 °C.

RNA isolation, library construction and sequencing

Total RNAs were isolated for each Se-treated and untreated samples using TRIZOL (Invitrogen) according to the manufacturer's instructions. Quality and quantity of RNAs were measured with a Nanodrop 2000 spectrophotometer (Nanodrop Technologies, Thermo Scientific, Wilmington, DE). The RNA samples were then pooled for treated and control samples, respectively, and then stored at −80 °C for further analysis. The library construc-

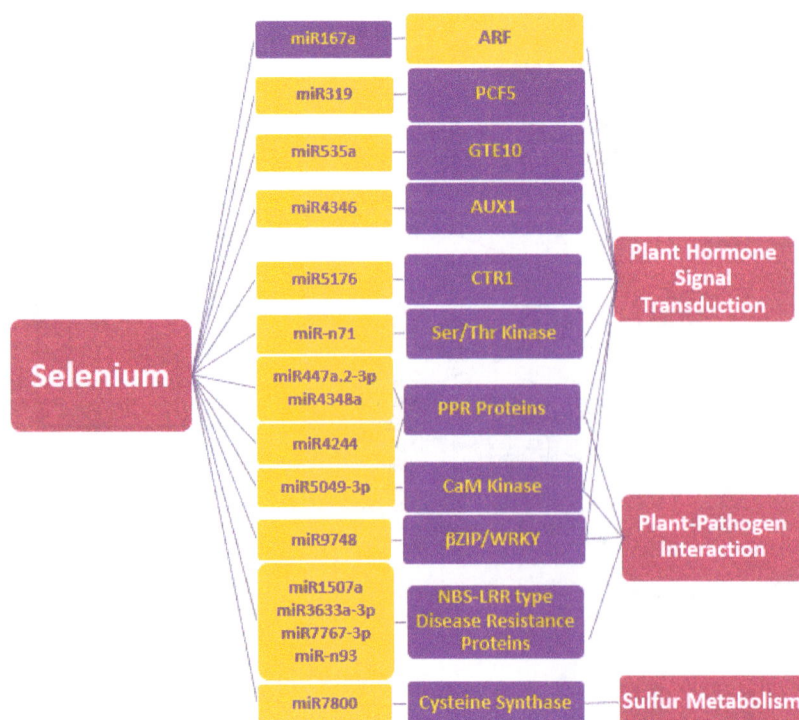

Figure 8 Possible mechanisms affected by Se-induced miRNAs.

tion and Illumina (Solexa) based-small RNA and degradome sequencing were carried out by the Beijing Genomics Institute (BGI, Shenzhen, China). Briefly, small RNAs were fractionated according to their sizes (18–30 nt long) and ligated to 5′ and 3′ oligonucleotide RNA adaptors. They were reverse-transcribed and amplified by PCR (Hafner et al., 2008). For degradome sequencing, the cleaved products were ligated to a single-stranded 5′ RNA adaptor. The ligation products were separated by oligo(dT) cellulose; after cDNAs were synthesized, MmeI digestion was performed (NEB). Ligation of a double-stranded adaptor to the 3′ end was performed followed by 21 cycles of PCR amplification (German et al., 2008). For both small RNA and degradome libraries, the purified PCR products were sequenced by SBS sequencing with a Solexa/Illumina genome analyser.

Small RNA and Degradome sequencing analyses

After sequencing, the vector sequences were removed first and the sequences between 18 and 30 nt were used for analyses as our previous reports (Xie et al., 2015a; Xie et al., 2015b). Briefly, miRNA length distribution was identified for clean, common and spesific sequences. Small RNA annotation was performed using tag2annotation software (BGI) to analyse the length distribution. The RNA-Seq data of A. chrysochlorus (unpublished data) obtained from our study was used as reference genome for mapping with the clean sequences by SOAP (Short Oligonucleotide Alignment Program) to determine the distribution on the genome (Li et al., 2008), as no available genome sequence information for this plant. The sequences that were matched up precisely were used for further analysis. The obtained small RNA sequences were compared to the GenBank (Benson et al., 2006) and Rfam database (11.0) (Gardner et al., 2009) by alignment and BLAST search to investigate rRNA, tRNA, snoRNA, snRNA and scRNA. The small RNAs were annotated with the priority rule [rRNAetc (in which GenBank > Rfam) > known miRNA > repeat > exon > intron] (Calabrese et al., 2007). To screen the known miRNAs, alignments were performed against the miRBase (Release 20) database (Griffiths-Jones et al., 2008) and differentially expressed known miRNAs among the samples that were identified previously. The small RNAs, which were unannotated, were screened for prediction of novel miRNAs using Mireap (BGI) with default parameters. The same method was used for degradome sequencing analysis, clean reads were generated and the alignments were performed against the A. chrysochlorus RNA-seq data, Rfam (11.0) database and GenBank database. The filtered reads were aligned to exons and introns of A. chrysochlorus mRNAs to investigate the fragments of degraded mRNAs. Then, clean tags were mapped to the reference genome (A. chrysochlorus RNA-seq data) using SOAP2.20 (Li et al., 2009) with allowing only two mismatches. 5′-position of sequences were chosen to predict miRNA cleavage sites using CleaveLand pipeline v3.0.1 (Addo-Quaye et al., 2009). Target (T-) plots were generated evaluating the position 10 or 11 of the miRNA.

Target prediction and functional classification based on GO and KEGG analyses

To predict potential mRNAs of A. chryschlorus, psRobot Small RNA Target Prediction Tool was used with default parameters (Wu et al., 2012). Glycine max (Soybean, JGI Glyma1.0 annotation of the chromosome-based Glyma1 assembly) and Medicago truncatula (Barrel medic, Release Mt3.0 from the Medicago Genome Sequence Consortium) were selected as

transcript libraries for target analysis. Gene Ontology (GO) provides a dictionary to render the characteristics of genes and their products. The results are classified by three ontologies in GO and these are molecular function, cellular component and biological process. Kyoto Encyclopedia of Genes and Genomes (KEGG) pathway analysis can also be used for the putative target genes. To determine the function of A. chryschlorus miRNAs, we used Blast2GO (http://www.blast2go.com) to determine the predicted target genes. As a first step, the mRNAs, which were targeted by miRNA, were aligned with BLASTX against nr database. As a second step, the hits were subjected to the GO and KEGG databases. Pathway enrichment and categorization were performed using GO (http://www.gene-ontology.org/) and KEGG databases (http://www.genome.jp/kegg/kegg1.html).

Acknowledgements

This work was kindly supported by the Research Fund of The Istanbul University with project number IRP-42927. O. Cakir received financial support from TUBITAK-BIDEB 2219-International Postdoctoral Research Fellowship Programme for this study. B. Candar-Cakir was supported partially by TUBITAK-BIDEB 2211/C and 2214/A programmes during her studies in U.S.A.

References

Addo-Quaye, C., Miller, W. and Axtell, M.J. (2009) CleaveLand: a pipeline for using degradome data to find cleaved small RNA targets. Bioinformatics, 25, 130–131.

Arı, Ş., Çakır, Ö. and Turgut-Kara, N. (2010) Selenium tolerance in Astragalus chrysochlorus: identification of a cDNA fragment encoding a putative Selenocysteine methyltransferase. Acta Physiol. Plant, 32, 1085–1092.

Barrera-Figueroa, B.E., Wu, Z. and Liu, R. (2012) Abiotic stress-associated microRNAs in plants: discovery, expression analysis, and evolution. Front. Biol. 8, 189–197.

Bartel, D.P. (2004) MicroRNAs: genomics, biogenesis, mechanism, and function. Cell, 116, 281–297.

Benchadi, W., Haba, H., Lavaud, C., Harakat, D. and Benkhaled, M. (2013) Secondary metabolites of Astragalus cruciatus link. and their chemotaxonomic significance. Rec. Nat. Prod. 7, 105–113.

Benson, D.A., Karsch-Mizrachi, I., Lipman, D.J., Ostell, J. and Wheeler, D.L. (2006) GenBank. Nucleic Acids Res. 34, D16–D20.

Birringer, M., Pilawa, S. and Flohe, L. (2002) Trends in selenium biochemistry. Nat. Prod. Rep. 19, 693–718.

Çakır, Ö. and Arı, Ş. (2013) Cloning and molecular characterization of selenocysteine methyltransferase (AchSMT) cDNA from Astragalus chrysochlorus. Plant Omics J. 6, 100–106.

Calabrese, J.M., Seila, A.C., Yeo, G.W. and Sharp, P.A. (2007) RNA sequence analysis defines Dicer's role in mouse embryonic stem cells. Proc. Natl Acad. Sci. USA, 104, 18097–18102.

Chen, L., Wang, T., Zhao, M., Tian, Q. and Zhang, W.H. (2012) Identification of aluminum-responsive microRNAs in Medicago truncatula by genome-wide high-throughput sequencing. Planta, 235, 375–386.

Dalmay, T. (2006) Short RNAs in environmental adaptation. Proc. Biol. Sci. 273, 1579–1585.

Ding, Y., Chen, Z. and Zhu, C. (2011) Microarray-based analysis of cadmium-responsive microRNAs in rice (Oryza sativa). J. Exp. Bot. 62, 3563–3573.

Freeman, J.L., Zhang, L.H., Marcus, M.A., Fakra, S., McGrath, S.P. and Pilon-Smits, E.A. (2006) Spatial imaging, speciation, and quantification of selenium in the hyperaccumulator plants Astragalus bisulcatus and Stanleya pinnata. Plant Physiol. 142, 124–134.

Freeman, J.L., Lindblom, S.D., Quinn, C.F., Fakra, S., Marcus, M.A. and Pilon-Smits, E.A. (2007) Selenium accumulation protects plants from herbivory by Orthoptera via toxicity and deterrence. *New Phytol.* **175**, 490–500.

Freeman, J.L., Tamaoki, M., Stushnoff, C., Quinn, C.F., Cappa, J.J., Devonshire, J., Fakra, S.C., Marcus, M.A., McGrath, S.P., Van Hoewyk, D. and Pilon-Smits, E.A. (2010) Molecular mechanisms of selenium tolerance and hyperaccumulation in *Stanleya pinnata. Plant Physiol.* **153**, 1630–1652.

Gardner, P.P., Daub, J., Tate, J.G., Nawrocki, E.P., Kolbe, D.L., Lindgreen, S., Wilkinson, A.C., Finn, R.D., Griffiths-Jones, S., Eddy, S.R. and Bateman, A. (2009) Rfam: updates to the RNA families database. *Nucleic Acids Res.* **37**, D136–D140.

German, M.A., Pillay, M., Jeong, D.H., Hetawal, A., Luo, S., Janardhanan, P., Kannan, V., Rymarquis, L.A., Nobuta, K., German, R., De Paoli, E., Lu, C., Schroth, G., Meyers, B.C. and Green, P.J. (2008) Global identification of microRNA-target RNA pairs by parallel analysis of RNA ends. *Nat. Biotechnol.* **26**, 941–946.

Griffiths-Jones, S., Saini, H.K., van Dongen, S. and Enright, A.J. (2008) miRBase: tools for microRNA genomics. *Nucleic Acids Res.* **36**, D154–D158.

Hafner, M., Landgraf, P., Ludwig, J., Rice, A., Ojo, T., Lin, C., Holoch, D., Lim, C. and Tuschl, T. (2008) Identification of microRNAs and other small regulatory RNAs using cDNA library sequencing. *Methods*, **44**, 3–12.

Hung, C.Y., Holliday, B.M., Kaur, H., Yadav, R., Kittur, F.S. and Xie, J. (2012) Identification and characterization of selenate- and selenite-responsive genes in a Se-hyperaccumulator *Astragalus racemosus. Mol. Biol. Rep.* **39**, 7635–7646.

Inal, B., Turktas, M., Eren, H., Ilhan, E., Okay, S., Atak, M., Erayman, M. and Unver, T. (2014) Genome-wide fungal stress responsive miRNA expression in wheat. *Planta*, **240**, 1287–1298.

Jones-Rhoades, M.W. and Bartel, D.P. (2004) Computational identification of plant microRNAs and their targets, including a stress-induced miRNA. *Mol. Cell*, **14**, 787–799.

Jones-Rhoades, M.W., Bartel, D.P. and Bartel, B. (2006) MicroRNAS and their regulatory roles in plants. *Annu. Rev. Plant Biol.* **57**, 19–53.

Kozomara, A. and Griffiths-Jones, S. (2011) miRBase: integrating microRNA annotation and deep-sequencing data. *Nucleic Acids Res.* **39**, D152–D157.

Kubachka, K.M., Meija, J., LeDuc, D.L., Terry, N. and Caruso, J.A. (2007) Selenium volatiles as proxy to the metabolic pathways of selenium in genetically modified *Brassica juncea. Environ. Sci. Technol.* **41**, 1863–1869.

Li, R., Li, Y., Kristiansen, K. and Wang, J. (2008) SOAP: short oligonucleotide alignment program. *Bioinformatics*, **24**, 713–714.

Li, R., Yu, C., Li, Y., Lam, T.W., Yiu, S.M., Kristiansen, K. and Wang, J. (2009) SOAP2: an improved ultrafast tool for short read alignment. *Bioinformatics*, **25**, 1966–1967.

Marsit, C.J., Eddy, K. and Kelsey, K.T. (2006) MicroRNA responses to cellular stress. *Cancer Res.* **66**, 10843–10848.

Murakami, Y., Yasuda, T., Saig, K., Urashima, T., Toyoda, H., Okanoue, T. and Shimotohno, K. (2006) Comprehensive analysis of microRNA expression patterns in hepatocellular carcinoma and non-tumorous tissues. *Oncogene*, **25**, 2537–2545.

Murashige, T. and Skoog, F. (1962) A revised medium for rapid growth and bioassays with tobacco tissue cultures. *Physiol. Plant.* **15**, 473–497.

Navarro, L., Dunoyer, P., Jay, F., Arnold, B., Dharmasiri, N., Estelle, M., Voinnet, O. and Jones, J.D. (2006) A plant miRNA contributes to antibacterial resistance by repressing auxin signaling. *Science*, **312**, 436–439.

Neuhierl, B. and Bock, A. (1996) On the mechanism of selenium tolerance in selenium-accumulating plants. Purification and characterization of a specific selenocysteine methyltransferase from cultured cells of *Astragalus bisculatus. Eur. J. Biochem.* **239**, 235–238.

Ozhuner, E., Eldem, V., Ipek, A., Okay, S., Sakcali, S., Zhang, B., Boke, H. and Unver, T. (2013) Boron stress responsive microRNAs and their targets in barley. *PLoS ONE*, **8**, e59543.

Pickering, I.J., Wright, C., Bubner, B., Ellis, D., Persans, M.W., Yu, E.Y., George, G.N., Prince, R.C. and Salt, D.E. (2003) Chemical form and distribution of

selenium and sulfur in the selenium hyperaccumulator *Astragalus bisulcatus. Plant Physiol.* **131**, 1460–1467.

Saini, A., Li, Y., Jagadeeswaran, G. and Sunkar, R. (2012) Role of microRNAs in plant adaptation to environmental stresses. In *MicroRNAs in Plant Development and Stress Responses, Signaling and Communication in Plants* (Sunkar, R., ed.), pp. 219–232. Berlin Heidelberg: Springer-Verlag.

Schiavon, M., Pilon, M., Malagoli, M. and Pilon-Smits, E.A. (2015) Exploring the importance of sulfate transporters and ATP sulphurylases for selenium hyperaccumulation-a comparison of *Stanleya pinnata* and *Brassica juncea* (Brassicaceae). *Front. Plant Sci.* **6**, 2.

Shrift, A. and Virupaksha, T.K. (1965) Seleno-amino acids in selenium-accumulating plants. *Biochim. Biophys. Acta*, **100**, 65–75.

Sors, T.G., Ellis, D.R. and Salt, D.E. (2005) Selenium uptake, translocation, assimilation and metabolic fate in plants. *Photosynth. Res.* **86**, 373–389.

Sors, T.G., Martin, C.P. and Salt, D.E. (2009) Characterization of selenocysteine methyltransferases from *Astragalus* species with contrasting selenium accumulation capacity. *Plant J.* **59**, 110–122.

Van Hoewyk, D., Takahashi, H., Inoue, E., Hess, A., Tamaoki, M. and Pilon-Smits, E.A. (2008) Transcriptome analyses give insights into selenium-stress responses and selenium tolerance mechanisms in *Arabidopsis. Physiol. Plant.* **132**, 236–253.

Wang, Y., Li, P., Cao, X., Wang, X., Zhang, A. and Li, X. (2009) Identification and expression analysis of miRNAs from nitrogen-fixing soybean nodules. *Biochem. Biophys. Res. Commun.* **378**, 799–803.

Whanger, P.D. (2002) Selenocompounds in plants and animals and their biological significance. *J. Am. Coll. Nutr.* **21**, 223–232.

Wu, H.J., Ma, Y.K., Chen, T., Wang, M. and Wang, X.J. (2012) PsRobot: a web-based plant small RNA meta-analysis toolbox. *Nucleic Acids Res.* **40**, W22–W28.

Xie, F.L., Huang, S.Q., Guo, K., Xiang, A.L., Zhu, Y.Y., Nie, L. and Yang, Z.M. (2007) Computational identification of novel microRNAs and targets in *Brassica napus. FEBS Lett.* **581**, 1464–1474.

Xie, F.L., Jones, D.C., Wang, Q.L., Sun, R.R. and Zhang, B.H. (2015) Small RNA sequencing identifies miRNA roles in ovule and fiber development. *Plant Biotechnol. J.* **13**, 355–369.

Xie, F., Wang, Q., Sun, R. and Zhang, B.H. (2015b) Deep sequencing reveals important roles of microRNAs in response to drought and salinity stress in cotton. *J. Exp. Bot.* **66**, 789–804.

Zeng, H.Q., Zhu, Y.Y., Huang, S.Q. and Yang, Z.M. (2010) Analysis of phosphorus-deficient responsive miRNAs and cis-elements from soybean (*Glycine max* L.). *J. Plant Physiol.* **167**, 1289–1297.

Zhang, B.H. (2015) MicroRNA: a new target for improving plant tolerance to abiotic stress. *J. Exp. Bot.* **66**, 1749–1761. doi:10.1093/jxb/erv013.

Zhang, B. and Wang, Q. (2015) MicroRNA-based biotechnology for plant improvement. *J. Cell. Physiol.* **230**, 1–15.

Zhou, Z.S., Huang, S.Q. and Yang, Z.M. (2008) Bioinformatic identification and expression analysis of new microRNAs from Medicago truncatula. *Biochem. Biophys. Res. Commun.* **374**, 538–542.

Zhou, Z.S., Zeng, H.Q., Liu, Z.P. and Yang, Z.M. (2012) Genome-wide identification of *Medicago truncatula* microRNAs and their targets reveals their differential regulation by heavy metal. *Plant Cell Environ.* **35**, 86–99.

4

Comparison of salt stress resistance genes in transgenic *Arabidopsis thaliana* indicates that extent of transcriptomic change may not predict secondary phenotypic or fitness effects

Zhulong Chan, Patrick J. Bigelow, Wayne Loescher and Rebecca Grumet*

Plant Breeding, Genetics and Biotechnology Program and Department of Horticulture, Plant and Soil Sciences Building, Michigan State University, East Lansing MI, USA

*Correspondence

email grumet@msu.edu
Accession numbers: Microarray data are available online in Gene Expression Omnibus (GEO) database (http://www.ncbi.nlm.nih.gov/geo/) under accessions number (GSE26983).

Keywords: abiotic stress resistance, risk assessment, environmental biosafety, CBF3/DREB1a, mannose-6-phosphate reductase, *SOS1*.

Summary

Engineered abiotic stress resistance is an important target for increasing agricultural productivity. There are concerns, however, regarding possible ecological impacts of transgenic crops. In contrast to the first wave of transgenic crops, many abiotic stress resistance genes can initiate complex downstream changes. Transcriptome profiling has been suggested as a comprehensive non-targeted approach to examine the secondary effects. We compared phenotypic and transcriptomic effects of constitutive expression of genes intended to confer salt stress tolerance by three different mechanisms: a transcription factor, CBF3/DREB1a; a metabolic gene, *M6PR*, for mannitol biosynthesis; and the Na⁺/H⁺ antiporter, *SOS1*. Transgenic *CBF3*, *M6PR* and *SOS1 Arabidopsis thaliana* were grown together in the growth chamber, greenhouse and field. In the absence of salt, M6PR and SOS1 lines performed comparably with wild type; CBF3 lines exhibited dwarfing as reported previously. All three transgenes conferred fitness advantage when subjected to 100 mM NaCl in the growth chamber. *CBF3* and *M6PR* affected transcription of numerous abiotic stress-related genes as measured by Affymetrix microarray analysis. *M6PR* additionally modified expression of biotic stress and oxidative stress genes. Transcriptional effects of *SOS1* in the absence of salt were smaller and primarily limited to redox-related genes. The extent of transcriptome change, however, did not correlate with the effects on growth and reproduction. Thus, the magnitude of global transcriptome differences may not predict phenotypic differences upon which environment and selection act to influence fitness. These observations have implications for interpretation of transcriptome analyses in the context of risk assessment and emphasize the importance of evaluation within a phenotypic context.

Introduction

Salt stress resulting from saline soils or irrigation water is a major factor limiting agricultural productivity worldwide (Yamaguchi and Blumwald, 2005; Shabala and Cuin, 2008; Munns and Tester, 2008). Increased irrigation, utilization of marginal crop land and increasing demand for food production are all anticipated to increase the rate of salinization, making salinity stress resistance an important goal for crop improvement. In recent years, genetic engineering of crops for environmental stress resistance has become increasingly important (Nickson, 2008; Beckie et al., 2010; Grumet et al., 2011). Field trials in the United States, for the crops engineered for resistances to drought, cold, heat and salt, increased from 23 in 2001 to 119 in 2010, and genetically engineered, drought-tolerant maize is approaching commercialization (USDA-APHIS records, http://www.isb.vt.edu/data.aspx; Edmeades, 2008).

Salt stress in plants is manifested as a combination of dehydration or osmotic-related stress effects owing to reduced water potential resulting from increased solute concentration and damage caused by toxic effects of excess sodium ions (Yamaguchi and Blumwald, 2005; Munns and Tester, 2008). Salt stress is also typically associated with oxidative stress (Hasegawa et al., 2000; Miller et al., 2010). Possible routes to counteract these negative effects include exclusion or sequestration of sodium ions or accumulation of compatible solutes or osmoprotectants. Compatible solutes or osmoprotectants have been suggested to osmotically balance stress-related decrease in water content, stabilize macromolecular structures and/or scavenge free radicals that accumulate in response to stress (Chen and Murata, 2002). Approaches to engineer salt stress resistance have included regulation of ion transport through introduction of Na⁺/H⁺ antiporters or H⁺ pumps (Apse et al., 1999; Gaxiola et al., 2001; Shi et al., 2003); synthesis of compatible solutes, e.g. mannitol (Zhifang and Loesher, 2003), proline (Kishore et al., 1995) or glycine betaine (Chen and Murata, 2008); or the introduction of transcription factors regulating expression of stress-responsive genes (e.g. Jaglo-Ottosen et al., 1998; Kasuga et al., 1999; Zheng et al., 2009).

While engineered salt stress resistance holds promise for agricultural productivity in impaired conditions, there has been considerable concern about the possible ecological impacts of release of transgenic crops. At the forefront are concerns about the risk of transgene escape into natural populations and potential effects on ecosystem balance (Conner *et al.*, 2003; Chandler and Dunwell, 2008; Craig *et al.*, 2008; Warwick *et al.*, 2009; Beckie *et al.*, 2010). The traits most likely to become established in natural environments are those that provide the greatest selective advantage (Hancock, 2003; Lu and Yang, 2009; Warwick *et al.*, 2009). Abiotic stress-related traits may fall in this category. Salt tolerance could provide a competitive advantage to recipient populations or allow a crop or wild relative to grow in areas that it could not previously colonize (Lu and Yang, 2009; Warwick *et al.*, 2009; Beckie *et al.*, 2010).

In addition to the selective advantages that may result from the primary intended effect of the transgene, e.g. ability to grow in saline environments, secondary changes in phenotype may also have reproductive or fitness effects. The first wave of transgenic crops primarily utilized genes whose protein product was directly responsible for the desired trait (e.g. Bt proteins confer insect resistance; herbicide resistance genes encode proteins that prevent binding of the herbicide or otherwise inactivate the herbicide; Carpenter *et al.*, 2002). These genes, as well as marker genes such as GUS or the kanamycin resistance gene, NPTII, have generally had minimal effects on fitness, except under the selective conditions (e.g. insect herbivory) for which they were developed (Crawley *et al.*, 2001; Pilson *et al.*, 2002; Snow *et al.*, 2003). They are largely inert with respect to other cellular functions as evidenced by minimal pleiotropic phenotypes and the results of global transcriptome and proteome studies (El Ouakfaoui and Miki, 2005; Ruebelt *et al.*, 2006; Cheng *et al.*, 2008; Zolla *et al.*, 2008; Little *et al.*, 2009). Indeed, transcriptome comparisons of single transgene differences vs. cultivar differences in wheat, rice, maize and soybean have shown greater differences among cultivars than as a result of transgene introduction (Baudo *et al.*, 2006; Batista *et al.*, 2008; Cheng *et al.*, 2008; Coll *et al.*, 2008). While introduction of a transgene, per se, may not cause extensive transcriptional modifications, the extent of changes is directly related to the nature of the introduced transgene and its biological function. It has been suggested that genes that are from distant biological sources or are novel to plants are less likely to interact with other plant processes than those that have specific plant-related functions (Miki *et al.*, 2009).

Many of the genes under consideration for abiotic stress resistance initiate subsequent changes within the cell that facilitate adaptive responses. They may cause the cell to produce compounds needed to survive, grow and respond to the environment. Such genes may encode transcription factors that regulate expression of other genes; signalling factors that initiate responses to perceived changes in the cellular environment; or metabolic pathway enzymes that result in the production of new cellular compounds. As a result of their downstream actions, these types of genes may have broader effects on plant metabolism, physiology and development, than genes for which the protein itself is the final product. Although the ability of a given gene to initiate a cascade of events can make it highly valuable for genetic engineering, such genes also have the potential to modify non-target phenotypes within the plant through pleiotropic or epistatic interactions (Wolfenbarger and Grumet, 2003; Little *et al.*, 2009; Miki *et al.*, 2009). These changes could, in turn, influence fitness of the recipient plant.

Therefore, different possible approaches to engineer salt stress resistance could have different secondary effects. In this work, we compared the phenotypic and transcriptomic effects of three types of genes intended to confer salt stress tolerance:

a regulatory gene, *CBF3/DREB1a*, coding for the C-repeat binding factor/drought-responsive element binding transcription factor (Jaglo-Ottosen *et al.*, 1998; Kasuga *et al.*, 1999); a metabolic gene, *M6PR*, coding for the mannose-6-phosphate reductase enzyme for mannitol biosynthesis (Zhifang and Loesher, 2003); and a membrane protein gene, *SOS1*, encoding a plasma membrane Na^+/H^+ antiporter (Shi *et al.*, 2003).

CBF/DREB1 genes encode a family of transcription factors that promote expression of a group of abiotic stress-responsive genes (Van Buskirk and Thomashow, 2006; Chinnusamy *et al.*, 2007). Transgenic *CBF/DREB1*-overexpressing *Arabidopsis* plants exhibit increased tolerance to freezing, drought and salinity stress (Jaglo-Ottosen *et al.*, 1998; Liu *et al.*, 1998; Gilmour *et al.*, 2004). Constitutive expression of CBF/DREB transcription factors in *Arabidopsis* leads to the expected increase in CBF/DREB target genes (the CBF/DREB regulon) (Seki *et al.*, 2001; Fowler and Thomashow, 2002; Maruyama *et al.*, 2004; Zhang *et al.*, 2004; Vogel *et al.*, 2005;). The induced genes include those that likely function in stress tolerance (e.g. LEA, dehydrin, antifreeze and galactinol/raffinose synthesis) as well as factors involved in signal transduction and gene regulation. The CBF/DREB-responding genes can be clustered into groups showing increased or decreased expression at different time periods following transfer to the cold, suggesting sequential induction by CBF, or activity of downstream CBF-induced transcription factors (Fowler and Thomashow, 2002; Vogel *et al.*, 2005). Constitutive *CBF* expression has been associated with growth reduction in the absence of stress and delayed reproductive development (Liu *et al.*, 1998; Kasuga *et al.*, 1999; Gilmour *et al.*, 2000; Achard *et al.*, 2008a).

M6PR is responsible for the conversion of mannose-6-phosphate to mannitol-1-phosphate, the first committed step in mannitol production in plants (Everard *et al.*, 1997; Zhifang and Loesher, 2003). Transgenic *Arabidopsis*-overexpressing *M6PR* showed increased resistance to salt stress as manifested by increased dry weight and seed yield and reduced inhibition of photosynthetic activity (Zhifang and Loesher, 2003; Sickler *et al.*, 2007; Chan *et al.*, 2011). *M6PR* did not confer resistance against drought stress (Sickler *et al.*, 2007). Laboratory analysis of *M6PR* transgenic *Arabidopsis* plants in the absence of salt stress did not show effects on growth, photosynthetic activity, time to bolting or seed set (Zhifang and Loesher, 2003; Sickler *et al.*, 2007; Chan *et al.*, 2011). Transcriptome analysis suggested that increased salt tolerance may be due, at least in part, to the expression of numerous stress-related genes prior to salt treatment (Chan *et al.*, 2011).

SOS1 is a plasma membrane–located sodium efflux carrier that functions to transport sodium ions out of the cell into the apoplast and so can reduce cytoplasmic sodium content (Qiu *et al.*, 2003; Shabala and Cuin, 2008). SOS1 acts in coordination with two other members of the SOS pathway, SOS2 and SOS3, to maintain cellular ion homeostasis (Shi *et al.*, 2003; Zhu, 2003). Transgenic *SOS1*-overexpressing *Arabidopsis* plants exhibited increased salt stress tolerance as indicated by higher per cent survival and less reduction in root growth, protein content, total chlorophyll and photosynthetic activity than the non-transgenic controls (Shi *et al.*, 2003). In the absence of salt stress, the transgenic plants did not show obvious differences in growth and development. To our knowledge, global transcriptional analyses have not been published for SOS transgenic plants.

Transcriptome or proteome profiling has been suggested as a comprehensive non-targeted approach to examine secondary effects resulting from the introduction of transgenes (Ruebelt et al., 2006; Batista et al., 2008; Chung et al., 2008; Zolla et al., 2008; Ricroch et al., 2011). These authors also have emphasized the importance of interpreting results within the context of naturally occurring variation that may result from genotypic or environmental factors. In this study, we sought to compare the transcriptome effects of the *CBF3*, *M6PR* and *SOS1* transgenes and, furthermore, to examine their phenotypic and fitness-associated effects when grown side by side in the growth chamber, greenhouse and field in the presence and absence of salinity stress. Our results indicate that each transgene influenced a different set of genes and that *CBF3* and *M6PR* had considerably greater effects on the transcriptome than *SOS1*. The magnitude of transcriptome effects, however, did not correlate with phenotypic and fitness effects.

Results

Growth and development of the CBF3, M6PR and SOS1 transgenic lines

To compare the growth, development and transcriptional effects of the three transgenes, three sets of lines, each including two independent transgenic lines and their respective wild-type (WT) parental controls [(1) CBF3 A30, CBF3 A40 and WS-WT; (2) M6PR M2, M6PR M5, and Col-WT; (3)SOS1 1-1, SOS1 7-6 and Col(gl)-WT], were grown together for their full life cycle in replicated experiments in the growth chamber, greenhouse and field. As has been reported previously for *CBF*-overexpressing plants in the growth chamber (Gilmour et al., 2000; Vogel et al., 2005), constitutive *CBF3* expression in *Arabidopsis* had negative effects on vegetative growth and reproductive development including delayed bolting and flowering (4 and 10 day delay to 50% flowering for A40 and A30, respectively; $P < 0.05$, Duncan's multiple range test; Figure S1), and reduced leaf number, rosette diameter, plant height, dry weight, and seed yield relative to the WS-WT parent genotype (Figure 1a; Table 1A; Figure S1). The CBF3 plants, especially line A30, which has a higher level of expression (data not shown), also exhibited significantly delayed flowering or maturation (Figure S1 and data not shown) and reduced dry matter in the greenhouse and seed production in the greenhouse and field (Figure 2).

The *M6PR* and *SOS1* genes had minimal effects on performance in the growth chamber. *M6PR* had somewhat positive effects on leaf number, dry weight and seed yield (Table 1A) and did not affect time to bolting or flowering (Figure S1). SOS1 1-1 exhibited some mild chlorosis, growth reduction and delay in flowering (Figure S1). M6PR and SOS1 plants exhibited some reduction in dry matter in the field, but not in the greenhouse; M6PR line M2 had increased seed yield in the field experiment (Figure 2).

Treatment with NaCl in the growth chamber experiments inhibited growth and development and caused progressive leaf injury symptoms of chlorosis and necrosis for all WT and transgenic genotypes (Figure 1; Table 1). Symptom development occurred more rapidly or to a greater extent on the WT lines than their transgenic counterparts as evidenced by CBF3 lines at 100 mM NaCl, M6PR lines at 100 mM and 200 mM NaCl, and SOS1 lines at 200 mM NaCl; 32 and 38 days after planting (DAP; $P = 0.05$, Duncan's multiple range). When treated with 100 mM NaCl, the transgenic lines (with the exception of CBF3 A30) had greater vegetative growth (stalk number, plant height,

dry weight) and seed yield relative to their wild-type counterparts (Figure 1, Table 1A,B; Figure S1). Both M6PR lines and the SOS1 7-6 line also showed increased bolting and flowering relative to their wild-type parents (Figure S1). The average reduction in dry matter (25%) and seed yield (18%) of the transgenic lines (except CBF3 A30) was approximately half that of the WT lines (51% and 36% for dry matter and seed yield, respectively). The 200 mM NaCl treatment caused severe growth reduction or death for all genotypes, although several vegetative measures, e.g. leaf number, rosette diameter, plant height or dry weight were greater for the transgenic plants than for WT parents (Table 1). Only a few plants, however, produced inflorescences, and none set seed, except a few SOS1 7-6 plants (Table 1; Figure S1).

Comparative transcriptome effects of the *CBF3*, *M6PR* and *SOS1* transgenes

Microarray analyses were conducted to compare transgene effects on gene expression in the presence and absence of salt stress. The microarray signal data from any pair of biological replicates were highly reproducible for all pairs of comparisons ($R^2 > 0.95$ for normalized signal data; Figure S2), indicating strong reproducibility between the experiments. Microarray results were also partially verified using quantitative real-time polymerase chain reaction (qRT-PCR) assay for a set of 21 transcripts representing different categories of relative transcript abundance (i.e. increased, decreased or essentially unchanged). The expression ratios measured by microarray and qRT-PCR were highly correlated ($R > 0.90$ for 252 transcript/salt/transgene combinations; Figure S3).

Different threshold parameters were examined to determine meaningful differences in gene expression levels (Table 2). To minimize potential statistical biases and avoid incorrectly declaring gene expression to be influenced by one transgene and not another, modest criteria (2-fold difference and $P \leq 0.05$), using both Bioconductor R package and GeneChip® Operating Software from Affymetrix as described in Materials and Methods, were used to declare differences. Comparable trends also were observed with more stringent criteria (3-fold cut-off or $P \leq 0.01$; Table 2).

Effects of the three transgenes on the transcriptome in the absence of salt stress

As expected, in the absence of stress, transgenic SOS1 and CBF3 plants showed increased expression of *SOS1* and *CBF3* transcripts, respectively (Table 3A). *M6PR* is not a native gene and cannot be detected by microarray, but the expression was verified by Northern blot analysis (Figure S4). The CBF3 lines also showed elevated expression of a large number of CBF-target genes as identified in previous studies, including *COR*, *RD*, *LT*, *ERD*, *ZAT* and dehydrin genes (Seki et al., 2001; Maruyama et al., 2004; Zhang et al., 2004; Vogel et al., 2005; Magome et al., 2008). About 70% (27/38) of *DREB1A* upregulated genes in Maruyama et al. (2004) and 90% (35/40) of *CBF* regulon genes in Zhang et al. (2004) were also significantly changed by the *CBF3* transgene in this experiment (Table 3C). Collectively, these results provide confidence in the microarray analyses.

In the absence of salt stress, *CBF3* and *M6PR* transgenes affected a much larger number of transcripts than the *SOS1* transgene (Table 2, Figure 3a). The large number of changes observed for CBF3 plants was consistent with its function as a

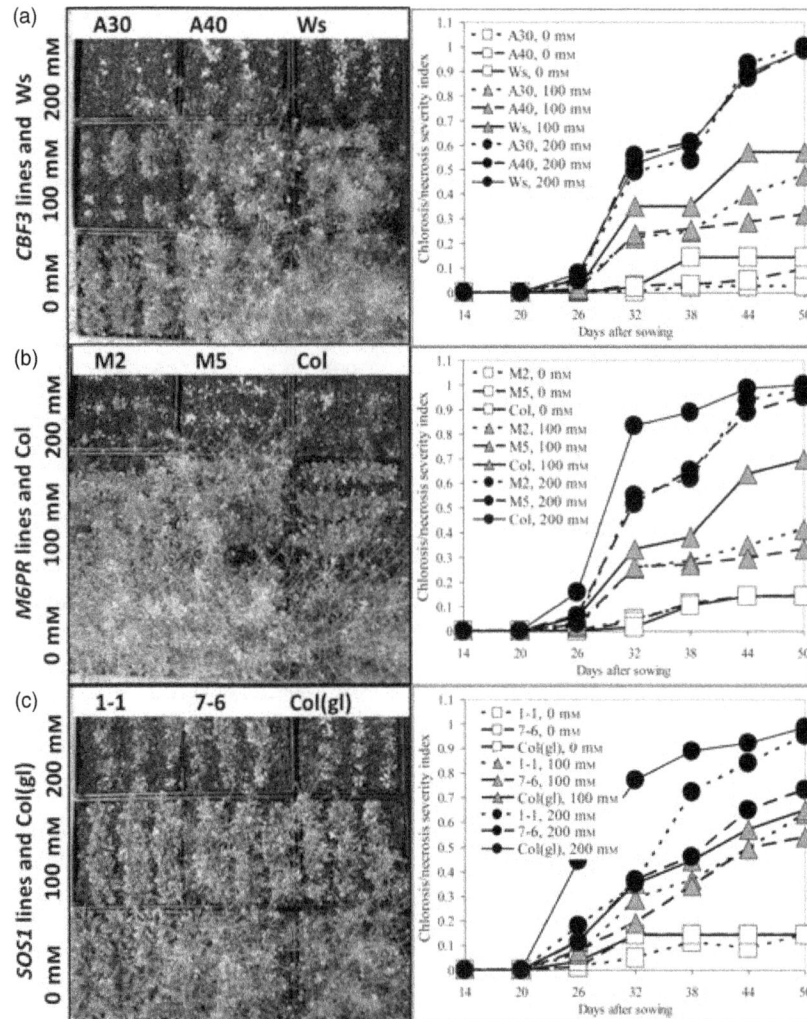

Figure 1 Growth and chlorosis/necrosis severity indices of transgenic and wild-type Arabidopsis in response to long-term salt stress in the growth chamber. (a) CBF3 and WS-WT; (b) M6PR and Col-WT; (c) SOS1 and Col(gl)-WT. The plants were photographed at 50 DAP. Chlorosis and necrosis were rated as: 0, no yellow or purple leaves; 1, older leaves turn yellow or purple; 3, younger leaves turn yellow or purple; 5, some leaves die; and 7, plants die. Chlorosis/necrosis severity indices were calculated as described in Experimental Procedures. Each value is the mean of three replicate trays, 18 plants/tray. White square, 0 mM NaCl; Grey triangle, 100 mM NaCl; Black circle, 200 mM NaCl.

transcription factor and with previous transcriptional analyses of CBF overexpressors (Seki et al., 2001; Maruyama et al., 2004; Zhang et al., 2004; Vogel et al., 2005; Magome et al., 2008). Gene expression changes included direct members of the CBF regulon as indicated earlier, as well as a variety of other genes previously shown to be targets of the CBF regulon (Fowler and Thomashow, 2002; Cook et al., 2004), including members of the stress-related categories of response to water deprivation, cold, osmotic stress and salt stress; numerous transport and minor carbohydrate metabolism genes, (e.g. pathway genes for the compatible solutes raffinose and trehalose); and numerous cell wall-associated genes (Table 3; Table S1; Table S2A,E,F). The CBF transgene influenced expression of several ABA-related genes, including up-regulation of several newly identified ABA receptors, [SNF1-related kinases (e.g. PYL5, SnRK2.2 and SnRK2.3)] and downstream ABA-responsive genes (Table S2D).

Expression of M6PR affected large number of the same genes as CBF3; approximately half (642, 49%; Figure 4A) of those up- or down-regulated were in common for the two transgenes.

These included the members of the stress-related categories of water deprivation, cold, osmotic stress and salt stress, ABC and potassium transport, minor CHO metabolism (including raffinose and trehalose), and ABA-related and cell wall-associated genes (Table 4; Table S1; Table S2A–F). However, the total number of transcripts affected was greater for M6PR plants than for CBF3 plants (1719 vs. 1350; Table 2, Figure 3A). In addition to the stress-related gene categories influenced by CBF3, M6PR strongly affected expression of biotic, oxidative and heat stress-related genes including several pathogenesis-related (PR) or putative resistance gene analogues, and glutathione, thioredoxin, glutaredoxin family genes (Table 4; Table S2B,C). M6PR also caused more extensive changes in cell wall-associated genes including up-regulation of several arabinogalactan and xyloglucan-related protein genes and down-regulation of cellulose synthases (Table S1).

In contrast to CBF3 and M6PR, many fewer transcripts were affected by the Na+/H+ antiporter SOS1 (Figure 3). While the general categories showing biological enrichment were similar

Table 1 Summary of transgene effects on growth parameters in the absence and presence of salt. The plants were harvested at 62 DAP. Leaf number, rosette diameter, stalk numbers and plant height were measured before harvest. Total dry weight and seed yield were measured after harvest. Each value is the mean of three replicate trays, 18 plants/tray. All data were analysed by ANOVA with SPSS 11.5 for windows (SPSS, Inc., Chicago, IL). Mean separations were performed by Duncan's multiple range test, $P < 0.05$. Detailed transgene effects on plant growth throughout the life cycle and on other growth parameters are shown in Figure S1

A. Growth relative to WT

	0 mM CBF3/Ws A30	A40	M6PR/Col M2	M5	SOS1/Col(gl) 1-1	7-6	100 mM CBF3/Ws A30	A40	M6PR/Col M2	M5	SOS1/Col(gl) 1-1	7-6	200 mM CBF3/Ws A30	A40	M6PR/Col M2	M5	SOS1/Col(gl) 1-1	7-6
Leaf number	0.64	0.81	1.05	1.17	0.93	1.10	0.86	1.04	1.55	1.64	1.26	1.62	0.63	0.77	1.19	1.26	1.90	3.22
Rosette diameter	0.86	0.86	1.08	1.10	1.00	1.01	0.78	0.86	1.03	1.18	1.12	1.06	0.51	0.85	1.13	1.15	0.94	1.31
Stalk number	0.89	1.00	1.05	0.98	0.90	0.98	0.57	1.38	1.47	1.47	1.29	1.52	0.00	2.65	/	/	/	/
Plant height	0.80	0.88	1.03	0.98	0.95	1.04	0.82	1.80	1.79	1.85	1.24	1.55	0.00	2.103	/	/	/	/
Dry weight	0.58	0.77	1.19	1.04	0.88	1.00	0.22	1.34	2.11	2.87	1.80	2.15	0.53	1.26	2.55	2.1	1.63	3.59
Seed yield	0.38	0.82	1.14	1.06	0.80	0.89	0.00	1.53	1.89	3.45	1.14	1.77	/	/	/	/	/	/

B. Growth relative to control plants without salt

	100 mM/0 mM CBF3/Ws Ws	A30	A40	M6PR/Col Col	M2	M5	SOS1/Col(gl) Col(gl)	1-1	7-6	200 mM/0 mM CBF3/Ws Ws	A30	A40	M6PR/Col Col	M2	M5	SOS1/Col(gl) Col(gl)	1-1	7-6
Dry weight	0.29	0.11	0.51	0.22	0.39	0.61	0.25	0.51	0.53	0.03	0.03	0.05	0.02	0.05	0.05	0.05	0.09	0.17
Seed yield	0.20	0.0	0.38	0.16	0.26	0.51	0.20	0.28	0.39	0.0	0.0	0.0	0.0	0.0	0.0	0.0	0.0	0.04

Grey background—significant negative effect when compared to WT (Duncan's multiple range, $P < 0.05$); Black background—significant positive effect when compared to WT; white—no significant difference.

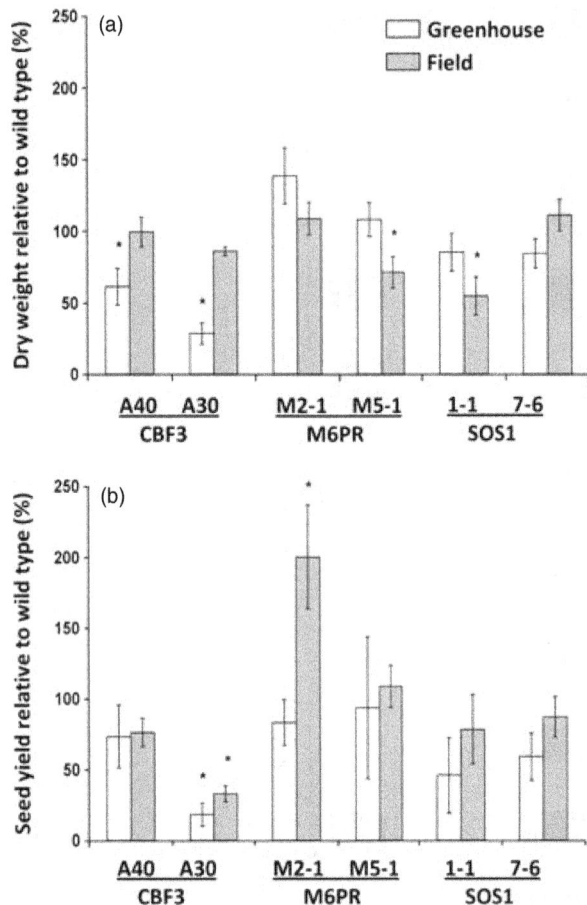

Figure 2 Dry matter accumulation (a) and seed yield (b) of greenhouse and field grown populations of transgenic CBF3, M6PR and SOS1 plants relative to wild-type parental populations. Plants were grown in the absence of salt stress. Values are expressed as per cent of wild type. Each value is the mean ± SE of five replicate trays. *Significant difference (ANOVA, Duncan's multiple range, $P < 0.05$) between transgenic and wild-type plants.

among all three transgenes (Table S3), very few of the specific genes affected by *SOS1* were in common with *CBF3* (7.7%) and *M6PR* (2.1%; Figure 4a). Only six expression changes were in common between all three. Microarray analysis indicated that *SOS1* overexpression influenced transcript levels of CBF3. However, the great majority of genes affected by CBF3 in the CBF3 transgenic lines were not affected in the SOS1 plants (Table 3). This may be due to the ~4-fold induction in SOS1 plants vs. ~200-fold induction for CBF3 plants. Of the stress-related categories, only response to oxidative stress was significantly overrepresented in the *SOS1*-influenced genes, but unlike M6PR plants, the majority of the affected redox-related genes in SOS1 plants were down-regulated, rather than up-regulated, especially in the glutaredoxin family (Table 4, Table S2C). These differences between *SOS1* effects relative to *CBF3* and *M6PR* are evident in the cluster analysis presented in Figure 5a.

Effects of the three transgenes on the transcriptome in the presence of salt stress

Imposition of salt stress affected a much smaller number of transcripts in the *CBF3* and *M6PR* plants than in the WT

parental WS and Col plants (Table 2, Figure 3c,d). Indeed, a substantial portion of the transcripts modulated by salt stress in Col and WS WT plants, including numerous abiotic stress, biotic stress, redox, cell wall, minor carbohydrate metabolism, and transport genes, was affected by the *CBF3* and *M6PR* transgenes prior to salt treatment [33% (449) for *CBF3* and 47% (817) for *M6PR*] (Figures 5a,c and 6; Table S2A–F).

The small number of salt-induced gene changes in the parental Col(gl) plants appears to be associated with transcriptional effects of the *gl* mutation leading to constitutive induction of many stress-associated genes (Zhulong Chan, Rebecca Grumet and Wayne Loescher, unpublished) that may also mask some of the effects of SOS1 overexpression. Only 5.2% of changes associated with *SOS1* overlapped with those by affected by salt (Figure 6). Unlike CBF3 and M6PR plants, the salt-stressed SOS1 plants did not exhibit a reduction in the number of gene expression differences relative to the parental Col(gl) plants in the absence of salt stress (Figure 3). Many disease resistance–related protein and glutaredoxin genes continued to be down-regulated in SOS1 plants relative to Col(gl)-WT, even in the presence of salt (Table S2A,B).

Pathways showed differences for all three transgenes relative to their WT parents in the presence of salt included hormone, secondary, and cell wall metabolism. Other pathways, like stress, transport, redox, minor CHO metabolism, S-assimilation and amino acid metabolism were enriched in salt-stressed CBF3 and SOS1 lines but not M6PR lines (Table S3; Table 4). Despite similar categories of genes, there was little overlap among the specific genes that differed in each transgene-WT comparison (Figure 5b). This is in contrast to the substantial overlap between transcripts affected by CBF3 and M6PR in the absence of salt stress; 47.5% without salt vs. 19.5% in the presence of 100 mm NaCl (Figure 4a,b). The transcriptional differences between CBF3 and WS-WT in the presence of salt continued to include a large number of CBF-target, ABA-related, and other abiotic stress-related genes, possibly reflecting continued effect of CBF3 overexpression in the presence of salt stress (Table 3, Table S2A–D). While salt treatment induced expression of many of these genes in both WS-WT and Col-WT plants, the level of increase was not great as was caused by the CBF3 overexpression.

As occurred in the absence of salt, a larger portion of overlap was found between the *CBF3*- and *M6PR*-affected transcripts, than for *CBF3* or *M6PR* with *SOS1* (Figure 4b). In the presence of salt stress, the SOS1 plants exhibited down-regulation of numerous disease resistance-related genes that did not occur in the absence of salt stress or in the other transgenic lines (Table S2B).

Discussion

Several recent studies have compared the magnitude of transcriptional or proteomic changes caused by a transgene with those observed following introgression of a specific trait, among cultivars resulting from conventional breeding, or as a result of environmental effects (e.g. Corpillo *et al.*, 2004; Baudo *et al.*, 2006; Ruebelt *et al.*, 2006; Albo *et al.*, 2007; Batista *et al.*, 2008; Cheng *et al.*, 2008; Coll *et al.*, 2008; Zolla *et al.*, 2008). The general conclusion from these studies is that fewer changes are observed for the transgene than by conventional breeding, and those that are observed, fall within the range of natural variation. These modest effects were attributed to the single

Table 2 Total numbers of changed transcripts by the three transgenes or salt stress at different cut-off values. All microarray data were normalized and analysed together using affylmGUI running on R package. The full list of genes is provided in Table S1

Comparisons	Fold change ≥2 and P-value <0.05		Fold change ≥3 and P-value <0.05		Fold change ≥2 and P-value <0.01	
	Up	Down	Up	Down	Up	Down
A. Transgene effects minus salt						
CBF3-0 mM_v_WS-0 mM	758	592	404	194	606	511
M6PR-0 mM_v_Col-0 mM	986	733	495	286	874	656
SOS1-0 mM_v_Col(gl)-0 mM	233	386	78	85	139	272
B. Transgene effects plus salt						
CBF3-100 mM_v_WS-100 mM	466	571	205	145	345	419
M6PR-100 mM_v_Col-100 mM	244	757	45	199	137	498
SOS1-100 mM_v_Col(gl)-100 mM	367	478	110	163	284	373
C. Salt effects on transgenic lines						
CBF3-100 mM_v_CBF3-0 mM	98	25	36	2	57	5
M6PR-100 mM_v_M6PR-0 mM	153	323	72	37	87	224
SOS1-100 mM_v_SOS1-0 mM	47	88	9	24	27	34
D. Salt effects on wild types						
WS-100 mM_v_WS-0 mM	774	319	327	80	572	231
Col-100 mM_v_Col-0 mM	1615	937	688	287	1423	843
Col(gl)-100 mM_v_Col(gl)-0 mM	80	58	29	19	48	38

gene change for the transgene vs. multiple changes that occur as a result of conventional breeding, even in a near-isogenic background, as well as the transcriptional flexibility widely exhibited by plants in response to variable environments. In most cases, the transgenes assessed encoded simple traits that were the direct product of the protein produced, such as endosperm seed storage protein in wheat (Baudo *et al.*, 2006); glyphosate tolerance in soybean (Cheng *et al.*, 2008); Bt protein in maize (Coll *et al.*, 2008); and selectable marker genes encoding kanamycin, biaphalos or glufosinate resistance (El Ouakfaoui and Miki, 2005; Abdeen and Miki, 2009; Miki *et al.*, 2009).

However, it is not unusual for introduction or deletion of a gene, especially those encoding transcription factors or proteins involved in signalling, to influence a cascade of gene expression changes, as has been noted for several stress response–related pathways (e.g. Vogel *et al.*, 2005; Perera *et al.*, 2008; Schramm *et al.*, 2008; Zhang *et al.*, 2011). Similarly, metabolic genes, such as chloroplast-targeted choline oxidase gene for glycine betaine synthesis introduced to engineer drought stress resistance in rice, can alter expression of many genes involved in stress responses, signal transduction, gene regulation, hormone signalling and cellular metabolism (Kathuria *et al.*, 2009). Thus, single gene modifications can have broad effects.

Here, we compared the phenotypic and transcriptomic effects of three alternate transgenic approaches to confer salt stress resistance with regard to implications for environmental risk assessment associated with genetically engineered crops. While it was anticipated that the transcription factor CBF3 would have the greatest effects, both the metabolic enzyme, M6PR, and ion transport protein, SOS1, could potentially affect a variety of cellular processes. The majority of published experiments have looked at short-term stress and early transcriptional responses and signalling. Here, we were particularly interested in long-term phenotypic impacts as observed throughout the life cycle, including effects on fecundity and

fitness, as well as long-term transcriptional adjustments. Long-term assessment is particularly important for salt stress, which most often results from saline soils or irrigation water, and so is less likely to be episodic than stresses such as cold, heat or drought.

Comparative phenotypic and fitness effects of the three transgenes

In the absence of salt stress, the transgenic M6PR and SOS1 lines performed comparably with their WT parental genotypes, indicating limited obvious secondary or fitness effects in the growth chamber or greenhouse. Reduced growth and development for CBF3 overexpressing plants, as has been observed previously in the growth chamber (Liu *et al.*, 1998; Kasuga *et al.*, 1999; Gilmour *et al.*, 2000; Achard *et al.*, 2008a), also was seen in these experiments, resulting in reduced fecundity in the growth chamber, greenhouse and field, especially for A30. Thus, consistent with reports of the transgene effects from growth chamber or agar plate studies in separate labs, significant negative effects on growth were observed for the CBF3 lines but not, or minimally, for M6PR and SOS1 lines (Shi *et al.*, 2003; Zhifang and Loesher, 2003; Sickler *et al.*, 2007; Chan *et al.*, 2011).

The dwarf phenotype in *CBF/DREB* overexpressing plants has been linked to changes in GA metabolism and response (Munns, 2002; Achard *et al.*, 2008a,b; Magome *et al.*, 2008). Increased expression of the negative regulator of GA response, *RGL3* (RGA-like protein 3; At5g17490), occurred in the CBF3 lines, but not in the M6PR or SOS1 lines. In contrast, in control and salt-stressed SOS1 plants and salt-stressed M6PR plants, there was an increase in *GA3ox1* (At1g15550) transcript for a key enzyme in production of the bioactive forms of GA, GA_1 and GA_4 (Yamaguchi, 2008). Another difference observed only in the CBF3 lines that may influence growth was altered expression of two guard cell localized potassium channels that func-

Table 3 Microarray data (log2 values) for overexpressed genes and selected CBF/DREB target genes ($P \leq 0.05$). Bold fonts indicate fold change ≥ 2.0. CBF/DREB target genes were chosen based on the study by Maruyama et al., 2004; Seki et al., 2001; Vogel et al., 2005; and Zhang et al., 2004

Probe Set ID	AGI	Description	CBF3 and wild type				M6PR and wild type				SOS1 and wild type			
			CBF3-0 mm vs. Ws-0 mm	CBF3-100 mm vs. Ws-100 mm	CBF3-100 mm vs. CBF3-0 mm	Ws-100 mm vs. Ws-0 mm	M6PR-0 mm vs. Col-0 mm	M6PR-100 mm vs. Col-100 mm	M6PR-100 mm vs. M6PR-0 mm	Col-100 mm vs. Col-0 mm	SOS1-0 mm vs. Col(gl)-0 mm	SOS1-100 mm vs. Col(gl)-100 mm	SOS1-100 mm vs. SOS1-0 mm	Col(gl)-100 mm vs. Col(gl)-0 mm
A. Over-expressed genes														
265252_at	At2g01980	SOS1	–	–	–	–	–	–	–	–	**1.85**	**1.87**	–	–
254066_at	At4g25480	CBF3/DREB1a	**6.84**	**6.87**	–	–	**-1.60**	–	**1.27**	–	**2.10**	**2.41**	–	–
B. CBF/DREB genes														
254074_at	At4g25490	CBF1/DREB1b		0.987				**-1.63**	0.87	**2.38**	–	–	–	–
254075_at	At4g25470	CBF2/DREB1c	**1.36**						**1.12**	**2.05**	**1.70**	**1.13**	–	–
C. CBF/DREB target genes														
264511_at	At1g09350	ATGOLS3	**6.19**	**6.17**					0.83		**1.14**	**1.37**		
266225_at	At2g28900	AtOEP16	**2.97**	**2.31**		0.72	0.83	0.81		**1.20**				
263789_at	At2g24560	GDSL-like Lipase	**4.16**	**3.80**	-0.69	-0.32								
260556_at	At2g43620	chitinase, putative	**2.58**	**2.68**			**1.41**			**2.33**				
254232_at	At4g23600	COR13/JR2	**1.05**		0.77	**1.77**			**1.30**	0.78				
263497_at	At2g42540	COR15a	**3.27**	**1.66**		**1.55**						0.65	0.63	
263495_at	At2g42530	COR15b	**5.57**	**4.14**	**1.00**	**1.44**	0.87		0.84	**1.78**			0.60	
256114_at	At1g16850	COR17	**5.27**	**3.40**		**2.87**			**1.50**	**2.23**			**1.14**	
262452_at	At1g11210	COR35	**3.32**	**3.72**			**1.67**			**2.65**				
265480_at	At2g15970	COR413-PM1	**2.40**	**1.80**		0.60						0.34		0.35
259789_at	At1g29395	COR414	**4.79**	**4.03**		0.59		**1.16**		**-1.16**				-0.73
253595_at	At4g30830	COR42	**1.04**	**1.61**	0.66									
259570_at	At1g20440	COR47/RD17	**4.41**	**2.26**		**2.23**	**1.14**			**1.06**				
246481_s_at	At5g15960	COR6.6/KIN2	**3.60**	**1.48**		**2.07**	0.61		**1.06**	**1.06**		**1.12**		
248337_at	At5g52310	COR78/RD29A	**5.48**	**2.82**		**2.97**				0.92				
267261_at	At2g23120	COR8.5	**2.47**	0.99		**1.24**					-0.48			
261749_at	At1g76180	Dehydrin	0.86	0.39		0.44								
252137_at	At3g50980	Dehydrin (Xero1)	**1.07**	**1.78**										
245523_at	At4g15910	Di21		**1.14**	0.92			**-1.52**	0.95	**2.30**		**1.31**	0.67	
259516_at	At1g20450	ERD10	**3.02**	**1.49**		**1.61**		-0.32		0.39		0.79	0.40	-0.50
256310_at	At1g30360	ERD4	**2.45**	**2.03**		0.42	**1.08**			0.52				
264787_at	At2g17840	ERD7	**3.23**	**2.38**		**1.55**	**1.07**			**1.72**				
265119_at	At1g62570	Flavin	**3.64**	**3.05**			-0.65	-0.78					0.95	
245427_at	At4g17550	Transporter-related	0.80	**1.56**			**-1.66**		0.66	-0.93		0.93		
247478_at	At5g62360	Invertase	**6.28**	**5.69**			**3.62**			**2.50**				

Table 3 Continued

Probe Set ID	AGI	Description	CBF3 and wild type				M6PR and wild type				SOS1 and wild type			
			CBF3-0 mm vs. Ws-0 mm	CBF3-100 mm vs. Ws-100 mm	CBF3-100 mm vs. CBF3-0 mm	Ws-100 mm vs. Ws-0 mm	M6PR-0 mm vs. Col-0 mm	M6PR-100 mm vs. Col-100 mm	M6PR-100 mm vs. M6PR-0 mm	Col-100 mm vs. Col-0 mm	SOS1-0 mm vs. Col(gl)-0 mm	SOS1-100 mm vs. Col(gl)-100 mm	SOS1-100 mm vs. SOS1-0 mm	Col(gl)-100 mm vs. Col(gl)-0 mm
247450_at	At5g62350	Invertase	1.86	0.88	–	0.96	0.33	–	0.27	0.61	0.36	0.29	–	–
259426_at	At1g01470	LEA protein	1.97	1.57	–	0.74	–	–	–	–	–	–	0.45	–
252102_at	At3g50970	LTI30/XERO2	8.55	6.11	–	2.46	2.01	–1.08	1.66	4.75	–	1.03	1.92	–
253627_at	At4g30650	LTI6A/RCI2A	5.03	2.82	–	1.98	1.23	–	–	2.26	–	–	–	–
254818_at	At4g12470	pEARLI 1-like	3.19	2.60	–	–	1.80	–	–	1.90	–	–	–	–
245807_at	At1g46768	RAP2.1	1.50	1.59	–	–	–	–	–	–	–	–	–	–
264415_at	At1g43160	RAP2.6	–	–	–	3.69	2.68	–	–	4.27	–	–	–	–
259364_at	At1g13260	RAV1	3.37	–	–	3.29	–	–	–	3.00	–	–	–1.46	–
252927_at	At4g39090	RD19A	2.64	1.47	–	1.53	1.27	–	–	0.69	–	–	–	–
253872_at	At4g27410	RD26	2.18	–	1.25	2.61	–	–	–	–	–	–	–	–
248352_at	At5g52300	RD29B	1.78	1.96	2.46	2.29	–	–	–	–	–	–	–	–
262440_at	At1g47710	Serpin, putative	1.71	1.33	–	0.58	–	–	0.60	0.51	–	0.37	0.47	–
264516_at	At1g10090	Similar to RXW8	1.59	2.31	–	–	–	–	–	–1.56	–	–	–	–
264989_at	At1g27200	Similar to zinc fing	3.13	2.64	–	0.57	–	0.60	–	–	–	–	–	–0.57
264654_s_at	At1g08500	Sugar transporter	1.85	2.41	–	–	0.73	0.70	–	–	–	–	–	–
252591_at	At3g45600	TETRASPANIN	0.92	1.06	–	–	–0.70	0.66	–	–0.99	–	–	–	–
262881_at	At1g64690	Transporter	2.84	2.20	–	0.84	–	–	–	–	–	–	0.63	–
261648_at	At1g27730	ZAT10	2.59	–	–	3.01	3.32	–	–	4.90	–	–	–1.65	–
247655_at	At5g59820	ZAT12	1.18	–1.16	–	3.19	1.57	–1.24	–	3.44	–	–	–0.9	–
245711_at	At5g04340	ZAT6/CZF2	2.76	2.36	–	1.28	–	–	–	2.51	–	–	–1.78	–

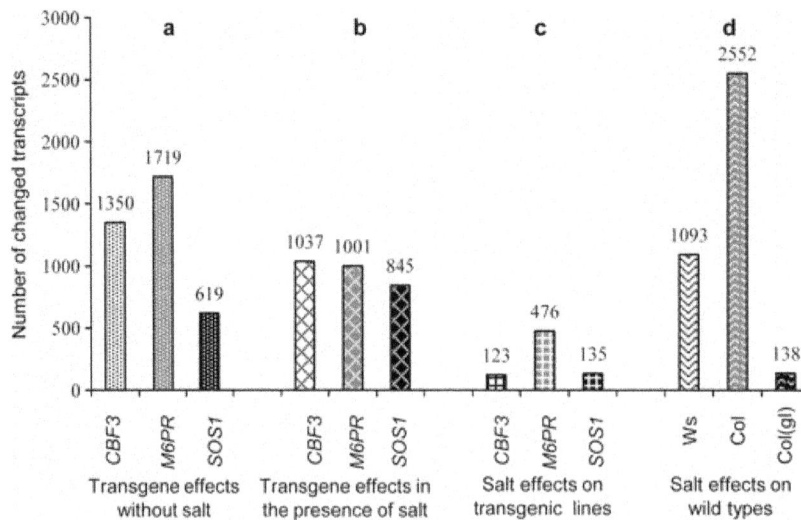

Figure 3 Total number of changed transcripts by three transgenes or salt stress. All microarray data were normalized and analysed together using affylmGUI running on R package. Transcripts level deemed significantly different were those with a fold change ≥2; a P-value ≤0.05, and a detection call of 'Present' in duplicate with the Affymetrix GCOS. The full list of genes is provided as Table S1.

tion reciprocally to drive stomatal closing and opening (Gambale and Uozumi, 2006; Ward et al., 2009). Similar to the observations by Vogel et al. (2005), the CBF-overexpressors exhibited 4-5-fold increased transcription of the guard cell outwardly rectifying potassium channel, Shaker-type GORK1 (At5g37500) gene and 4-5-fold decreased transcription of the inwardly rectifying KAT1 potassium channel (At5g46240) gene (Table S2E), thereby potentially increasing stomatal closure, while decreasing rate of water loss, photosynthetic capacity, and growth.

With the exception of the more severely dwarfed line, CBF3 A30, all transgenic lines exhibited a significant fitness advantage relative to their wild-type parents when subjected to moderate (100 mM) salt stress throughout their life cycle verifying that all three transgenes can confer tolerance to the long-term salt stress imposed in these experiments. Seed production ranged from 114% to 345% of salt-stressed WT genotypes. SOS1 conferred the greatest salt tolerance as measured by reduced salt injury effects and greater survival at 200 mM NaCl. M6PR plants, however, had the greatest fecundity at 100 mM NaCl. The M6PR plants also exhibited enhanced seed yield in the field, possibly reflecting better adaptation to environmental stresses that can be experienced in field conditions. Preliminary results of direct competition experiments between each transgenic line and corresponding parental genotype in field tests also showed strong negative fitness effects for CBF3 plants and somewhat negative effects for SOS1, while M6PR had somewhat positive effects (Bigelow et al., 2010). Thus, relative fitness advantages or disadvantages caused by the transgenes varied depending on the presence, absence and level of stress.

Comparative transcriptome effects of the three transgenes

In contrast to the minimal effects of M6PR on phenotype and fitness in absence of salt stress, the global transcriptome effects of M6PR were at least as great as those of CBF3 and included many changes in common. Among the changes induced by

M6PR was strong activation of three recently identified ABA receptor genes (PYL4, PYL5 and PYL6; Ma et al., 2009; Park et al., 2009) (Table S2C) and down-regulation of two ABA signalling inhibitor genes, type 2C protein phosphatases (PP2C), ABI1 and ABI2 (Table S2C). Increased ABA signalling may contribute to the broad range of stress-related gene expression and commonality in many expression responses between M6PR and CBF3 plants.

There were also numerous gene expression changes in the M6PR plants not seen in the CBF3 plants, especially with respect to biotic stress and oxidative stress-related genes including many disease resistance-related proteins, and glutaredoxin and thioredoxin family protein genes. As many pathogenic fungi produce mannitol during the infection process (Vogele et al., 2005; Cheng et al., 2009), the endogenous mannitol production may be perceived by the M6PR plants as a signal of pathogen attack to stimulate expression of biotic stress-related genes. Work of the past decade has led to increasing recognition of extensive crosstalk between biotic and abiotic stress responses, including ABA- and reactive oxygen-mediated signalling (Garg and Manchanda, 2009; Klinger et al., 2010). Indeed, several of the disease-resistance-related genes whose expression was up-regulated by M6PR were also induced by salt treatment of the WT Col plants. Similarly, if mannitol is perceived as a sign of pathogen attack, the resultant defences may include abiotic responses in common with those induced by CBF3.

Transcriptional effects of the SOS1 transgene in the absence of salt stress were considerably smaller than for CBF3 or M6PR. These results are consistent with the apparent independence of the SOS signalling pathway from CBF, ABA and MYC/MYB pathways as was observed for sos2 and sos3 mutants of Arabidopsis (Kamei et al., 2005). The small number of gene expression changes may also be influenced by the Col(gl) background (Zhulong Chan, Rebecca Grumet and Wayne Loescher, unpublished) leading to a partial masking of SOS1 effects. The SOS1 transgene, did however, have substantial effects on oxidative stress or redox-related genes, resulting in down-regulation of numerous transcripts. Direct interplay between the SOS pathway and redox signalling has been observed by interaction between SOS2 and the redox signalling pathway proteins, nucleoside diphosphate kinase 2 (NDPK2) and catalases, and between SOS1 and RCD1, a regulator of oxidative stress (Katiyar-Agarwal et al., 2006; Verslues et al., 2007). Mutants of sos2 had increased sensitivity to oxidative stress (Zhu et al., 2007). However, consistent with reduced expression of redox-related genes in the SOS1 overexpressors, sos1 mutants had increased tolerance to oxidative stress induced by methyl viologen, indicating that SOS1 expression can act to make plants more sensitive to oxidative stress (Katiyar-Agarwal et al., 2006; Chung et al., 2008).

The relative impacts of the transgenes on global transcription changed in the presence of salt. While the differences between the transgenic CBF3 and M6PR plants and their WT parents were greater without salt, the differences for SOS1 plants relative to WT parents were greater in the presence of salt, both in terms of numbers of genes affected and level of induction or repression (Figure 4). Similarly, significant enrichment for modified expression of genes associated with response to osmotic stress and salt stress only occurred for SOS1 plants when subjected to salt stress (Table 4). These observations are consistent with studies showing stabilization of the SOS1 protein and increased ion exchange activity in response to salinity (Qiu et al., 2003, 2004; Chung et al., 2008). Similar results were observed with transgenic Arabidopsis overexpressing the drought-related transcription factor, ABF3, wherein extensive transcriptional differences were only observed after application of drought stress (Abdeen et al., 2010). Minimal changes in gene expression in the absence of drought stress were attributed to lack of activation of ABF3 by SnRK2-mediated phosphorylation that is normally induced in response to abscisic acid.

Relationship between transcriptome and phenotype

The transcriptome data show that transgenes intended to confer salt stress tolerance can have extensive and variable effects on the transcriptome. The three transgenes affected different pathways or groups of pathways consistent with their different functions as summarized in Figure 7. The global transcriptional differences as measured by number of genes affected by each transgene, however, did not correlate with changes in pheno-

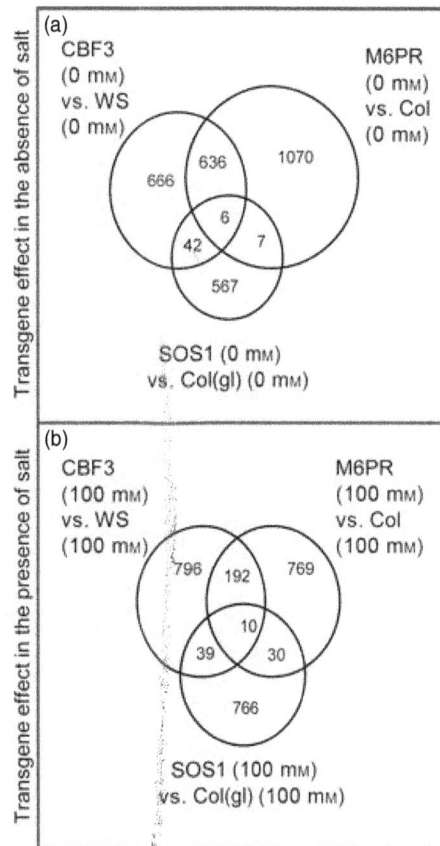

Figure 4 Venn diagrams showing overlapping transcripts (P-value ≤0.05 and fold change ≥2.0) affected by the three transgenes in the absence (a) and presence (b) of salt stress (100 mM NaCl).

type (Figure 8). Furthermore, while the effects of the transgenes on plant growth and effects on global gene expression varied in response to salt stress, they did not vary in parallel. In the absence of salinity, despite a range of transcriptional differences, performance differences [as measured by average difference in dry weight or seed yield between the transgenic and WT counterparts (with the exception of CBF A30)] were rel-

Table 4 Stress-related GO term enrichment analysis. Term enrichment analysis was performed using AmiGO software

GO Terms	CBF3-0 mM vs. Ws-0 mM		M6PR-0 mM vs. Col-0 mM		SOS1-0 mM vs. Col(gl)-0 mM		CBF3-100 mM vs. Ws-100 mM		M6PR-100 mM vs. Col-100 mM		SOS1-100 mM vs. Col(gl)-100 mM	
	FC	P-value	FC	P-value	FC	P-value	FC	P-value	FC	P-value	FC	P-value
GO:0009414 response to water deprivation	5.12	0.000	3.09	0.000	–	–	6.22	0.000	–	–	–	–
GO:0009409 response to cold	4.50	0.000	2.87	0.000	–	–	4.89	0.000	–	–	–	–
GO:0006970 response to osmotic stress	3.15	0.000	2.59	0.000	–	–	3.76	0.000	–	–	2.87	0.000
GO:0009651 response to salt stress	3.10	0.000	2.56	0.000	–	–	3.52	0.000	–	–	3.00	0.000
GO:0006979 response to oxidative stress	–	–	2.72	0.000	3.68	0.001	2.85	0.003	–	–	3.63	0.000
GO:0009408 response to heat	–	–	2.78	0.048	–	–	5.96	0.000	–	–	–	–
GO:0009607 response to biotic stimulus	2.42	0.000	2.26	0.000	2.58	0.000	2.90	0.000	–	–	2.84	0.000

FC: enriched fold change was calculated as frequency of transcripts from the functional category relative to total changed transcripts/background frequency of that functional category in the Arabidopsis genome. Colour scales represent fold enrichment ▮ >6.0 ▮ 5.0–6.0 ▮ 4.0–5.0 ▮ 3.0–4.0 ▮ 2.0–3.0 ▮ <2.0 .

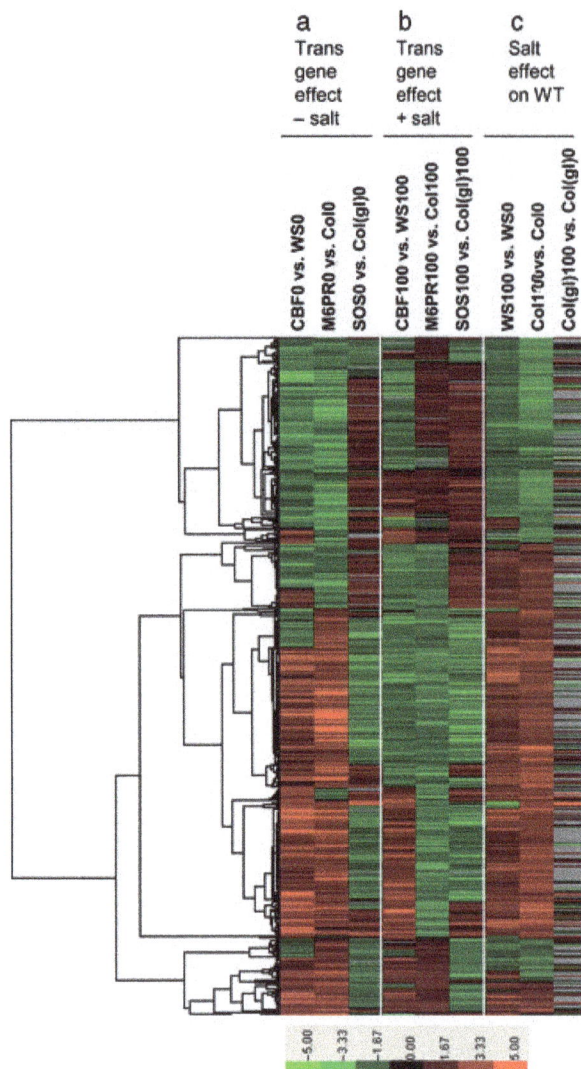

Figure 5 Cluster analysis of transcripts with expression levels significantly affected (*P*-value ≤0.05) by transgenes or salt. Red, black and green scales indicate fold change for genes with significant changes. Red, up-regulation; green, down-regulation. Gray, transcription levels were not significantly changed for that comparison. Hierarchical cluster analysis was performed with Cluster 3.0 software. Resulting tree figures were displayed using the software package, Java Treeview. The detailed gene IDs and fold changes are listed in Table S1.

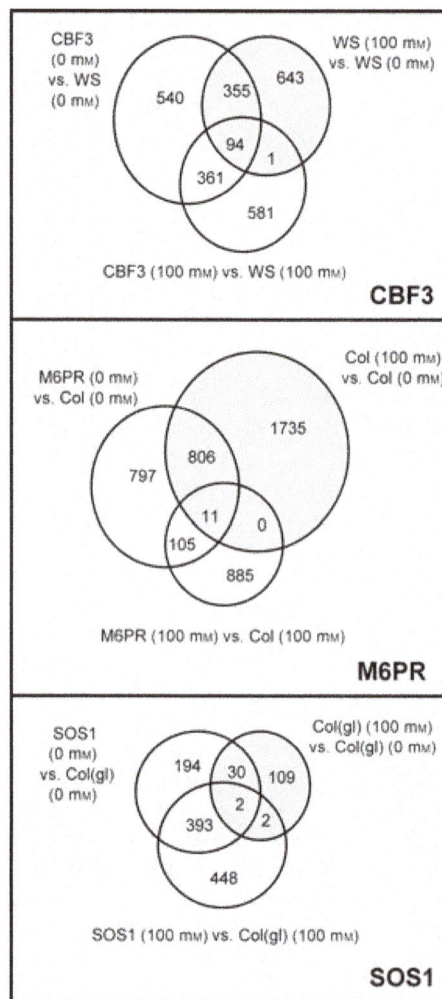

Figure 6 Venn diagrams showing overlapping transcripts affected by salt stress (shaded) and the *CBF3*, *M6PR* and *SOS1* transgenes in the presence and absence of salt (*P*-value ≤0.05 and fold change ≥2.0). Comparisons are indicated around the circle.

atively modest (average transgene effect was 85%–110% of WT seed yield). In the presence of salt stress, the range of transcriptional differences between the transgenic lines was quite small, but the performance differences, relative to each other and to WT plants, were considerably increased (average transgene effect was 146%–267% of WT seed yield).

These results suggest that depending on genotype and environment, extensive transcriptional changes may serve different functions. They could facilitate adaptive expression of fitness-associated traits, or they may reflect response to injury. They may also be an adaptive response to buffer the effects of genetic perturbation. Evolutionarily conserved buffering systems modifying gene expression have been observed across organisms and are hypothesized as an adaptive mechanism to minimize potential negative impacts of mutation on fitness (Boerjan

and Vuylsteke, 2009; Fu *et al.*, 2009). Studies in *Arabidopsis* have shown that only a handful out of thousands of expression differences are observed at the phenotype level, indicating that much of the genetic variation in gene expression is hidden by non-linearity in response functions (Fu *et al.*, 2009). It was proposed that such robust system properties serve to keep traits within acceptable limits, thereby preventing dysfunction of the organism.

Lack of correspondence between magnitude of transcriptional differences and performance differences indicates that extent of global transcriptome differences may not predict phenotypic differences upon which environment and selection act in influencing fitness and fecundity. These observations have implications for the use of global gene expression data for purposes of risk assessment. The sorts of changes identified, however, may provide guidance for risk assessment analyses. For example, given the transcriptional changes for biotic stress-related genes, do the M6PR plants show altered disease responses? Collectively these observations emphasize the importance of evaluation of the transcriptomic effects of transgene within a phenotypic context.

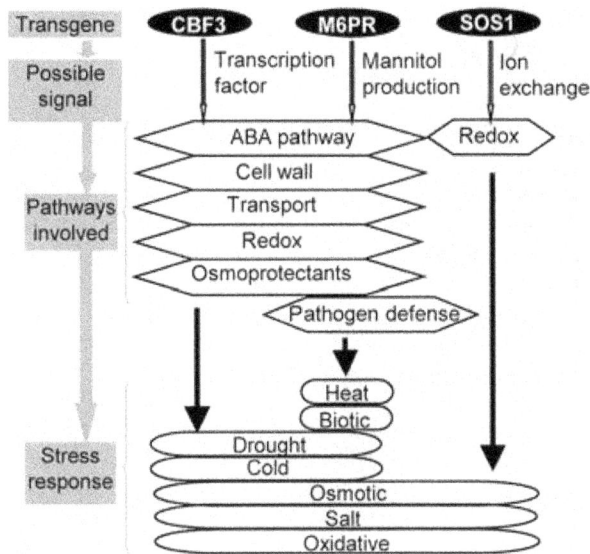

Figure 7 Model of relative transgene effects of *CBF3*, *M6PR* and *SOS1* on Arabidopsis gene expression and stress responses.

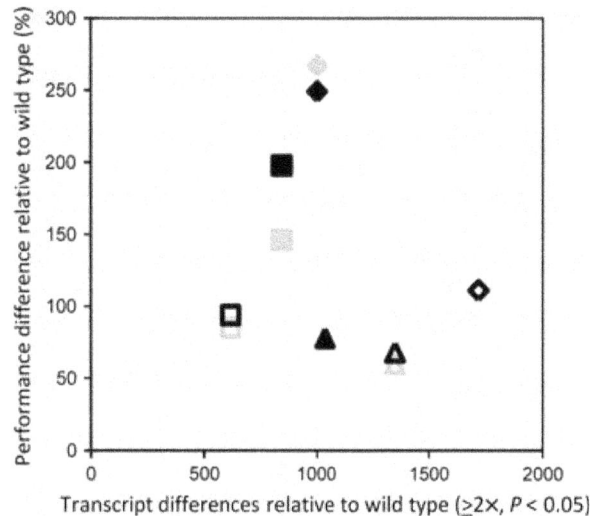

Figure 8 Lack of relationship between magnitude of transcriptional effects of the CBF3 (triangles), M6PR (diamonds) and SOS1 (squares) transgenes and effect of the transgenes on vegetative (dry weight, black symbols) and reproductive (seed production, grey symbols) performance in the growth chamber. Open symbols, 0 mM NaCl; closed symbols, 100 mM NaCl. Performance values are averaged over the two lines for each transgene. Magnitude of transcriptional effects was number of significantly changed transcript levels for the transgenic lines vs. wild type. Performance difference refers to above-ground dry weight or seed yield for the transgenic lines relative to wild type, expressed as a per cent of wild type.

Experimental procedures

Plant materials

Arabidopsis thaliana L. (Heynh) plants overexpressing three abiotic stress resistance genes under the control of CaMV 35S promoter, as well as their wild-type parents, were used in this experiment. Two lines were used for each transgene. A30 and A40 transgenic lines overexpressing C-repeat/DRE binding factors (*CBF3*) and Wassilewskija (Ws) background were kindly provided by Michael F. Thomashow (Gilmour *et al.*, 2000). M2 and M5 lines overexpressing celery mannose-6-phosphate reductase (*M6PR*) in the Columbia (Col) background were produced by Zhifang and Loesher (2003). Two plasma membrane Na^+/H^+ antiporter (*SOS1*) transgenic lines (#1-1 and #7-6) and Columbia-glabrous (*gl1-1*) (Col(gl)) background were generously provided by Huazhong Shi (Shi *et al.*, 2003). All transgenic and wild-type lines were verified for the presence and expression of the relevant transgenes by Southern and Northern blot analyses prior to initiation of the experiments (data not shown).

Growth conditions and salt treatment in the growth chamber

Seed production, planting and growth conditions were as described by Chan *et al.* (2011). Salt treatment was initiated at 14 DAP (6 true leaf stage). Plants subjected to salt stress were sub-irrigated to field capacity with NaCl solution dissolved in ½ strength Hoagland solution and then sprayed with the same concentration of NaCl solution from the top, ensuring adequate leaching and preventing excess salinity. The concentrations of NaCl supplementation were increased stepwise by 50 mM every 2 days for each line, to the indicated maximum (0, 100, or 200 mM). Plants were then watered every 2 days at the indicated concentrations. The pots were rotated in the growth chamber everyday to minimize the effect of environment. All genotype–salt combinations were grown together in the growth chamber at the same time.

Measurement of growth parameters including bolting and flowering time, leaf number, rosette diameter, plant height,

stalk number, dry weight, and seed yield were taken as described by Chan *et al.* (2011). Chlorosis/necrosis severity indices, leaf numbers and rosette diameters were measured every 6 days. Chlorosis/necrosis severity was rated as follows: 0, no yellow or purple leaves; 1, older leaves turn yellow or purple; 3, younger leaves turn yellow or purple; 5, some leaves die; and 7, plants die. Severity indices were calculated analogous to the disease severity index of Piccinni *et al.* (2000) as follows: Σ (number of plants with each score × score value)]/(total number of plants × highest score). The plants were photographed at 50 DAP and harvested at 62 DAP when most of them reached maturity. The complete experiment was repeated three times. All data were analysed with SPSS 11.5 for windows (SPSS, Inc., Chicago, IL). Mean separations were performed by Duncan's multiple range test. Differences at $P < 0.05$ were considered to be significant.

Growth conditions and parameters in the greenhouse and field

Seed from verified growth chamber grown plants of all transgenic and wild-type lines were counted into 5 replicate batches per line for the greenhouse and field experiments. Each batch contained approximately 180 seeds. All batches of seeds were stratified as described previously, mixed with sterile sand and randomly scattered onto 26 × 26 × 6 cm pots filled with a standard planting medium (Baccto, Houston, TX) mixed with 2.1 kg/m³ Osmocote Classic 14-14-14 slow release fertilizer (The Scotts Miracle-Gro Company, Marysville, OH). All plant populations were germinated and grown in the greenhouse with supplemental lighting providing 12 h light/12 h dark. Pots were sub-irrigated as required. Greenhouse populations were rotated biweekly to minimize location effects.

At the 6 true leaf stage, the pots for the field experiment were moved from the greenhouse and placed into anchored 52 × 26 × 6 cm trays placed atop weed barrier plastic in the field. Trays were spaced every 0.6 m and watered as needed by trickle hose to allow for sub-irrigation of the pots. The plants were maintained in the field until they approached senescence and then returned to the greenhouse to complete senescence and dry down. Total above-ground dry weight and seed yield were measured after harvest in the greenhouse. All data were analysed with SAS 9.2 for windows (SAS Institute Inc., Cary, NC). Statistical tests were performed as described before.

Plant growth and salt treatment for microarray experiment

Seeds of three transgenic lines and their wild-type plants were sowed as described previously with two replicate pots for each genotype and salt combination (0 or 100 mm). Two replications were performed in different growth chambers on different dates with 36 plants/replicate pot for each genotype and salt combination. Plants were grown at 23/18 °C in the growth chamber 10-h light/14-h dark cycle at 350 μmol/m^2/s and 70% relative humidity. Salt treatments were initiated at 14 DAP and applied as described previously. Sampling was performed at 20 DAP by collecting fully developed but not senescent leaves (about 0.5 cm width × 1.5 cm length) from at least 15 seedlings/treatment.

RNA isolation, GeneChip® hybridization and microarray analysis

RNA isolation and GeneChip® hybridization for microarray experiments was performed as described by Chan et al. (2011). Total RNA was extracted and purified from leaves of at least 15 plants per genotype and salt treatment combination. Two biological replicates from different growth chambers were prepared for each genotype and salt combination. To minimize the transgene position effects, equal amounts of total RNAs from the two lines for each transgene (A30 and A40 for CBF3, M2 and M5 for M6PR, and 1-1 and 7-6 for SOS1) were pooled for biotin labelling.

The reproducibility of the microarray experiments was characterized by comparing each set of data generated from the duplicated experiment with Affymetrix GCOS software. Raw signal data from two biological replicates were compared, and a correlation coefficient was calculated between the duplicate experiments. All biological replicates had a coefficient of determination (R^2) larger than 0.91 (Figure S2). All the Affymetrix data files produced with Affymetrix GCOS software (*.CEL files) were analysed using Bioconductor, a public source software for the analyses of genomic data rooted in the statistical computing environment R (Gentleman et al., 2004). The data were normalized by robust multiarray normalization of probe-level data with RMA and analysed using affylmGUI running on R software (Wettenhall et al., 2006). To determine meaningful differences between samples, modest threshold parameters were applied in this study to minimize any potential statistical biases. Transcript levels deemed significantly different were those with (i) a fold change larger than 2; (ii) a P-value smaller than 0.05; and (iii) a detection call of 'Present' in duplicate with the Affymetrix GCOS. Microarray data are available online in Gene Expression Omnibus (GEO) database (http://www.ncbi.nlm.nih.gov/geo/) under accessions number (GSE26983). The M6PR-Col WT microarray data were previously deposited to the Gene Expression Omnibus (GEO) online database (http://www.ncbi.nlm.nih.gov/geo/) under accession number GSE18217 and published by Chan et al. (2011).

Quantitative real-time PCR

Total RNAs extraction, cDNA synthesis and PCR amplification were performed as described by Chan et al. (2011). All reactions were run in duplicates, and the average values were calculated. Quantification was performed with at least two independent experiments. The housekeeping F-actin gene (At3g05520) was used as endogenous control. Relative expression levels of target genes and SD values were calculated using the $2^{-\Delta\Delta CT}$ method (Livak and Schmittgen, 2001).

Twenty genes with at least one significant sample difference from nine comparisons based on microarray data were selected for qRT-PCR analyses, along with a single gene that did not (At3g63490). Log2 values for each replicate and their averages and standard errors were calculated. Primers used for real-time PCR are listed in Table S4.

Biological enrichment and metabolic pathway analyses

All transcripts with P-value ≤0.05 and fold change ≥2 were loaded and annotated in the Classification SuperViewer Tool w/Bootstrap web database (http://bar.utoronto.ca/ntools/cgi-bin/ntools_classification_superviewer.cgi) (Provart and Zhu, 2003). MapMan was used as the classification source to assign functional categories for each gene (Thimm et al., 2004). The absolute values and normalized frequency relative to the Arabidopsis genomic set of each functional category were then calculated as described by Chan et al. (2011). For GO term enrichment analysis, all transcripts with P-value ≤0.05 and fold change ≥2 were loaded in 'Term enrichment' using AmiGO software (http://amigo.geneontology.org) (Carbon et al., 2009). Enriched fold change of each functional category was calculated as following: enriched fold change = sample frequency of each category in this experiment/background frequency of each category in the Arabidopsis genome. Hierarchical cluster analyses was performed on selected sets of genes using the CLUSTER program (http://bonsai.ims.u-tokyo.ac.jp/~mdehoon/software/cluster/) (deHoon et al., 2004) by the uncentred matrix and complete linkage method. Resulting tree figures were displayed using the software package, Java Treeview (http://jtreeview.sourceforge.net/).

Acknowledgements

This work was supported in part by USDA-BRAG #2005-39454-16516 and USDA-NNF #2005-38420-15789 for P. Bigelow. We thank Dr Sunchung Park for his assistance with analysis of microarray data and Ms. Jean Bronson for her help with the field experiments. We also thank Drs Jim Hancock and Michael Thomashow for their helpful reviews of the manuscript.

References

Abdeen, A. and Miki, B. (2009) The pleiotropic effects of the bar gene and glufosinate on the Arabidopsis transcriptome. Plant Biotechnol. J. 7, 266–282.

Abdeen, A., Schnell, J. and Miki, B. (2010) Transcriptome analysis reveals absence of unintended effects in drought tolerant transgenic plants overexpressing the transcription factor, ABF3. BMC Genomics 11, 69.

Achard, P., Gong, F., Cheminant, S., Alioua, M., Hedden, P. and Genschik, P. (2008a) The cold-inducible CBF1 factor-dependent signaling pathway modulates the accumulation of the growth-repressing DELLA proteins via its effect on gibberellin metabolism. *Plant Cell* **20**, 2117.

Achard, P., Renou, J.P., Berthome, R., Harberd, N.P. and Genschik, P. (2008b) Plant DELLAs restrain growth and promote survival of adversity by reducing the levels of reactive oxygen species. *Curr. Biol.* **18**, 656–660.

Albo, A.G., Mila, S., Digilo, G., Motto, M., Aime, S. and Corpillo, D. (2007) Proteomic analysis of a genetically modified maize flour carrying *CRY1AB* gene and comparison to the corresponding wild-type. *Maydica* **52**, 443–445.

Apse, M.P., Aharon, G.S., Snedden, W.A. and Blumwald, E. (1999) Salt tolerance conferred by overexpression of a vacuolar Na$^+$/H$^+$ antiporter in Arabidopsis. *Science* **285**, 1256–1258.

Batista, R., Saibo, N., Lurenco, T. and Oliveira, M.M. (2008) Microarray analyses reveal that plant mutagenesis may induce more transcriptomic changes than transgene insertion. *Proc. Natl Acad. Sci. USA* **105**, 3640–3645.

Baudo, M.M., Lyons, R., Powers, S., Pastouri, G.M., Edwards, K.J., Holdsworth, M.J. and Shewry, P.R. (2006) Transgenesis has less impact on the transcriptome of wheat grain than conventional breeding. *Plant Biotechnol. J.* **4**, 369–380.

Beckie, H.J., Hall, L.M., Simard, M.J., Leeson, J.Y. and Willnborg, C.J. (2010) A framework for postrelease environmental monitoring of second-generation crops with novel traits. *Crop Sci.* **50**, 1587–1604.

Bigelow, P., Loescher, W. and Grumet, R. (2010) The competitive fitness of abiotic stress tolerance enhancing transgenes under field conditions. *Amer. Soc. Plant Biol.* (abstract) P07106. Http://abstracts.aspb.org/pb2010/public/P07?P07106.html.

Boerjan, W. and Vuylsteke, M. (2009) Integrative genetical genomics in Arabidopsis. *Nature Genet.* **41**, 144–145.

Carbon, S., Ireland, A., Mungall, C.J., Shu, S., Marshall, B., Lewis, S. and Ami, G.O. (2009) Hub Web Presence Working Group 2009. AmiGO: online access to ontology and annotation data. *Bioinformatics* **25**, 288–289.

Carpenter, J., Felsot, A., Goode, T., Hammig, M., Onstad, D. and Sankula, S. (2002) *Comparative Environmental Impacts of Biotechnology-Derived and Traditional Soybean, Corn, and Cotton Crops.* Ames, Iowa: Council Agric. Sci. Technol.

Chan, Z., Grumet, R. and Loescher, W.H. (2011) Global gene expression analysis of transgenic, mannitol-producing and salt tolerant Arabidopsis thaliana indicates widespread changes in expression of abiotic- and biotic-stress related genes. *J. Exp. Bot.* **62**, 4787–4803.

Chandler, S. and Dunwell, J.M. (2008) Gene flow, risk assessment and the environmental release of transgenic plants. *Crit. Rev. Plant Sci.* **27**, 25–49.

Chen, T.H.H. and Murata, N. (2002) Enhancement of tolerance of abiotic stress by metabolic engineering of betaines and other compatible solutes. *Curr. Opin. Plant Biol.* **5**, 250–257.

Chen, T.H.H. and Murata, N. (2008) Glycinebetaine: an effective protectant against abiotic stress in plants. *Trends Plant Sci.* **13**, 499–505.

Cheng, K.C., Beaulier, J., Iquira, E., Belzile, F.J., Fortin, M.G. and Stromvik, M.V. (2008) Effect of transgenes on global gene expression in soybean is within the natural range of variation of conventional cultivars. *J. Agric. Food. Chem.* **56**, 3057–3067.

Cheng, F.Y., Zamski, E., Guo, W.W., Pharr, D.M. and Williamson, J.D. (2009) Salicylic acid stimulates secretion of the normally symplastic enzyme mannitol dehydrogenase: a possible defense against mannitol secreting fungal pathogens. *Planta* **230**, 1093–1103.

Chinnusamy, V., Zhu, J. and Zhu, J.K. (2007) Cold stress regulation of gene expression in plants. *Trends Plant Sci.* **12**, 444–451.

Chung, J.S., Zhu, J.K., Bressan, R.A., Hasegawa, P.M. and Shi, H.H. (2008) Reactive oxygen species mediate Na$^+$-induced *SOS1* mRNA stability in Arabidopsis. *Plant J.* **53**, 554–565.

Coll, A., Nadal, A., Palaudelmas, M., Messenguer, J., Mele, E., Puigomenech, P. and Pla, M. (2008) Lack of repeatable differential expression patterns between MON810 and comparable commercial varieties of maize. *Plant Molec. Biol.* **68**, 105–117.

Conner, A.J., Glare, T.R. and Nap, J. (2003) The release of genetically modified crops into the environment, Part II. Overview of ecological risk assessment. *Plant J.* **33**, 19–46.

Cook, D., Fowler, S., Fiehn, O. and Thomashow, M.F. (2004) A prominent role for the CBF cold response pathway in configuring the low temperature metabolome of Arabidopsis. *Proc. Natl Acad. Sci. USA* **101**, 15243–15248.

Corpillo, D., Gardini, G., Vaira, A.M., Basso, M., Aime, S., Accotto, G.R. and Fasono, M. (2004) Proteomics as a tool to improve investigation of substantial equivalence in genetically modified organisms: the case of a virus-resistant tomato. *Proteomics* **4**, 193–200.

Craig, W., Tepfer, M., Degrassi, G. and Ripandelli, D. (2008) An overview of general features of risk assessments of genetically modified crops. *Euphytica* **164**, 853–880.

Crawley, M.J., Brown, S.L., Hails, S.R., Kohn, D.D. and Rees, M. (2001) Transgenic crops in natural habitats. *Nature* **409**, 682–683.

Edmeades, G.O. (2008) *ISAAA Brief 39.* http://www.isaaa.org/resources/publications/briefs/39/pressrelease/default.html.

El Ouakfaoui, S. and Miki, B. (2005) The stability of the *Arabidopsis* transcriptome in transgenic plants expressing the marker genes *nptII* and *uidA*. *Plant J.* **41**, 791–800.

Everard, J.D., Cantini, C., Grumet, R., Plummer, J. and Loescher, W.H. (1997) Molecular cloning of mannose-6-phosphate reductase and its developmental expression in celery. *Plant Physiol.* **113**, 1427–1435.

Fowler, S. and Thomashow, M.F. (2002) Arabidopsis transcriptome profiling indicates that multiple regulatory pathways are activated during cold acclimation in addition to the CBF cold response pathway. *Plant Cell* **14**, 1675–1690.

Fu, H., Keurentjes, J.J.B., Bouwmeester, H., America, T., Verstappen, F.W.A., Ward, J.L., Beale, M.H., de Vos, R.C.H., Dijkstra, M., Scheltema, R.A., Johannes, F., Kornneef, M., Vreugdenhil, D., Breitling, R. and Jansen, R.C. (2009) System-wide molecular evidence for phenotypic buffering in *Arabidopsis*. *Nature Genet.* **41**, 166–167.

Gambale, F. and Uozumi, N. (2006) Properties of *Shaker*-type potassium channels in higher plants. *J. Membrane Biol.* **210**, 1–19.

Garg, N. and Manchanda, G. (2009) ROS generation in plants: boon or bane? *Plant Biosystems* **143**, 81–96.

Gaxiola, R.A., Li, J.S., Undurraga, S., Dang, L.M., Allen, G.J., Alper, S.L. and Fink, G.R. (2001) Drought and salt tolerant plants result from overexpression of the AVP H$^+$ pump. *Proc. Natl Acad. Sci. USA* **98**, 11444–11449.

Gentleman, R., Carey, V., Bates, D., Bolstad, B., Dettling, M., Dudoit, S., Ellis, B., Gautier, L., Ge, Y., Gentry, J., Hornik, K., Hothorn, T., Huber, W., Iacus, S., Irizarry, R., Leisch, F., Li, C., Maechler, M., Rossini, A., Sawitzki, G., Smith, C., Smyth, G., Tierney, L., Yang, Y. and Zhang, J. (2004) Bioconductor: open software development for computational biology and bioinformatics. *Genome Biol.* **5**, R80.

Gilmour, S.J., Sebolt, A.M., Salazar, M.P., Everard, J.D. and Thomashow, M.F. (2000) Overexpression of the Arabidopsis CBF3 transcriptional activator mimics multiple biochemical changes associated with cold acclimation. *Plant Physiol.* **124**, 1854–1865.

Gilmour, S.J., Fowler, S.G. and Thomashow, M.F. (2004) Arabidopsis transcriptional activators CBF1, CBF2, and CBF3 have matching functional activities. *Plant Molec. Biol.* **54**, 767–781.

Grumet, R., Wolfenbarger, L. and Ferenczi, A. (2011) Future possible genetically engineered crops and traits and potential environmental concerns. In *Environmental Safety of Genetically Engineered Crops* (Grumet, R., Hancock, J., Maredia, K. and Weebadde, C., eds), pp. 47–57. East Lansing: Michigan State University Press.

Hancock, J.F. (2003) A framework for assessing the risk of transgenic crops. *Bioscience* **53**, 5412–5519.

Hasegawa, P.M., Bressan, R.A., Zhu, J.K. and Bohnert, H.J. (2000) Plant cellular and molecular responses to high salinity. *Annu. Rev. Plant Physiol. Plant Molec. Biol.* **51**, 463–499.

deHoon, M.J.L., Imoto, S., Nolan, J. and Miyano, S. (2004) Open source clustering software. LINK http://bioinformatics.oupjournals.org. *Bioinformatics* **20**, 1453–1454.

Jaglo-Ottosen, K.R., Gilmour, S.J., Sarka, D.G., Schabenberger, O. and Thomashow, M.F. (1998) Arabidopsis *CBF* overexpression induces *COR* genes and enhances freezing tolerance. *Science* **280**, 104–106.

Kamei, A., Seki, M., Umezawa, T., Ishida, J., Satou, M., Akiyama, K., Zhu, J.K. and Shinozaki, K. (2005) Analysis of gene expression profiles in

Arabidopsis salt overly sensitive mutants *sos2-1* and *sos3-1*. *Plant Cell Env.* **28**, 1267–1275.

Kasuga, M., Liu, Q., Miura, S., Yamaguchi-Shinozaki, K. and Shinozaki, K. (1999) Improving plant drought, salt, and freezing tolerance by gene transfer of a single stress-inducible transcription factor. *Nat. Biotechnol.* **17**, 287–291.

Kathuria, K., Giri, J., Karaba, N., Nataraja, K.N., Murata, N., Udayakumar, M. and Tyagi, A.K. (2009) Glycinebetaine-induced water-stress tolerance in *codA*-expressing transgenic indica rice is associated with up-regulation of several stress responsive genes. *Plant Biotechnol. J.* **7**, 512–526.

Katiyar-Agarwal, S., Zhu, J., Kim, K., Agarwal, M., Fu, X., Huang, A. and Zhu, J.K. (2006) The plasma membrane Na$^+$/H$^+$ antiporter SOS1 interacts with RCD1 and functions in oxidative stress tolerance in Arabidopsis. *Proc. Natl Acad. Sci. USA* **103**, 18816–18821.

Kishore, P.B.K., Hong, Z., Miao, G.H., Hu, C.A. and Verma, D.P.S. (1995) Overexpression of Δ-pyrroline-5-carboxylate synthase increases proline production and confers osmotolerance in transgenic plants. *Plant Physiol.* **108**, 1387–1394.

Klinger, J.P., Batelli, G. and Zhu, J.K. (2010) ABA receptors: the START of a new paradigm in phytohormone signaling. *J. Exp. Bot.* **61**, 3199–3210.

Little, H.A., Grumet, R. and Hancock, J.F. (2009) Modified ethylene signaling as an example of engineering for complex traits: Secondary effects and implications for risk assessment. *HortScience* **44**, 94–101.

Liu, Q., Kasuag, M., Sakuma, Y., Abe, H., Miura, S., Yamaguchi-Shinozaki, K. and Shinozaki, K. (1998) Two transcription factors, DREB1 and DREB2 with and EREBP/AP2 binding domain separate two cellular signal transduction pathways in drought and low-temperature responsive gene expression, respectively, in Arabidopsis. *Plant Cell* **10**, 1391–1406.

Livak, K.J. and Schmittgen, T.D. (2001) Analyses of relative gene expression data using real-time quantitative PCR and the 2$^{-\Delta\Delta CT}$ method. *Methods* **25**, 402–408.

Lu, B.R. and Yang, C. (2009) Gene flow from genetically modified rice to its wild relatives: assessing potential ecological consequences. *Biotechnol. Adv.* **6**, 1083–1091.

Ma, Y., Szostkiewicz, I., Korte, A., Moes, D., Yang, Y., Christmann, A. and Grill, E. (2009) Regulators of PP2C phosphatase activity function as abscisic acid sensors. *Science* **324**, 1064–1068.

Magome, H., Yamaguchi, S., Hanada, A., Kamiya, Y. and Oda, K. (2008) The DDF1 transcriptional activator upregulates expression of a gibberellin-deactivating gene, GA2ox7, under high-salinity stress in Arabidopsis. *Plant J.* **56**, 613–626.

Maruyama, K., Sakuma, Y., Kasuga, M., Ito, Y., Seki, M., Goda, H., Shimada, Y., Yoshida, S., Shinozaki, K. and Yamaguchi-Shinozaki, K. (2004) Identification of cold-inducible downstream genes of the Arabidopsis DREB1A/CBF3 transcriptional factor using two microarray systems. *Plant J.* **38**, 982–993.

Miki, B., Abdeen, A., Manabe, Y. and MacDonald, P. (2009) Selectable marker genes and unintended changes to the plant transcriptome. *Plant Biotechnol. J.* **7**, 211–218.

Miller, G., Suzuki, N., Ciftci-Yilmaz, S. and Mittler, R. (2010) Reactive oxygen species homeostatis and signaling during drought and salinity stresses. *Plant Cell Env.* **33**, 453–467.

Munns, R. (2002) Comparative physiology of salt and water stress. *Plant Cell Env.* **25**, 239–250.

Munns, R. and Tester, M. (2008) Mechanisms of salinity tolerance. *Annu. Rev. Plant Biol.* **59**, 651–681.

Nickson, T.E. (2008) Planning environmental risk assessment for genetically modified crops: problem formulation for stress tolerant crops. *Plant Physiol.* **147**, 494–502.

Park, S.Y., Fung, .P., Nishimura, N., Jensen, D.R., Fujii, H., Zhao, Y., Lumba, S., Santiago, J., Rodrigues, A., Chow, T.F., Alfred, S.E., Bonetta, D., Finkelstein, R., Provart, N.J., Desveaux, D., Rodriguez, P.L., McCourt, P., Zhu, J.-K., Schroeder, J.I., Volkman, B.F. and Cutler, S.R. (2009) Abscisic acid inhibits Type 2C protein phosphatases via the PYR/PYL family of START proteins. *Science* **324**, 1068–1071.

Perera, I.Y., Hung, D.Y., Moore, C.D., Stevenson-Paulik, J. and Boss W, F. (2008) Transgenic Arabidopsis plants expressing the type 1 inositol

5-phosphatase exhibit increased drought tolerance and altered abscisic acid signaling. *Plant Cell* **20**, 2876–2893.

Piccinni, G., Rush, C.M., Vaughn, K.M. and Lazar, M.C. (2000) Lack of relationship between susceptibility to common root rot and drought tolerance among several closely related wheat lines. *Plant Dis.* **84**, 25–28.

Pilson, D., Snow, A., Rieseberg, L. and Alexander, H. (2002) *Fitness and Populational Effects of Gene Flow from Transgenic Sunflower to Wild Helianthus annuus*. Gene Flow Workshop, The Ohio State University, Columbus.

Provart, N. and Zhu, T. (2003) A browser-based functional classification superviewer for Arabidopsis genomics. *Curr. Comput. Molec. Biol.* **2003**, 271–272.

Qiu, Q.S., Barkla, B.J., Vera-Estrella, R., Zhu, J.K. and Schumaker, K.S. (2003) Na$^+$/H$^+$ exchange activity in the plasma membrane of Arabidopsis. *Plant Physiol.* **132**, 1041–1052.

Qiu, Q.S., Guo, Y., Quintero, F.J., Pardo, J.M., Schumaker, K.S. and Zhu, J.K. (2004) Regulation of vacuolar Na$^+$/H$^+$ exchange in *Arabidopsis thaliana* by the salt-overly-sensitive (SOS) pathway. *J. Biol. Chem.* **279**, 207–215.

Ricroch, A.E., Berge, J.B. and Kuntz, M. (2011) Evaluation of genetically engineered crops using transcriptomic, proteomic, and metabolomic profiling techniques. *Plant Physiol.* **155**, 1752–1761.

Ruebelt, M.C., Reynolds, T.L., Schmuke, J.J., Astwood, J.D., Della Penna, D., Engel, K.H. and Jany, K.D. (2006) Application of two-dimensional gel electrophoresis to interrogate alterations in the proteome of genetically modified crops. 3. Assessing unintended effects. *J. Agric. Food. Chem.* **54**, 2169–2177.

Schramm, F., Larkindale, J., Kiehlmann, E., Ganguli, A., Englich, G., Vierling, E. and von Koskull-Döring, P. (2008) A cascade of transcription factor *DREB2A* and heat stress transcription factor *HsfA3* regulates the heat stress response of *Arabidopsis*. *Plant J.* **53**, 264–274.

Seki, M., Narusaka, M., Abe, H., Kasuga, M., Yamaguchi-Shinozaki, K., Carninci, P., Hayashizaki, Y. and Shinozaki, K. (2001) Monitoring the expression pattern of 1300 Arabidopsis genes under drought and cold stresses by using a full-length cDNA microarray. *Plant Cell* **13**, 61–72.

Shabala, S. and Cuin, T.A. (2008) Potassium transport and plant salt tolerance. *Physiol. Plant.* **133**, 651–669.

Shi, H., Lee, B., Wu, S.J. and Zhu, J.K. (2003) Overexpression of a plasma membrane Na$^+$/H$^+$ antiporter gene improves salt tolerance in *Arabidopsis thaliana*. *Nat. Biotech.* **21**, 81–85.

Sickler, C.M., Edwards, G.E., Kiirats, O., Gao, Z. and Loescher, W. (2007) Response of mannitol-producing *Arabidopsis thaliana* to abiotic stress. *Funct. Plant Biol.* **34**, 382–391.

Snow, A.A., Pilson, D., Rieseberg, L.H., Paulsen, M.J., Pleskac, N., Reagon, M.R., Wolf, D.E. and Selbo, S.M. (2003) A Bt transgene reduces herbivory and enhances fecundity in wild sunflowers. *Ecol. App.* **13**, 279–286.

Thimm, O., Blasing, O., Gibon, Y., Nagel, A., Meyer, S., Kruger, P., Selbig, J., Muller, L.A., Rhee, S.Y. and Stitt, M. (2004) MAPMAN: a user-driven tool to display genomics data sets onto diagrams of metabolic pathways and other biological processes. *Plant J.* **37**, 914–939.

Van Buskirk, H.A. and Thomashow, M.F. (2006) Arabidopsis transcription factors regulating cold acclimation. *Physiologia Plant.* **126**, 72–80.

Verslues, P.E., Batelli, G., Grillo, S., Agius, F., Kim, Y.S., Zhu, J., Agarwal, M., Katiyar-Agarwal, S. and Zhu, J.K. (2007) Interaction of SOS2 with nucleoside diphosphate kinase 2 and catalases reveals a point of connection between salt stress and H$_2$O$_2$ signaling in *Arabidopsis thaliana*. *Molec. Cell Biol.* **27**, 7771–7780.

Vogel, J.T., Zarka, D.G., Van Buskirk, H.A., Fowler, S.G. and Thomashow, M.F. (2005) Roles of the CBF2 and ZAT12 transcription factors in configuring the low temperature transcriptome of Arabidopsis. *Plant J.* **41**, 195–211.

Vogele, R.T., Hahn, M., Lohaus, G., Link, T., Heiser, I. and Mendgen, K. (2005) Possible roles for mannitol and mannitol dehydrogenase in the biotrophic plant pathogen *Uromyces fabae*. *Plant Physiol.* **137**, 190–198.

Ward, J.M., Maser, P. and Schroeder, J.I. (2009) Plant ion channels: gene families, physiology, and functional genomic analyses. *Annu. Rev. Physiol.* **71**, 59–82.

Warwick, S.I., Beckie, H.J. and Hall, L.M. (2009) Gene flow, invasiveness, and ecological impact of genetically modified crops. *Ann. N. Y. Acad. Sci.* **1168**, 72–99.

Wettenhall, J.M., Simpson, K.M., Satterley, K. and Smyth, G.K. (2006) affylmGUI: a graphical user interface for linear modeling of single channel microarray data. *Bioinformatics* **22**, 897–899.

Wolfenbarger, L.L. and Grumet, R. (2003) Executive summary. In *Proceedings of a Workshop on: Criteria for Field Testing of Plants with Engineered Regulatory, Metabolic, and Signaling Pathways*, June 3–4, 2002 (Wolfenbarger, L.L., ed), pp. 5–12. Blacksburg, Virginia: Information Systems for Biotechnology.

Yamaguchi, S. (2008) Gibberellin metabolism and its regulation. *Annu. Rev. Plant Biol.* **59**, 225–251.

Yamaguchi, T. and Blumwald, E. (2005) Developing salt-tolerant crop plants: challenges and opportunities. *Trends Plant Sci.* **10**, 615–620.

Zhang, X., Fowler, S., Cheng, H., Lou, Y., Rhee, S.Y., Stockinger, E.J. and Thomashow, M.F. (2004) Freezing sensitive tomato has a functional CBF cold response pathway, but a CBF regulon that differs from that of freezing tolerant Arabidopsis. *Plant J.* **39**, 905–991.

Zhang, X., Wang, L., Meng, H., Wen, H.T., Fan, Y.L. and Zhao, J. (2011) Maize ABP9 enhances tolerance to multiple stresses in transgenic Arabidopsis by modulating ABA signaling and cellular levels of reactive oxygen species. *Plant Molec. Biol.* **75**, 365–378.

Zheng, X.N., Chen, B., Lu, G.J. and Han, B. (2009) Overexpression of a NAC transcription factor enhances rice drought and salt tolerance. *Biochem. Biophys. Res. Comm.* **379**, 985–989.

Zhifang, G. and Loesher, W. (2003) Expression of a celery mannose 6-phosphate reductase in *Arabidopsis thaliana* enhances salt tolerance and induces biosynthesis of both mannitol and a mannitol dimer. *Plant Cell Env.* **26**, 275–283.

Zhu, J.-K. (2003) Regulation of ion homeostasis under salt stress. *Curr. Opin. Plant Biol.* **6**, 441–445.

Zhu, J., Fu, X., Koo, Y.D., Zhu, J.K., Jenny Jr, F.E.., Adams, M.W.W., Zhu, Y., Shi, H., Yun, D.-J., Hasegawa, P.M. and Bressan, R.A. (2007) An enhancer mutant of *Arabidopsis salt overly sensitive 3* mediates both ion homeostasis and the oxidative stress response. *Mol. Cell. Biol.* **27**, 5214–5224.

Zolla, L., Antonioli, P. and Righetti, P.G. (2008) Proteomics as a complementary tool for identifying unintended side effects occurring in transgenic maize seeds as a result of genetic modifications. *J. Proteome Res.* **5**, 1850–1861.

5

Oral immunogenicity of porcine reproductive and respiratory syndrome virus antigen expressed in transgenic banana

Hui-Ting Chan[1], Min-Yuan Chia[2], Victor Fei Pang[2], Chian-Ren Jeng[2], Yi-Yin Do[1,]* and Pung-Ling Huang[1,3,]*

[1]Department of Horticulture and Landscape Architecture, National Taiwan University, Taiwan, Republic of China
[2]Graduate Institute of Veterinary Medicine, National Taiwan University, Taiwan, Republic of China
[3]Graduate Institute of Biotechnology, Chinese Culture University, Taiwan, Republic of China

*Correspondence
email pungling@ntu.edu.tw; yiyindo@ntu.edu.tw

Keywords: porcine reproductive and respiratory syndrome virus, glycoprotein 5, transgenic banana, oral vaccine, oral immunogenicity.

Summary

Porcine reproductive and respiratory syndrome virus (PRRSV) is a persistent threat of economically significant influence to the swine industry worldwide. Recombinant DNA technology coupled with tissue culture technology is a viable alternative for the inexpensive production of heterologous proteins in planta. Embryogenic cells of banana cv. 'Pei chiao' (AAA) have been transformed with the ORF5 gene of PRRSV envelope glycoprotein (GP5) using Agrobacterium-mediated transformation and have been confirmed. Recombinant GP5 protein levels in the transgenic banana leaves were detected and ranged from 0.021%–0.037% of total soluble protein. Pigs were immunized with recombinant GP5 protein by orally feeding transgenic banana leaves for three consecutive doses at a 2-week interval and challenged with PRRSV at 7 weeks postinitial immunization. A vaccination-dependent gradational increase in the elicitation of serum and saliva anti-PRRSV IgG and IgA was observed. Furthermore, significantly lower viraemia and tissue viral load were recorded when compared with the pigs fed with untransformed banana leaves. The results suggest that transgenic banana leaves expressing recombinant GP5 protein can be an effective strategy for oral delivery of recombinant subunit vaccines in pigs and can open new avenues for the production of vaccines against PRRSV.

Introduction

Porcine reproductive and respiratory syndrome virus (PRRSV) is a positive-strand RNA virus of the family Arteriviridae (Dea et al., 2000). PRRSV was first observed in the North American countries in 1987 and in the European countries in 1990 (Gilbert et al., 1997). PRRS was first reported in Taiwanese pigs in 1993 (Chang et al., 1993). The disease caused by PRRSV is characterized by reproductive failure in gilts and sows and by a respiratory disease in young pigs. PRRSV genome contains nine open reading frames (ORFs) that code for the viral replicase (ORF1a and 1b), four membrane-associated glycoproteins (ORF2a to 5), two unglycosylated proteins (ORF2b and 6) and the nucleocapsid protein (ORF7) (Meng et al., 1994). ORF5 encodes a major envelope protein (GP5) containing immunologically important domains associated with virus neutralization. The ectodomain of GP5 comprises approximately the first 40 residues of the mature protein and contains a variable number of N-glycosylation sites, and it has been characterized that linear neutralizing epitopes were located within the residues 37–45 in this region (Ostrowski et al., 2002; Plagemann, 2004; Wang et al., 2008; Wissink et al., 2003). The ORF5 nucleotide sequence of a Taiwanese PRRSV isolate, MD-001, revealed a 53.7%–54.7% amino acid identity with the European viruses and 82.6%–89.1% identity to the North American and Asian viruses (Cheuh and Lee, 2001). Current vaccines against PRRSV, including both live attenuated and killed vaccines, have been widely used, but live PRRSV vaccines have the potential to revert to virulence (Opriessnig

et al., 2002). This problem can be overcome if a vaccine consists of 'subunit' proteins of the pathogen that can stimulate immune responses (Mason et al., 2002).

Plant-derived subunit vaccines lack contamination with animal pathogens, are heat stable and may be engineered to contain multiple antigens (Rigano et al., 2009). Plant-based antigens induced the synthesis of antigen-specific mucosal IgA and serum IgG when delivered orally to mice and humans (Sala et al., 2003). Plants offer general advantages in terms of production scale and economy, product safety and ease of storage and distribution (Ma et al., 2005). Vegetable and fruit crops would be ideal host systems for the production of oral vaccines. The technology for harvesting and processing plant material on a large scale is available, and the purification requirement can be eliminated when the plant tissue containing the recombinant antigen is used as feed (Daniell et al., 2001; Kumar et al., 2007).

Potential plant species used for pharmaceutical protein production include tobacco, lettuce, carrot, tomato, maize, potato, alfalfa, soybean, banana and rice (Peterson and Arntzen, 2004). However, banana is considered an ideal host for producing oral vaccine antigens. A number of plant-based PRRSV oral vaccines carrying GP5 as antigen have been reported in tobacco (Chia et al., 2010, 2011) and potato (Chen and Liu, 2011). Other PRRSV antigens making use of M protein or N protein have also been developed in corn (Hu et al., 2012) and soybean (Vimolmangkang et al., 2012). Banana is grown extensively throughout the year in the tropics and subtropics and can be consumed raw without any modification. Furthermore, bananas are clonally propagated through suckers and pose no risk for gene containment with segregation in field (Mason et al., 2002). From an ecological point of view, the transgenic banana provides an added benefit of having low risk in the agricultural environment when used for the purpose of pharmaceutical protein production. In this paper, we present the results of a study of the recombinant GP5 protein produced by transgenic banana and delivered as a feed additive for the evaluation of immunogenicity. We demonstrate that the pigs orally immunized with transgenic banana showed a specific anti-GP5 response and displayed a significant reduction in serum and tissue viral loads. We found that the banana vaccine is a potent oral immunogen.

Results

Transformation and genetic analysis of transgenic banana

A binary plasmid pGKU-35PRRSV of 11 647 bp (Figure 1) encoding PRRSV ORF5 (from Taiwan MD-001 strain) with HDEL driven by the cauliflower mosaic virus (CaMV) 35S promoter was constructed. To allow transcriptional activation of OFR5 gene, the 3′-flanking region of Mh-UBQ1 (GenBank accession no. JQ782419) was used as a matrix-attachment region (MARs) because of its abundance of adenosine and thymidine, the major

characteristic of MARs. Banana cultivar 'Pei Chiao' (AAA genome) embryogenic suspension cells were transformed by co-cultivation with Agrobacterium tumefaciens strain LBA4404 harbouring the binary plasmid pGKU-35PRRSV. The transformed cells developed into somatic embryos after 1–2 months of selection on the banana embryo induction medium containing 50 mg/L geneticin. The embryos were then regenerated into plantlets on embryo germination medium, and the germinated embryos were transferred to MS medium containing 1 mg/L benzyladenine, 0.1 mg/L gibberellic acid and 100 mg/L geneticin for complete plantlet development (Figure 2a–c). Transgenic banana plants were grown to maturity in the transgenic glasshouse (Figure 2d –f), and both the root and leaf of transgenic plantlets tested positive for the GUS histochemical staining assay (Figure 3).

Southern hybridization was used to confirm the stable integration of the ORF5 gene into the banana genome. Analysis of the five transgenic plants showed the presence of transgene, and the different hybridization patterns confirmed that the five transgenic banana plants were independent transgenic lines. No signal was observed in genomic DNA of the untransformed plant. The copy number in the transgenic plants ranged from 1 to 2 based on the number of bands in each hybridization pattern (Figure 4).

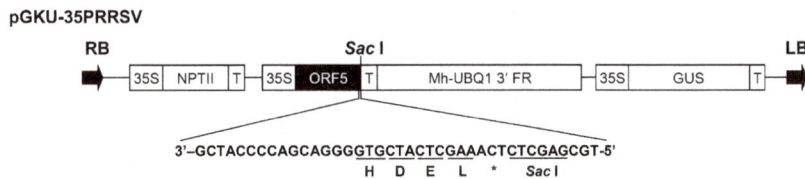

Figure 1 Schematic representation of T-DNA region of pGKU-35PRRSV for GP5 expression. 35S, cauliflower mosaic virus 35S promoter; NPTII, neomycin phosphotransferase II gene; T, nopaline synthase terminator; ORF5, PRRSV glycoprotein gene; HDEL, ER retention signal; Mh-UBQ1 3′ FR, Mh-UBQ1 3′-flanking region; GUS, the β-glucuronidase gene; LB, left border of T-DNA region; RB, right border of T-DNA region.

Figure 2 Somatic embryogenesis and regeneration of transgenic banana transformed with plasmid pGKU-35PRRSV. (a) Suspension cells were transformed and selected after 7 days. (b) Somatic embryo formation on selection medium containing 50 mg/L G418 (bar = 0.1 mm). (c) Shoot formation and rooting on selection medium containing 100 mg/L G418. (d–f) Transgenic banana plants in pots in the transgenic glasshouse.

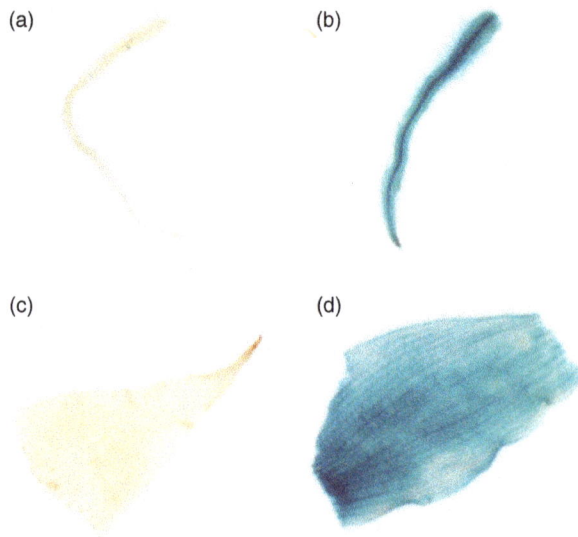

Figure 3 Histochemical GUS activity of transgenic banana 'Pei Chiao' transformed with plasmid pGKU-35PRRSV. (a, c) Negative control; (b, d) Root and leaf of plantlets transformed with pGKU-35PRRSV.

GP5 expression in transgenic plants

GP5 protein levels in the leaf samples of the five transgenic banana plants were determined by indirect enzyme-linked immunosorbent assay (ELISA). The amounts of GP5 protein in the transgenic plants were quantified by comparison with a known quantity of purified bacterial recombinant GP5 protein. The quantitative ELISA analysis showed that the GP5 protein level in the banana leaf samples ranged from 208.2 ± 24.3 to 365.1 ± 8.9 ng/mg (0.021%–0.037%) of total soluble protein (TSP), equivalent to 154–257 ng/g FW of banana leaves (Table 1).

PRRSV-specific lymphocyte proliferation

Twelve pigs were divided into two groups and were orally administrated with 50 g of fresh leaves from transgenic or untransformed banana three times on days 1, 15 and 29. The antigen dosage in 50 g banana leaves is approximately between 7.7 and 12.9 µg per dose. PRRSV-specific lymphoproliferative cell-mediated immune responses were determined in the peripheral blood mononuclear cells (PBMCs) collected from the immunized pigs at 1, 3, 5 and 7 weeks postinitial immunization (WPI). It can be inferred from Figure 5a that the PRRSV-specific lymphocyte proliferative responses showed a significant increase after the second oral vaccination ($P < 0.05$). A further gradual inoculation frequency–dependent increase in stimulation index was observed throughout the study.

PRRSV-specific serum IgG and saliva IgA responses

Pigs orally administrated with 50 g of fresh banana leaf (transgenic and untransformed) three times on days 1, 15 and 29 were examined to check whether the GP5 protein in the transgenic leaves had the ability to induce cellular and humoral immune responses. Significant increase ($P < 0.05$) in anti-PRRSV GP5 IgG antibody response was observed in pigs fed with leaves from transgenic plants compared to pigs fed with untransformed banana leaves as early as 3 WPI, and the levels further increased following subsequent booster (Figure 5b). From Figure 5c, it can be observed that anti-PRRSV-specific saliva IgA responses were elicited after two booster doses. Three oral doses with no adjuvant during a 6-week period stimulated strong serum and saliva anti-PRRSV responses that continued to increase up to 7 WPI. In general, oral vaccination frequency–dependent

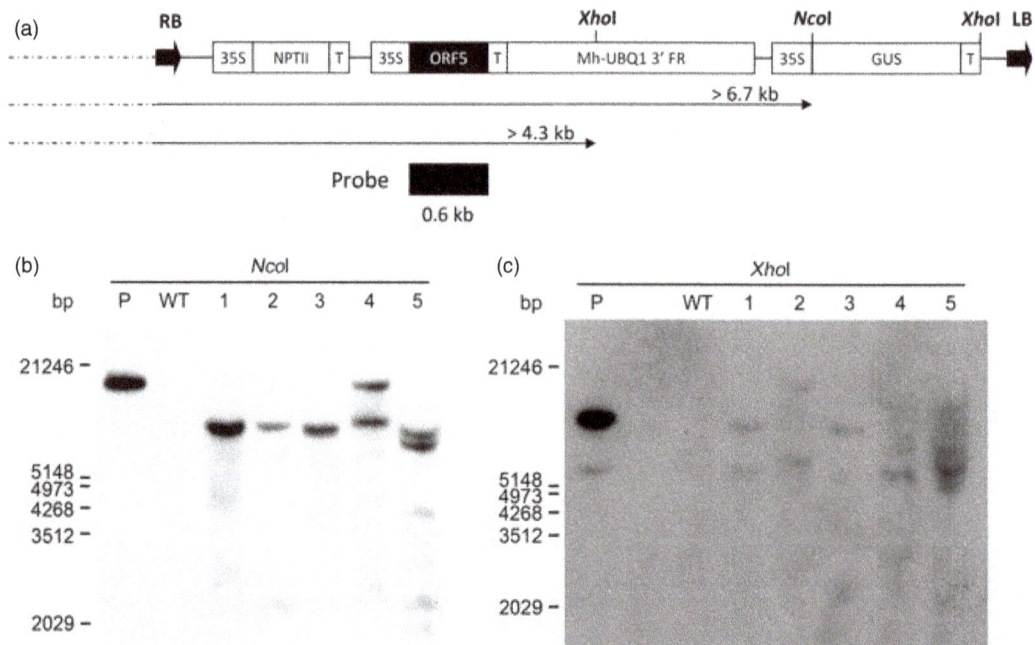

Figure 4 Southern blot analysis of individual banana plants transformed with plasmid pGKU-35PRRSV (P). The probe used for Southern blot analysis (0.6 kb) is marked by a filled box, and the expected hybridized fragments are marked by thin lines (a). Genomic DNA (20 µg) from nontransformed plant (WT) and transformed plants (lines 1–5) was digested with NcoI (b) and XhoI (c), and the plasmid (pGKU-35PRRSV) was used as a positive control. Band sizes are given in bps.

Table 1 Expression levels of recombinant GP5 protein in transgenic banana leaves

Line no.	TSP (ng/mg)	FW (ng/g)	% of TSP
1	208.2 ± 24.3	157.9 ± 18.4	0.021
2	255.2 ± 13.3	179.9 ± 9.4	0.026
3	234.3 ± 25.1	154.2 ± 16.5	0.023
4	365.1 ± 8.9	257.6 ± 6.3	0.037
5	312.8 ± 32.4	234.6 ± 24.3	0.031

FW, fresh weight.

gradational elicitation could be observed in both serum anti-PRRSV IgG and saliva-specific IgA.

PRRSV-specific neutralizing antibody levels

To determine whether the plant expressed GP5 proteins induced PRRSV-specific NAs in pigs, serum samples were evaluated at 3, 5 and 7 WPI as well as 1, 2 and 3 weeks postchallenge (WPC). As shown in Table 2, PRRSV-specific NAs were detected as early as 5 WPI in pigs. At 7 WPI, all pigs fed with transgenic banana leaves developed neutralizing antibody (NA) titres equal to or greater than 2 $\log_2$. By 3 WPC, those pigs had titres of 2–4 $\log_2$, respectively. On the contrary, pigs immunized with nontransformed plants had no detectable anti-PRRSV NAs at the designated time points.

Viraemia and tissue viral load after PRRSV challenge

The serum PRRSV loads were determined by real-time quantitative reverse transcription polymerase chain reaction (real-time qRT-PCR) based on the *ORF7* sequence, and the results are summarized in Figure 6. During the 3 weeks' post-PRRS virus challenge, pigs orally immunized with transgenic banana leaves displayed significantly lower viral loads in the serum than pigs fed on leaves from untransformed plants. To further evaluate the viral distribution in tissues, all pigs were necropsied at 3 WPC, and the viral loads in the lungs, tonsils, spleen and bronchial lymph nodes (LN) were determined by real-time qRT-PCR. As shown in Figure 7, although PRRSV RNA could still be detected in the selected tissues of all pigs immunized with transgenic banana leaves, the levels ranged from 15 to 24 copies/mg and were significantly lower than the viral load detected in pigs fed with leaves of untransformed banana plants, ranging from 156 to 203 copies/mg.

Discussion

In this study, we report a new generation of PRRSV vaccine, the recombinant GP5 subunit vaccine produced by transgenic banana. Pigs administered with this banana-based oral vaccine developed specific humoral and cell-mediated immune response against PRRSV as pigs administered with other plant-based oral vaccines did. GP5 was expressed as a 2–3 times higher protein level in transgenic banana (approximately 0.021%–0.037% TSP) than in transgenic tobacco (approximately 0.0108%–0.0155% TSP) leaves (Chia *et al.*, 2011). Pigs immunized with GP5-expressed banana developed 1 : 4–1 : 8 NA titres with further increases to 1 : 4–1 : 16 at 3 WPC, a similar level to that found in tobacco-immunized pigs (Chia *et al.*, 2011), indicating an anamnestic immune response after challenge. It is known that NA titre of 1 : 16 by passive immunization can clear PRRSV infection and prevent reproductive failure (Ostrowski *et al.*, 2002). PRRSV

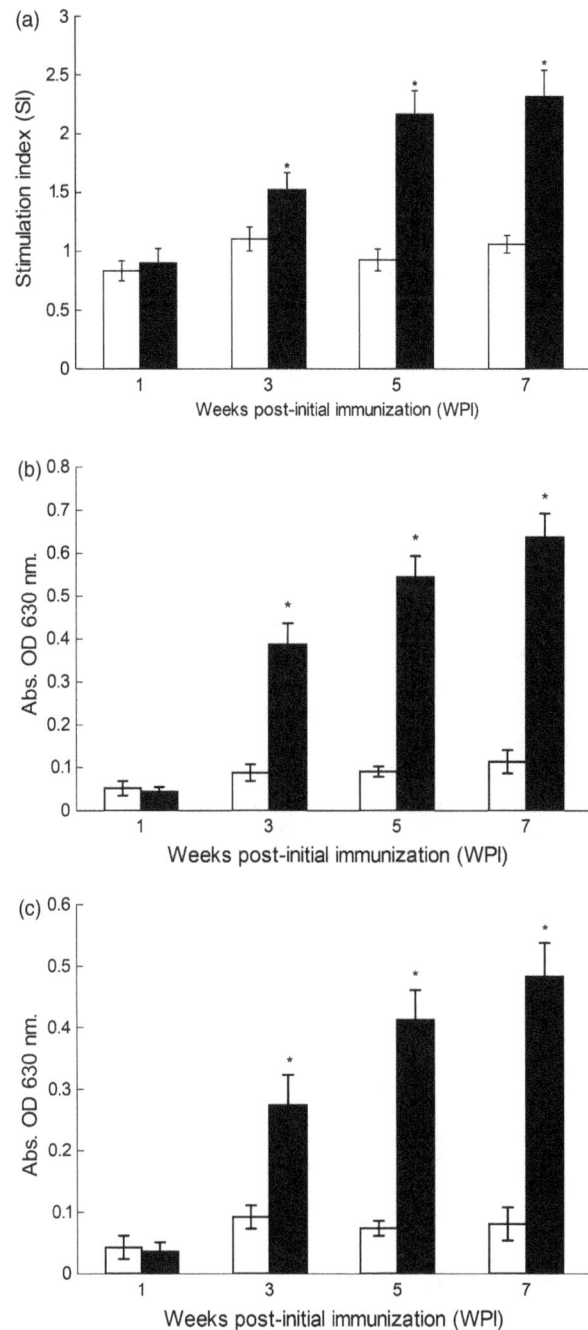

Figure 5 Change in PRRSV-specific lymphocyte proliferative responses (a), serum PRRSV-specific IgG (b) and saliva PRRSV-specific IgA (c) with time in pigs immunized with transgenic banana plants. Pigs with six in each group (*n* = 6) were inoculated orally with 50 g of PRRSV GP5 transgenic banana leaves (filled boxes) or wild-type banana leaves (open boxes), at a 2-week interval for 3 times. The peripheral blood mononuclear cells (PBMCs) were isolated at 1, 3, 5, 7 weeks postinitial immunization (WPI) and were stimulated with PRRSV strain MD-001 at 10^4 TCID$_{50}$ in hexaplicate. After 72 h of stimulation, MTS were added and the OD value was determined. Serum and saliva samples were collected at 1, 3, 5, 7 WPI. The anti-PRRSV-specific IgG antibody and the anti-PRRSV-specific IgA antibody titres were determined at a single 32-fold and 20-fold dilution in triplicate, respectively. Data are expressed as mean ± SD, and the vertical bars represent group standard deviations. An asterisk indicates a statistically significant (*P* < 0.05, Duncan's multiple range test) difference in comparison with control.

Table 2 Changes in the titres of serum anti-PRRSV-neutralizing antibodies (NAs) at various designated time points in pigs orally immunized with transgenic banana plants followed by PRRSV challenge

	NA titre ($\log_2$)* at weeks postinitial immunization					
Pig number	3	5	7	8 (1)[†]	9 (2)	10 (3)
WT						
1	–[‡]	–	–	–	–	–
2	–	–	–	–	–	–
3	–	–	–	–	–	–
4	–	–	–	–	–	–
5	–	–	–	–	–	–
6	–	–	–	–	–	–
GP5						
1	–	–	1	2	2	2
2	–	1	2	2	3	3
3	–	1	2	3	3	3
4	–	2	3	3	3	4
5	–	2	2	3	3	4
6	–	2	3	3	4	4

WT: nontransformed banana plant; GP5:PRRSV GP5 expressed transgenic banana plant; groups of six pigs were immunized orally with banana plants 3 times at 2-week intervals, followed by intranasal challenge with 2 mL of PRRSV MD-001 strain at 2×10^5 TCID$_{50}$ at 7 weeks postinitial immunization.

*Serum samples were collected at various designated time points and titrated ($\log_2$) individually in MARC-145 cells for the levels of anti-PRRSV NAs determined as the reciprocal of the highest dilution that inhibited 90% of positive foci in fluorescent assay.

[†]Weeks postinitial immunization (weeks postchallenge).

[‡]The NA titre was less than 1 : 2.

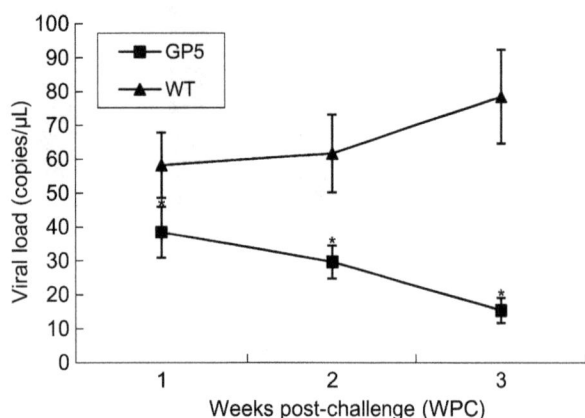

Figure 6 Changes in serum viral load with time in pigs orally immunized with transgenic banana plants followed by PRRSV challenge. Groups of six pigs were fed on transgenic banana or nontransformed leaves 3 times at 2-week intervals, followed by intranasal challenge with 2 mL of PRRSV MD-001 strain at 2×10^5 TCID$_{50}$ at 7 weeks postinitial immunization (WPI). Serum samples were collected at 1, 2 and 3 weeks postchallenge (WPC), and the PRRSV loads were determined by real-time qRT-PCR as RNA copies per microlitre of serum. Data are expressed as mean ± SD, and the vertical bars represent group standard deviations. An asterisk indicates a statistically significant ($P < 0.05$, Duncan's multiple range test) difference in comparison with control.

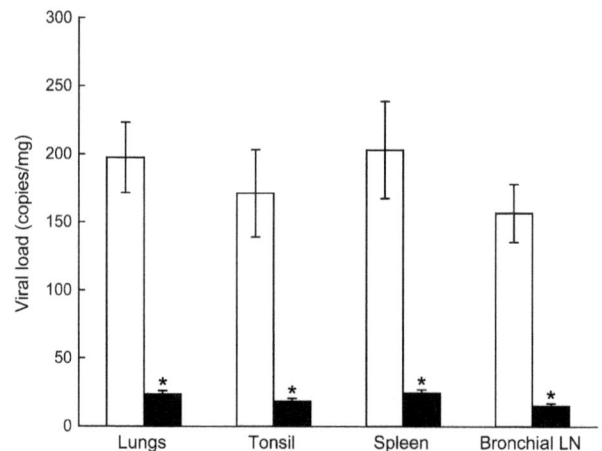

Figure 7 Tissue viral loads in pigs orally immunized with transgenic banana plants followed by PRRSV challenge. Groups of six pigs were fed on transgenic banana (filled boxes) or nontransformed (open boxes) leaves 3 times at 2-week intervals, followed by intranasal challenge with 2 mL of PRRSV MD-001 strain at 2×10^5 TCID$_{50}$ at 7 weeks postinitial immunization (WPI). The lungs, tonsils, spleen and bronchial lymph nodes (LN) were collected at 3 weeks postchallenge. The PRRSV load in each tissue was determined by real-time qRT-PCR as RNA copies per milligram of tissue. Data are expressed as mean ± SD, and the vertical bars represent group standard deviations. An asterisk indicates a statistically significant ($P < 0.05$, Duncan's multiple range test) difference in comparison with control.

enters the host via the mucosa of gastrointestinal, respiratory and reproductive tracts. Therefore, it is possible to activate the mucosal immunity by subunit oral vaccine administration to prevent PRRSV infection (Hyland et al., 2004). Moreover, the fact that no clinical symptoms were observed in pigs orally immunized with transgenic banana before PRRSV challenge indicates that vaccines are safe for pigs by oral administration.

Using banana-derived subunit vaccines can not only induce immune protection but also overcome problems associated with current PRRSV vaccines. Widely used PRRSV vaccines are now made from live attenuated and killed viruses (Murtaugh and Genzow, 2011), have the potential to revert to virulence (Jiang et al., 2006; Opriessnig et al., 2002) and have other drawbacks in terms of low levels of immunogenicity, antigenic variability between species, and possible transfer of genetic materials to wild-type strains (Floss et al., 2007). Many studies have confirmed the increase in immunogenicity by plant-based oral vaccines. The heat-labile enterotoxin (LT) B subunit (LT-B) of enterotoxigenic Escherichia coli (ETEC) produced in transgenic potato and corn has demonstrated partial protection against oral challenge with bacterial LT in animal studies (Lamphear et al., 2002; Mason et al., 1998). Plant-based human and animal recombinant subunit vaccines such as ETEC vaccine (Tacket et al., 1998, 2004), Norwalk virus vaccine (Mason et al., 1996; Tacket et al., 2000) and hepatitis B (HB) vaccine have shown efficacy and safety in preclinical and clinical studies under Phase 1 (Yusibov et al., 2011). Dow AgroSciences has developed a plant cell production system based on the use of a transformed plant cell line to introduce the target Newcastle disease virus subunit vaccine and has gained the first commercial licence for plant-made veterinary vaccines (Ling et al., 2010).

Modulation on gene expression in plants to enhance the accumulation levels of target proteins, the limitation for generating a protective immune response in plants, could be achieved by using some components in the expression system, such as codon usage, promoter, leader and polyadenylation signals (Sala et al., 2003). To optimize the accumulation of GP5, several components, including strong CaMV 35S promoter, endoplasmic reticulum (ER) retention signal HDEL and MARs, were assembled in our expression vector. In our results, GP5 protein expression level in banana leaves reached 154–257 ng/g FW, which was 6.8–15.8 times higher than that of HB surface antigen, 9.76–38 ng/g FW, driven by Arabidopsis UBQ3 promoter or banana EFE promoter (Kumar et al., 2005). Using ER retention signal could enhance the integration and retention of target protein in the ER, increase expression levels typically varying from 0.35% to 2% TSP (Fischer and Emans, 2000; Matsushima et al., 2003; Wright et al., 2001), enhance the accumulation of soluble recombinant proteins and prevent the degradation of these proteins in the cytoplasm (Lindh et al., 2009). The PRRSV structural proteins GP5 and M accumulate in the ER of infected cells, where they form disulphide-linked heterodimers, and constitute the basic protein matrix of virion envelope (Cruz et al., 2010). The PRRSV GP5-M heterodimer may be anchored in membranes, both proteins exposed a short N-terminal ectodomain on the virion surface, being involved in receptor recognition (Cruz et al., 2010). MARs, A/T-rich DNA sequences in eukaryotic genomes, are thought to influence the expression and stability of flanking genes, and assays of the transient expression of transgenes suggested that MAR elements might be involved in the structure and organization of the chromatin (Fukuda and Nishikawa, 2003) into independent and transcriptionally active loop domains (De Bolle et al., 2007). The Mh-UBQ1 3'-flanking region applied in this study was suggested as MARs because of its high AT content (64.49%; GenBank accession no. JQ782419), and hence used in plasmid construction to sustain the stability of the ORF5 gene. Mh-UBQ1 3'-flanking region also contains MAR-associated motifs, including MAR consensus sequence (AATATTTTTT; Cockerill and Garrard, 1986), Saccharomyces cerevisiae ARS consensus sequence (WTTTAYRTTTW; Amati and Gasser, 1988), Drosophila melanogaster topoisomerase II cleavage sites (GTNWAYATTNATNNR; Sander and Hsieh, 1985) and vertebrate topoisomerase II cleavage sites (RNYNNCNNGYNGKTNYNY; Spitzner and Muller, 1988). Previous studies have confirmed that MARs increase or stabilize the expression of flanking transgenes and tend to reduce or prevent the occurrence of transgene silencing (Li et al., 2008; Sidorenko et al., 2003).

Anti-PRRSV immunity was raised by GP5-expressed banana based on the evaluation results of viral load and tissue pathology of PRRSV-challenged pigs. Because PRRSV RNA could be detected in the lungs, tonsils, spleen and bronchial LN, and because PRRSV could also be detected in the serum at 3 WPC, decreased viraemia and limited tissue viral load reflect that the GP5-expressed banana could effectively reduce PRRSV replication activity. We have demonstrated previously that pigs fed on transgenic tobacco developed PRRSV-specific mucosal as well as systemic immunity to PRRSV challenge (Chia et al., 2011). Studies have demonstrated that GP5 and M proteins formed disulphide-linked heterodimers in the PRRSV particle and are associated with the development of neutralizing antibodies and cellular immunity (Jiang et al., 2006). The biological function of LT-B is dependent on the pentamer structure formation binding with G_{M1}-ganglioside receptors on the surface of intestinal epithelial cells (de Haan et al., 1998). Therefore, LT-B can act as an adjuvant and carrier for genetically linked foreign proteins (Kim et al., 2010).

To enhance immunogenicity, the new PRRSV oral vaccines may require the incorporation of other PRRSV antigens or other mucosal adjuvants.

In this study, we demonstrated the feasibility of banana oral vaccine. Banana is cultivated widely in developing countries where vaccines are most desperately needed and can be consumed raw by infants, while risk potential for outcrossing and biosafety considerations for banana are quite low due to its clonal propagation (Kumar et al., 2005). For preliminary evaluation of oral immunogenicity and the ease of mixing with feedstuff, we have chosen leaf as the plant material for oral administration. In addition, high crude protein content, known to be the most critical factor in oral vaccines in eliciting immunogenicity, in banana leaves varied from 1.53% to 14.96% (Yang et al., 2012) and in banana fruits from 2.1% to 4.88% (Gibert et al., 2009). Although protein content of potato tubers is higher (10.9%–13.8%; Ekin, 2011), their consumption in cooked form may result in poor immunogenicity because of the denaturation of the vaccine. Tomato fruits are edible but contain relatively low in protein content (0.26%–0.47%; Adebooye et al., 2006). Tobacco may provide the highest crude protein content in leaves (7–69 mg/g of dry weight; Szedljak et al., 2007), but the toxic alkaloids such as nicotine are not appropriate for oral delivery (Mason et al., 2002). Moreover, banana has a high yield of total biomass with a maximum residual biomass (13–20 metric tons dry matter/ha) in pseudostem and leaves, but tobacco plants can only produce 2.2 metric tons of biomass per hectare (Fischer and Emans, 2000). Also, removal of the lower fully expanded banana leaves during 8–12 occasions does not affect the fruit yield (Meyreles and Preston, 1979). Thus, the banana leaves can be used as animal feed throughout the year (Ffoulkes et al., 1977; Kimambo and Muya, 1991). Further studies aimed at enhancing expression in banana should include the application of tissue- or stage-specific promoter; for example, fruit-specific or ripening promoter is thought to be used for enhancing the expression in banana fruits. Higher protein expression levels could also be increased by using 5'-UTR translational enhancer, codon optimization, or expressing foreign proteins in chloroplasts (Sala et al., 2003).

Plant-based antigens can be fed directly to animals and are also safe due to no chance of animal pathogen contamination. Low costs for the application of the plant-derived vaccines can be achieved either by oral delivery of crude plant material or by the development of simple purification methods (Floss et al., 2007). Moreover, plant-derived subunit vaccines can be engineered to contain multiple antigens, such as those combined with subunits of cholera toxin, for the protection of animals against multiple infectious diseases. Using plant-based oral vaccines could revolutionize the vaccine industry by eliminating the significant costs incurred in complex production systems, such as fermentation, purification, cold storage, transportation and sterile delivery (Daniell et al., 2009).

Finally, this study demonstrates that pigs immunized with transgenic GP5-expressed banana leaves showed a specific anti-PRRSV response when their sera and saliva were analysed, demonstrating that GP5 proteins expressed in banana plants were able to elicit the production of the mucosal and systemic immune responses against PRRSV infection in pigs. These results support the foreseeable potential of developing novel edible plant-based oral vaccines for important future prevention of epidemic diseases. Our study showed promising results for banana-based vaccine in animal systems and suggest potential applications to human banana vaccines.

Experimental procedures

Binary plasmid construction and banana transformation

The ORF5 gene of PRRSV GP5 antigen with ER retention signal (HDEL) was PCR-amplified using the primers PRF: 5'-CAGA-TGCATTGGGGAACTGCTTG-3' (forward) and PRR: 5'-TGCGAGCTCTCAAAGCTCATCGTGGGGACGACCCCATCG-3' (reverse) from the PRRSV strain MD-001 (accession no. AF121131) isolated from the lungs of a PRRSV-infected pig (Chueh et al., 1998). The amplified product was cloned into an intermediate plasmid between the CaMV 35S promoter and nopaline synthase terminator after the digestion with NsiI and SacI. The expression cassette was then released by PstI digestion and inserted into the plant binary vector pGKU, which was modified from pGreen (Hellens et al., 2000) and contains β-glucuronidase (GUS) gene and kanamycin resistance gene (npt II) driven by the CaMV 35S promoter to obtain plasmid pGKU-35PRRSV. All cloning and cell transformation were performed in Escherichia coli JM109-competent cells as described by Sambrook et al. (1989). The resultant binary plant expression vector was electroporated into A. tumefaciens strain LBA4404 (Shen and Forde, 1989), which then served to transform the embryogenic cells of banana cultivar 'Pei Chiao' (AAA genome), as described by Hsu et al. (2008). The embryogenic cells were co-cultivated with A. tumefaciens strain LBA4404 harbouring the binary plasmid pGKU-35PRRSV. The cells were transferred to SH medium (Schenk and Hildebrandt, 1972) supplemented with zeatin (0.05 mg/L), isopentenyl adenine (0.2 mg/L), kinetin (0.1 mg/L), naphthalene acetic acid (0.2 mg/L), cefotaxime (250 mg/L) and geneticin (G418; Sigma, St. Louis, MO; 50 mg/L) after co-cultivation for somatic embryo development. The germinating embryos were transferred to MS medium (Murashige and Skoog, 1962) supplemented with benzyladenine (1 mg/L), gibberellic acid (0.1 mg/L) and geneticin (100 mg/L) for plantlet development. The fully developed plants were hardened in the transgenic glasshouse and then used for molecular analysis and pig feed trial experiments.

Histochemical GUS analysis

GUS histochemical analysis was performed on leaf and root samples collected from transgenic and nontransformed (control) plants as described by Hiei et al. (1994). Plant samples were incubated with GUS-staining solution [50 mM NaH_2PO_4 (pH 7.0), 1 mM 5-bromo-4-chloro-3-indoxyl-β-D-glucuronic acid] overnight at 37 °C. After staining, the sample was treated with 70% ethanol for 12 to 24 h at 37 °C to remove chlorophyll.

Southern blot analysis of transgenic plants

Genomic DNA was extracted from nontransformed and putative transgenic plants as described by Dellaporta et al. (1983). Twenty micrograms of each genomic DNA samples was digested with NcoI and XhoI, respectively, electrophoresed on a 0.7% (w/v) agarose gel and then transferred to a Hybond-N membrane (Amersham Pharmacia Biotech, Buckinghamshire, UK). A positive control (pGKU-35PRRSV) and a negative control (DNA from nontransformed banana leaves) were included in this experiment. Membranes were hybridized overnight at 65 °C in hybridization solution (6 × SSPE, 5 × Denhardt's reagent, 0.5% SDS, 250 μg/mL salmon sperm DNA and 10% dextran sulphate). This DNA was hybridized with ^{32}P-labelled random-primed (Promega, Madison, WI) probe and washed twice with wash buffer I (2 × SSPE and

0.1% SDS) at room temperature for 15 min and twice with wash buffer II (1 × SSPE and 0.1% SDS) at 65 °C for 15 min. The results were visually displayed by exposure to X-ray film (Kodak, Rochester, NY).

GP5 protein quantification in transgenic banana

The expression levels of GP5 in transgenic banana leaves were measured by quantitative ELISA as described by Chia et al. (2010). One gram of freeze-dried leaves from untransformed and transgenic banana plants was homogenized by grinding in mortar and pestle on ice in extraction buffer (50 mM Tris–HCl, pH 8.0, 1 mM EDTA, 10 mM β-mercaptoethanol, 4 mM dithiothreitol, 2 mM phenylmethylsulfonyl fluoride and 10 μg/mL leupeptin), and the TSP was extracted as described by Donald and Jackson (1994). The leaf homogenate contained approximately 0.7–0.75 mg of TSP as determined by Quant-iTTM Protein Assay Kit (Invitrogen, Carlsbad, CA). The 96-well ELISA plates were coated with 10 μg of TSP in triplicate and incubated overnight at 37 °C. The plates were washed three times with PBS (8 g/L NaCl, 0.2 g/L KH_2PO_4, 2.9 g/L $Na_2HPO_4 \cdot 12H_2O$, 0.2 g/L KCl, 0.2 g/L NaN_3, pH 7.4) containing 0.05% Tween-20 (Sigma) (PBST) and incubated with a 1 : 1000 dilution of mouse anti-GP5 monoclonal antibody (Ingenasa, Madrid, Spain) for 1 h at 37 °C. After three washes with PBST, the plates were incubated with a 1 : 1000 dilution of biotinylated goat anti-mouse IgG ABC kit (Promega) for 1 h at 37 °C and washed three times with PBST. The plates were developed with ρ-nitrophenylphosphate (pNPP, Sigma) substrate for 1 h, and absorbance/optical density (OD) at 405 nm was measured on Multiskan Ascent plate reader (Thermo Scientific, Waltham, MA)/ELISA reader (SpectraMax M5, Molecular Devices, Sunnyvale, CA). The amount of GP5 expressed in the transgenic banana plant was estimated based on the known amount of purified bacterial GP5.

Immunization and PRRSV challenge in pigs

Twelve six-week-old, crossbred, specific-pathogen-free (SPF) PRRSV-seronegative pigs were randomly grouped into experimental and control groups with six pigs in each group. Oral immunization was achieved by feeding 50 g fresh sliced leaves of either transgenic banana line no. 4 (GP5 reaching 12.9 μg per dose) or untransformed banana thoroughly mixed with 500 g of commercial feedstuff three times in a 2-week period. Care was taken to ensure that every pig received the designated dose by feeding individually in isolation until the mixed feedstuff was completely consumed. The animals were housed in a facility equipped with a filtered in-flow and out-flow air system. After the completion of each oral immunization, pigs were returned to regular water and normal commercial feed. All study procedures and animal care were conducted in accordance with the guidelines and under the supervision of the Institutional Committee on Animal Care and Use, National Taiwan University. PBMCs, serum and saliva samples were collected at designated times (1, 3, 5 and 7 WPI) for the measurement of lymphocyte proliferation and PRRSV-specific antibody responses. All pigs were challenged intranasally with 2 mL of PRRSV MD-001 strain at 2×10^5 50% tissue culture infection dose ($TCID_{50}$) at 7 WPI. All pigs were then monitored daily for clinical symptoms and killed by electrocution at 3 WPC.

Lymphocyte proliferation assay

Lymphocyte proliferation assay was performed using PBMCs from immunized pigs, and swine PBMCs were isolated as described by

Summerfield *et al.* (1998). The PBMCs were plated in 96-well round-bottom plates at a concentration of 2×10^5 cells/well (100 µL/well) and incubated with or without PRRSV (10^4 TCID$_{50}$/well) in hexaplicate for 72 h at 37 °C in 5% CO_2. Concanavalin A (Sigma) 5 µg/mL was used as a positive control. After incubation, 20 µL of MTS [3-(4,5-dimethylthiazol-2-yl)-5-(3-carboxymethoxyphenyl)-2-(4-sulfophenyl)-2H-tetrazolium] (Promega) was added to each well, and then the absorbance at 490 nm was detected by an ELISA reader. The stimulation index (SI) was calculated as the ratio of the average OD$_{490}$ value of wells containing PRRSV-stimulated cells to the average OD$_{490}$ value of wells containing nonstimulated cells.

Serological examination

Sera and whole saliva samples collected periodically from all control and experimental animals were tested to measure the response of anti-PRRSV IgG and IgA antibodies as described by Chia *et al.* (2010). For the analysis of the sera, 96-well flat-bottom plates (Costar) were coated overnight with purified PRRSV at the protein concentration of 1.86 µg/well. Following PBST wash, 32-fold diluted serum samples in PBS were added to PRRSV-coated wells in triplicate. After 1 h of reaction at room temperature, the wells were washed three times with PBST and filled with 100 µL of 1 : 1000 diluted horseradish peroxidase–conjugated goat anti-pig IgG (Bethyl Laboratories Inc., Montgomery, TX) as secondary antibody and 3,3′,5,5′-tetramethylbenzidine (TMB) (in 120 mM citrate buffer and 0.005% hydrogen peroxide) (Sigma) was used for colour development. For the detection of the levels of anti-PRRSV IgA and IgG antibodies in the samples of saliva, approximately 2–3 mL of whole saliva was collected by placing a known weight of gauze under the tongue of each pig for 30–45 s until saturated with saliva. The saliva-saturated gauze was then placed in a 50-mL centrifuge tube and a certain amount of PBS was added to reach a 10-fold dilution. After centrifugation at 3000 × g for 15 min, the supernatant was collected and treated with protease inhibitors (10 µM leupeptin, 1 µg/mL aprotinin, 50 µM PMSF and 5 µM bestatin). Following further twofold dilution in PBS to reach the finalized 20-fold dilution, the levels of anti-PRRSV IgA and IgG antibodies in the samples of saliva were measured as that described for the serum anti-PRRSV antibodies, except the secondary antibody used was horseradish peroxidase–conjugated goat anti-pig IgA or IgG (Bethyl Laboratories). NA titres were measured as described by Chia *et al.* (2010). Serum and saliva samples were heat-inactivated at 56 °C for 30 min prior to performing the neutralization assay. Fifty microlitres of twofold serially diluted serum or saliva samples was mixed with an equal volume of PRRSV at 10^2 TCID$_{50}$ in Dulbecco's modified Eagle medium (DMEM) (Gibco Laboratories, Grand Island, NY) supplemented with 10% heat-inactivated foetal bovine serum (HIFBS), 2 mM L-glutamine, 100 µg/mL streptomycin and 100 IU/mL penicillin in 96-well microtitration plates and incubated for 1 h at 37 °C in 5% CO_2. The mixtures were then transferred to 96-well flat-bottom plates containing confluent MARC-145 cells which had been seeded for 48 h. After 72-h incubation at 37 °C in a humidified atmosphere containing 5% CO_2, the cells were fixed with a solution of 50% methanol and 50% acetone for 10 min. After extensive washing with PBS, the expression of N protein of PRRSV was detected with an anti-N protein monoclonal antibody (South Dakota State University, South Dakota, USA) (Chang *et al.*, 2005) followed by incubation with fluorescein isothiocyanate–conjugated goat anti-mouse IgG (Leinco, St. Louis, MO). The NA titres were expressed as the reciprocal of the highest dilution that inhibited 90% of the positive foci present in the corresponding control wells with the addition of pooled serum from anti-PRRSV antibody-free SPF pigs. Each sample was run in triplicate.

Real-time quantitative reverse transcription polymerase chain reaction (real-time qRT-PCR)

Serum samples and tissue samples collected at 1, 2 and 3 WPC from all animals were subjected to analysis of viraemia by two-step real-time qRT-PCR as described by Chia *et al.* (2011). Total RNA from each serum sample collected was purified by RNeasy® Mini Kit (Qiagen). Tissue samples (1 g) from the lungs, tonsils, spleen and bronchial lymph nodes were collected and homogenized mechanically in DMEM to make a 10% suspension. Total RNA of tissue samples was extracted from supernatants by using TRIzol® reagent (Invitrogen). Each RNA sample was diluted to equal concentration and was then reversely transcribed by the oligo (dT)$_{12-18}$ primer (Invitrogen) and the SuperScript™ II reverse transcriptase (Invitrogen). A 107-bp region (2901–3007) of PRRSV *ORF7* was amplified and detected by LightCycler® TaqMan Master Mix (Hoffmann-La Roche Inc., Nutley, NJ) The fluorescent signals were determined each cycle at the end of the extension step. A standard curve was generated by 10-fold serially diluted plasmid standard (PRRSV *ORF7* gene insertion into PCR-Script™ SK vector, Stratagene Inc., La Jolla, CA) from 10^0–10^9 copies/µL, where the threshold of sensitivity of the assay was 10^0 copy/µL. The reaction was considered positive if the cycle threshold level was obtained at ≤ 39 cycles. Positive and negative reference samples were tested along with the unknown samples, and each sample was tested in triplicate.

Statistical analysis

The data were analysed using analysis of variance (ANOVA) (Statistical Analysis System; SAS for windows 6.12; SAS Institute Inc., Cary, NC) and Duncan's multiple range test. A *P*-value of less than 0.05 was considered statistically significant.

Acknowledgements

We are grateful to Prof. Mark D. Barnes (Department of Natural Resources, Chinese Culture University) and Dr. Raghu Rajasekara for critical reading of the manuscript. This work was supported by grants 97AS-1.2.1-AD-U1(18) and 98AS-1.2.1-ST-a2 of The National Science and Technology Program for Agricultural Biotechnology supported by the Council of Agriculture, Republic of China.

References

Adebooye, O.C., Adeoye, G.O. and Tijani-Eniola, H. (2006) Quality of fruits of three varieties of tomato (*Lycopersicon esculentum* (L.) Mill) as affected by phosphorus rates. *J. Agron.* **5**, 396–400.

Amati, B.B. and Gasser, S.M. (1988) Chromosomal ARS and CEN elements bind specifically to the yeast nuclear scaffold. *Cell*, **54**, 967–978.

De Bolle, M.F., Butaye, K.M., Goderis, I.J., Wouters, P.F., Jacobs, A., Delaure, S.L., Depicker, A. and Cammue, B.P. (2007) The influence of matrix attachment regions on transgene expression in *Arabidopsis thaliana* wild type and gene silencing mutants. *Plant Mol. Biol.* **63**, 533–543.

Chang, C.C., Chung, W.B., Lin, W.M., Yang, P.C., Weng, C.N., Chiu, Y.T., Chang, W.F. and Chu, R.M. (1993) Porcine reproductive and respiratory syndrome (PRRS) in Taiwan I: viral isolation. *J. Chin. Soc. Vet. Sci.* **19**, 268–276.

Chang, H.W., Jeng, C.R., Liu, J.J., Lin, T.L., Chang, C.C., Chia, M.Y., Tsai, Y.C. and Pang, V.F. (2005) Reduction of porcine reproductive and respiratory syndrome virus (PRRSV) infection in swine alveolar macrophages by porcine circovirus 2 (PCV2)-induced interferon-alpha. *Vet. Microbiol.* **108**, 167–177.

Chen, X. and Liu, J. (2011) Generation and immunogenicity of transgenic potato expressing the GP5 protein of porcine reproductive and respiratory syndrome virus. *J. Virol. Methods*, **173**, 153–158.

Cheuh, L.L. and Lee, K.H. (2001) Sequence analysis of two membrane-associated protein genes of a porcine reproductive and respiratory syndrome virus, Taiwan MD001 strain. *J. Chin. Soc. Vet. Sci.* **27**, 80–88.

Chia, M.Y., Hsiao, S.H., Chan, H.T., Do, Y.Y., Huang, P.L., Chang, H.W., Tsai, Y.C., Lin, C.M., Pang, V.F. and Jeng, C.R. (2010) Immunogenicity of recombinant GP5 protein of porcine reproductive and respiratory syndrome virus expressed in tobacco plant. *Vet. Immunol. Immunopathol.* **135**, 234–242.

Chia, M.Y., Hsiao, S.H., Chan, H.T., Do, Y.Y., Huang, P.L., Chang, H.W., Tsai, Y.C., Lin, C.M., Pang, V.F. and Jeng, C.R. (2011) Evaluation of the immunogenicity of a transgenic tobacco plant expressing the recombinant fusion protein of GP5 of porcine reproductive and respiratory syndrome virus and B subunit of *Escherichia coli* heat-labile enterotoxin in pigs. *Vet. Immunol. Immunopathol.* **140**, 215–225.

Chueh, L.L., Lee, K.H., Wang, F.I., Pang, V.F. and Weng, C.N. (1998) Sequence analysis of the nucleocapsid protein gene of the porcine reproductive and respiratory syndrome virus Taiwan MD-001 strain. *Adv. Exp. Med. Biol.* **440**, 795–799.

Cockerill, P.N. and Garrard, W.T. (1986) Chromosomal loop anchorage of the kappa immunoglobulin gene occurs next to the enhancer in a region containing topoisomerase II sites. *Cell*, **44**, 273–282.

Cruz, J.L.G., Zúñiga, S., Bécares, M., Sola, I., Ceriani, J.E., Juanola, S., Plana, J. and Enjuanes, L. (2010) Vectored vaccines to protect against PRRSV. *Virus Res.* **154**, 150–160.

Daniell, H., Streatfield, S.J. and Wycoff, K. (2001) Medical molecular farming: production of antibodies, biopharmaceuticals and edible vaccines in plants. *Trends Plant Sci.* **6**, 219–226.

Daniell, H., Singh, N.D., Mason, H. and Streatfield, S.J. (2009) Plant-made vaccine antigens and biopharmaceuticals. *Trends Plant Sci.* **14**, 669–679.

Dea, S., Gagnon, C.A., Mardassi, H., Pirzadeh, B. and Rogan, D. (2000) Current knowledge on the structural proteins of porcine reproductive and respiratory syndrome (PRRS) virus: comparison of the North American and European isolates. *Arch. Virol.* **145**, 659–688.

Dellaporta, S.L., Wood, J. and Hicks, J.B. (1983) A plant DNA minipreparation: version II. *Plant Mol. Biol. Rep.* **1**, 19–21.

Donald, R.G. and Jackson, A.O. (1994) The barley stripe mosaic virus gamma b gene encodes a multifunctional cysteine-rich protein that affects pathogenesis. *Plant Cell*, **6**, 1593–1606.

Ekin, Z. (2011) Some analytical quality characteristics for evaluating the utilization and consumption of potato (*Solanum tuberosum* L.) tubers. *Afr. J. Biotechnol.* **10**, 6001–6010.

Ffoulkes, D., Espejo, S., Marie, D., Delpeche, M. and Preston, T.R. (1977) The banana plant as cattle feed: composition and biomass production. *Trop. Anim. Prod.* **3**, 45–50.

Fischer, R. and Emans, N. (2000) Molecular farming of pharmaceutical proteins. *Transgenic Res.* **9**, 279–299.

Floss, D.M., Falkenburg, D. and Conrad, U. (2007) Production of vaccines and therapeutic antibodies for veterinary applications in transgenic plants: an overview. *Transgenic Res.* **16**, 315–332.

Fukuda, Y. and Nishikawa, S. (2003) Matrix attachment regions enhance transcription of a downstream transgene and the accessibility of its promoter region to micrococcal nuclease. *Plant Mol. Biol.* **51**, 665–675.

Gibert, O., Dufour, D., Giraldo, A., Sánchez, T., Reynes, M., Pain, J.P., González, A., Fernández, A. and Díaz, A. (2009) Differentiation between cooking bananas and dessert bananas. 1 Morphological and compositional characterization of cultivated Colombian Musaceae (Musa sp.) in relation to consumer preferences. *J. Agric. Food Chem.* **57**, 7857–7869.

Gilbert, S.A., Larochelle, R., Magar, R., Cho, H.J. and Deregt, D. (1997) Typing of porcine reproductive and respiratory syndrome viruses by a multiplex PCR assay. *J. Clin. Microbiol.* **35**, 264–267.

de Haan, L., Verweij, W.R., Feil, I.K., Holtrop, M., Hol, W.G., Agsteribbe, E. and Wilschut, J. (1998) Role of GM1 binding in the mucosal immunogenicity and adjuvant activity of the *Escherichia coli* heat-labile enterotoxin and its B subunit. *Immunology*, **94**, 424–430.

Hellens, R.P., Edwards, E.A., Leyland, N.R., Bean, S. and Mullineaux, P.M. (2000) pGreen: a versatile and flexible binary Ti vector for *Agrobacterium*-mediated plant transformation. *Plant Mol. Biol.* **42**, 819–832.

Hiei, Y., Ohta, S., Komari, T. and Kumashiro, T. (1994) Efficient transformation of rice (*Oryza sativa* L.) mediated by *Agrobacterium* and sequence analysis of the boundaries of the T-DNA. *Plant J.* **6**, 271–282.

Hsu, S.T., Liao, Y.W. and Huang, P.L. (2008) Establishment of *Agrobacterium*-mediated transformation system in a cultivated banana (Musa 'Pei Chiao', AAA group). *J. Taiwan Soc. Hort. Sci.*, **54**, 173–181.

Hu, J., Ni, Y., Drymanb, B.A., Meng, X.J. and Zhang, C. (2012) Immunogenicity study of plant-made oral subunit vaccine against porcine reproductive and respiratory syndrome virus (PRRSV). *Vaccine*, **30**, 2068–2074.

Hyland, K., Foss, D.L., Johnson, C.R. and Murtaugh, M.P. (2004) Oral immunization induces local and distant mucosal immunity in swine. *Vet. Immunol. Immunopathol.* **102**, 329–338.

Jiang, W., Jiang, P., Li, Y., Tang, J., Wang, X. and Ma, S. (2006) Recombinant adenovirus expressing GP5 and M fusion proteins of porcine reproductive and respiratory syndrome virus induce both humoral and cell-mediated immune responses in mice. *Vet. Immunol. Immunopathol.* **113**, 169–180.

Kim, T.G., Kim, B.G., Kim, M.Y., Choi, J.K., Jung, E.S. and Yang, M.S. (2010) Expression and immunogenicity of enterotoxigenic *Escherichia coli* heat-labile toxin B subunit in transgenic rice callus. *Mol. Biotechnol.* **44**, 14–21.

Kimambo, A.E. and Muya, H.M.H. (1991) Rumen degradation of dry matter and organic matter of different parts of the banana plant. *Livest Res. Rural Dev.* **3**, 35–40.

Kumar, G.B., Ganapathi, T.R., Revathi, C.J., Srinivas, L. and Bapat, V.A. (2005) Expression of hepatitis B surface antigen in transgenic banana plants. *Planta*, **222**, 484–493.

Kumar, G.B., Ganapathi, T.R. and Bapat, V.A. (2007) Production of hepatitis B surface antigen in recombinant plant systems: an update. *Biotechnol. Prog.* **23**, 532–539.

Lamphear, B.J., Streatfield, S.J., Jilka, J.M., Brooks, C.A., Barker, D.K., Turner, D.D., Delaney, D.E., Garcia, M., Wiggins, B., Woodard, S.L., Hood, E.E., Tizard, I.R., Lawhorn, B. and Howard, J.A. (2002) Delivery of subunit vaccines in maize seed. *J. Control Release*, **85**, 169–180.

Li, J.Y., Brunner, A.M., Meilan, R. and Strauss, S.H. (2008) Matrix attachment region elements have small and variable effects on transgene expression and stability in field-grown Populus. *Plant Biotechnol. J.* **6**, 887–896.

Lindh, I., Wallin, A., Kalbina, I., Savenstrand, H., Engstrom, P., Andersson, S. and Strid, A. (2009) Production of the p24 capsid protein from HIV-1 subtype C in *Arabidopsis thaliana* and *Daucus carota* using an endoplasmic reticulum-directing SEKDEL sequence in protein expression constructs. *Protein Expr. Purif.* **66**, 46–51.

Ling, H.Y., Pelosi, A. and Walmsley, A.M. (2010) Current status of plant-made vaccines for veterinary purposes. *Expert Rev. Vaccines*, **9**, 971–982.

Ma, J.K., Chikwamba, R., Sparrow, P., Fischer, R., Mahoney, R. and Twyman, R.M. (2005) Plant-derived pharmaceuticals–the road forward. *Trends Plant Sci.* **10**, 580–585.

Mason, H.S., Ball, J.M., Shi, J.J., Jiang, X., Estes, M.K. and Arntzen, C.J. (1996) Expression of Norwalk virus capsid protein in transgenic tobacco and potato and its oral immunogenicity in mice. *Proc. Natl Acad. Sci. USA*, **93**, 5335–5340.

Mason, H.S., Haq, T.A., Clements, J.D. and Arntzen, C.J. (1998) Edible vaccine protects mice against *Escherichia coli* heat-labile enterotoxin (LT): potatoes expressing a synthetic LT-B gene. *Vaccine*, **16**, 1336–1343.

Mason, H.S., Warzecha, H., Mor, T. and Arntzen, C.J. (2002) Edible plant vaccines: applications for prophylactic and therapeutic molecular medicine. *Trends Mol. Med.* **8**, 324–329.

Matsushima, R., Kondo, M., Nishimura, M. and Hara-Nishimura, I. (2003) A novel ER-derived compartment, the ER body, selectively accumulates a beta-glucosidase with an ER-retention signal in *Arabidopsis*. *Plant J.* **33**, 493–502.

Meng, X.J., Paul, P.S. and Halbur, P.G. (1994) Molecular cloning and nucleotide sequencing of the 3'-terminal genomic RNA of the porcine reproductive and respiratory syndrome virus. *J. Gen. Virol.* **75**, 1795–1801.

Meyreles, L. and Preston, T. R. (1979) Studies on leaf production in the banana plant. *Trop. Anim. Prod.* **4**, 302.

Murashige, T. and Skoog, F. (1962) A revised medium for rapid growth and bioassays with tobacco tissue cultures. *Physiol. Plant.* **15**, 473–497.

Murtaugh, M.P. and Genzow, M. (2011) Immunological solutions for treatment and prevention of porcine reproductive and respiratory syndrome (PRRS). *Vaccine*, **29**, 8192–8204.

Opriessnig, T., Halbur, P.G., Yoon, K.J., Pogranichniy, R.M., Harmon, K.M., Evans, R., Key, K.F., Pallares, F.J., Thomas, P. and Meng, X.J. (2002) Comparison of molecular and biological characteristics of a modified live porcine reproductive and respiratory syndrome virus (PRRSV) vaccine (ingelvac PRRS MLV), the parent strain of the vaccine (ATCC VR2332), ATCC VR2385, and two recent field isolates of PRRSV. *J. Virol.* **76**, 11837–11844.

Ostrowski, M., Galeota, J.A., Jar, A.M., Platt, K.B., Osorio, F.A. and Lopez, O.J. (2002) Identification of neutralizing and nonneutralizing epitopes in the porcine reproductive and respiratory syndrome virus GP5 ectodomain. *J. Virol.* **76**, 4241–4250.

Peterson, R.K. and Arntzen, C.J. (2004) On risk and plant-based biopharmaceuticals. *Trends Biotechnol.* **22**, 64–66.

Plagemann, P.G. (2004) GP5 ectodomain epitope of porcine reproductive and respiratory syndrome virus, strain Lelystad virus. *Virus Res.* **102**, 225–230.

Rigano, M.M., Manna, C., Giulini, A., Vitale, A. and Cardi, T. (2009) Plants as biofactories for the production of subunit vaccines against bio-security-related bacteria and viruses. *Vaccine*, **27**, 3463–3466.

Sala, F., Manuela Rigano, M., Barbante, A., Basso, B., Walmsley, A.M. and Castiglione, S. (2003) Vaccine antigen production in transgenic plants: strategies, gene constructs and perspectives. *Vaccine*, **21**, 803–808.

Sambrook, J., Fritsch, E.F. and Maniatis, T. (1989) Plasmid vectors. In *Molecular Cloning: A Laboratory Manual*, 2nd edn. Vol. I, pp. 1.1–1.110. Cold Spring Harbor: Cold Spring Harbor Laboratory Press.

Sander, M. and Hsieh, T.S. (1985) *Drosophila* topoisomerase II double-stranded DNA cleavage: analysis of DNA sequence homology at the cleavage site. *Nucleic Acids Res.* **13**, 1057–1072.

Schenk, R.U. and Hildebrandt, A.C. (1972) Medium and techniques for induction and growth of monocotyledonous plant cell cultures. *Can. J. Bot.* **50**, 199–204.

Shen, W.J. and Forde, B.G. (1989) Efficient transformation of *Agrobacterium* spp. by high voltage electroporation. *Nucleic Acids Res.* **17**, 8385.

Sidorenko, L., Bruce, W., Maddock, S., Tagliani, L., Li, X., Daniels, M. and Peterson, T. (2003) Functional analysis of two matrix attachment region (MAR) elements in transgenic maize plants. *Transgenic Res.* **12**, 137–154.

Spitzner, J.R. and Muller, M.T. (1988) A consensus sequence for cleavage by vertebrate DNA topoisomerase II. *Nucleic Acids Res.* **16**, 5533–5556.

Summerfield, A., Knotig, S.M. and McCullough, K.C. (1998) Lymphocyte apoptosis during classical swine fever: implication of activation-induced cell death. *J. Virol.* **72**, 1853–1861.

Szedljak, I., Szántainé Kőjegyi, K. and Kosáry, J. (2007) Preliminary biochemical studies on a model growing of different tobacco plant (*Nicotiana tabacum* L.) cultivars. *Intl. J. Hort. Sci.* **13**, 83–87.

Tacket, C.O., Mason, H.S., Losonsky, G., Clements, J.D., Levine, M.M. and Arntzen, C.J. (1998) Immunogenicity in humans of a recombinant bacterial antigen delivered in a transgenic potato. *Nat. Med.* **4**, 607–609.

Tacket, C.O., Mason, H.S., Losonsky, G., Estes, M.K., Levine, M.M. and Arntzen, C.J. (2000) Human immune responses to a novel norwalk virus vaccine delivered in transgenic potatoes. *J. Infect. Dis.* **182**, 302–305.

Tacket, C.O., Pasetti, M.F., Edelman, R., Howard, J.A. and Streatfield, S. (2004) Immunogenicity of recombinant LT-B delivered orally to humans in transgenic corn. *Vaccine*, **22**, 4385–4389.

Vimolmangkang, S., Gasic, K., Soria-Guerra, R., Rosales-Mendoza, S., Moreno-Fierros, L. and Korban, S.S. (2012) Expression of the nucleocapsid protein of porcine reproductive and respiratory syndrome virus in soybean seed yields an immunogenic antigenic protein. *Planta*, **235**, 513–522.

Wang, C., Lee, F., Huang, T.S., Pan, C.H., Jong, M.H. and Chao, P.H. (2008) Genetic variation in open reading frame 5 gene of porcine reproductive and respiratory syndrome virus in Taiwan. *Vet. Microbiol.* **131**, 339–347.

Wissink, E.H., van Wijk, H.A., Kroese, M.V., Weiland, E., Meulenberg, J.J., Rottier, P.J. and van Rijn, P.A. (2003) The major envelope protein, GP5, of a European porcine reproductive and respiratory syndrome virus contains a neutralization epitope in its N-terminal ectodomain. *J. Gen. Virol.* **84**, 1535–1543.

Wright, K.E., Prior, F., Sardana, R., Altosaar, I., Dudani, A.K., Ganz, P.R. and Tackaberry, E.S. (2001) Sorting of glycoprotein B from human cytomegalovirus to protein storage vesicles in seeds of transgenic tobacco. *Transgenic Res.* **10**, 177–181.

Yang, J., Tan, H., Zhai, H., Wang, Q., Zhao, N., Cai, Y., Li, M. and Zhou, H. (2012) Research on chemical composition and ensiling characteristics of banana stems and leaves. *Adv. Materials Res.* **4347–4353**, 1647–1651.

Yusibov, V., Streatfield, S.J. and Kushnir, N. (2011) Clinical development of plant-produced recombinant pharmaceuticals: vaccines, antibodies and beyond. *Hum. Vaccine*, **7**, 313–3.

6

Improving rice tolerance to potassium deficiency by enhancing *OsHAK16p:WOX11*-controlled root development

Guang Chen, Huimin Feng, Qingdi Hu, Hongye Qu, Aiqun Chen, Ling Yu and Guohua Xu*

State Key Laboratory of Crop Genetics and Germplasm Enhancement, MOA Key Laboratory of Plant Nutrition and Fertilization in Lower-Middle Reaches of the Yangtze River, Nanjing Agricultural University, Nanjing, China

*Correspondence

email ghxu@njau.edu.cn

Summary

Potassium (K) deficiency in plants confines root growth and decreases root-to-shoot ratio, thus limiting root K acquisition in culture medium. A WUSCHEL-related homeobox (WOX) gene, *WOX11*, has been reported as an integrator of auxin and cytokinin signalling that regulates root cell proliferation. Here, we report that ectopic expression of *WOX11* gene driven by the promoter of *OsHAK16* encoding a low-K-enhanced K transporter led to an extensive root system and adventitious roots and more effective tiller numbers in rice. The *WOX11*-regulated root and shoot phenotypes in the *OsHAK16p:WOX11* transgenic lines were supported by K-deficiency-enhanced expression of several *RR* genes encoding type-A cytokinin-responsive regulators, *PIN* genes encoding auxin transporters and *Aux/IAA* genes. In comparison with WT, the transgenic lines showed increases in root biomass, root activity and K concentrations in the whole plants, and higher soluble sugar concentrations in roots particularly under low K supply condition. The improvement of sugar partitioning to the roots by the expression of *OsHAK16p:WOX11* was further indicated by increasing the expression of *OsSUT1* and *OsSUT4* genes in leaf blades and several *OsMSTs* genes in roots. Expression of *OsHAK16p:WOX11* in the rice grown in moderate K-deficient soil increased total K uptake by 72% and grain yield by 24%–32%. The results suggest that enlarging root growth and development by the expression of *WOX11* in roots could provide a useful option for increasing K acquisition efficiency and cereal crop productivity in low K soil.

Keywords: inducible promoter, *Oryza sativa*, potassium deficiency, root growth, sugar partition, *WOX11*.

Introduction

Potassium (K) is an essential macronutrient playing the crucial role in a number of biochemical and physiological processes for plant growth and development (Amrutha *et al.*, 2007; Gierth and Mäser, 2007; Yang *et al.*, 2009). Due to strong adsorption and fixation by soil particles and low mobility in soil, K availability for plant roots is limited in large areas of agricultural land in the world (e.g. 3/4 of paddy soils in China, 2/3 of the wheatbelt in Southern Australia) (Rengel and Damon, 2008). Therefore, increasing the efficiency of plant K acquisition and utilization is important for agricultural production.

K deficiency affects shoot and root growth (Cakmak *et al.*, 1994a; Rengel and Damon, 2008). In roots, it impairs both lateral root initiation and development (Armengaud *et al.*, 2004; Shin and Schachtman, 2004). It also seems to have a depressive effect on primary root growth (Gruber *et al.*, 2013; Kim *et al.*, 2010; Rengel and Damon, 2008). In *Arabidopsis*, a major trade-off between main root (MR) and lateral root elongation results in two extreme strategies of morphological adaptation to low K (Kellermeier *et al.*, 2013). The strategy I is for maintaining MR growth but compromising lateral root elongation, while the strategy II is for arresting MR elongation in favour of lateral branching. Interestingly, both the strategies allow the plants to maintain elongation of at least certain root parts, and as a result, no differences in shoot growth were observed in low K supply

condition (Kellermeier *et al.*, 2013). In rice, K deficiency decreased root growth and root-to-shoot ratio by reducing soluble sugar contents in the roots (Cai *et al.*, 2012).

Potassium enhances phloem loading of carbohydrates through a direct effect on the loading mechanism (e.g. maintenance of the transmembrane pH gradient) and an indirect effect via an increase in osmotic potential of phloem sap and, thus, the rate of mass flow in the sieve tubes (Gajdanowicz *et al.*, 2011; Lalonde *et al.*, 2003; Lang, 1983; Peel and Rogers, 1982). K deficiency reduces photo-assimilate production and phloem transport of assimilates, thus changing metabolite concentrations in vegetative plant organs (Cakmak *et al.*, 1994a,b). Accumulation of sugars in mature leaves is the consequence of inhibited export of sugars from the leaves (Cai *et al.*, 2012; Hermans *et al.*, 2006) and smaller assimilate supply to sink organs such as growing roots (Cakmak *et al.*, 1994a,b), buds (Pettigrew, 2008), developing leaves (Gerardeaux *et al.*, 2010) and fleshy fruits (e.g. tomato) (Kanai *et al.*, 2007), thus reducing their growth. Sucrose from source leaves is translocated to sink organs through a process that involves the activity of sucrose transporters (SUTs) at specific points (Ayre, 2011; Eom *et al.*, 2012; Scofield *et al.*, 2007). In sink organs, monosaccharides produced through hydrolysis of the unloaded sucrose by apoplastic invertase are imported into cells by monosaccharide transporters (MSTs) (Lim *et al.*, 2006; Slewinski, 2011). Thus, coordinated regulation of SUTs and MSTs is crucial for controlling the whole-plant carbon distribution, particularly in sink tissues such as developing roots (Bihmidine et al., 2013; Lim et al., 2006; Patrick and Offler, 2001).

A low plant K status triggers or up-regulates expression of some high-affinity K transporters and K channels and activates signalling cascades (Ashley et al., 2006; Li et al., 2014a; Wang and Tsay, 2011; Wang and Wu, 2013; Yang et al., 2014). The KT/ HAK/KUP family is a major K transporter family present in bacteria, fungi and plants (Gierth and Mäser, 2007). In plants, a number of KT/KUP/HAK genes have been subsequently identified from different plant species and the transcription of many genes, such as AtHAK5, CaHAK1, HvHAK1, LeHAK5 and ThHAK5, is up-regulated by K starvation (Alemán et al., 2009; Fulgenzi et al., 2008; Gierth et al., 2005; Nieves-Cordones et al., 2007; Qi et al., 2008; Wang et al., 2002). In rice (Oryza sativa cv. Nipponbare), 27 OsHAK genes were identified in the genome (Gupta et al., 2008; Yang et al., 2009). Among them, OsHAK1, OsHAK5, OsHAK7 and OsHAK16 transcript levels were found to be significantly increased in rice roots under K starvation condition (Bañuelos et al., 2002; Horie et al., 2011; Okada et al., 2008; Yang et al., 2014). In addition, a Shaker K channel OsAKT1 has recently been shown to play a critical role in K uptake in rice roots (Li et al., 2014a).

WUSCHEL-related homeobox (WOX) genes are developmental regulators (Wang et al., 2014). There are at least 15 Arabidopsis and 13 rice WOX family members, and some of these WOX genes play important roles in regulating cell division and differentiation during root system development in both monocot and dicot plants (Cheng et al., 2014; Cho et al., 2013; Yoo et al., 2013). For example, in Arabidopsis, WOX2 regulates cell fates during basal embryonic root formation (Breuninger et al., 2008). WOX5 is essential for stem cell maintenance (Stahl et al., 2009). Mutants of WOX8/9 display reduced cell division and lethality during embryo and seedling developmental stages (Wang et al., 2014; Wu et al., 2005). In rice, OsWOX3A (nal2/3) mutants produced fewer lateral roots (Cho et al., 2013). However, root hair growth was markedly increased in nal2/3, suggesting that OsWOX3A is involved in root hair development as well as lateral root development (Yoo et al., 2013). Loss-of-function mutation or down-regulation of OsWOX11 reduced the number and the growth rate of crown roots, whereas overexpression of the gene induced precocious crown root growth and dramatically increased the root biomass by producing crown roots at the upper stem nodes and the base of florets (Zhao et al., 2009).

The aim of this study was to test whether conditionally enlarging the root system and development of adventitious roots could improve K uptake efficiency at low K supply level. We used the promoter of OsHAK16, a K-deficiency up-regulated gene encoding a K transporter, to drive the expression of OsWOX11 gene (WOX11 as abbreviation thereafter in this article) in rice roots. We found that K deficiency increased the expression of WOX11 mainly in the roots of the transgenic plants. Expression of OsHAK16p:WOX11 (HAK16p:WOX11 as abbreviation thereafter in this article) very significantly improved root development and activity, elevated soluble sugar concentration in roots and young leaves, and increased effective tiller numbers, total K uptake and grain yield under low K supply conditions. Alteration of several RR genes encoding type-A cytokinin-responsive regulators, PIN genes encoding auxin transporters and Aux/IAA genes, and SUTs and MSTs genes in the transgenic lines confirmed the WOX11-regulated root and shoot phenotype and increased sugar distribution into the roots. The results suggest that application of the promoter of low K up-regulated transporter genes in

combination with the WOX11 gene in roots could be considered in molecular breeding for improving the tolerance of crops, at least rice, to K deficiency.

Results

K-deficiency-enhanced OsHAK16 expression in roots

It was reported that transcription of OsHAK16 in rice roots was up-regulated by K starvation (Okada et al., 2008), while its expression in the shoot was uncertain. In this study, we detected that the expression of OsHAK16 gene in the roots was about fourfold higher under low K (0.1 mM, −K) than under normal K (1 mM, +K) supply condition; however, there was no significant difference of OsHAK16 expression between the K-deficient and K-sufficient shoots (Figure 1a).

To investigate the effect of K deficiency on OsHAK16 promoter activity, the transgenic rice expressing the GUS reporter gene driven by OsHAK16 promoter (1846 bp) was generated and named as HAK16p:GUS. HAK16p:GUS transgenic lines were cultured in +K and −K solution for 7 days. Compared with +K supply, −K increased the GUS activity by 2.5-fold in roots, while not significant in shoot (Figure 1b). The GUS staining showed that OsHAK16 was expressed throughout the K-deficient primary and lateral roots (Figure 1c–e). Its expression was detected moderately at epidermis and cortices, strongly at the root stele vascular tissue and emerging lateral root primordium (Figure 1f, g), while slightly at the leaf blades (Figure 1h).

Generation of the transgenic rice expressing HAK16p: WOX11

It was reported that K deficiency resulted in a stunted root growth and further limited K uptake in rice (Cai et al., 2012). WOX11 functions in stimulating the emergence and growth of adventitious roots; however, ubiquitin promoter:WOX11 transgenic rice showed a large root system but poorly developed shoots; thus, some of the transgenic plants died before producing seeds (Zhao et al., 2009). To counteract the effect of K starvation on the development of root system and thereby improve rice tolerance to low K, we manipulated expression of WOX11 gene by transforming HAK16 promoter:WOX11 construct into rice.

We obtained 16 independent HAK16p:WOX11 transgenic lines in T$_0$ generation by Southern blotting analysis. Moreover, five null segregants were detected by both GUS staining and Southern blotting analysis in T$_1$ generation (data not shown). The phenotypes of roots of two untransformed lines and eight HAK16p: WOX11 transgenic lines in T$_0$ generation were presented in Figure S1a. As no significant difference was observed between WT and the null segregants in T$_1$ generation for the expression of WOX11, root growth (root biomass, root-to-shoot ratio, number of adventitious roots and total root length), root activity (the active absorbing surface percentage), K concentration in both root and shoot, shoot architecture and yield at both K-deficient and K-sufficient conditions (Figures S1b,c, S2 and S3), we used WT as the only negative control in detailed analyses in hydroponics, soil and field experiments for the transgenic lines in T$_2$ generation. In addition, five independent T$_1$ transgenic lines showed similar but distinct difference with WT and the null segregants in shoot architecture and yield (Figure S3), and we selected three single-copy transgenic lines (L1, L2 and L3) in T$_2$ generation for further detailed analysis (Figure 2a).

In comparison with WT, the transgenic lines showed higher expression of WOX11 in the roots, particularly at low K (0.1 mM)

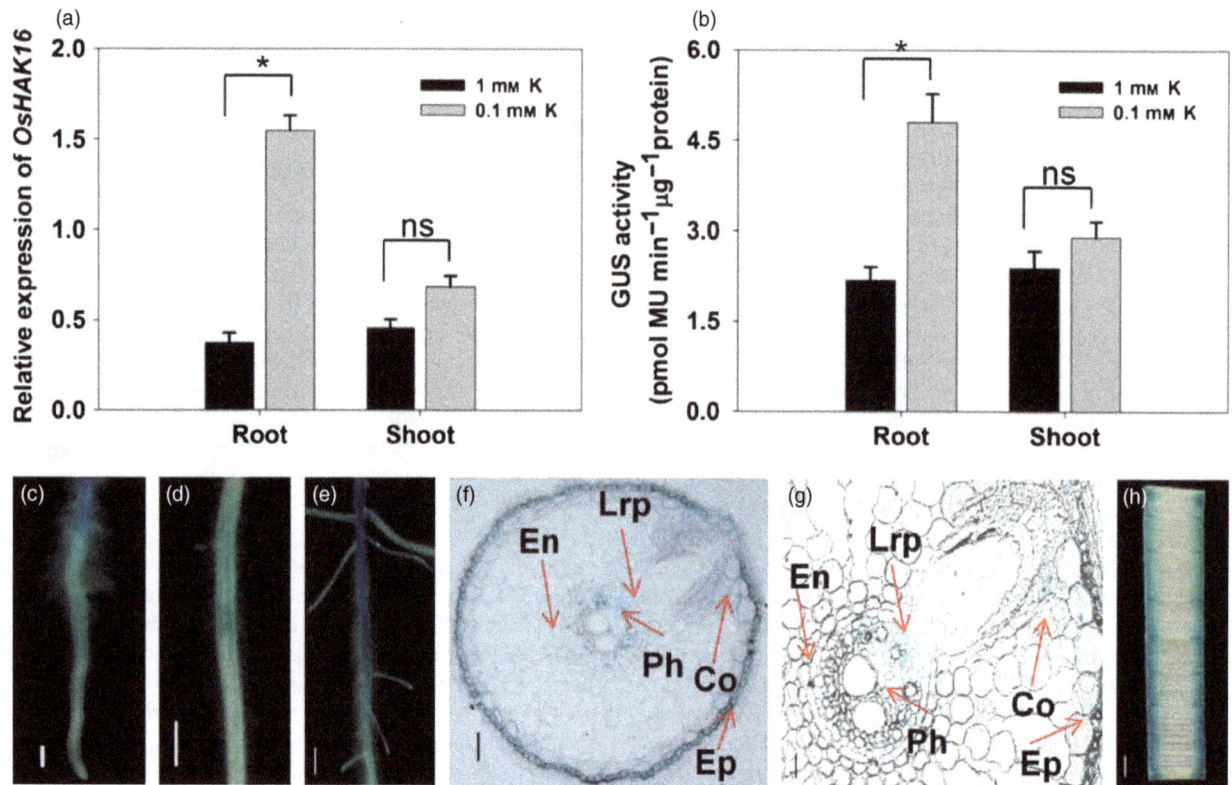

Figure 1 Effect of K deficiency on the expression of *OsHAK16* in wild type and GUS activity in *HAK16p:GUS* plants. (a), Expression of *OsHAK16* in response to K deficiency in wild type. Rice seedlings were supplied with IRRI nutrient solution containing 1 mm K for 14 days and then transferred to 1 or 0.1 mm K for 3 days. Total RNA was extracted from root and shoot. *OsActin* was used as an internal control. (b) quantification of GUS activity. (c–h) GUS staining in *OsHAK16p:GUS* transgenic rice seedlings. Seedlings were cultured with 1 mm K for 3 weeks and then treated with 0.1 mm K for 7 days. (c), root tip; (d), root hair zone; (e), root lateral branching zone. (f) and (g), root cross sections. (h) leaf blade. GUS staining was detected in epidermis (Ep), cortex (Co), endodermis (En), phloem (Ph) and lateral root primordium (Lrp). (c–e and h), Bars = 1 mm. (f) Bar = 50 μm. (g) Bar = 20 μm. Error bars indicate SE ($n = 3$) of three biological replicates. Significant differences between 1 and 0.1 mm K supply are indicated with asterisks ($P < 0.05$, one-way ANOVA), and ns indicates nonsignificant differences at that level of significance.

condition (Figure 2b). In WT, *WOX11* transcript was not regulated by K status, but in *HAK16:WOX11* plants, K deficiency increased *WOX11* transcripts in the roots by two–threefold (Figure 2b). Expression of *WOX11* in the shoots was only at marginal detectable level in both WT and the transgenic lines (Figure 2b).

Decreasing K supply from 1 to 0.1 mm resulted in smaller roots for WT plants, whereas the transgenic lines expressing *HAK16p: WOX11* maintained much larger root system than WT irrespective of K supply (Figure 2c,d). In addition, the shoot growth had no difference for WT and *HAK16p:WOX11* transgenic plants in the sterilized solid medium containing either 1 or 0.1 mm K (Figure S4). The root phenotypes were inheritable from T_0 (Figure S1a) to T_1 (Figure S1b–d) and T_2 generation (Figures 2c,d and S4).

To confirm that the changed phenotype was caused by ectopic expression of *WOX11* in the transgenic plants, we detected the location of one-copy T-DNA insertion in these three lines. The T-DNA containing *OsHAK16* promoter was inserted in the chromosomes 4, 3 and 7 in the genome of L1, L2 and L3, respectively (Figure S5). There was no putative gene in the insertion site for all three lines. This direct genetic evidence further supported the contribution of enhanced *WOX11* expression to the altered phenotypes of the transgenic rice in Figures 2 and 8 and S1.

Effects of *HAK16p:WOX11* expression on root growth and K uptake at seedling stage

Rice seeds of WT and *HAK16p:WOX11* transgenic lines were germinated and grown in sterilized solid 0.4% Phytagel containing IRRI nutrient solution with different levels of K (0.1, 0.5 and 1 mm K) for 14 days. In comparison with WT, the transgenic lines produced a larger root system with more root biomass, number of adventitious roots and total root length (Figure 3a–c). As K concentration decreased from 1 to 0.1 mm in the medium, the root growth was decreased in WT, but increased in the transgenic lines which resulted in largest difference at 0.1 mm K level between them (Figure 3a–c). Remarkably, expression of *HAK16p: WOX11* not only increased root growth, but also the root activity as indicated by increase in the active absorbing surface percentage (Figure 3d).

Expression of *HAK16p:WOX11* increased not only root growth, but also K concentration in both roots and shoots (Figure 4a,b). Increasing K supply in the medium improved total K accumulation in both roots and shoots of WT, but decreased K accumulation in the transgenic lines (Figure 4c,d). Nevertheless, the transgenic lines acquired much more K from the medium than WT, particularly in the roots (Figure 4c). Expression of *HAK16p: WOX11* increased total K accumulation in the shoots was about

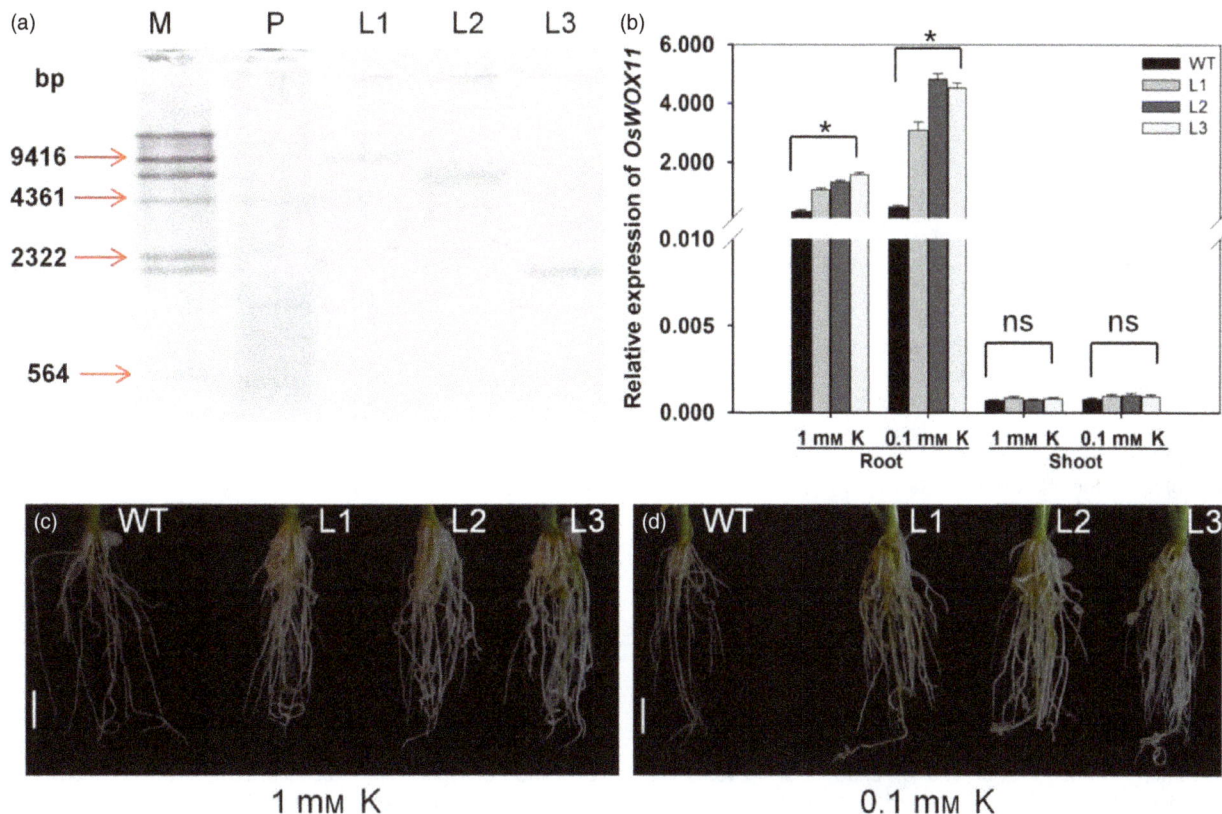

Figure 2 Characterization of *HAK16p:WOX11* transgenic lines. (a) Southern blot analysis of the transgene copy number. Genomic DNA of T1 generation of *HAK16p:WOX11* transgenic lines (L1, L2, L3) was digested with the restriction enzyme *Hind*III and *Bam*HI. The hygromycin probe was used for hybridization. M, marker; P, positive control. (b) real-time quantitative RT-PCR analysis of endogenous *OsWOX11* in *HAK16p:WOX11* transgenic lines and wild type (WT). RNA was extracted from both roots and shoots. (c and d) root phenotypes of *HAK16p:WOX11* compared with WT supplied with different concentration of K. Rice seedlings were supplied with IRRI nutrient solution containing 1 mm K (b, c) and 0.1 mm K (b, d) for 14 days after germination. Values are shown with mean ± SE (*n* = 3). Significant differences between WT and transgenic lines are indicated with asterisks (*P* < 0.05, one-way ANOVA), and ns indicates nonsignificant differences at that level of significance. Bars = 2 cm.

80% at 0.1 mm K supply level, while such increase was much less at high K (1 mm) condition (Figure 4d).

To learn whether the improved root activity (Figure 3d) by the presence of *HAK16p:WOX11* affects the K uptake rate of unit weight roots, we quantified the specific absorption rate of K (SARK) by calculating the net K uptake rate during the days 7–14 of growth supplied with 0.1, 0.5 and 1 mm K (Figure 4e), respectively, and transcriptional expression of one K channel (*OsAKT1*) and two K transporter genes (*OsHAK1* and *OsHAK5*) in the roots under both 1 and 0.1 mm K supply conditions (Figure S6a–c). The SARK was slightly lower in the transgenic lines compared with WT at K-sufficient condition (Figure 4e). As K concentration decreased from 1 to 0.1 mm in the medium, the SARK was decreased in WT, but slightly increased in the transgenic lines (Figure 4e). In addition, the expression of *OsAKT1* was higher in the transgenic lines than WT irrespective K supply levels (Figure S6a), while expression of *OsHAK1* and *OsHAK5* was up-regulated by *HAK16p:WOX11* only at K-deficient roots (Figure S6b,c).

Effects of *HAK16p:WOX11* expression on auxin- and cytokinin-responsive genes and auxin efflux carrier genes

It has been reported that constitutively overexpression or RNA interference knockdown of *WOX11* gene altered expressions of some auxin- and cytokinin-responsive genes (Zhao *et al.*, 2009). To confirm whether the phenotype derived from the expression of *HAK16p:WOX11* construct was involved hormone signalling, we quantified the change of the genes possibly involved in auxin and cytokinin signalling and transportation in comparison with WT at both K-sufficient (1 mm K, +K) and K-deficient (0.1 mm K, −K) conditions. All six putative cytokinin type-A *RR* genes (Du *et al.*, 2007; To and Kieber, 2008) in the roots were down-regulated by enhanced expression of *WOX11* (Figure 5a,b). *RR2* and *RR8* were much more sensitive than *RR*1, *RR*4, *RR*6 and *RR*7 to the expression of *WOX11* gene. Notably, expression of *RR8* gene in the roots of *HAK16:WOX11* transgenic line was suppressed greatly at low K supply condition (Figure 5b).

All four putative auxin-responsive *Aux/IAA* genes, *IAA11/13/23/31*, were up-regulated in the transgenic lines in comparison with their expression in WT (Figure 5c,d). Among them were *IAA23* and *IAA11* the most and least sensitive to the K-deficiency-enhanced expression of *WOX11* (Figure 5c,d). Interestingly, among four putative auxin efflux carriers *PIN* genes, *PIN2* and *PIN10a* were up-regulated, while *PIN5a* and *PIN9* were extremely sensitive and down-regulated by *WOX11* expression (Figure 5e,f). The alteration of all these genes expression caused by *HAK16p:WOX11* was several-fold larger at 0.1 mm K than at 1.0 mm K supply condition (Figure 5a–f), supporting the role of *HAK16* promoter in driving *WOX11*

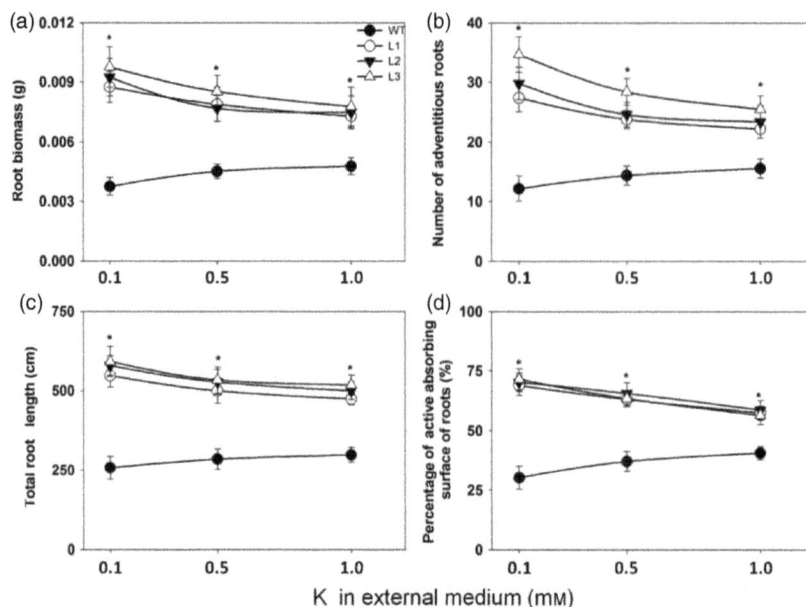

Figure 3 Root growth of *HAK16p:WOX11* transgenic lines compared with WT at young seedling stage. Rice seeds (WT and *HAK16p: WOX11* transgenic lines) were germinated and grown in sterilized solid medium (IRRI nutrient solution with 0.4% Phytagel) containing 0.1, 0.5 and 1 mM K for 14 days. Root architecture of the young seedlings was analysed. (a) root biomass. (b) number of adventitious roots. (c) total root length. (d) proportion of active root surface area. The values are means ± SE of six replicates. Significant differences between WT and transgenic lines are indicated with asterisks ($P < 0.05$, one-way ANOVA).

expression and the role of *WOX11* in regulating expression of the tested genes in rice.

Based on the previous reports on the involvement of *WOX11* in cytokinin- and auxin-coordinated crown root initiation and development (Coudert *et al.*, 2010; Zhao *et al.*, 2009), we modified the hypothetical pathway of *OsWOX11* in regulating the crown root growth by showing the either up-regulated or down-regulated genes belonging to *RR*, *IAA* and *PIN* families, respectively, detected in this study (Figure 5g).

Effects of *HAK16p:WOX11* expression on plant growth, K uptake, soluble sugar content and distribution under solution culture condition

To evaluate whether the transgenic lines could continuously grow better than WT at relatively later stage, *HAK16p:WOX11* transgenic and WT plants were cultured in solution for 9 weeks. Two-week-old seedlings were grown first in IRRI solution for 4 weeks and then transferred to the nutrient solution containing 1 or 0.1 mM K for 3 weeks. Total root biomass of the transgenic lines on the average was 48% at 1 mM K level and 150% at 0.1 mM K level more than that of WT, respectively (Figure 6a). In contrast, there was no significant difference of the shoot biomass between WT and the transgenic lines under 1 mM K condition (Figure 6b). However, expression of *HAK16p:WOX11* significantly in rice increased shoot biomass in 0.1 mM K solution (Figure 6b), demonstrating that K became the limiting factor for the growth of aboveground parts under this K supply level. In addition, K concentrations in the roots, culms and sheaths, mature blades and developmental blades were all significantly higher in the transgenic lines than in WT, particularly at the low K condition (Figure 6c,d). The total K accumulated in the roots and shoots of three transgenic lines were 140% and 25% higher than that of WT at 1 mM K solution (Figure 6e), while these values were increased to 350% and 78% when grown at 0.1 mM K solution (Figure 6f). The differences of both the root and shoot growth and K acquisition between the transgenic lines and WT were further enlarged by decreasing K supply from 0.1 to 0.02 mM (Figure S7), demonstrating that the ectopic expression of *HAK16p:WOX11* could play more pronounced role in improving rice tolerance to K stress under more severe deficiency condition.

In plants, photosynthetically produced sugar is moved from source leaves to heterotrophic sink organs including roots for supporting the growth and development of plants and the carbon storage process (Kühn and Grof, 2010). K starvation suppressed the export of sugar from source leaves; as a result, the growth of rice plants was impaired (Cai *et al.*, 2012). To learn whether the increased K concentration in the different parts of transgenic lines was beneficial to sugar transporting efficiently from source to sink organs, we measured the soluble sugar concentrations in different parts of the plants including mature leaf blade (sugar source), culm and sheath (transport phloem network), and developmental (young) leaf blade and roots (sugar sink). In comparison with normal K (1 mM) supply, K limitation (0.1 mM) decreased soluble sugar concentration in all parts of WT, whereas the soluble sugar concentration of root, culm and sheath, and developing leaf blade in the low K supplied transgenic lines were similar or even higher than those in normal K supplied WT (comparison of Figure 7a,b).

Interestingly, expression of *HAK16p:WOX11* increased soluble sugar concentration much more pronounced in the two sink organs, roots and developmental leaf blades than in source organ (mature leaf blades), particularly under low K condition (Figure 7a,b). The results suggested enhanced distribution of soluble sugar from the source leaf blades to the sink organs by elevated expression of *HAK16p:WOX11* due to improved K status at low K supply level (Figure 6c,d).

It was reported that sugar transporters play a pivotal role in the membrane transport of sugars and their distribution throughout the plant to ensure that all sink tissues receive an adequate supply of sugars for growth and development (Williams *et al.*, 2000). Sucrose is loaded as a major carbon photo-assimilate into the phloem apoplastically by sucrose transporters (SUTs) and unloaded in sink tissues, where it is converted by cell wall invertases (CINs) into hexose that is imported into cells by monosaccharide transporters (MSTs) (Lim *et al.*, 2006). To further elucidate whether the enhanced root growth and K acquisition affected the sugar transportation, the transcript level of *SUT1* and *SUT4* in the source leaves (mature leaf blades) and the expression of *MSTs* in the sink roots of the transgenic and WT plants were quantitatively detected at both the K-deficient and K-sufficient

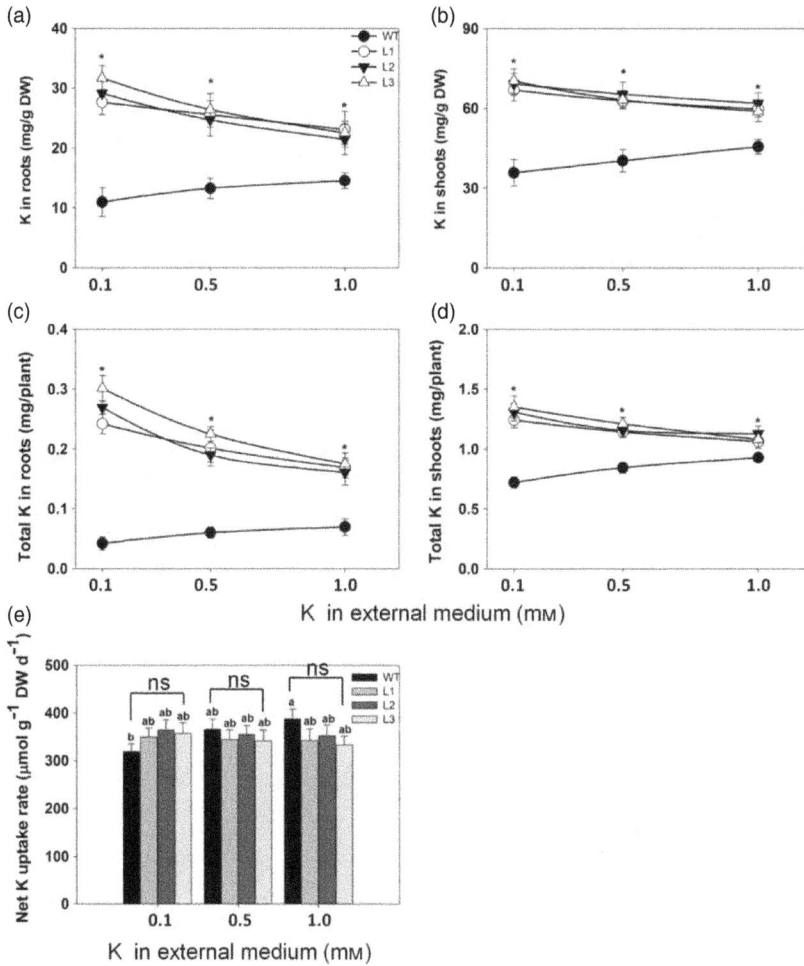

Figure 4 K accumulation and net K uptake rates in WT and *HAK16p:WOX11* transgenic lines at young seedling stage. Rice seedlings were cultured as described in Figure 3. (a) K concentration in roots. (b) K concentration in shoots. (c) total K in roots. (d) total K in shoots. The K concentration multiplied by dry weight of roots or shoots represents the total K in roots or shoots. Data are means of 6 replications ± SE. Significant differences between WT and transgenic lines are indicated with asterisks ($P < 0.05$, one-way ANOVA). (e) net K uptake rates during days 7–14 of growth in sterilized solid medium containing 0.1, 0.5 and 1 mM K. Values are shown with mean ± SE ($n = 6$). Means with the same letter are not significantly different at $P < 0.05$, according to two-way ANOVA followed by Tukey's test and ns indicates nonsignificant differences between WT and transgenic lines at $P < 0.05$, one-way ANOVA. DW, dry weight.

conditions. In comparison with WT, the transgenic lines showed increase in *SUT1* and *SUT4* expression in the sugar source blades by 39% and 47% on the average at 1 mM K level, respectively (Figure 7c). K deficiency suppressed expression of both *SUT1* and *SUT4* in the source leaf blades of WT (comparison of Figure 7c,d), supporting their potential role in exporting soluble sugar to sink organs, while such suppression was slight in the source leaf blades of the transgenic lines resulting in the threefold–fourfold increase in their expression (Figure 7d). All of the four tested *MST* genes (*MST1*, *MST3*, *MST4* and *MST5*) showed higher expression in the roots at the transgenic lines than at WT, while K deficiency did not significantly change their expression abundance (Figure 7e,f). These data indicated that the *HAK16p:WOX11* transgenic lines had the evident enhancement of soluble sugar allocation from source blades to sink roots, particularly by stimulating sugar export (loading to phloem) from source leaves at K-starved condition.

Effects of *HAK16p:WOX11* expression on K uptake, growth and grain yield under soil culture condition

We further evaluated the tolerance of *HAK16p:WOX11* transgenic rice in both T_1 and T_2 generations to low K stress under soil culture condition. We first planted WT together with the null segregants and transgenic lines of T_1 generation in paddy field with no K-fertilized clay-loam soil (Figure S3a). The transgenic lines showed significantly lower height, but more effective tiller number and shoot biomass (Figure S3b–d), their grain yield per

plant on the average was 32% higher than that of WT and the null segregants (Figure S3e).

The transgenic lines of T_2 generation and WT were then grown in the pots filled with the paddy soil containing 70 mg/kg exchangeable K as moderate K-deficient treatment and adding 0.2 g fertilizer K/kg soil as K-sufficient treatment. As shown in Figure 8(a,b) and Table 1, similar to its T_1 generation, ectopic expression of *HAK16p:WOX11* in rice increased effective tiller numbers and decreased height very significantly. Such change was more significant at no K-fertilized soil. In comparison with WT, the transgenic lines showed increase in total shoot biomass, number of grains per plant and grain yield only by 5%, 8% and 7% grown in the K-fertilized soil, while they had 16%, 24% and 24% increase in the none K-fertilized soil (Table 1).

As expected, the transgenic lines of *HAK16p:WOX11* produced much more roots and larger shoot biomass than WT (Figure 8c–h). Similar to the effect under agar and solution culture condition (Figures 2c,d; 3a; and 6a), K deficiency expanded root system of the transgenic lines in contrast to that observed in WT (Figure 8e–f). The shoot biomass of transgenic lines remained the same as WT in the K-fertilized soil, while it was significantly higher than WT in the non-K-fertilized soil (Figure 8h).

Ectopic expression of *HAK16p:WOX11* increased K concentration significantly only in both roots and culm in the K-fertilized soil (Figures 9a and S8a), while it increased K concentration in all the vegetative organs including roots, culm, leaf sheath and blade in

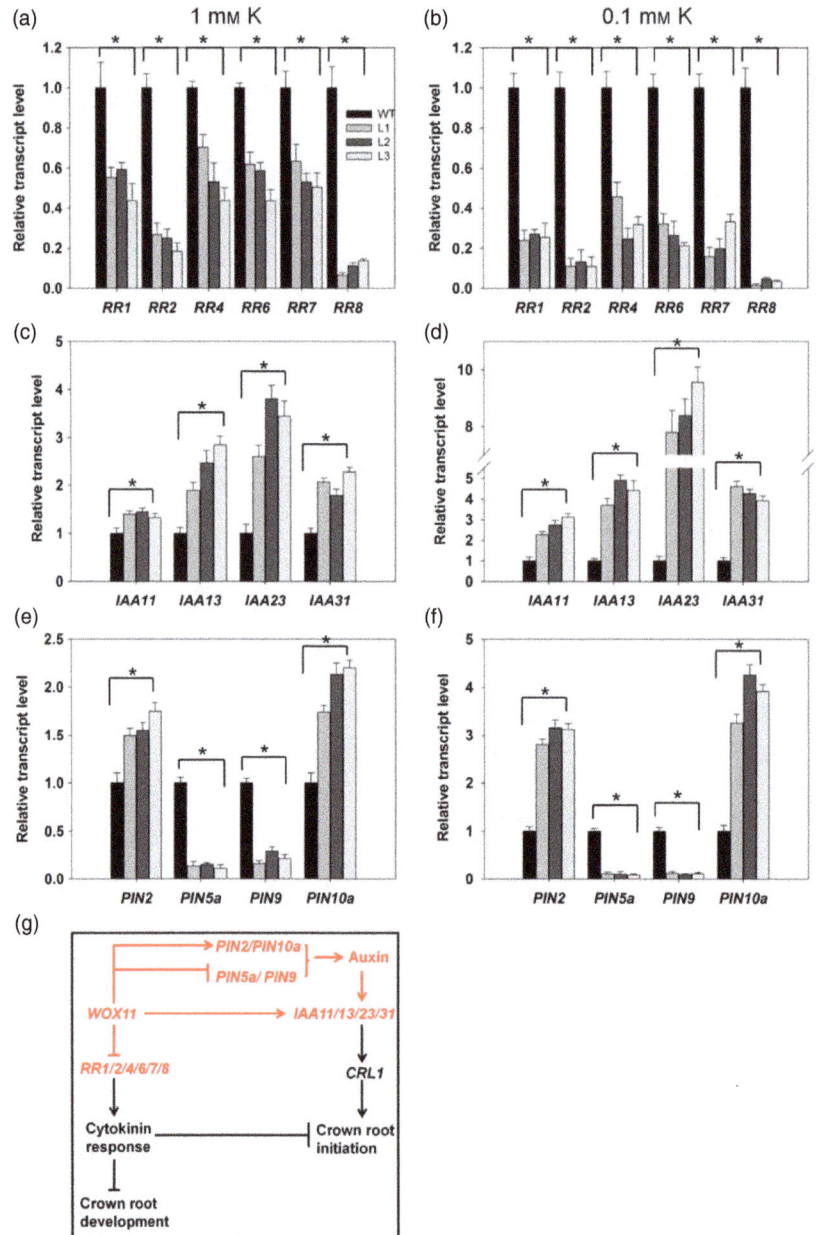

Figure 5 Expression level of hormone-responsive and auxin efflux carrier genes in WT and transgenic lines. Rice seeds (WT and *HAK16p: WOX11* transgenic lines) were germinated and grown in sterilized solid medium (IRRI nutrient solution with 0.4% Phytagel) containing 1 mM K (a, c, e) and 0.1 mM K (b, d, f) for 14 days. RNA was extracted from roots, and qRT-PCR was used to detect the transcript level of all genes. A and B, transcript expression of cytokinin type-A *RR* genes (*RR1, RR2, RR4, RR6, RR7* and *RR8*). (c and d) transcript expression of auxin-responsive Aux/IAA genes (*IAA11, IAA13, IAA23* and *IAA31*). (e and f) transcript expression of auxin efflux carrier genes (PIN2, *PIN5a, PIN9* and *PIN10a*). The PCR signals were normalized with *OsActin* transcripts. Transcript levels from WT were set at 1. Data are means ± SE of three biological replicates. Significant differences between WT and transgenic lines are indicated with asterisks (*P* < 0.05, one-way ANOVA). (g) hypothetical gene regulatory network controlling crown root initiation and development in rice. Arrows represent the positive regulatory action of one element of the network on another one. A line ending with a trait represents the negative regulatory action of one element of the network on another one.

the low K soil (Figures 9b and S8b). The transgenic lines accumulated total K per plant were about 30% more grown in the K-sufficient soil (Figures 9c and S8c) and 72%–80% more in the moderate K-deficient soil (Figures 9d and S8d).

The expression of *HAK16p:WOX11* increased soluble sugar concentration in the roots as a sink organ at both tillering and mature stage, while its effect on the sugar concentration in the shoot was tissue and growth stage dependent (Figures 9e,f and S8e,f). In comparison with WT, the transgenic lines contained higher soluble sugar in the developmental blades (sink organ) and the low-K-supplied culm and sheath at tillering stage (Figure S8e, f), while they had lower soluble sugar in the culms, and low-K-supplied leaf blades (Figure 9e,f).

To determine whether other elements changed in face of the enhancement of root growth by the expression of *HAK16p: WOX11*, we also measured N and P acquisition in the rice plants at both tillering and mature stages in different K supply soil (Figures S9 and S10). Unlike remarkable improvement of K

concentration in different organs by the gene transformation, there were only slight increases in total N and P concentration in the roots under both K-sufficient and K-deficient conditions, and in the culm and sheath under K-deficient condition, there was no significant change of total N and P concentration in the other organs (Figures S9 and S10). Total accumulation of both N and P was increased in the transgenic lines (Figures S9 and S10), indicating that expression of *WOX11* under the control of the *HAK16* promoter enhances the accumulation of other elements probably as a consequence of enhanced root growth.

Discussion

Preventing root growth retardation caused by K deficiency could enhance the adaptation of plant to K limitation

The strong ability of the roots to take up K from soil is a prerequisite for plant survival under low K condition (Rengel and

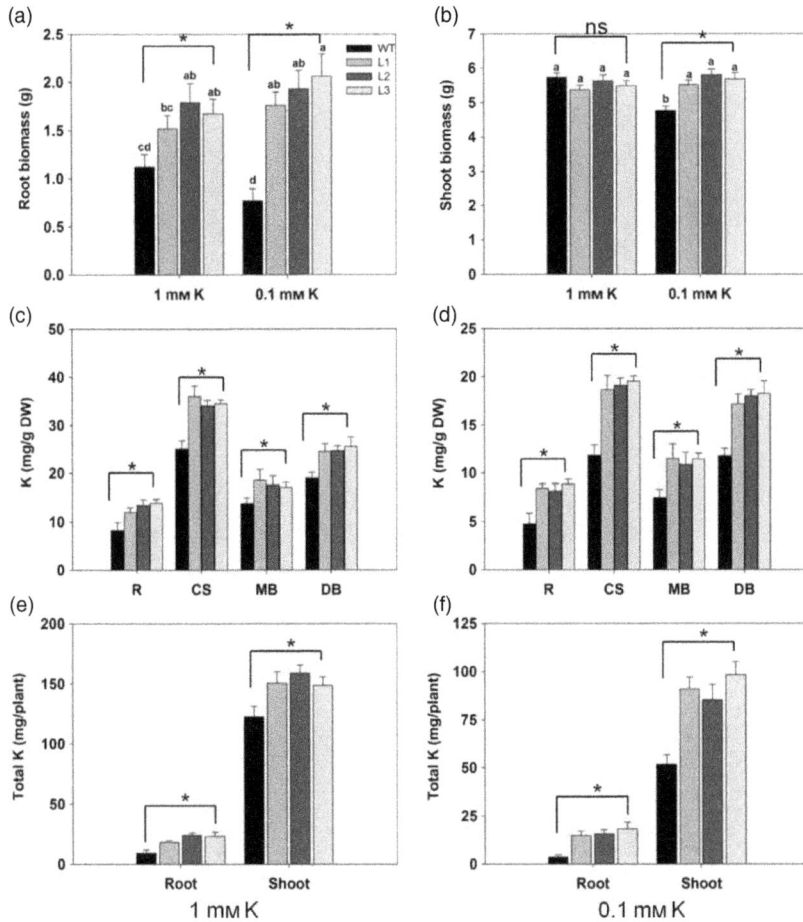

Figure 6 Root and shoot growth, K accumulation in WT and *HAK16p:WOX11* transgenic rice at tillering stage. Two-week-old rice seedlings (WT and *HAK16p:WOX11* transgenic lines) were grown in IRRI solution for 4 weeks and then transferred to nutrient solution containing 1 or 0.1 mm K for 3 weeks. (a) root biomass under indicated K conditions. (b) shoot biomass under indicated K conditions. (c and d) K concentrations in roots (R), culms and leaf sheaths (CS), mature blades (MB) and developmental blades (DB) under 1 mm K (c) and 0.1 mm K (d) conditions. e and f, total K in roots and shoots under 1 mm K (e) and 0.1 mm K (f) conditions. The values are mean ± SE of five replicates. (a and b) means with the same letter are not significantly different at $P < 0.05$, according to two-way ANOVA followed by Tukey's test. a–f, Significant differences between WT and transgenic lines are indicated with asterisks ($P < 0.05$, one-way ANOVA). DW, dry weight.

Damon, 2008; Shin, 2014). The most efficient strategy to improve root acquisition of the nutrients is directed at increasing root system to obtain a large root surface area that is in contact with the soil solution. Enhancing root growth for larger root system and increasing root-to-shoot ratio is a common response of plant roots to the deficiency of the nutrients in soil, including nitrogen, phosphorus, magnesium (see the review of Hermans *et al.*, 2006; references therein) and iron (Santi and Schmidt, 2009). However, unlike deficiency of other major nutrients, K deficiency commonly resulted in root growth retardation and smaller root-to-shoot biomass ratio (Cai *et al.*, 2012; Drew, 1975; Hermans *et al.*, 2006) although in some cases, the root-to-shoot ratio at different K levels remains constant or slightly increase (Moriconi *et al.*, 2012; White, 1993). K in plants plays multiple roles in maintaining both root and shoot growth including the regulation of cell cycle (Sano *et al.*, 2007) and execution of cell death programmes (Peters and Chin, 2007). As K is the most abundant cation in plant cells for maintaining the balance with anions and osmotic potential, lack of K in the root may prevent the cell expansion due to decrease in the turgor pressure for root growth (Dolan and Davies, 2004; Jordan-Meille and Pellerin, 2008; Mengel and Arneke, 1982), as well as by charge imbalance (Walker *et al.*, 1998).

In this study, we forced rice roots to expand the growth by ectopic expression of *OsWOX11* gene driven by the promoter of K-deficient enhanced K transporter gene *OsHAK16* in roots. Using different culture mediums including solid agar, nutrient solution and soil, we detected that the expression of *HAK16p:*

WOX11 could increase total root biomass and active absorbing surface area ratio by about 120%–130% on the average under low K supply condition (Figures 3a,d; 6a; and 8g). Limitation of K supply increased K upward transport from root to shoot and aggregated K-deficiency-induced root retardation (Chérel *et al.*, 2014; Gruber *et al.*, 2013), which was consistency with our observation for WT in this study (Figures 3, 4, 6, 8, 9a–d, and S8a–d). It is notable that K concentration in the low-K-supplied roots of transgenic lines was maintained higher than that of high-K-supplied WT roots under both solution culture (Figure 6c,d) and soil culture conditions (Figures 9a,b and S8a,b). These data suggest that maintaining high K concentration in the roots is prerequisite for enlarging root system through the transgenic approach.

In comparison with WT, the transgenic lines expressing *HAK16p:WOX11* showed increase in total K acquisition by 95%–110% and 25%–35% at 0.1 and 1 mm K supply level, respectively, in agar and solution culture mediums (Figures 4c,d and 6e,f). As no significant difference was observed between WT and the *HAK16p:WOX11* transgenic lines irrespective of K supply levels for the root-specific absorption rate of K (SARK) (Figure 4e), such improvement of total K acquisition was mainly due to enlarged root growth and activity (Figures 2, 3, 6 and 8; S1,S2, S4,S7). It is interesting that expression of *OsAKT1*, *OsHAK1* and *OsHAK5* was strongly up-regulated by the expression of *WOX11* in the transgenic lines (Figure S6). The data suggest that WOX11 might function in regulating the genes globally involved in K capture.

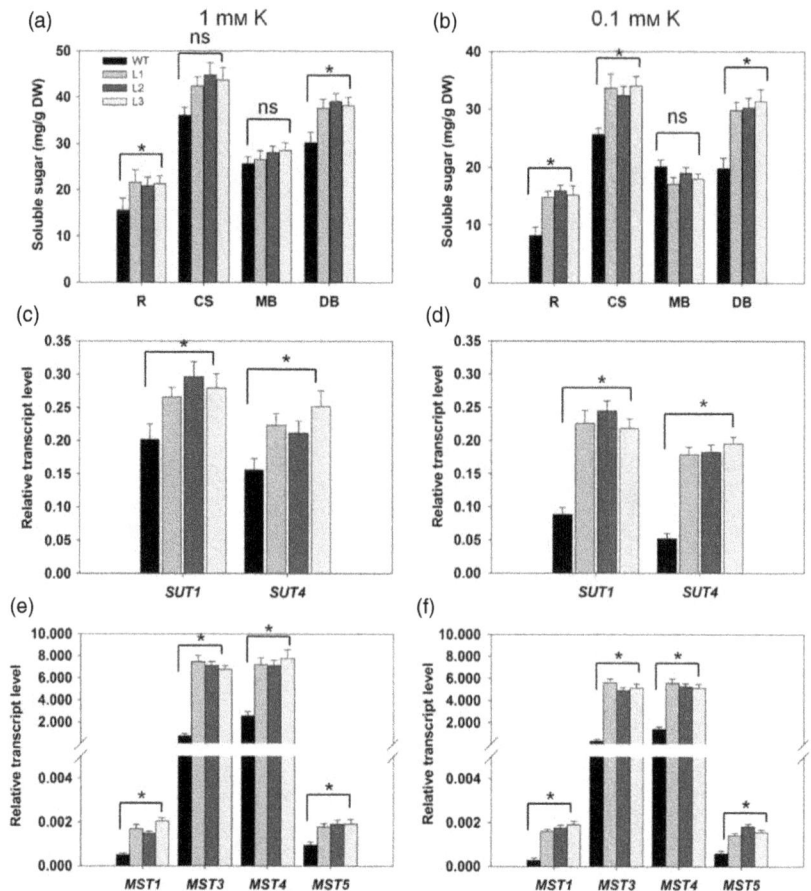

Figure 7 Soluble sugar concentrations and transcripts of sugar transport related genes (*SUTs* and *MSTs*) in WT and *HAK16p:WOX11* transgenic rice at tillering stage. Rice plants were grown under the conditions as described in the legends of Figure 6. (a and b) Soluble sugar concentrations in roots (R), culms and leaf sheaths (CS), mature blades (MB) and developmental blades (DB) of WT and transgenic lines under 1 mM K (a) and 0.1 mM K (b) conditions. Values are shown with mean ± SE (n = 5). Expression level of *SUT* genes (c and d) and *MST* genes (e and f) was tested by qRT-PCR in WT and *HAK16p:WOX11* transgenic lines under 1.0 mM K (c and e) or 0.1 mM K (d and f) conditions. Total RNA was isolated from the source blades and roots. *OsActin* was used as a control. Data represent mean ± SE of three replicates. Significant differences between WT and transgenic lines are indicated with asterisks ($P < 0.05$, one-way ANOVA), and ns indicates nonsignificant differences at that level of significance. DW, dry weight.

Table 1 Comparison of agronomic traits and yields between WT and *HAK16p:WOX11* transgenic rice in different K supply soil

Treatment	Line	Plant height (cm)	Effective tiller number per plant	Shoot biomass (g/plant)	1000-grain weight (g)	Number of grains per plant	Grain yield (g/plant)
+K	WT	73.3 ± 1.6 a	24.3 ± 2.5 b	23.6 ± 1.1 a	19.1 ± 0.3 a	941.3 ± 3.9 b	17.7 ± 0.6 a
	L1	53.6 ± 3.1 c	28.3 ± 1.5 a	24.4 ± 2.2 a	18.6 ± 0.2 a	1002.2 ± 7.4 a	18.6 ± 1.1 a
	L2	52.4 ± 2.4 c	29.0 ± 1.7 a	25.2 ± 1.6 a	18.8 ± 0.2 a	1025.6 ± 4.6 a	19.4 ± 0.8 a
	L3	50.8 ± 1.7 c	30.3 ± 2.1 a	24.9 ± 1.4 a	18.5 ± 0.3 a	1021.5 ± 5.5 a	19.1 ± 0.7 a
*Difference		−21.1 (−28.7%)	4.9 (20.2%)	1.2 (5.2%)	−0.5 (−2.5%)	75.1 (8.0%)	1.3 (7.5%)
−K	WT	66.5 ± 1.9 b	19.7 ± 1.2 c	20.2 ± 1.2 b	18.5 ± 0.4 a	755.9 ± 4.8 c	13.9 ± 1.2 b
	L1	52.1 ± 1.5 c	26.7 ± 2.3 ab	23.4 ± 1.8 a	18.4 ± 0.3 a	941.7 ± 6.1 b	17.2 ± 0.9 a
	L2	50.6 ± 2.5 c	26.3 ± 1.5 ab	23.1 ± 1.3 a	18.5 ± 0.2 a	925.4 ± 7.2 b	16.9 ± 0.7 a
	L3	50.5 ± 2.2 c	27.7 ± 2.6a b	23.9 ± 1.9 a	18.3 ± 0.3 a	947.8 ± 6.7 b	17.6 ± 1.3 a
*Difference		−15.4 (−23.3%)	7.2 (36.5%)	3.3 (16.3%)	−0.1 (−0.5%)	182.4 (24.1%)	3.3 (23.7%)

Numbers are presented as mean ± SE. The number of observations in each mean is 5.

*Difference between WT and three transgenic lines on the average. Parentheses data are the decreasing or increasing percentage relative to WT by enhanced expression of *WOX11*. Means in the same column followed by the same letter are not significantly different at $P < 0.05$, according to two-way ANOVA followed by Tukey's test.

The transgenic lines grown in the moderate K-deficient soil showed significant higher K concentration in culm, leaf sheath and blade (Figure 9b), and their total K accumulation per plant at mature stage was 72% more than WT (Figure 9d). It is noticeable that the ectopic expression of *HAK16p:WOX11* increased shoot biomass by 16%–24% and grain yield by 24%–32% for the rice grown in the moderate K-deficient soil (Figure S3d,e; Table 1), while it did not significantly improve the accumulation of shoot and grain weight when K fertilizer was applied in the soil (Table 1). In the solution culture, we have shown that decreasing K supply from 0.1 to 0.02 mM enlarged the difference of rice growth and K acquisition between the transgenic lines and WT (comparison of Figures 6 and S7). As the intensity of *WOX11* gene expression in the *HAK16p:WOX11* transgenic lines was related to the degree of K deficiency, it is worth to be tested whether the transgenic plants expressing *HAK16p:WOX11* could generate much larger root system and higher K acquisition efficiency and gain highest

Figure 8 Phenotypes of WT and *HAK16p: WOX11* transgenic rice at mature stage in different K supply soil. WT and *HAK16:WOX11* transgenic rice plants were grown in pot with 200 mg fertilizer K/kg (+K) soil and 0 mg fertilizer K/kg (−K) soil until harvest. (a) phenotype of plants in +K soil. (b) phenotype of plants in −K soil. (c-h) root and shoot growth at active tillering stage. (c–f) root growth performance in −K soil. Bars = 10 cm. (g) root biomass grown in the indicated soil. (h) shoot biomass grown in the indicated soil. The values are mean ± SE of five replicates. Means with the same letter are not significantly different at $P < 0.05$, according to two-way ANOVA followed by Tukey's test. Significant differences between WT and transgenic lines are indicated with asterisks ($P < 0.05$, one-way ANOVA). DW, dry weight.

increase in grain yield if they were grown in less available K soil for the rice plant.

Enhanced ectopic expression of *WOX11* driven by OsHAK16 promoter increased soluble sugar allocation from source to sink organs to maintain large root system

The nutritional status of plants may considerably change the partitioning of dry matter and carbohydrates between shoots and roots (Cakmak *et al.*, 1994b). In K-deficient plants, high shoot/root dry weight ratios are accompanied by low carbohydrate allocation to roots (Cakmak *et al.*, 1994a). Increased K supply stimulates export of carbohydrates from source leaves by affecting various steps of sucrose transport, particularly loading and utilization (Cakmak *et al.*, 1994b; Conti and Geiger, 1982).

In the present study, soluble sugar concentration was significantly higher in the sink organs including roots, culm (for sugar transport) and developing leaf blade of the transgenic lines compared with that of WT under both hydroponic and soil culture condition (Figures 7 a,b and 8 e,f), while it was higher in roots but lower in culm and leaf blade grown in the low K soil at mature stage (Figure 9f). One reason for this is that phloem transport of photosynthates from source organs to sink organs requires adequate K (Cakmak *et al.*, 1994b; Deeken *et al.*, 2002; Pilot *et al.*, 2003). We observed that K concentrations were significantly higher in all the vegetative organs of transgenic plants than WT under low K supply conditions (Figures 6d; 9b; and S8b). The enlarged effect of K on sugar transport is attributed to the stimulation of sink activity (Conti and Geiger, 1982), and improved ATP status of plants (Mengel and Viro, 1974) and efflux of sugars into the apoplast from source leaves (Huber and Moreland, 1981), as well as to the osmotic pressure in the sieve tubes and thus the flow rate of photosynthates into sink organs (Hayaishi and Chino, 1990). Another reason for this may be that enhancing sugar translocation to the root driven by an increased root demand. Root biomass (sink capacity) and growth rate (sink activity) control the rate at which photosynthetic products move from leaves (Venkateswarlu and Visperas, 1987). The transgenic rice plants transformed with *HAK16p:WOX11* construct showed larger root system (Figures 2 c,d; 3a; 6a; and 8c–g), which might accelerate sugar translocation rate. The third reason for this may

be that the increased expression of sugar transporters. Sugar transport proteins (SUT) play a crucial role in the cell-to-cell and long-distance distribution of sugars throughout the plant (Ayre, 2011; Kühn and Grof, 2010; Lim *et al.*, 2006; Williams *et al.*, 2000). In rice, *OsSUT1* is involved in phloem loading of sucrose retrieved from the apoplasm for long-distance pathway of sucrose transport (Scofield *et al.*, 2007). We observed that the expression of both *SUT1* and *SUT4* genes in source leaf blades of the transgenic plants was much higher than that of WT, particularly under low K stress conditions (Figure 7c,d), which enhancing the export of carbohydrates from source leaves in the transgenic plants by increasing the expression of *SUT* genes.

Plants appear to have several monosaccharide and disaccharide transporters (MSTs and SUTs) to coordinate sugar transport in diverse tissues, at different developmental stages and under varying environmental conditions (Lim *et al.*, 2006; Williams *et al.*, 2000). Although most MSTs do not directly participate in long-distance transport, their indirect roles greatly impact carbon allocation and transport flux to the heterotrophic tissues of the plant (Slewinski, 2011). In rice, a number of MSTs have been reported, such as *OsMST4* (Wang *et al.*, 2007), *OsMST5* (Ngampanya *et al.*, 2003) and *OsMST6* (Wang *et al.*, 2008). We found that the transcript level of four *MSTs* in the sink roots of transgenic plants was evidently higher compared with WT irrespective of external K levels (Figure 7e,f). Therefore, we speculated that, in the transgenic plants due to the up-regulated expression of *MSTs*, the sugar unloading might be extremely enhanced, increasing the sugar concentration gradients between source and organs driving source-to-sink transport. It is possible that the *SUTs* and *MSTs* genes were indirectly regulated by enhanced expression of *WOX11*. One way that this might occur, for example, is that the enhanced *WOX11* expression led to stimulate root growth and was accompanied by increased K and sugar concentrations, which served as signals to regulate the expression of *SUTs* and *MSTs* genes. Consequently, K accumulation and sugar partitioning indicated by both hydroponic and pot experiments in transgenic plants were significantly increased during the whole growth stages, which resulted in vigorous growth and improvement in grain yield, particularly under K deficiency (Table 1).

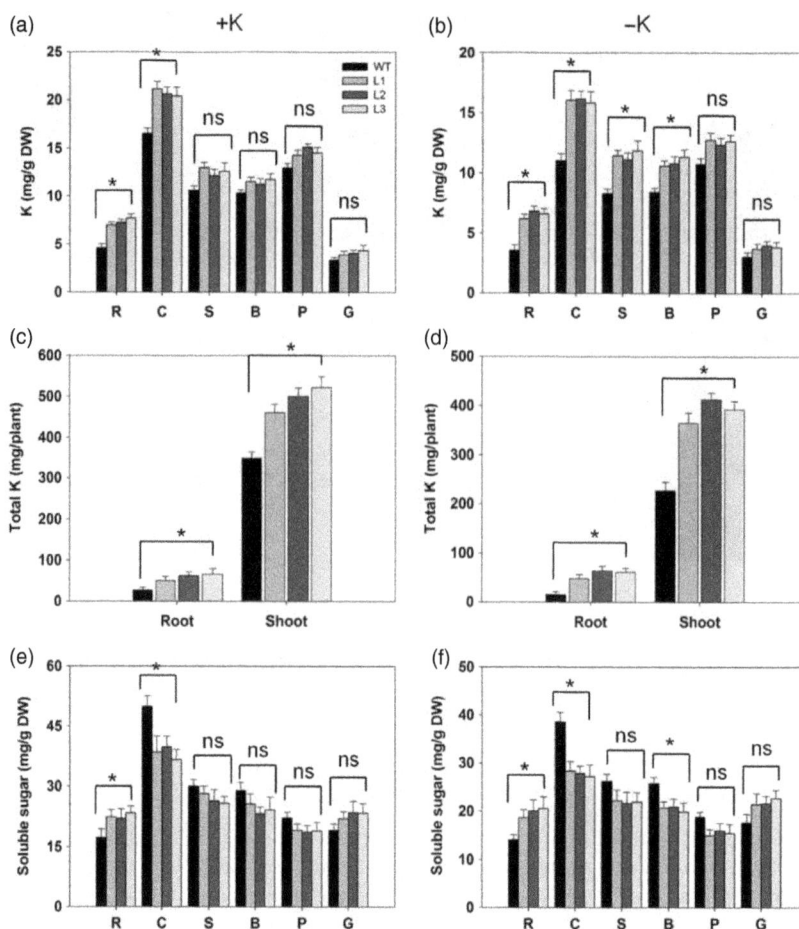

Figure 9 K accumulation and soluble sugar of WT and *HAK16p:WOX11* transgenic rice at mature stage in different K supply soil. Rice plants (WT and *HAK16:WOX11* transgenic rice) were grown under +K (a, c, e) and −K soil (b, d, f) conditions as described in the legend of Figure 8. (a and b) K concentrations in different organs. (c and d) total K in roots and shoots. (e and f) soluble sugar in different organs. Roots (R), clums (C), leaf sheaths (S), blades (B), panicle axes (P) and rice grains (G). Data represent mean ± SE of five replicates. Significant differences between WT and transgenic lines are indicated with asterisks ($P < 0.05$, one-way ANOVA), and ns indicates nonsignificant differences at that level of significance. DW, dry weight.

Enhanced ectopic expression of *WOX11* driven by *OsHAK16* promoter altered both signalling and transportation of cytokinin and auxin to promote root development

Plant root architecture is regulated by the integration of endogenous factors including phytohormones with environmental stimuli, for example, the availability of water and nutrients (Petricka *et al.*, 2012). Auxin and cytokinin are known to be two of the major players in regulating root development (Gao *et al.*, 2014; Růžička *et al.*, 2009). It has been reported that *OsWOX11* gene in rice is likely to function in a part of a homeostatic feedback loop of both cytokinin and auxin signalling to promote crown root development (Zhao *et al.*, 2009). In this work, we detected that expression of *HAK16p:WOX11* affected the expression of several cytokinin- and auxin-responsive genes in rice (Figure 5a–d).

Type-A *RR* genes function as redundant negative regulators of cytokinin signalling (To *et al.*, 2004). Zhao *et al.* (2009) have shown that *WOX11* directly represses one of the type-A *RR2* genes, which is expressed in crown root primordial, and suggested that activation of RR2 may have a negative effect on the expression of other type-A *RR* genes. In addition, Hirose *et al.* (2007) have shown that overexpression of *RR6* gene results in a poorly developed root system and shoot dwarf. We found that all six type-A *RR* genes including *RR2* and *RR6* detected in this study were largely suppressed by ectopic expression of *HAK16p: WOX11* in the roots, in particular under low K (0.1 mM) supply condition (Figure 5a,b). Interestingly, *RR8* gene was most sensi-

tive to be regulated by *WOX11*, and it was almost completely suppressed in the transgenic plants under K deficiency condition (Figure 5b). In contrast, all the four putative auxin-responsive *Aux/IAA* genes were up-regulated; in particular, *IAA23* was most responsive to enhanced expression of *WOX11* in the K-deficient transgenic plants (Figure 5d), which is consistency with the observation in the *WOX11* constitutively expressed plants (Zhao *et al.*, 2009).

The auxin spatial distribution largely depends on the polar localization of the PIN-FORMED (PIN) auxin efflux carrier family members, which plays a critical role in plant growth and development (Wang *et al.*, 2009). Some of these *PIN* genes in rice, such as *PIN1b* (Xu *et al.*, 2005), *PIN2* (Chen *et al.*, 2012) and *PIN3t* (*PIN10a*) (Zhang *et al.*, 2012), have been found to be involved in polar auxin transport, representing altered root growth performance. In addition, the expressions of *PIN2*, *PIN5b* and *PIN9* were altered in the *osgnom1* mutants, which exhibited reduced numbers of lateral roots and partial loss of gravitropism (Liu *et al.*, 2009). We observed that the expression of *PIN2*, *PIN5a*, *PIN9* and *PIN10a* genes was altered which also correlated with the expression of *WOX11* (Figure 5e,f). Interestingly, overexpression of *PIN2* (Chen *et al.*, 2012) or *PIN3t* (*PIN10a*) (Zhang *et al.*, 2012) in rice altered auxin distribution, resulting in more tillers and a larger tiller angle which was similar to that phenotype of *HAK16p:WOX11* transgenic rice plants generated in this study (Figure 8a,b, Table 1). These data suggest that the phenotype of enlarged root system, short height and more tillers by the expression of *HAK16:WOX11* in rice may partially link to the allocation of auxins from shoots to roots. Our data confirmed that

WOX11 was a positive regulator of auxin-responsive IAA genes and could activate or suppress some auxin efflux carrier genes represented with red pathways in Figure 5g.

Using K-deficiency-enhancing promoter in roots decreased the detrimental effects of ectopic *WOX11* overexpression on rice shoot development

A number of experiments have shown that strongly constitutive overexpression of a transcription factor (TF) may frequently lead to negative developmental phenotypes in transgenic plants (Kasuga *et al.*, 1999; Morran *et al.*, 2011). Stress-inducible promoters have been used to overcome these negative defects (Kasuga *et al.*, 1999; Kovalchuk *et al.*, 2013; Morran *et al.*, 2011). *WOX11* is expressed in rapid cell division regions of both root and shoot meristems, and constitutive overexpression of *WOX11* under the control of a maize ubiquitin promoter produced not only an extensive root system, but also large number of stem nodes-borne roots, resulting in poorly developed shoots and reduced grain yield (Zhao *et al.*, 2009).

To use *WOX11* gene properly in extending root development and meanwhile to overcome its defect in generating aberrant shoot phenotype, we searched the promoters of the genes which were expressed mainly in K-deficient roots. In this study, we detected that the one of *KT/HAK/KUP* genes putatively encoding high-affinity K transporters in rice (Bañuelos *et al.*, 2002; Horie *et al.*, 2011; Okada *et al.*, 2008; Yang *et al.*, 2014), *OsHAK16*, was expressed mainly in K-starved roots (Figure 1). We demonstrated that the *OsHAK16* promoter was strongly activated by K deficiency in rice roots, mainly at the root tips, cell division zones of primary and lateral roots (Figure 1a–c). Interestingly, *OsHAK16* expression was detected at the root stele vascular tissue and emerging lateral root primordium (Figure 1d,e), co-localized with distribution and transport route of auxin (Friml, 2003; Wang *et al.*, 2009). We found that ectopic expression of *WOX11* gene under the control of *OsHAK16* promoter did not generate stem node-borne crown roots under both K-deficient and K-sufficient conditions (Figures S3a and Figure 8 a,b), demonstrating that the basal activity or tissue specific localization of *OsHAK16* might be not strongly enough to drive *WOX11* expression in the rice shoot. The results also suggested that the promoters of other K-starvation-induced root genes could also be considered for control of WOX11 expression in improving rice tolerance to K deficiency.

Experimental procedures

Plant growth conditions

For hydroponic experiments, the protocol of seed sterilization and the basal IRRI nutrient solution composition for seedling growth were described previously (Li *et al.*, 2006). Four-week-old seedlings with uniform size and vigour were transplanted into holes in a lid placed over the top of pots (eight holes in a lid and one seedling per hole). Two wild-type plant and six transgenic plants (two seedlings per line) were grown in each pot, and every treatment contained three pots. All pots were filled with 5 L of a nutrient solution (1.25 mM NH_4NO_3, 1 mM $CaCl_2 \cdot 2H_2O$, 1 mM $MgSO_4 \cdot 7H_2O$, 0.5 mM Na_2SiO_3, 20 μM NaFeEDTA, 20 μM H_3BO_3, 9 μM $MnCl_2 \cdot 4H_2O$, 0.32 μM $CuSO_4 \cdot 5H_2O$, 0.77 μM $ZnSO_4 \cdot 7H_2O$ and 0.39 μM $Na_2MoO_4 \cdot 2H_2O$, pH 5.5), supplemented with 0.3 mM KH_2PO_4 plus 0.35 mM K_2SO_4 (K-sufficient solution; +K) or 0.2 mM NaH_2PO_4 plus 0.1 mM KH_2PO_4 (K-deficient solution; −K). The hydroponic experiments were conducted in a phytotron

with a 14-h-light (30 °C)/10-h-dark (22 °C) photoperiod and relative humidity at approximately 70%. The nutrient solutions were replaced every 2 days for all the treatments. The assay of transcriptional expression of the sugar transporter genes by qRT-PCR, K concentration and soluble sugar concentration were carried out for the plants in 3 weeks after the treatment. The soil experiment was performed with five replications for each treatment in a glasshouse. One WT or one transgenic rice plant was grown in each pot filled with 7.5 kg of air-dried loam soil containing 70 mg/kg (−K) exchangeable K extracted by 1 M neutral ammonium acetate or adding 200 mg/kg K (KCl) to the soil (+K).

Assay of qRT-PCR

Entire root tissues of each WT and the transgenic plant at 2–3 weeks after the treatment were used for the isolation of total RNA. Triplicate quantitative assays for each of triple biological replicates were performed with SYBR Premix Ex Taq™ II (Perfect Real Time) kit (TaKaRa Biotechnology, Dalian, China) on the Step One Plus Real-Time PCR Systems (Applied Biosystems, Bio-Rad, CA). The amplification of *OsActin* was used as an internal control to normalize all data, and the expression levels were calculated using the relative quantification method (Li *et al.*, 2014b). All of the primers used for qRT-PCR are listed in Table S1.

Construction of an *OsHAK16* promoter fusion with GUS, *HAK16p:WOX11* vector and generation of transgenic rice

A 1846-bp fragment of upstream *OsHAK16* encoding region was amplified from genomic DNA of rice (*Oryza sativa* L. ssp. *Japonica* cv. Nipponbare) for driving GUS reporter expression using the specific primers listed in Table S2. The restriction sites (*Hind*III and *Bam*HI) were incorporated into the primers to facilitate cloning into the expression vector. The PCR product was first cloned into the pMD19-T vector (TaKaRa) and confirmed by restriction enzyme digestion and DNA sequencing. Afterwards, the cloned fragment in the T vector was cut and inserted into upstream of the 5′ end of the GUS reporter gene in the binary vector 1300GN (Jia *et al.*, 2011; Tang *et al.*, 2012). For generating the transgenic rice lines expressing with *HAK16p:WOX11* construct, the native promoter of *OsHAK16* was used to replace the ubiquitin promoter in the expression vector pTCK303 (Ai *et al.*, 2009). The open reading frame of *WOX11* was amplified using the specific primers listed in Table S3 from genomic cDNA. The PCR product was digested with *Bam*HI and *Sac*I and ligated into pTCK303 expression vector, and then, the pTCK303 vector was digested with *Hind*III and *Bam*HI to cut off the ubiquitin promoter. The putative promoter of *OsHAK16*, 1846-bp regions, in the immediate upstream of the translation start codon was amplified by PCR from the genomic DNA, using the primers listed in Table S3. The PCR product was digested with *Hind*III and *Bam*HI, and ligated into the fragment of pTCK303 vector without ubiquitin promoter. The above constructs were transferred to *Agrobacterium tumefaciens* strain EHA105 by electroporation and then transformed into rice as described previously (Ai *et al.*, 2009).

Assay of GUS staining and activity

The histochemical analysis of GUS staining in rice was examined as described previously (Ai *et al.*, 2009). To investigate subcellular expression patterns of *OsHAK16*, the stained root materials were embedded into spur resin and sectioned. The sections (8 μm thick) were transferred onto a slide and visualized with an Olympus BX51T stereomicroscope with a colour CCD camera (Olympus, http://www.olympus-global.com). The GUS activity analysis was performed as described previously (Feng et al., 2011). Protein concentrations were determined by G250 colorimetric method. The 4-methylumbelliferone fluorescence intensity was measured by a universal Microplate spectrometer (SpectraMax M5; Molecular Devices, Sunnyvale, CA).

Measurement of root number, length and biomass and the proportion of active root surface

For the root morphology observation, rice seeds were germinated and grown in sterilized solid medium (IRRI nutrient solution with 0.4% Phytagel) containing 0.1, 0.5 and 1 mM K for 14 days. The numbers of adventitious and lateral roots were directly counted as described by Song et al. (2011) to detect the difference in elongation and initiation of roots among the plant materials. Root analysis machine (WinRhizoV4.0b; Regent instrument Inc., Quebec, Canada) was used to scan the roots of different treatments (Song et al., 2011). Then, the roots were placed in an oven at 105 °C for 30 min to inactivate the enzymes and finally dried to a constant weight at 70 °C for getting the root biomass. Six individual plants of each line were measured for getting the mean and standard error of each root parameter per plant.

Root activity was evaluated by measuring the proportion of active absorbing surface area of plant roots with methylene blue ($C_{16}H_{18}N_3SCl\cdot3H_2O$) staining with six biological replicates, as described in detail by Cai et al. (2012) with some modifications. The rice roots were immersed in three beakers containing methylene blue solution in sequence, for 1.5 min in each beaker. Afterwards, 1 mL solution from each beaker was extracted and diluted to 10 mL. Absorbance of the diluted solutions at 660 nm (A660) was determined by a universal Microplate spectrometer (SpectraMax M5; Molecular Devices, Sunnyvale, CA). Mass of methylene blue absorbed by roots was determined according to a standard curve.

Measurement of K, N and P concentrations and soluble sugar concentration

Samples for the transgenic and WT plants from different K supply conditions were assayed separately. The measurement of K concentration in the solution of digested plants was performed by an ICP emission spectrometer (Optima 2100DV; PerkinElmer Inc., Shelton, CT, USA) as described previously (Cai et al., 2012). Total N concentration in plants was determined by the Kjeldahl method (Li et al., 2006). Total P concentration in the plants was measured using the molybdate blue method described by Chen et al. (2007).

Measurement of soluble sugar concentrations in different tissues of WT and transgenic lines was conducted in five biological replicates based on the method of Hansen and Møller (1975), as described in detail by Ding et al. (2006) with some modifications. Briefly, about 0.1 g of each dry weight of different tissues was extracted with 4 mL 80% ethanol for 30 min at 80 °C. The extraction was repeated two times. Supernatants were filtrated and collected to a volume of 10 mL. The reaction mixtures, containing 0.1–1 mL extraction buffer, 0.9–0 mL deionized water and 5 mL of 0.1% (w/v) anthrone with 80% H_2SO_4, were incubated in boiled water for 10 min. After cooling, the soluble sugar content was determined by a universal Microplate spectrometer (SpectraMax M5; Molecular Devices, Sunnyvale, CA) at 620 nm using sucrose as a standard.

Measurement of net K uptake rates

The measurement of net K uptake rates was performed according to the procedure described by Yang et al. (2014) with some modifications. Rice seeds were germinated and grown in sterilized solid medium (IRRI nutrient solution with 0.4% Phytagel) containing 0.1 or 1 mM K for 14 days. Plants were harvested at two time points: on day 7 (7 days after the treatment) and 14 after the germination (end of the treatment), for the determination of their dry weight and K concentrations. The net K uptake rates into the plants were calculated according to the equation (Yang et al., 2014): Net Uptake Rate = $(C_2-C_1)/[(t_2-t_1) * (R_2+R_1)/2]$, where C is the total K content, R is the root dry weight, and t is the time at each of the two harvests; the subindexes 1 and 2 indicate the start and end of the period for which the uptake rate is calculated (t_2-t_1 = 14 days) and $(R_2+R_1)/2$ = the mean root dry weight.

Southern blot and T-DNA insertion site analysis

The independent transgenic lines transformed with the HAK16p: WOX11 construct, namely L1, L2 and L3, were determined by Southern blot analysis following the procedures described previously (Jia et al., 2011). Briefly, genomic DNA was extracted from leaves of WT and transgenic plants using the SDS method, and 60 µg of genomic DNA was digested with the restriction enzyme HindIII and BamHI overnight at 37 °C. The digested DNA was separated on a 1% (w/v) agarose gel, transferred to a Hybond-N⁺nylon membrane and hybridized with the coding sequence of the hygromycin-resistant gene used as the hybridization probe. Thermal asymmetric interlaced (tail) PCR (Genome Walking Kit; http://www.takara.com.cn) was used to detect the T-DNA insertion site of the transgenic lines.

Statistical analysis

Data were analysed by ANOVA using the SPSS 10 program (SPSS Inc., Chicago, IL). Different letters on the histograms or after the mean value between the transgenic plants and WT and/or different treatments indicate their statistically difference at $P \leq 0.05$.

Acknowledgements

This work was funded by National Natural Science Foundation of China (31361140357), National Program on R&D of Transgenic Plants, the Fundamental Research Funds for the Central Universities (KYTZ201404), 111 Project (B12009) and Innovative Research Team Development Plan of the Ministry of Education of China (IRT1256), Israeli Dead Sea Works Ltd, PAPD in Jiangsu Higher Education Institutions.

References

Ai, P., Sun, S., Zhao, J., Fan, X., Xin, W., Guo, Q., Yu, L., Shen, Q.R., Wu, P., Miller, A.J. and Xu, G. (2009) Two rice phosphate transporters, OsPht1;2 and OsPht1; 6, have different functions and kinetic properties in uptake and translocation. Plant J. **57**, 798–809.

Alemán, F., Nieves-Cordones, M., Martínez, V. and Rubio, F. (2009) Differential regulation of the HAK5 genes encoding the high-affinity K⁺ transporters of Thellungiella halophila and Arabidopsis thaliana. Environ. Exp. Bot. **65**, 263–269.

Amrutha, R.N., Sekhar, P.N., Varshney, R.K. and Kishor, P.B. (2007) Genome-wide analysis and identification of genes related to potassium transporter families in rice (Oryza sativa L.). Plant Sci. **172**, 708–721.

Armengaud, P., Breitling, R. and Amtmann, A. (2004) The potassium-dependent transcriptome of Arabidopsis reveals a prominent role of jasmonic acid in nutrient signaling. *Plant Physiol.* **136**, 2556–2576.

Ashley, M.K., Grant, M. and Grabov, A. (2006) Plant responses to potassium deficiencies: a role for potassium transport proteins. *J. Exp. Bot.* **57**, 425–436.

Ayre, B.G. (2011) Membrane-transport systems for sucrose in relation to whole-plant carbon partitioning. *Mol. Plant*, **4**, 377–394.

Bañuelos, M.A., Garciadeblas, B., Cubero, B. and Rodríguez-Navarro, A. (2002) Inventory and functional characterization of the HAK potassium transporters of rice. *Plant Physiol.* **130**, 784–795.

Bihmidine, S., Hunter, C.T. III, Johns, C.E., Koch, K.E. and Braun, D.M. (2013) Regulation of assimilate import into sink organs: update on molecular drivers of sink strength. *Front. Plant Sci.* **4**, 177.

Breuninger, H., Rikirsch, E., Hermann, M., Ueda, M. and Laux, T. (2008) Differential expression of WOX genes mediates apical-basal axis formation in the Arabidopsis embryo. *Dev. Cell*, **14**, 867–876.

Cai, J., Chen, L., Qu, H., Lian, J., Liu, W., Hu, Y. and Xu, G. (2012) Alteration of nutrient allocation and transporter genes expression in rice under N, P, K, and Mg deficiencies. *Acta Physiol. Plant*, **34**, 939–946.

Cakmak, I., Hengeler, C. and Marschner, H. (1994a) Partitioning of shoot and root dry matter and carbohydrates in bean plants suffering from phosphorus, potassium and magnesium deficiency. *J. Exp. Bot.* **45**, 1245–1250.

Cakmak, I., Hengeler, C. and Marschner, H. (1994b) Changes in phloem export of sucrose in leaves in response to phosphorus, potassium and magnesium deficiency in bean plants. *J. Exp. Bot.* **45**, 1251–1257.

Chen, A.Q., Hu, J., Sun, S.B. and Xu, G.H. (2007) Conservation and divergence of both phosphate- and mycorrhiza-regulated physiological responses and expression patterns of phosphate transporters in solanaceous species. *New Phytol.* **173**, 817–831.

Chen, Y., Fan, X., Song, W., Zhang, Y. and Xu, G. (2012) Over-expression of OsPIN2 leads to increased tiller numbers, angle and shorter plant height through suppression of OsLAZY1. *Plant Biotechnol. J.* **10**, 139–149.

Cheng, S., Huang, Y., Zhu, N. and Zhao, Y. (2014) The rice WUSCHEL-related homeobox genes are involved in reproductive organ development, hormone signaling and abiotic stress response. *Gene*, **549**, 266–274.

Chérel, I., Lefoulon, C., Boeglin, M. and Sentenac, H. (2014) Molecular mechanisms involved in plant adaptation to low K+ availability. *J. Exp. Bot.* **65**, 833–848.

Cho, S.H., Yoo, S.C., Zhang, H., Pandeya, D., Koh, H.J., Hwang, J.Y., Kim, G.T. and Paek, N.C. (2013) The rice narrow leaf2 and narrow leaf3 loci encode WUSCHEL-related homeobox 3A (OsWOX3A) and function in leaf, spikelet, tiller and lateral root development. *New Phytol.* **198**, 1071–1084.

Conti, T.R. and Geiger, D.R. (1982) Potassium nutrition and translocation in sugarbeet. *Plant Physiol.* **70**, 168–172.

Coudert, Y., Périn, C., Courtois, B., Khong, N.G. and Gantet, P. (2010) Genetic control of root development in rice, the model cereal. *Trends Plant Sci.* **15**, 219–226.

Deeken, R., Geiger, D., Fromm, J., Koroleva, O., Ache, P., Langenfeld-Heyser, R., Sauer, N., May, S.T. and Hedrich, R. (2002) Loss of the AKT2/3 potassium channel affects sugar loading into the phloem of Arabidopsis. *Planta*, **216**, 334–344.

Ding, Y., Luo, W. and Xu, G. (2006) Characterisation of magnesium nutrition and interaction of magnesium and potassium in rice. *Ann. Appl. Biol.* **149**, 111–123.

Dolan, L. and Davies, J. (2004) Cell expansion in roots. *Curr. Opin. Plant Biol.* **7**, 33–39.

Drew, M.C. (1975) Comparison of the effects of a localised supply of phosphate, nitrate, ammonium and potassium on the growth of the seminal root system, and the shoot, in barley. *New Phytol.* **75**, 479–490.

Du, L., Jiao, F., Chu, J., Jin, G., Chen, M. and Wu, P. (2007) The two-component signal system in rice (*Oryza sativa* L.): a genome-wide study of cytokinin signal perception and transduction. *Genomics*, **89**, 697–707.

Eom, J.S., Choi, S.B., Ward, J.M. and Jeon, J.S. (2012) The mechanism of phloem loading in rice (*Oryza sativa*). *Mol. Cells*, **33**, 431–438.

Feng, H., Yan, M., Fan, X., Li, B., Shen, Q., Miller, A.J. and Xu, G. (2011) Spatial expression and regulation of rice high-affinity nitrate transporters by nitrogen and carbon status. *J. Exp. Bot.* **62**, 2319–2332.

Friml, J. (2003) Auxin transport-shaping the plant. *Curr. Opin. Plant Biol.* **6**, 7–12.

Fulgenzi, F.R., Peralta, M.L., Mangano, S., Danna, C.H., Vallejo, A.J., Puigdomenech, P. and Santa-María, G.E. (2008) The ionic environment controls the contribution of the barley HvHAK1 transporter to potassium acquisition. *Plant Physiol.* **147**, 252–262.

Gajdanowicz, P., Michard, E., Sandmann, M., Rocha, M., Corrêa, L.G.G., Ramírez-Aguilar, S.J., Gomez-Porras, J.L., González, W., Thibaud, J.B., van Dongen, J.T. and Dreyer, I. (2011) Potassium (K+) gradients serve as a mobile energy source in plant vascular tissues. *Proc. Natl Acad. Sci. USA*, **108**, 864–869.

Gao, S., Fang, J., Xu, F., Wang, W., Sun, X., Chu, J., Cai, B., Feng, Y. and Chu, C. (2014) CYTOKININ OXIDASE/DEHYDROGENASE4 integrates cytokinin and auxin signaling to control rice grown root formation. *Plant Physiol.* **165**, 1035–1046.

Gerardeaux, E., Jordan-Meille, L., Constantin, J., Pellerin, S. and Dingkuhn, M. (2010) Changes in plant morphology and dry matter partitioning caused by potassium deficiency in *Gossypium hirsutum* (L.). *Environ. Exp. Bot.* **67**, 451–459.

Gierth, M. and Mäser, P. (2007) Potassium transporters in plants–Involvement in K+ acquisition, redistribution and homeostasis. *FEBS Lett.* **581**, 2348–2356.

Gierth, M., Mäser, P. and Schroeder, J.I. (2005) The potassium transporter AtHAK5 functions in K+ deprivation-induced high-affinity K+ uptake and AKT1 K+ channel contribution to K+ uptake kinetics in *Arabidopsis* roots. *Plant Physiol.* **137**, 1105–1114.

Gruber, B.D., Giehl, R.F., Friedel, S. and von Wirén, N. (2013) Plasticity of the Arabidopsis root system under nutrient deficiencies. *Plant Physiol.* **163**, 161–179.

Gupta, M., Qiu, X., Wang, L., Xie, W., Zhang, C., Xiong, L., Lian, X. and Zhang, Q. (2008) KT/HAK/KUP potassium transporters gene family and their whole-life cycle expression profile in rice (*Oryza sativa*). *Mol. Genet. Genomics*, **280**, 437–452.

Hansen, J. and Møller, I.B. (1975) Percolation of starch and soluble carbohydrates from plant tissue for quantitative determination with anthrone. *Anal. Biochem.* **68**, 87–94.

Hayaishi, H. and Chino, M. (1990) Chemical composition of phloem sap from the uppermost internode of the rice plant. *Plant Cell Physiol.* **134**, 308–315.

Hermans, C., Hammond, J.P., White, P.J. and Verbruggen, N. (2006) How do plants respond to nutrient shortage by biomass allocation? *Trends Plant Sci.* **11**, 610–617.

Hirose, N., Makita, N., Kojima, M., Kamada-Nobusada, T. and Sakakibara, H. (2007) Overexpression of a type-A response regulator alters rice morphology and cytokinin metabolism. *Plant Cell Physiol.* **48**, 523–539.

Horie, T., Sugawara, M., Okada, T., Taira, K., Kaothien-Nakayama, P., Katsuhara, M., Shinmyo, A. and Nakayama, H. (2011) Rice sodium-insensitive potassium transporter, OsHAK5, confers increased salt tolerance in tobacco BY2 cells. *J. Biosci. Bioeng.* **111**, 346–356.

Huber, S.C. and Moreland, D.E. (1981) Co-transport of potassium and sugars across the plasmalemma of mesophyll protoplasts. *Plant Physiol.* **67**, 163–169.

Jia, H., Ren, H., Gu, M., Zhao, J., Sun, S., Zhang, X., Chen, J., Wu, P. and Xu, G. (2011) The phosphate transporter gene OsPht1; 8 is involved in phosphate homeostasis in rice. *Plant Physiol.* **156**, 1164–1175.

Jordan-Meille, L. and Pellerin, S. (2008) Shoot and root growth of hydroponic maize (*Zea mays* L.) as influenced by K deficiency. *Plant Soil*, **304**, 157–168.

Kanai, S., Ohkura, K., Adu-Gyamfi, J.J., Mohapatra, P.K., Nguyen, N.T., Saneoka, H. and Fujita, K. (2007) Depression of sink activity precedes the inhibition of biomass production in tomato plants subjected to potassium deficiency stress. *J. Exp. Bot.* **58**, 2917–2928.

Kasuga, M., Liu, Q., Miura, S., Yamaguchi-Shinozaki, K. and Shinozaki, K. (1999) Improving plant drought, salt, and freezing tolerance by gene transfer of a single stress-inducible transcription factor. *Nat. Biotechnol.* **17**, 287–291.

Kellermeier, F., Chardon, F. and Amtmann, A. (2013) Natural variation of Arabidopsis root architecture reveals complementing adaptive strategies to potassium starvation. *Plant Physiol.* **161**, 1421–1432.

Kim, M.J., Ciani, S. and Schachtman, D.P. (2010) A peroxidase contributes to ROS production during Arabidopsis root response to potassium deficiency. *Mol. Plant*, **3**, 420–427.

Kovalchuk, N., Jia, W., Eini, O., Morran, S., Pyvovarenko, T., Fletcher, S., Bazanova, N., Harris, J., Beck-Oldach, K., Shavrukov, Y., Langridge, P. and

Lopato, S. (2013) Optimization of TaDREB3 gene expression in transgenic barley using cold-inducible promoters. *Plant Biotechnol. J.* **11**, 659–670.

Kühn, C. and Grof, C.P. (2010) Sucrose transporters of higher plants. *Curr. Opin. Plant Biol.* **13**, 287–297.

Lalonde, S., Tegeder, M., Throne-Holst, M., Frommer, W.B. and Patrick, J.W. (2003) Phloem loading and unloading of sugars and amino acids. *Plant, Cell Environ.* **26**, 37–56.

Lang, A. (1983) Turgor-regulated translocation. *Plant, Cell Environ.* **6**, 683–689.

Li, B., Xin, W., Sun, S., Shen, Q. and Xu, G. (2006) Physiological and molecular responses of nitrogen-starved rice plants to re-supply of different nitrogen sources. *Plant Soil*, **287**, 145–159.

Li, J., Long, Y., Qi, G.N., Li, J., Xu, Z.J., Wu, W.H. and Wang, Y. (2014a) The OsAKT1 channel is critical for K⁺ uptake in rice roots and is modulated by the rice CBL1-CIPK23 complex. *Plant Cell*, **26**, 3387–3402.

Li, Y.T., Gu, M., Zhang, X., Zhang, J., Fan, H.M., Li, P.P., Li, Z.F. and Xu, G.H. (2014b) Engineering a sensitive visual tracking reporter system for real-time monitoring phosphorus deficiency in tobacco. *Plant Biotechnol. J.* **12**, 674–684.

Lim, J.D., Cho, J.I., Park, Y.I., Hahn, T.R., Choi, S.B. and Jeon, J.S. (2006) Sucrose transport from source to sink seeds in rice. *Physiol. Plant.* **126**, 572–584.

Liu, S., Wang, J., Wang, L., Wang, X., Xue, Y., Wu, P. and Shou, H. (2009) Adventitious root formation in rice requires OsGNOM1 and is mediated by the OsPINs family. *Cell Res.* **19**, 1110–1119.

Mengel, K. and Arneke, W.W. (1982) Effect of potassium on the water potential, the pressure potential, the osmotic potential and cell elongation in leaves of *Phaseolus vulgaris*. *Physiol. Plantarum*, **54**, 402–408.

Mengel, K. and Viro, M. (1974) Effect of potassium supply on the transport of photosynthates to the fruits of tomatoes (*Lycopersicon esculentum*). *Physiol. Plantarum* **30**, 295–300.

Moriconi, J.I., Buet, A., Simontacchi, M. and Santa-María, G.E. (2012) Near-isogenic wheat lines carrying altered function alleles of the Rht-1 genes exhibit differential responses to potassium deprivation. *Plant Sci.* **185**, 199–207.

Morran, S., Eini, O., Pyvovarenko, T., Parent, B., Singh, R., Ismagul, A., Eliby, S., Shirley, N., Langridge, P. and Lopato, S. (2011) Improvement of stress tolerance of wheat and barley by modulation of expression of DREB/CBF factors. *Plant Biotechnol. J.* **9**, 230–249.

Ngampanya, B., Sobolewska, A., Takeda, T., Toyofuku, K., Narangajavana, J., Ikeda, A. and Yamaguchi, J. (2003) Characterization of rice functional monosaccharide transporter, OsMST5. *Biosci. Biotech. Biochem.* **67**, 556–562.

Nieves-Cordones, M., Martínez-Cordero, M., Martínez, V. and Rubio, F. (2007) An NH₄⁺-sensitive component dominates high-affinity K⁺ uptake in tomato plants. *Plant Sci.* **172**, 273–280.

Okada, T., Nakayama, H., Shinmyo, A. and Yoshida, K. (2008) Expression of OsHAK genes encoding potassium ion transporters in rice. *Plant Biotech.* **25**, 241–245.

Patrick, J.W. and Offler, C.E. (2001) Compartmentation of transport and transfer events in developing seeds. *J. Exp. Bot.* **52**, 551–564.

Peel, A.J. and Rogers, S. (1982) Stimulation of sugar loading into sieve elements of willow by potassium and sodium salts. *Planta*, **154**, 94–96.

Peters, J. and Chin, C.K. (2007) Potassium loss is involved in tobacco cell death induced by palmitoleic acid and ceramide. *Arch. Biochem. Biophys.* **465**, 180–186.

Petricka, J.J., Winter, C.M. and Benfey, P.N. (2012) Control of Arabidopsis root development. *Annu. Rev. Plant Biol.* **63**, 563–590.

Pettigrew, W.T. (2008) Potassium influences on yield and quality production for maize, wheat, soybean and cotton. *Physiol. Plant.* **133**, 670–681.

Pilot, G., Gaymard, F., Mouline, K., Chérel, I. and Sentenac, H. (2003) Regulated expression of Arabidopsis Shaker K⁺ channel genes involved in K⁺ uptake and distribution in the plant. *Plant Mol. Biol.* **51**, 773–787.

Qi, Z., Hampton, C.R., Shin, R., Barkla, B.J., White, P.J. and Schachtman, D.P. (2008) The high affinity K⁺ transporter AtHAK5 plays a physiological role in planta at very low K⁺ concentrations and provides a caesium uptake pathway in *Arabidopsis*. *J. Exp. Bot.* **59**, 595–607.

Rengel, Z. and Damon, P.M. (2008) Crops and genotypes differ in efficiency of potassium uptake and use. *Physiol. Plantarum*, **133**, 624–636.

Růžička, K., Šimášková, M., Duclercq, J., Petrášek, J., Zažímalová, E., Simon, S., Friml, J., Montagu, M.C.E.V. and Benková, E. (2009) Cytokinin regulates root meristem activity via modulation of the polar auxin transport. *Proc. Natl Acad. Sci. USA*, **106**, 4284–4289.

Sano, T., Becker, D., Ivashikina, N., Wegner, L.H., Zimmermann, U., Roelfsema, M.R., Nagata, T. and Hedrich, R. (2007) Plant cells must pass a K⁺ threshold to re-enter the cell cycle. *Plant J.* **50**, 401–413.

Santi, S. and Schmidt, W. (2009) Dissecting iron deficiency-induced proton extrusion in Arabidopsis roots. *New Phytol.* **183**, 1072–1084.

Scofield, G.N., Hirose, T., Aoki, N. and Furbank, R.T. (2007) Involvement of the sucrose transporter, OsSUT1, in the long-distance pathway for assimilate transport in rice. *J. Exp. Bot.* **58**, 3155–3169.

Shin, R. (2014) Strategies for improving potassium use efficiency in plants. *Mol. Cells*, **37**, 435–502.

Shin, R. and Schachtman, D.P. (2004) Hydrogen peroxide mediates plant root cell response to nutrient deprivation. *Proc. Natl Acad. Sci. USA*, **101**, 8827–8832.

Slewinski, T.L. (2011) Diverse functional roles of monosaccharide transporters and their homologs in vascular plants: a physiological perspective. *Mol. Plant*, **4**, 641–662.

Song, W., Makeen, K., Wang, D., Zhang, C., Xu, Y., Zhao, H., Tu, E., Zhang, Y., Shen, Q. and Xu, G. (2011) Nitrate supply affects root growth differentially in two rice cultivars differing in nitrogen use efficiency. *Plant Soil*, **343**, 357–368.

Stahl, Y., Wink, R.H., Ingram, G.C. and Simon, R. (2009) A signaling module controlling the stem cell niche in Arabidopsis root meristem. *Curr. Biol.* **19**, 909–914.

Tang, Z., Fan, X., Li, Q., Feng, H., Miller, A.J., Shen, Q. and Xu, G. (2012) Knockdown of a rice stelar nitrate transporter alters long-distance translocation but not root influx. *Plant Physiol.* **160**, 2052–2063.

To, J.P. and Kieber, J.J. (2008) Cytokinin signaling: two-components and more. *Trends Plant Sci.* **13**, 85–92.

To, J.P., Haberer, G., Ferreira, F.J., Deruère, J., Mason, M.G., Schaller, G.E., Alonso, J.M., Ecker, J.R. and Kieber, J.J. (2004) Type-A Arabidopsis response regulators are partially redundant negative regulators of cytokinin signaling. *Plant Cell*, **16**, 658–671.

Venkateswarlu, B. and Visperas, R.M. (1987) *Source-sink relationships in crop plants*.

Walker, D.J., Black, C.R. and Miller, A.J. (1998) The role of cytosolic potassium and pH in the growth of barley roots. *Plant Physiol.* **118**, 957–964.

Wang, Y.Y. and Tsay, Y.F. (2011) Arabidopsis nitrate transporter NRT1. 9 is important in phloem nitrate transport. *Plant Cell*, **23**, 1945–1957.

Wang, Y. and Wu, W.H. (2013) Potassium transport and signaling in higher plants. *Annu. Rev. Plant Biol.* **64**, 451–476.

Wang, Y.H., Garvin, D.F. and Kochian, L.V. (2002) Rapid induction of regulatory and transporter genes in response to phosphorus, potassium, and iron deficiencies in tomato roots. Evidence for cross talk and root/rhizosphere-mediated signals. *Plant Physiol.* **130**, 1361–1370.

Wang, Y., Xu, H., Wei, X., Chai, C., Xiao, Y., Zhang, Y., Chen, B., Xiao, G., Ouwerkerk, P.B.F., Wang, M. and Zhu, Z. (2007) Molecular cloning and expression analysis of a monosaccharide transporter gene OsMST4 from rice (*Oryza sativa* L.). *Plant Mol. Biol.* **65**, 439–451.

Wang, Y., Xiao, Y., Zhang, Y., Chai, C., Wei, G., Wei, X., Xu, H., Wang, M., Ouwerkerk, P.B.F. and Zhu, Z. (2008) Molecular cloning, functional characterization and expression analysis of a novel monosaccharide transporter gene OsMST6 from rice (*Oryza sativa* L.). *Planta*, **228**, 525–535.

Wang, J.R., Hu, H., Wang, G.H., Li, J., Chen, J.Y. and Wu, P. (2009) Expression of PIN genes in rice (*Oryza sativa* L.): tissue specificity and regulation by hormones. *Mol. Plant*, **2**, 823–831.

Wang, W., Li, G., Zhao, J., Chu, H., Lin, W., Zhang, D., Wang, Z. and Liang, W. (2014) DWARF TILLER1, a WUSCHEL-related homeobox transcription factor, is required for tiller growth in rice. *PLoS Genet.* **10**, e1004154.

White, P.J. (1993) Relationship between the development and growth of rye (*Secale cereale* L.) and the potassium concentration in solution. *Ann. Bot.* **72**, 349–358.

Williams, L.E., Lemoine, R. and Sauer, N. (2000) Sugar transporters in higher plants—a diversity of roles and complex regulation. *Trends Plant Sci.* **5**, 283–290.

Wu, X., Dabi, T. and Weigel, D. (2005) Requirement of homeobox gene STIMPY/WOX9 for Arabidopsis meristem growth and maintenance. *Curr. Biol.* **15**, 436–440.

Xu, M., Zhu, L., Shou, H. and Wu, P. (2005) A PIN1 family gene, OsPIN1, involved in Auxin-dependent adventitious root emergence and tillering in rice. *Plant Cell Physiol.* **46**, 1674–1681.

Yang, Z., Gao, Q., Sun, C., Li, W., Gu, S. and Xu, C. (2009) Molecular evolution and functional divergence of HAK potassium transporter gene family in rice (*Oryza sativa* L.). *J. Genet. Genomics*, **36**, 161–172.

Yang, T., Zhang, S., Hu, Y., Wu, F., Hu, Q., Chen, G., Cai, J., Wu, T., Moran, N., Yu, L. and Xu, G. (2014) The role of OsHAK5 in potassium acquisition and transport from roots to shoots in rice at low potassium supply levels. *Plant Physiol.* **166**, 945–959.

Yoo, S.C., Cho, S.H. and Paek, N.C. (2013) Rice WUSCHEL-related homeobox 3A (OsWOX3A) modulates auxin-transport gene expression in lateral root and root hair development. *Plant Signal. Behav.* **8**, e25929.

Zhang, Q., Li, J., Zhang, W., Yan, S., Wang, R., Zhao, J., Yujing Li, Y., Qi, Z., Zongxiu Sun, Z. and Zhu, Z. (2012) The putative auxin efflux carrier OsPIN3t is involved in the drought stress response and drought tolerance. *Plant J.*, **72**, 805–816.

Zhao, Y., Hu, Y., Dai, M., Huang, L. and Zhou, D.X. (2009) The WUSCHEL-related homeobox gene WOX11 is required to activate shoot-borne crown root development in rice. *Plant Cell*, **21**, 736–748.

7

Enhanced resistance to soybean cyst nematode *Heterodera glycines* in transgenic soybean by silencing putative CLE receptors

Xiaoli Guo[1], Demosthenis Chronis[2,†], Carola M. De La Torre[1,†], John Smeda[1,‡], Xiaohong Wang[2,3] and Melissa G. Mitchum[1,*]

[1]*Division of Plant Sciences and Bond Life Sciences Center, University of Missouri, Columbia, MO, USA*
[2]*Robert W. Holley Center for Agriculture and Health, US Department of Agriculture, Agricultural Research Service, Ithaca, NY, USA*
[3]*Department of Plant Pathology and Plant-Microbe Biology, Cornell University, Ithaca, NY, USA*

*Correspondence
email goellnerm@missouri.
edu

†Equal contributors.

‡Present address: Department of Plant Breeding & Genetics, Cornell University, 320 Bradfield Hall, Ithaca, NY 14853.

Keywords: CLAVATA, CLE, cyst nematode, *Heterodera*, soybean, transgenic.

Summary

CLE peptides are small extracellular proteins important in regulating plant meristematic activity through the CLE-receptor kinase-WOX signalling module. Stem cell pools in the SAM (shoot apical meristem), RAM (root apical meristem) and vascular cambium are controlled by CLE signalling pathways. Interestingly, plant-parasitic cyst nematodes secrete CLE-like effector proteins, which act as ligand mimics of plant CLE peptides and are required for successful parasitism. Recently, we demonstrated that Arabidopsis CLE receptors CLAVATA1 (CLV1), the CLAVATA2 (CLV2)/CORYNE (CRN) heterodimer receptor complex and RECEPTOR-LIKE PROTEIN KINASE 2 (RPK2), which transmit the CLV3 signal in the SAM, are required for perception of beet cyst nematode *Heterodera schachtii* CLEs. Reduction in nematode infection was observed in *clv1*, *clv2*, *crn*, *rpk2* and combined double and triple mutants. In an effort to develop nematode resistance in an agriculturally important crop, orthologues of Arabidopsis receptors including CLV1, CLV2, CRN and RPK2 were identified from soybean, a host for the soybean cyst nematode *Heterodera glycines*. For each of the receptors, there are at least two paralogues in the soybean genome. Localization studies showed that most receptors are expressed in the root, but vary in their level of expression and spatial expression patterns. Expression in nematode-induced feeding cells was also confirmed. *In vitro* direct binding of the soybean receptors with the HgCLE peptide was analysed. Knock-down of the receptors in soybean hairy roots showed enhanced resistance to SCN. Our findings suggest that targeted disruption of nematode CLE signalling may be a potential means to engineer nematode resistance in crop plants.

Introduction

Soybean cyst nematode (SCN, *Heterodera glycines*) is the most damaging pathogen of soybean, causing more than $1 billion dollars in yield losses annually in the US (Koenning and Wrather, 2010). A combination of nonhost crop rotation and the use of SCN-resistant cultivars are the primary means of managing SCN population levels in the field. Nevertheless, an overdependence on resistant cultivars coupled with genetic variation in SCN field populations has led to population shifts (Mitchum *et al.*, 2007; Niblack *et al.*, 2008) that necessitate the development of alternative resistance strategies. Mapping quantitative trait loci (QTL) associated with SCN resistance and the recent cloning of the resistance (*R*) genes at two major QTLs, *Rhg1* and *Rhg4*, promises to accelerate our understanding of SCN resistance (Cook *et al.*, 2012; Liu *et al.*, 2012). However, susceptibility (*S*) genes offer an alternative strategy for conferring resistance to nematodes. *S* genes encode proteins targeted by pathogen effectors to promote disease development (Gawehns *et al.*, 2013), and therefore, the inactivation or modification of these genes has the potential to reduce parasitic success and may be stacked with natural resistance.

Soybean cyst nematode is an obligate, sedentary endoparasite that induces the formation of a feeding site called a syncytium in soybean roots. The syncytium is a large, elaborate complex of fused plant cells with multiple, enlarged nuclei, highly condensed cytoplasm and thickened cell walls with ingrowths, from which the nematode ingests nutrients for its own benefit (Mitchum *et al.*, 2008, 2012). Nematode effector proteins secreted from the oesophageal glands are delivered into the host root through a protrusible stylet and are pivotal for syncytium development and maintenance (Davis *et al.*, 2008; Mitchum *et al.*, 2013).

The CLAVATA3/ENDOSPERM SURROUNDING REGION (CLE)-like peptides (Mitchum *et al.*, 2012) represent one of the best-studied classes of stylet-secreted nematode effectors. Nematode *CLE* genes have been cloned from several different cyst nematodes including SCN (*HgCLEs*) (Gao *et al.*, 2003; Wang *et al.*, 2001, 2005, 2010b), the potato cyst nematode (PCN) *Globodera rostochiensis* (*GrCLEs*) (Lu *et al.*, 2009) and the beet cyst nematode (BCN) *Heterodera schachtii* (*HsCLEs*) (Patel *et al.*, 2008; Wang *et al.*, 2011). *CLEs* are expressed in the dorsal oesophageal gland cell throughout the parasitic life stages (Lu *et al.*, 2009; Wang *et al.*, 2005, 2010b) and are required for nematode parasitism. Previous studies have shown that RNAi silencing of *CLE* gene expression in SCN and BCN leads to a reduction in nematode infection (Bakhetia *et al.*, 2007; Patel *et al.*, 2008). Overexpression of CLE proteins and exogenous peptide treatment demonstrated that nematode CLEs are biologically active mimics of plant CLEs (Lu *et al.*, 2009; Wang *et al.*, 2005, 2010b, 2011). *In planta* immunolocalization studies found that nematode CLEs are delivered into the syncytium cytoplasm of host roots, but the peptides need to be redirected out to the apoplast where they can interact with extracellular plant CLE receptors to function (Wang *et al.*, 2010b). Genetic and biochemical analysis confirmed that the variable domain of the HgCLE protein is sufficient for trafficking (Wang *et al.*, 2010a,b); however, the underlying mechanism for peptide processing and trafficking remains elusive. Plant CLE processing proteases were found to be involved in GrCLE1 processing, which indicates that plant factors are essential for nematode CLE peptide mimicry (Guo *et al.*, 2011).

Plant CLE peptides are involved in stem cell maintenance in the shoot apical meristem (SAM), the root apical meristem (RAM) and vascular meristem (Betsuyaku *et al.*, 2011; Miyawaki *et al.*, 2013; Simon and Stahl, 2006). In Arabidopsis, the CLE-receptor-WOX module is a common pathway of CLE signalling to tightly control stem cell pools. Leucine-rich repeat (LRR) receptor-like kinase (RLK) family members are the predominant players in CLE peptide recognition. In the SAM, the CLV3 peptide signal is perceived by three receptor complexes, CLAVATA1 (CLV1), CLAVATA2 (CLV2)/CORYNE (CRN) and RECEPTOR-LIKE PROTEIN KINASE 2 (RPK2)/TOADSTOOL 2 (TOAD2) and the transmitted signal leads to the transcriptional repression of *WUSCHEL* (*WUS*), which encodes a homeodomain transcription factor required for stem cell proliferation (Brand *et al.*, 2000; Clark *et al.*, 1993; Kayes and Clark, 1998; Mayer *et al.*, 1998; Miwa *et al.*, 2008; Müller *et al.*, 2008; Schoof *et al.*, 2000). In the distal region of the RAM, CLV1 cooperates with receptor-like kinase, ARABIDOPSIS CRINKLY4 (ACR4) to perceive CLE40 and restrict expression of *WUS*-related homeobox 5 (*WOX5*) in the quiescent centre (QC) (Sarkar *et al.*, 2007; Stahl *et al.*, 2009, 2013). Similar to the CLV3-WUS and CLE40-WOX5 pathways in SAM and RAM, the TDIF (tracheary element differentiation inhibitory factor)–TDR (TDIF receptor) pathway acts in the vascular meristem to activate *WOX4* expression to promote procambium proliferation (Etchells *et al.*, 2013; Fisher and Turner, 2007; Hirakawa *et al.*, 2008, 2010; Ito *et al.*, 2006).

The Arabidopsis–BCN model pathosystem has been used to investigate whether similar pathways are utilized during nematode CLE signalling. Reduced infection on several Arabidopsis mutants including *clv1*, *clv2*, *crn/sol2* and *rpk2/TOAD2* and combined double or triple mutants revealed that synergistic interaction of CLV1, CLV2/CRN and RPK2 pathways is required for BCN infection and syncytium development (Replogle *et al.*, 2011, 2013). All of the receptors were found to be expressed in the feeding cells and both nematode development and syncytium size were compromised in the mutants, especially the triple mutant. Biochemical studies demonstrated GrCLE1 and HsCLE2 binding to Arabidopsis CLV1 and CLV2 providing the first direct evidence that host plant CLE receptors can perceive nematode CLE signals (Guo *et al.*, 2010, 2011). Therefore, the identified receptors could serve as vulnerable target points for engineering nematode resistance in crop plants.

Here, we identified potential CLE receptors from soybean and investigated their role in SCN CLE signalling by biochemical and functional analysis. This is the first time nematode CLE signalling components have been characterized in a crop plant and demonstrates the potential for translating basic knowledge from the model plant Arabidopsis to engineer nematode resistance in agronomically important crop plants.

Results

Cloning, sequence analysis and expression of potential soybean CLE receptors

We focused on CLV1, CLV2, CRN and RPK2 CLE receptors in this study. The soybean homologues of these Arabidopsis receptors were identified from the Williams 82 genome sequence through BLAST analysis. Due to genome duplication in soybean (Schmutz *et al.*, 2010), at least two paralogues were found for each of the receptors and named after the Arabidopsis homologues.

A phylogenetic tree was generated using receptor protein sequences from Arabidopsis, the top six hits from soybean and homologous sequences from *Lotus japonicus* (Figure 1a). At the start of this study, the soybean counterparts of CLV1 were annotated in the Glyma1.0 assembly as Glyma11g12190 (GmCLV1A) and Glyma12g04390 (GmCLV1B) corresponding to the sequences described by Wong *et al.* (2013). However, the *CLV1A* gene region on chromosome 11 was recently re-annotated in the Glyma1.1 and Glyma2.0 assembly (Wm82.a2.v1). Consequently, Glyma11g12190 was replaced with Glyma.11g114100 (alias Glyma11g12186) and Glyma.11g114200 (alias Glyma11g12193), which we have named as *CLV1A* and *CLV1C*, respectively. A diagram representing the domain structure of the soybean receptors and Arabidopsis homologues is shown in Figure 1b. Similar to Arabidopsis homologues, GmCLV1A, GmCLV1B, GmRPK2A and GmRPK2B contain an extracellular domain with leucine-rich repeats (eLRRs) and a putative kinase domain. GmCLV1C, GmCLV2A and GmCLV2B only contain eLRRs, while GmCRNA and GmCRNB only contain a putative kinase domain.

Soybean CLE receptors are expressed in root tissues and nematode feeding sites

To determine whether CLE receptors are expressed in root tissues and nematode feeding sites, promoter–β-glucuronidase (GUS) constructs were generated and transformed into soybean hairy roots for localization studies. Five receptor gene promoters were cloned (Figure 2a). For the others, we were unable to amplify the promoter sequences after several attempts, probably because of the high A-T content in the regions. Approximately 2-kb upstream from the start codon was used, except for GmCLV2A and GmCLV1C, in which case the maximum length to the stop codon of the previous gene was used. Transgenic soybean hairy roots expressing *pGmCLV2A::GUS*, *pGmCRNA::GUS*, *pGmCRNB::GUS*, *pGmRPK2B::GUS* and *pGmCLV1C::GUS* constructs were inoculated with infective second-stage juveniles (J2) of SCN and evaluated at 3 and 7 dpi (days-post-inoculation) and compared to their respective noninoculated controls. *pGmCLV2A::GUS* expression was predominantly found in the root vasculature (Figure S1a). At 3 dpi, *GUS* expression was observed in the feeding site and was more evident by 7 dpi (Figure 2b). Expression within the syncytium was confirmed in a longitudinal section through an SCN-infected soybean hairy root (Figure 2c). *pGmCRNA::GUS* showed expression in root tips (Figure S1b), and *GUS* expression was observed in nematode feeding sites as early as 3 dpi (Figure 2d). *pGmCRNB::GUS* showed expression in root

(a)

(b)

Figure 1 Cloning and sequence analysis of soybean CLE receptors. (a) Phylogenetic tree of plant CLE receptors. Putative CLE receptors from soybean and Lotus were obtained with BLAST using Arabidopsis CLE receptor sequences as query. The top 6 soybean hits for each receptor were included. The accession numbers of the top two or three soybean hits were identified and named after the Arabidopsis homologues as follows: GmCLV2A (Glyma.09g251800, Alias Glyma09g38720), GmCLV2B (Glyma.18g240800, Alias Glyma18g47610), GmCRNA (Glyma.08g257700, Alias Glyma08g28900), GmCRNB (Glyma.18g282100, Alias Glyma18g51820), GmRPK2A (Glyma.13g056200, Alias Glyma13g06210), GmRPK2B (Glyma.19g030400, Alias Glyma19g03710), GmCLV1A (Glyma.11g114100, Alias Glyma11g12186), GmCLV1B (Glyma.12g040000, Alias Glyma12g04390) and GmCLV1C (Glyma.11g114200, Alias Glyma11g12193). Amino acid sequences were aligned using MUSCLE algorithm of the phylogenetic software MEGA6 (Tamura *et al.*, 2013). Bootstrap consensus phylogenetic tree of CLE receptor sequences inferred from 1000 replicates using the neighbour-joining method. (b) Schematic representation of domain structure of soybean CLE receptors and their Arabidopsis homologues.

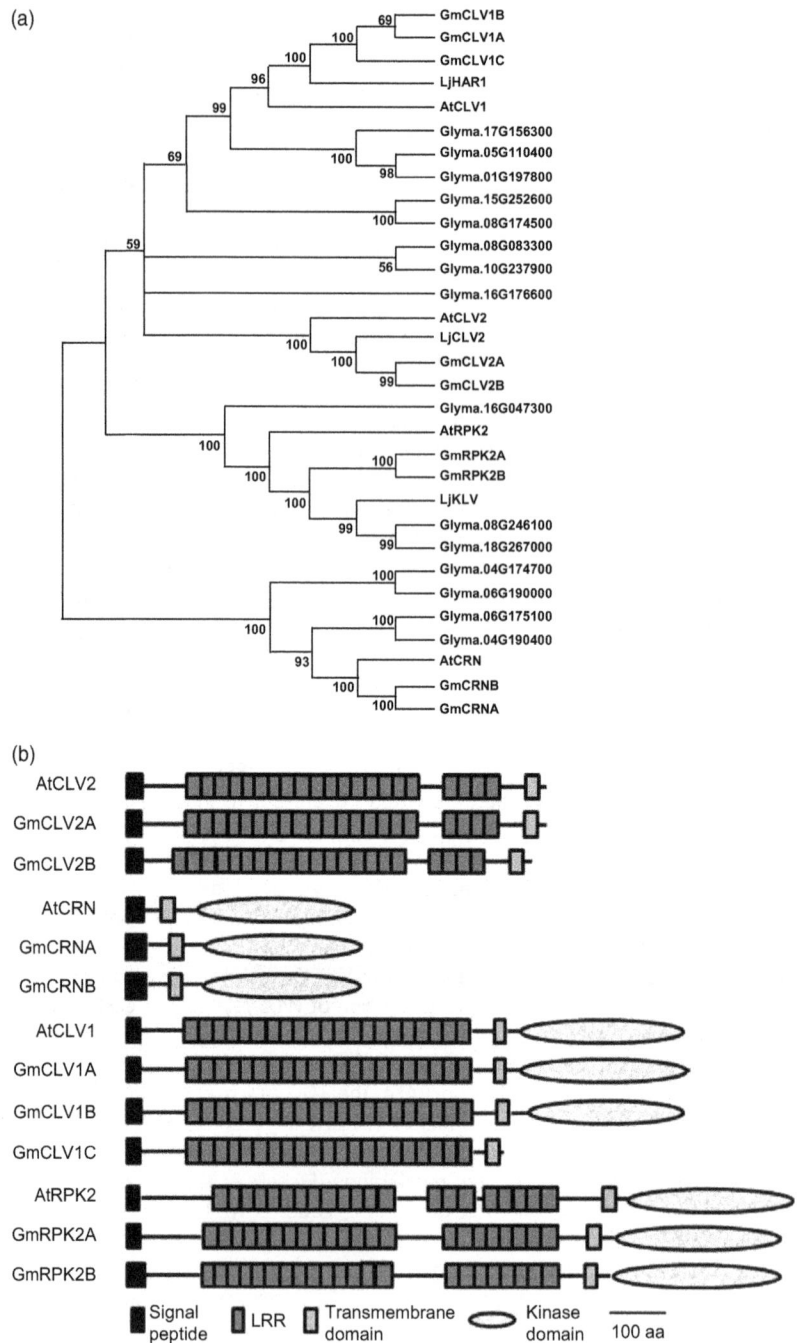

tips and vascular tissues (Figure S1c,d). Higher levels of *GUS* expression were observed in response to nematode infection compared to uninfected roots (Figure 2e). *pGmRPK2B::GUS* also showed strong expression in root tips and vascular tissues (Figure S1e,f). Expression was observed in nematode feeding sites as early as 3 dpi but was more evident at 7 dpi (Figure 2f). For *pGmCLV1C::GUS*, no positive GUS staining could be detected with or without nematode inoculation.

RNA-seq expression profiles for the receptors in uninfected soybean root tissues were examined from Soybase (http://soybase.org/), a publicly available database (Figure S1g). Reads were used to represent the expression level. A low level of expression of each receptor was detected in root tissues examined, consistent with the promoter–GUS analysis described

herein. Additionally, a syncytium-specific RNA-seq data set generated using laser-capture microdissection (LCM) was also analysed. The data represent the transcripts expressed in W82 syncytia 5 dpi. We found reads for all the receptors in the syncytia sample confirming the promoter–GUS expression in feeding sites (Figure S1h).

Silencing soybean CLE receptors reduces nematode infection

The function of the soybean receptors in nematode parasitism was analysed using the hairy root system. For RNAi, 300- to 400-bp hairpin transcripts targeting each receptor were driven by the nematode-inducible zinc finger transcription factor promoter (pZF) (Kandoth *et al.*, 2011; Liu *et al.*, 2012). Due

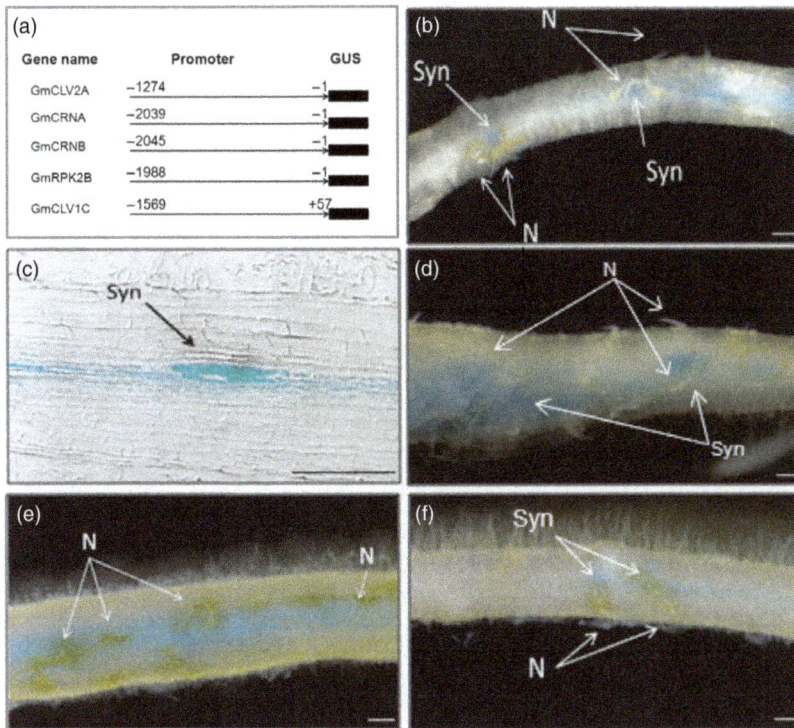

Figure 2 Expression of soybean CLE receptors. (a) Diagram showing the length of the five promoters evaluated and their corresponding names. (b–f) Promoter–GUS expression pattern in W82 transgenic hairy root lines 7 d after soybean cyst nematode (SCN) infection. (b), and (c), GmCLV2A. (d), GmCRNA. (e), GmCRNB. (f), GmRPK2B. (c), 10 μm root section showing pGmCLV2A::GUS expression in a syncytium. N, nematode; Syn, syncytium. Bars = 200 μm.

to high similarity between paralogues, at least two genes were targeted by the silencing constructs. For GmCLV1Ri, a C-terminal fragment was used which can target GmCLV1A and GmCLV1B, but not GmCLV1C due to the loss of the sequence corresponding to the kinase domain in the latter. For GmCLV1CRi, the N-terminal sequence was used which targets GmCLV1A, GmCLV1B and GmCLV1C because of high sequence similarity between the genes. A statistically significant reduction in nematode development was observed on hairy roots transformed with CLV2Ri, CRNRi and CLV1CRi compared to the GUSRi controls (Table 1). No significant differences were found for CLV1Ri and RPK2Ri. Enhanced resistance of CLV2Ri lines was further confirmed using the constitutive promoter p15 (promoter for Glyma15g06130 which encodes an arabinogalactan protein) (Kandoth et al., 2011). We observed a 32% reduction in nematode infection with p15::CLV2Ri (Figure 3a). The relative expression level of GmCLV2 from 14 independent hairy roots was examined using qRT-PCR, which shows the variation in knock-down of gene expression in individual hairy roots (Figure S2a). Because of the variability in gene silencing among root lines, the percent reduction may be lower than what could be achieved by selecting only lines that had high levels of silencing. Taken together, these data provide evidence of a role for GmCLV2, GmCRN and GmCLV1 in SCN infection.

In Arabidopsis, CLV1, CLV2 and RPK2 work synergistically during BCN parasitism of Arabidopsis (Replogle et al., 2013). Therefore, we tested the effect of silencing multiple soybean receptors on SCN infection. GmCLV2-GmRPK2 and GmCLV2-GmCLV1-GmRPK2 were targeted by 2geneRi and 3geneRi constructs, respectively. We observed a statistically significant reduction of 33% and 35% for the 2geneRi and 3geneRi constructs, respectively, but this was not significant when compared with CLV2Ri gene silencing (Figure 3). Silencing of GmCLV2 was confirmed in independent hairy roots (Figure S2b); however, due to the low expression levels of GmRPK2 and

Table 1 Enhanced nematode resistance in soybean receptor-knock-down transgenic hairy roots

Genotype	n	Mean cyst no.	SE Mean	SD	P value	Reps
CLV2Ri	66	6.00	0.49	3.96	<0.01	4
GUSRi	76	8.50	0.61	5.30		
CRNRi	42	5.12	0.44	2.87	< 0.01	2
GUSRi	50	8.52	0.79	5.57		
CLV1CRi	93	6.15	0.42	4.05	0.03	5
GUSRi	104	7.90	0.50	5.07		
CLV1Ri	30	6.33	0.74	4.03	0.18	2
GUSRi	44	7.82	0.70	4.62		
RPK2Ri	52	7.98	0.70	5.04	0.22	3
GUSRi	68	9.18	0.73	6.04		

Nematode-inducible promoter pZF (Kandoth et al., 2011) was used to drive 300–400-bp hairpin transcripts. At least two biological replicates were conducted for each construct. P-value was calculated using ANOVA and Tukey's test. Cyst number in the table was shown as mean number from all the replicates. Genotype was indicated as GUSRi and the individual receptor Ri. n, number of hairy root lines used in all replicates. Rep means the number of biological replicates performed for each receptor construct. SE Mean and SD are standard error and standard deviation, respectively.

GmCLV1 combined with variable silencing levels among independent hairy root lines, we could not confirm silencing of these genes by qRT-PCR.

We also investigated the function of CLE receptors using a virus-induced gene silencing approach in soybean (Kandoth et al., 2013). We cloned a CLV2-RPK2 fragment into the bean pod mottle virus (BPMV) RNA2 vector (pBPMV-IA-V1) (Zhang et al., 2010). The BPMV vector containing a GFP fragment insert was used as a control. Plasmids were bombarded into soybean to generate virus-infected tissue. Lyophilized infected leaf tissue was ground and used as inoculum for VIGS experiments. Soybean

(a)

(b)

Figure 3 Enhanced nematode resistance in soybean receptor-silenced hairy roots. Nematode reproduction was reduced in *CLV2* RNAi (CLV2Ri) (a), *CLV2-RPK2* RNAi (2geneRi) and *CLV2-CLV1-RPK2* RNAi (3geneRi) (b) W82-transgenic hairy roots. Constitutive promoter p15 was used. Two biological replicates were performed for each construct. Error bars = standard error. *P*-value was calculated using ANOVA and Tukey's test. *$P \leq 0.05$.

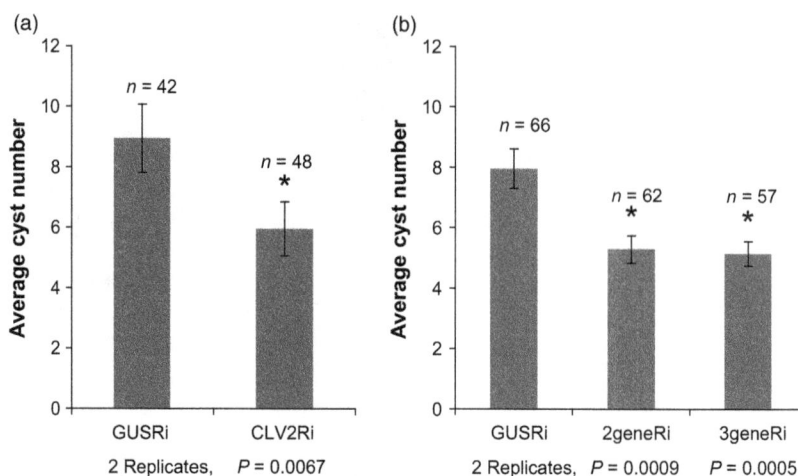

unifoliate leaves were rub-inoculated with virus-infected tissue. Three weeks after virus infection, *GmCLV2* transcripts were silenced to 50% in the second trifoliate leaves of CLV2RPK2VIGS plants; however, no silencing was detected in the roots (Figure S3a,b). Consequently, no significant difference for nematode reproduction was detected in the VIGS experiment likely due to unsuccessful silencing of the target genes in the roots (Figure S3c). Interestingly, we observed a lesion-mimic phenotype on unifoliate leaves of the CLV2RPK2VIGS plants which was not present on GFPVIGS inoculated plants (Figure S3d).

HgCLE peptide binds soybean CLE receptors

Two *HgCLE* genes, *HgCLE1* and *HgCLE2*, have been cloned and encode proteins with an identical 12-amino acid CLE motif peptide called HgCLE herein (Wang *et al.*, 2010b). To test whether the soybean receptors can bind to the HgCLE peptide, we assessed binding ability *in vitro*. The BCN HsCLE2 peptide, which represents a CLE peptide from a nematode that cannot develop on soybean, was also included for comparison. We cloned six candidate soybean receptors into pBIN61-eGFP-HA vector which adds an eGFP-HA tag at the C-terminus of the protein. For CLV1, we focused on the full-length LRR-RLKs GmCLV1A and GmCLV1B. For GmCLV2 and GmRPK2, we cloned GmCLV2A, GmCLV2B, GmRPK2A and GmRPK2B. For optimum protein expression, receptor proteins were expressed without the kinase domains using agrobacterium-mediated transient expression in *Nicotiana benthamiana* leaves. Specific binding of iodine radio-labelled HgCLE peptide to each receptor was tested. The LRR-RLK BAK1 from Arabidopsis was used as a negative control and exhibited no specific binding to the HgCLE peptide. We compared the ability of HgCLE and HsCLE2 peptides to compete with iodine radio-labelled HgCLE peptide for binding to individual receptor proteins. Specific binding was calculated by subtracting binding activity with competitor peptide from total binding without peptide. Both the HgCLE peptide and the HsCLE2 peptide showed higher binding affinity to GmCLV1A, GmCLV2A, GmCLV2B, GmRPK2A and GmRPK2B compared with GmCLV1B (Figure 4a). Although binding for the HsCLE2 peptide was detected, it was not as strong when compared with HgCLE peptide binding.

We further determined the binding kinetics of HgCLE to GmCLV1A, GmCLV2A, GmCLV2B and GmRPK2B. The receptors had similar binding affinity for HgCLE peptide with K_d values of 34.93 nM for GmCLV1A, 30.84 nM for GmCLV2A, 29.04 nM for

GmCLV2B and 32.69 nM for GmRPK2B (Figure 4b–e). Subcellular localization of C-terminal GFP:HA tagged GmCLV2A and GmCLV2B in *N. benthamiana* leaves was also investigated. LRR-RLKs are localized to the plasma membrane in order to function as receptors. As expected, the GmCLV2 receptors colocalized with the plasma membrane marker (Figure S4).

Discussion

The use of resistant soybean cultivars is the primary method of managing levels of SCN in the field. However, the limited genetic base of nematode resistance in soybean coupled with the increase in virulent SCN populations necessitates the development of novel resistance strategies. Our prior studies identified CLV1, CLV2-CRN and RPK2 as the host targets for BCN CLE effector proteins in the model plant Arabidopsis (Replogle *et al.*, 2013). These findings encouraged us to begin translating this knowledge into crop plants to test the feasibility of targeting these genes as a means to engineer novel cyst nematode resistance. Therefore, in this study, we focused our efforts on identifying soybean receptors with a role in SCN infection by expression, *in vitro* binding and gene silencing analyses.

As the soybean cultivar Williams 82 genome sequence is publicly available, we identified the putative orthologous sequences of the Arabidopsis receptors of HsCLE peptides in soybean. In this study, we focused our work on CLV1, CLV2, CRN and RPK2. Due to soybean genome duplication, at least two close homologues were identified for each receptor. Phylogenetic analysis showed the receptors from soybean and Arabidopsis also grouped with receptors from Lotus. Interestingly, recent studies found that legume CLE peptides and CLE receptors are important in the regulation of nodulation in rhizobium–legume interactions. Legumes have a systemic negative feedback regulatory system, called autoregulation of nodulation (AON) to prevent excess nodulation. AON is controlled by two long-distance signals, a root-derived infection signal and a shoot-derived inhibition signal. *Lotus japonicas HYPERNODULATION ABERRANT ROOT FORMATION 1* (*HAR1*), *KLAVIER* (*KLV*) and *LjCLV2* encode a CLV1-like, a RPK2-like and CLV2-like LRR-RLK, respectively (Krusell *et al.*, 2002, 2011; Miyazawa *et al.*, 2010; Nishimura *et al.*, 2002). GmCLV1B, also named GmNARK, is similar to CLV1 (Searle *et al.*, 2003). Knockout or knock-down of the receptors in legumes causes hypernodulation in response to *Rhizobium*, suggesting

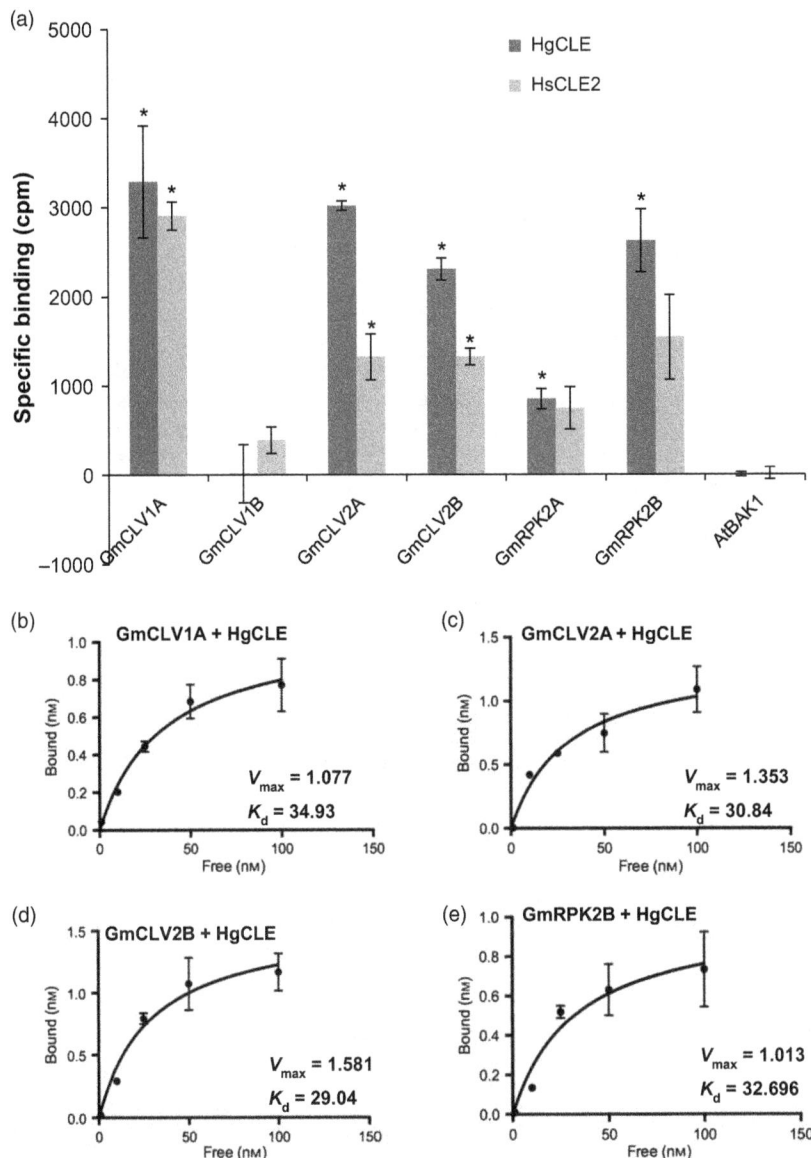

Figure 4 Receptor binding of HgCLE peptide. (a) Receptors were tested for iodine-labelled HgCLE peptide binding with and without excess unlabelled HgCLE and HsCLE2 peptide as competitor. AtBAK1 was used as negative control. Specific binding of individual CLE peptides to each receptor was calculated by subtracting binding activity with competitor from total binding when 10 μM competitive peptide was present. Data represent the mean ± SE (n = 3). Asterisks indicate $P \leq 0.05$ when compared with AtBAK1. (b–e) Binding kinetics of HgCLE peptide with GmCLV1A, GmCLV2A, GmCLV2B and GmRPK2B. Nonlinear regression analysis was performed, and the equilibrium dissociation constant (K_d) and V_{max} were calculated.

possible roles in shoot-derived inhibition during AON. Two CLE peptides from *L. japonicus*, *CLE ROOT SIGNAL1* and *2* (*LjCLE-RS1* and *LjCLE-RS2*) and three CLE peptides from soybean, *Glycine max RHIZOBIA-INDUCED CLE 1/2* (*GmRIC1/2*) and *NITRATE-INDUCED CLE 1* (*GmNIC1*) are induced by *Rhizobium* inoculation or nitrate treatment and considered as putative root-derived infection signals (Okamoto *et al.*, 2009; Reid *et al.*, 2011). In *Medicago truncatula*, a similar module was observed (Elise *et al.*, 2005; Mortier *et al.*, 2010). The CLE peptides are transmitted via the xylem from the root to the shoot where they are perceived by the receptor complex to repress excess nodulation (Okamoto *et al.*, 2013). Other than their role in nodulation, little is known about CLE peptide and receptor function in legumes.

Because nematode CLEs are delivered locally into selected cells, the receptors need to be expressed at the site of infection in order to function. We analysed the expression of putative CLE receptors in soybean, in particular in roots and nematode-induced feeding cells. For this, we cloned the promoters corresponding to five receptors including *GmCLV1C*, *GmCLV2A*, *GmCRNA*, *GmCRNB* and *GmRPK2B*, upstream of the *GUS* reporter gene, and evaluated expression in transformed soybean hairy roots. Our

data confirmed the expression of four of the five putative receptors in feeding cells upon nematode infection, which supports a role in nematode CLE peptide perception. Consistently, RNA-seq data indicated a low level of expression of the receptors in roots and syncytia.

To further demonstrate the role of the receptors in SCN CLE peptide perception, we carried out loss-of-function studies. A hairpin RNAi-silencing approach in soybean hairy roots was used. We previously identified several nematode-inducible gene promoters to specifically target gene silencing in nematode feeding cells and used these to design our silencing vectors (Kandoth *et al.*, 2011; Liu *et al.*, 2012). The gene silencing constructs were designed to silence similar genes to reduce the possibility of functional redundancy. We observed a statistically significant reduction in nematode infection on hairy roots transformed with CLV2Ri, CRNRi and CLV1CRi constructs. A consistent effect was observed when silencing *GmCLV2* using the constitutive promoter p15. These data indicate that *GmCLV2*, *GmCRN* and *GmCLV1* in soybean play an important role in SCN infection. Additionally, the targeted silencing of these genes under the control of a nematode-inducible promoter may be utilized for engineering novel resistance in crop plants, especially as silencing of these genes may have potential effects on plant growth and development. We also tested the silencing constructs targeting multiple receptors using the p15 promoter. Statistically significant reductions in infection were observed compared with controls, but we did not observe a significant additive effect compared with CLV2Ri single gene silencing. This might be due to the low silencing efficiency of multiple genes in individual hairy root lines and/or the high level of variability in nematode infection rates in the hairy root system. Thus, by selecting for high levels of silencing in whole transgenic soybean or taking advantage of targeted gene mutagenesis, much greater levels of reduction may be achieved than what we observed in hairy roots.

Previously, effective gene silencing in soybean has also been achieved by virus-induced gene silencing (VIGS) (Kandoth et al., 2013; Liu et al., 2012). Thus, we evaluated the role of the receptors in whole plant nematode infection using VIGS. Although efficient gene silencing in soybean roots using this approach has been achieved (Juvale et al., 2012; Liu et al., 2012; Zhang et al., 2009), we were unable to silence the receptor genes in this tissue. On the contrary, we did observe significant gene knock-down of *GmCLV2* in soybean leaves indicating that the lack of silencing in the roots was not due to construct design. The results indicate that significant variation in silencing efficiency for certain genes in the roots needs to be considered when using a VIGS approach in soybean. Interestingly, we observed stunted plants and a lesion-mimic phenotype on the leaves of these plants, suggesting that these genes may play a role in modulating plant defences against the BPMV virus and other pathogens. However, the molecular basis of the above-ground phenotype and whether the phenotype may have affected virus-induced gene silencing in the roots is unclear. Plant disease resistance-related and plant stress responsive genes were found to be up-regulated in the *crn* mutant (Miwa et al., 2008). This finding and the results described herein support our prior speculation that nematode CLE activation of the CLV2/CRN signalling pathway may lead to a suppression of plant defences to promote parasitism (Replogle et al., 2011).

Recently, nematode CLE peptides including HsCLE2 (identical to AtCLE5/6) and GrCLE1 were shown to interact with Arabidopsis receptors (Guo et al., 2010, 2011). To test for a potential interaction of the SCN HgCLE peptide with the candidate soybean receptors, we performed *in vitro* receptor binding assays. Six candidate receptors in soybean were cloned and used in agrobacterium-mediated transient expression in *N. benthamiana* leaves for the binding assays. Using ^{125}I-labelled HgCLE peptide, we confirmed that HgCLE could bind specifically at some level to the entire set of soybean receptors tested. Although *GmCLV1B* is expressed in soybean roots, the low binding to the HgCLE peptide indicates that this receptor may not have a significant role in HgCLE perception. *GmCLV1B* expression in the leaf plays a primary role in long-distance communication with root-derived CLE signals (Reid et al., 2011; Searle et al., 2003). In Lotus, the GmCLV1B homologue HAR1 binds directly to root-derived CLE glycopeptides, which plays a major role in regulation of nodulation (Okamoto et al., 2013). Although the function of the other soybean receptors remains to be determined, we demonstrate here that SCN may co-opt these receptors for the establishment of feeding sites.

Genetic engineering to improve plant disease resistance is a strategy that has the potential to complement classical and molecular breeding technologies. In addition to the use of R

genes, interference of *S* genes targeted by microbial pathogens offers an alternative approach to control pathogen infection. Stacking of *R* and *S* genes under the control of tissue-specific or pathogen-inducible promoters holds promise for the development of elite crop plants. As the number of host targets of nematode effectors identified in plants continues to increase (Mitchum et al., 2013), the probability of identifying targets with the potential to be used as *S* genes is improving. Several host targets of nematode effectors have been identified in the model plant Arabidopsis; however, none of these have been advanced to testing in crop plants. Here, we identified the soybean orthologues of Arabidopsis receptors targeted by nematode CLE effector proteins and demonstrated the potential for targeting these *S* genes to improve crop plant resistance to nematodes. Future work is needed to determine whether the silencing of these genes has any effect on other traits, such as yield or plant vigour in whole transgenic soybean plants. Such pleiotropic effects will need to be determined before the soybean receptors described herein have applicability as *S* genes. Nevertheless, advances in technology continue to offer novel methods to better target gene silencing to specific cell types and tissues or directly edit the target gene to modify it in such a way that pleiotropic effects may be reduced without compromising nematode resistance.

Experimental procedures

Plant and nematode material

The soybean cultivar Williams 82 (W82) which is susceptible to SCN (*Heterodera glycines* Ichinohe) was used in this study. The SCN inbred line PA3 (HG type 0) was mass selected and propagated according to standard procedures (Niblack et al., 1993) on W82 at the University of Missouri.

Phylogenetic analysis

Putative CLE receptors from soybean and Lotus were obtained with BLAST using Arabidopsis CLE receptor sequences as query. Full-length amino acid sequences were aligned using the MUSCLE algorithm of the phylogenetic analysis software MEGA6 (Tamura et al., 2013). A bootstrap consensus tree inferred from 1000 replicates using the neighbour-joining method was constructed. Evolutionary distances were calculated using a *P*-distance model with complete gap deletion.

Hairy root silencing analysis

To generate silencing constructs, a 364-bp fragment for *GmCLV2*, a 432-bp fragment for *GmCRN*, a 415-bp fragment for *GmRPK2*, a 371-bp fragment for *GmCLV1* and a 352-bp fragment for *GmCLV1C* were amplified from soybean Williams 82 root cDNA and then cloned into pDONR/zeo gateway cloning vector (Invitrogen Life Technologies, Applied Biosystem, Carlsbad, CA, USA) and sequenced. The fragments were subcloned into gateway RNAi binary vectors pZF-RNAi and p15-RNAi under the control of nematode-inducible promoter pZF and constitutive promoter p15, respectively. To generate *CLV2-RPK2* RNAi (2geneRi) and *CLV2-CLV1-RPK2*RNAi (3geneRi) vectors, a 231-bp *CLV2* fragment, a 207-bp *CLV1A* fragment and a 198-bp *RPK2* fragment were PCR amplified and used in overlap PCR to amplify *CLV2-RPK2* and *CLV2-CLV1-RPK2* fragments. Then, those two fragments were cloned into pDONR/zeo gateway vector and subsequently moved into p15-RNAi binary vector. Transgenic W82 hairy roots were generated using a GFP-selectable marker and infected with approximately 400 sterile infective J2s as described previously (Liu *et al.*, 2012). After 30 days, cysts were counted using a stereoscope. The experiments were conducted at least two times with at least 16 independent hairy roots for each treatment. The results were combined and analysed by ANOVA using Minitab with the general linear model and Tukey's simultaneous test for significant difference.

Histochemical promoter–GUS assays

The promoter sequences for the receptors were downloaded from Phytozome (www.phytozome.net). Approximately 1- to 2-kb promoter fragments upstream of the start codon or the maximum length to the stop codon of the previous gene were amplified with primers listed in Table S1 using W82 genomic DNA as template. The PCR products were cloned into the cloning vector pGEX-T easy (Promega, Madison, WI, USA) and sequenced. The promoter fragments were then subcloned into pBI101.1 using *SalI* and *BamHI*. After PCR confirmation, the plasmids were transformed into *Agrobacterium rhizogenes* K599. Transformed soybean hairy roots were selected using kanamycin and infected with 300 infective J2s. Root pieces were vacuum-infiltrated with GUS staining solution (100 mM Tris, pH 7.0, 50 mM NaCl, 1 mM 5-bromo-4-chloro-3-indolyl-b-glucuronic acid, 1.0 mM potassium ferricyanide, pH 7.0, and 0.06% [v/v] Triton X-100) twice for 10 min each and incubated overnight at 37 °C. The reaction was stopped with 70% ethanol. Stained roots were visualized with a stereoscope, and sections were performed according to Liu *et al.* (2012).

RNA extraction and quantitative PCR analysis

To check the silencing level of the target gene, gene-specific primers were designed and qRT-PCR was performed. Total RNA was extracted from root tissues using the RNeasy plant mini kit (Qiagen, Germantown, MD, USA), and cDNA was synthesized using superscript III reverse transcriptase (Invitrogen) according to the manufacturer's instructions. Quantitative PCR was performed using Applied Biosystem (Life Technologies, Applied Biosystem, Carlsbad, CA, USA) 7500 real-time PCR system. A soybean gene *SKP16* (Glyma12 g05510, Glyma.12 g051100) was used as an endogenous control. Relative gene expression level was determined using $\Delta\Delta C_t$ method compared with the internal control.

Virus-induced gene silencing

Bean pod mottle virus vectors used in this study were previously made by cloning RNA1 (essential for viral replication and maintenance; pBPMV-IA-R1M) and RNA2 (pBPMV-IA-V1) in plasmid vectors (Zhang *et al.*, 2009, 2010). A CLV2-RPK2 fragment was amplified as described for RNAi vector construction with primers containing *BamHI* restriction sites. The product was digested with *BamHI* and ligated into the pBPMV-IA-V1 vector to generate the CLV2RPK2VIGS construct. The BPMV vector with GFP fragment insert was used as a control. VIGS was performed as described (Kandoth *et al.*, 2013). To generate infected tissues, plasmid DNA encoding RNA1 was cobombarded into soybean leaves with either RNA2 CLV2RPK2VIGS vector or GFP control vector using a Bio-Rad® PDS-1000/He biolistic transformation system (Bio-Rad, Hercules, CA, USA). Infected leaf tissue was collected and lyophilized after 3–4 weeks. Hundred-milligram leaf tissue was ground in 2 mL 0.05 M potassium phosphate buffer (pH 7.0) and used as inoculum for VIGS experiments. Three weeks after rub-inoculation with virus inoculum, leaves were collected for RNA analysis and the plants were inoculated with SCN eggs (1500 eggs/plant). Cysts were counted at 35-day postinoculation.

Overexpresssion vector construction

Receptor gene fragments without the sequence encoding the kinase domain were amplified from W82 cDNA using gene-specific primers (Table S1). The PCR products were first cloned into pGEM-T easy vector and sequenced. The fragments were digested with *XbaI* and *BamHI* and subsequently cloned into pBIN61-eGFP-HA binary vector. The receptor proteins were transiently expressed in *N. benthamiana* leaves as described (Chronis *et al.*, 2013), and proteins were extracted 2 days after agro-infiltration.

Peptide binding assay

All the peptides used in the binding assays were synthesized by Selleckchem (Selleckchem.com). HgCLE peptide (RLSPSGPDPHHH) was iodine-labelled using the Bolton-Hunter method (Phoenix Pharmaceutical, Burlingame, CA, USA). Unlabelled HgCLE and HsCLE2 (RVSPGGPDPQHH) peptide were used as cold competitors. *Agrobacterium tumefaciens* C58C1 transformed with receptor constructs was used for transient expression of the receptor: GFP:HA fusion protein in *N. benthamiana* leaves (Chronis *et al.*, 2013). Two days after infiltration, leaves were evaluated for expression using a Leica DM5500B microscope; positively expressing leaves were collected. Leaves were ground in liquid nitrogen and homogenized in GTEN buffer (Sacco *et al.*, 2007) containing 10 mM dithiothreitol and 2% protease inhibitors (Sigma-Aldrich). The resulting slurry was passed through miracloth (Calbiochem) and centrifuged at 5000 g for 20 min at 4 °C. The supernatant was then centrifuged at 100 000 g for 1 h at 4 °C to pellet microsomal fractions. Aliquots of microsomal proteins (500 µg) were suspended in 250 µl of binding buffer (50 mM MES-KOH, 100 mM sucrose, pH 5.5) containing 100 nM ^{125}I-HgCLE in the presence or absence of competitor peptides (10 µM) and then incubated on ice for 30 min. Bound and free ^{125}I-HgCLE were separated by layering the reaction mixture onto 900 µl of wash buffer (50 mM MES-KOH, 500 mM sucrose, pH 5.5) and centrifuged for 15 min at 20 800 g. Radioactivity associated with the pellets was counted using an automatic gamma counter (Wizard). Specific binding was calculated by subtracting binding activity with competitor from total binding (without competitor). For binding kinetics assay, 500 µg of microsomal protein fractions was incubated with ^{125}I-HgCLE (0–100 nM), and saturation curves were generated using GraphPad Prism (GraphPad Software Inc, La Jolla, CA, USA).

Subcellular localization

pBIN61-eGFP-HA vectors carrying CLV2A and CLV2B receptor sequences were used to localize the CLV2 protein in *N. benthamiana*. *Agrobacterium* strain GV3101 containing the p19 protein was used to suppress gene silencing. Bacterial cultures containing each of the receptor GFP constructs, a PM-RFP marker plasmid and the p19 plasmid were mixed and infiltrated into *N. benthamiana*. The epidermal cell layer of the leaf was observed using a Zeiss LSM510 META confocal scanning microscope (Zeiss, Oberkochen, Germany) 2 days after infiltration. GFP was excited at 488 nm, and emission detected with 500–550 nm. RFP was excited at 543 nm, and emission detected with 565–615 nm.

Acknowledgements

We gratefully acknowledge funding by grants from USDA-AFRI to M.G.M and X.W., a MU-HHMI C^3 Undergraduate Research

Fellowship to J.S., and funding from the Daniel F. Millikan Endowment to C.C. We thank Robert Heinz for nematode population maintenance.

References

Bakhetia, M., Urwin, P.E. and Atkinson, H.J. (2007) qPCR analysis and RNAi define pharyngeal gland cell-expressed genes of *Heterodera glycines* required for initial interactions with the host. *Mol. Plant Microbe Interact.* **20**, 306–312.

Betsuyaku, S., Sawa, S. and Yamada, M. (2011) The function of the CLE peptides in plant development and plant-microbe interactions. *The Arabidopsis Book*, **9**, e0149.

Brand, U., Fletcher, J.C., Hobe, M., Meyerowitz, E.M. and Simon, R. (2000) Dependence of stem cell fate in Arabidopsis on a feedback loop regulated by CLV3 activity. *Science*, **289**, 617–619.

Chronis, D., Chen, S., Lu, S., Hewezi, T., Carpenter, S.C.D., Loria, R., Baum, T.J. and Wang, X. (2013) A ubiquitin carboxyl extension protein secreted from a plant-parasitic nematode *Globodera rostochiensis* is cleaved in planta to promote plant parasitism. *Plant J.* **74**, 185–196.

Clark, S.E., Running, M.P. and Meyerowitz, E.M. (1993) CLAVATA1, a regulator of meristem and flower development in Arabidopsis. *Development*, **119**, 397–418.

Cook, D.E., Lee, T.G., Guo, X., Melito, S., Wang, K., Bayless, A.M., Wang, J., Hughes, T.J., Willis, D.K., Clemente, T.E., Diers, B.W., Jiang, J., Hudson, M.E. and Bent, A.F. (2012) Copy number variation of multiple genes at *Rhg1* mediates nematode resistance in soybean. *Science*, **338**, 1206–1209.

Davis, E.L., Hussey, R.S., Mitchum, M.G. and Baum, T.J. (2008) Parasitism proteins in nematode–plant interactions. *Curr. Opin. Plant Biol.* **11**, 360–366.

Elise, S., Etienne-Pascal, J., de Fernanda, C.-N., Gérard, D. and Julia, F. (2005) The *Medicago truncatula SUNN* gene encodes a CLV1-like leucine-rich repeat receptor kinase that regulates nodule number and root length. *Plant Mol. Biol.* **58**, 809–822.

Etchells, J.P., Provost, C.M., Mishra, L. and Turner, S.R. (2013) WOX4 and WOX14 act downstream of the PXY receptor kinase to regulate plant vascular proliferation independently of any role in vascular organisation. *Development*, **140**, 2224–2234.

Fisher, K. and Turner, S. (2007) PXY, a receptor-like kinase essential for maintaining polarity during plant vascular-tissue development. *Curr. Biol.* **17**, 1061–1066.

Gao, B.L., Allen, R., Maier, T., Davis, E.L., Baum, T.J. and Hussey, R.S. (2003) The parasitome of the phytonematode *Heterodera glycines*. *Mol. Plant Microbe Interact.* **16**, 720–726.

Gawehns, F., Cornelissen, B.J.C. and Takken, F.L.W. (2013) The potential of effector-target genes in breeding for plant innate immunity. *Microb. Biotech.* **6**, 223–229.

Guo, Y., Han, L., Hymes, M., Denver, R. and Clark, S.E. (2010) CLAVATA2 forms a distinct CLE-binding receptor complex regulating Arabidopsis stem cell specification. *Plant J.* **63**, 889–900.

Guo, Y., Ni, J., Denver, R., Wang, X. and Clark, S.E. (2011) Mechanisms of molecular mimicry of plant CLE peptide ligands by the parasitic nematode *Globodera rostochiensis*. *Plant Physiol.* **157**, 476–484.

Hirakawa, Y., Shinohara, H., Kondo, Y., Inoue, A., Nakanomyo, I., Ogawa, M., Sawa, S., Ohashi-Ito, K., Matsubayashi, Y. and Fukuda, H. (2008) Non-cell-autonomous control of vascular stem cell fate by a CLE peptide/receptor system. *Proc. Natl Acad. Sci. USA*, **105**, 15208–15213.

Hirakawa, Y., Kondo, Y. and Fukuda, H. (2010) TDIF peptide signaling regulates vascular stem cell proliferation via the *WOX4* homeobox gene in *Arabidopsis*. *Plant Cell*, **22**, 2618–2629.

Ito, Y., Nakanomyo, I., Motose, H., Iwamoto, K., Sawa, S., Dohmae, N. and Fukuda, H. (2006) Dodeca-CLE peptides as suppressors of plant stem cell differentiation. *Science*, **313**, 842–845.

Juvale, P.S., Hewezi, T., Zhang, C., Kandoth, P.K., Mitchum, M.G., Hill, J.H., Whitham, S.A. and Baum, T.J. (2012) Temporal and spatial bean pod mottle virus-induced gene silencing in soybean. *Mol. Plant Pathol.* **13**, 1140–1148.

Kandoth, P.K., Ithal, N., Recknor, J., Maier, T., Nettleton, D., Baum, T.J. and Mitchum, M.G. (2011) The soybean *rhg1* locus for resistance to the soybean cyst nematode *Heterodera glycines* regulates the expression of a large number of stress- and defense-related genes in degenerating feeding cells. *Plant Physiol.* **155**, 1960–1975.

Kandoth, P., Heinz, R., Yeckel, G., Gross, N., Juvale, P., Hill, J., Whitham, S., Baum, T. and Mitchum, M. (2013) A virus-induced gene silencing method to study soybean cyst nematode parasitism in *Glycine max. BMC Res. Notes*, **6**, 255.

Kayes, J.M. and Clark, S.E. (1998) CLAVATA2, a regulator of meristem and organ development in Arabidopsis. *Development*, **125**, 3843–3851.

Koenning, S.R. and Wrather, J.A. (2010) Suppression of soybean yield potential in the continental United States from plant diseases estimated from 2006 to 2009. *Plant Health Prog.* http://dx.doi.org/10.1094/PHP-2010-1122-1001-RS.

Krusell, L., Madsen, L.H., Sato, S., Aubert, G., Genua, A., Szczyglowski, K., Duc, G., Kaneko, T., Tabata, S., de Bruijn, F., Pajuelo, E., Sandal, N. and Stougaard, J. (2002) Shoot control of root development and nodulation is mediated by a receptor-like kinase. *Nature*, **420**, 422–426.

Krusell, L., Sato, N., Fukuhara, I., Koch, B.E.V., Grossmann, C., Okamoto, S., Oka-Kira, E., Otsubo, Y., Aubert, G., Nakagawa, T., Sato, S., Tabata, S., Duc, G., Parniske, M., Wang, T.L., Kawaguchi, M. and Stougaard, J. (2011) The *Clavata2* genes of pea and *Lotus japonicus* affect autoregulation of nodulation. *Plant J.* **65**, 861–871.

Liu, S., Kandoth, P.K., Warren, S.D., Yeckel, G., Heinz, R., Alden, J., Yang, C., Jamai, A., El-Mellouki, T., Juvale, P.S., Hill, J., Baum, T.J., Cianzio, S., Whitham, S.A., Korkin, D., Mitchum, M.G. and Meksem, K. (2012) A soybean cyst nematode resistance gene points to a new mechanism of plant resistance to pathogens. *Nature*, **492**, 256–260.

Lu, S.-W., Chen, S., Wang, J., Yu, H., Chronis, D., Mitchum, M.G. and Wang, X. (2009) Structural and functional diversity of CLAVATA3/ESR (CLE)-like genes from the potato cyst nematode *Globodera rostochiensis*. *Mol. Plant Microbe Interact.* **22**, 1128–1142.

Mayer, K.F.X., Schoof, H., Haecker, A., Lenhard, M., Jürgens, G. and Laux, T. (1998) Role of WUSCHEL in regulating stem cell fate in the Arabidopsis shoot meristem. *Cell*, **95**, 805–815.

Mitchum, M.G., Wrather, J.A., Heinz, R.D., Shannon, J.G. and Danekas, G. (2007) Variability in distribution and virulence phenotypes of *Heterodera glycines* in Missouri during 2005. *Plant Dis.* **91**, 1473–1476.

Mitchum, M.G., Wang, X. and Davis, E.L. (2008) Diverse and conserved roles of CLE peptides. *Curr. Opin. Plant Biol.* **11**, 75–81.

Mitchum, M.G., Wang, X., Wang, J. and Davis, E.L. (2012) Role of nematode peptides and other small molecules in plant parasitism. *Annu. Rev. Phytopathol.* **50**, 175–195.

Mitchum, M.G., Hussey, R.S., Baum, T.J., Wang, X., Elling, A.A., Wubben, M. and Davis, E.L. (2013) Nematode effector proteins: an emerging paradigm of parasitism. *New Phytol.* **199**, 879–894.

Miwa, H., Betsuyaku, S., Iwamoto, K., Kinoshita, A., Fukuda, H. and Sawa, S. (2008) The receptor-like kinase SOL2 mediates CLE signaling in Arabidopsis. *Plant Cell Physiol.* **49**, 1752–1757.

Miyawaki, K., Tabata, R. and Sawa, S. (2013) Evolutionarily conserved CLE peptide signaling in plant development, symbiosis, and parasitism. *Curr. Opin. Plant Biol.* **16**, 598–606.

Miyazawa, H., Oka-Kira, E., Sato, N., Takahashi, H., Wu, G.-J., Sato, S., Hayashi, M., Betsuyaku, S., Nakazono, M., Tabata, S., Harada, K., Sawa, S., Fukuda, H. and Kawaguchi, M. (2010) The receptor-like kinase KLAVIER mediates systemic regulation of nodulation and non-symbiotic shoot development in *Lotus japonicus*. *Development*, **137**, 4317–4325.

Mortier, V., Den Herder, G., Whitford, R., Van de Velde, W., Rombauts, S., D'haeseleer, K., Holsters, M. and Goormachtig, S. (2010) CLE peptides control *Medicago truncatula* nodulation locally and systemically. *Plant Physiol.* **153**, 222–237.

Müller, R., Bleckmann, A. and Simon, R. (2008) The receptor kinase CORYNE of Arabidopsis transmits the stem cell-limiting signal CLAVATA3 independently of CLAVATA1. *Plant Cell*, **20**, 934–946.

Niblack, T.L., Heinz, R.D., Smith, G.S. and Donald, P.A. (1993) Distribution, density, and diversity of *Heterodera-glycines* in Missouri. *J. Nematol.* **25**, 880–886.

Niblack, T.L., Colgrove, A.L., Colgrove, K. and Bond, J.P. (2008) Shift in virulence of soybean cyst nematode is associated with use of resistance from PI 88788. *Plant Health Prog.* http://dx.doi.org/10.1094/PHP-2008-0118-1001-RS.

Nishimura, R., Hayashi, M., Wu, G.-J., Kouchi, H., Imaizumi-Anraku, H., Murakami, Y., Kawasaki, S., Akao, S., Ohmori, M., Nagasawa, M., Harada, K. and Kawaguchi, M. (2002) HAR1 mediates systemic regulation of symbiotic organ development. *Nature*, **420**, 426–429.

Okamoto, S., Ohnishi, E., Sato, S., Takahashi, H., Nakazono, M., Tabata, S. and Kawaguchi, M. (2009) Nod factor/nitrate-induced *CLE* genes that drive HAR1-mediated systemic regulation of nodulation. *Plant Cell Physiol.* **50**, 67–77.

Okamoto, S., Shinohara, H., Mori, T., Matsubayashi, Y. and Kawaguchi, M. (2013) Root-derived CLE glycopeptides control nodulation by direct binding to HAR1 receptor kinase. *Nat. Commun.* http://www.nature.com/ncomms/2013/130812/ncomms3191.html

Patel, N., Hamamouch, N., Chunying, L., Hussey, R., Mitchum, M., Baum, T., Wang, X. and Davis, E.L. (2008) Similarity and functional analyses of expressed parasitism genes in *Heterodera schachtii* and *Heterodera glycines*. *J. Nematol.* **40**, 299–310.

Reid, D.E., Ferguson, B.J. and Gresshoff, P.M. (2011) Inoculation- and nitrate-induced CLE peptides of soybean control NARK-dependent nodule formation. *Mol. Plant Microbe Interact.* **24**, 606–618.

Replogle, A., Wang, J., Bleckmann, A., Hussey, R.S., Baum, T.J., Sawa, S., Davis, E.L., Wang, X., Simon, R. and Mitchum, M.G. (2011) Nematode CLE signaling in Arabidopsis requires CLAVATA2 and CORYNE. *Plant J.* **65**, 430–440.

Replogle, A., Wang, J., Paolillo, V., Smeda, J., Kinoshita, A., Durbak, A., Tax, F.E., Wang, X., Sawa, S. and Mitchum, M.G. (2013) Synergistic interaction of CLAVATA1, CLAVATA2, and RECEPTOR-LIKE PROTEIN KINASE 2 in cyst nematode parasitism of Arabidopsis. *Mol. Plant Microbe Interact.* **26**, 87–96.

Sacco, M.A., Mansoor, S. and Moffett, P. (2007) A RanGAP protein physically interacts with the NB-LRR protein Rx, and is required for Rx-mediated viral resistance. *Plant J.* **52**, 82–93.

Sarkar, A.K., Luijten, M., Miyashima, S., Lenhard, M., Hashimoto, T., Nakajima, K., Scheres, B., Heidstra, R. and Laux, T. (2007) Conserved factors regulate signalling in *Arabidopsis thaliana* shoot and root stem cell organizers. *Nature*, **446**, 811–814.

Schmutz, J., Cannon, S.B., Schlueter, J., Ma, J., Mitros, T., Nelson, W., Hyten, D.L., Song, Q., Thelen, J.J., Cheng, J., Xu, D., Hellsten, U., May, G.D., Yu, Y., Sakurai, T., Umezawa, T., Bhattacharyya, M.K., Sandhu, D., Valliyodan, B., Lindquist, E., Peto, M., Grant, D., Shu, S., Goodstein, D., Barry, K., Futrell-Griggs, M., Abernathy, B., Du, J., Tian, Z., Zhu, L., Gill, N., Joshi, T., Libault, M., Sethuraman, A., Zhang, X.-C., Shinozaki, K., Nguyen, H.T., Wing, R.A., Cregan, P., Specht, J., Grimwood, J., Rokhsar, D., Stacey, G., Shoemaker, R.C. and Jackson, S.A. (2010) Genome sequence of the palaeopolyploid soybean. *Nature*, **463**, 178–183.

Schoof, H., Lenhard, M., Haecker, A., Mayer, K.F.X., Jürgens, G. and Laux, T. (2000) The stem cell population of Arabidopsis shoot meristems is maintained by a regulatory loop between the *CLAVATA* and *WUSCHEL* genes. *Cell*, **100**, 635–644.

Searle, I.R., Men, A.E., Laniya, T.S., Buzas, D.M., Iturbe-Ormaetxe, I., Carroll, B.J. and Gresshoff, P.M. (2003) Long-distance signaling in nodulation directed by a CLAVATA1-like receptor kinase. *Science*, **299**, 109–112.

Simon, R. and Stahl, Y. (2006) Plant cells CLEave their way to differentiation. *Science*, **313**, 773–774.

Stahl, Y., Wink, R.H., Ingram, G.C. and Simon, R. (2009) A signaling module controlling the stem cell niche in Arabidopsis root meristems. *Curr. Biol.* **19**, 909–914.

Stahl, Y., Grabowski, S., Bleckmann, A., Kühnemuth, R., Weidtkamp-Peters, S., Pinto Karine, G., Kirschner Gwendolyn, K., Schmid Julia, B., Wink René, H., Hülsewede, A., Felekyan, S., Seidel Claus, A.M. and Simon, R. (2013) Moderation of Arabidopsis root stemness by CLAVATA1 and ARABIDOPSIS CRINKLY4 receptor kinase complexes. *Curr. Biol.* **23**, 362–371.

Tamura, K., Stecher, G., Peterson, D., Filipski, A. and Kumar, S. (2013) MEGA6: molecular evolutionary genetics analysis version 6.0. *Mol. Biol. Evol.* **30**, 2725–2729.

Wang, X.H., Allen, R., Ding, X.F., Goellner, M., Maier, T., de Boer, J.M., Baum, T.J., Hussey, R.S. and Davis, E.L. (2001) Signal peptide-selection of cDNA cloned directly from the esophageal gland cells of the soybean cyst nematode *Heterodera glycines*. *Mol. Plant Microbe Interact.* **14**, 536–544.

Wang, X.H., Mitchum, M.G., Gao, B.L., Li, C.Y., Diab, H., Baum, T.J., Hussey, R.S. and Davis, E.L. (2005) A parasitism gene from a plant-parasitic nematode with function similar to CLAVATA3/ESR (CLE) of *Arabidopsis thaliana*. *Mol. Plant Pathol.* **6**, 187–191.

Wang, J., Joshi, S., Korkin, D. and Mitchum, M.G. (2010a) Variable domain I of nematode CLEs directs post-translational targeting of CLE peptides to the extracellular space. *Plant Signal. Behav.* **5**, 1633–1635.

Wang, J., Lee, C., Replogle, A., Joshi, S., Korkin, D., Hussey, R., Baum, T.J., Davis, E.L., Wang, X. and Mitchum, M.G. (2010b) Dual roles for the variable domain in protein trafficking and host-specific recognition of *Heterodera glycines* CLE effector proteins. *New Phytol.* **187**, 1003–1017.

Wang, J., Replogle, A.M.Y., Hussey, R., Baum, T., Wang, X., Davis, E.L. and Mitchum, M.G. (2011) Identification of potential host plant mimics of CLAVATA3/ESR (CLE)-like peptides from the plant-parasitic nematode *Heterodera schachtii*. *Mol. Plant Pathol.* **12**, 177–186.

Wong, C.E., Singh, M.B. and Bhalla, P.L. (2013) Spatial expression of CLAVATA3 in the shoot apical meristem suggests it is not a stem cell marker in soybean. *J. Exp. Bot.* **64**, 5641–5649.

Zhang, C., Yang, C., Whitham, S.A. and Hill, J.H. (2009) Development and use of an efficient DNA-based viral gene silencing vector for soybean. *Mol. Plant Microbe Interact.* **22**, 123–131.

Zhang, C., Bradshaw, J.D., Whitham, S.A. and Hill, J.H. (2010) The development of an efficient multipurpose bean pod mottle virus viral vector set for foreign gene expression and RNA silencing. *Plant Physiol.* **153**, 52–65.

8

The *QQS* orphan gene of Arabidopsis modulates carbon and nitrogen allocation in soybean

Ling Li* and Eve Syrkin Wurtele*

Department of Genetics, Development and Cell Biology, Iowa State University, Ames, IA, USA

*Correspondence

emails liling@iastate.edu and
mash@iastate.edu

Accession numbers: Sequence data from
this article can be found in The Arabidopsis
Genome Information Resource under the
following accession numbers: *QQS*
(At3g30720).

Keywords: *QQS*, orphan, carbon and
nitrogen allocation, protein, starch,
Glycine max.

Summary

The genome of each species contains as high as 8% of genes that are uniquely present in that species. Little is known about the functional significance of these so-called species specific or orphan genes. The *Arabidopsis thaliana* gene Qua-Quine Starch (*QQS*) is species specific. Here, we show that altering *QQS* expression in Arabidopsis affects carbon partitioning to both starch and protein. We hypothesized *QQS* may be conserved in a feature other than primary sequence, and as such could function to impact composition in another species. To test the potential of *QQS* in affecting composition in an ectopic species, we introduced *QQS* into soybean. Soybean T1 lines expressing *QQS* have up to 80% decreased leaf starch and up to 60% increased leaf protein; T4 generation seeds from field-grown plants contain up to 13% less oil, while protein is increased by up to 18%. These data broaden the concept of *QQS* as a modulator of carbon and nitrogen allocation, and demonstrate that this species-specific gene can affect the seed composition of an agronomic species thought to have diverged from Arabidopsis 100 million years ago.

Introduction

The ability to optimize protein productivity of plant-based foods could have far-ranging impacts to both world health and to sustainability (Godfray *et al.*, 2010; Heitschmidt *et al.*, 1996; Pimentel and Pimentel, 2003). Dietary protein is essential for animals, whereas photosynthetic organisms can biosynthesize all amino acids required for protein synthesis. Over four billion of the seven billion people on our planet obtain the majority of their dietary protein from plants (Pimentel and Pimentel, 2003; Young and Pellett, 1994). However, for many people, protein intake is insufficient, and its deficiency results in mental retardation, stunting of growth, and greatly increased susceptibility to disease, predominantly affecting children (Gomes *et al.*, 2009; Muller and Krawinkel, 2005; Victora *et al.*, 2010).

Because plants are solar-powered heterotrophs in the food web, consumption of plant-derived proteins has far less impact on the environment than consumption of animal protein sources, especially considering the earth's dwindling water resources (Pimentel and Pimentel, 2003). Thus, increasing the use of plants as a protein source, rather than animals, would have a major ecological significance.

Plant composition is determined by a metabolic network that mediates the conversion of imported, photosynthetically-derived carbon and nitrogen into protein, oil and carbohydrate. Many of the pathways by which plants synthesize or degrade protein, oil and starch have been delineated; however, much less is understood about the mechanisms that integrate these pathways and that regulate carbon and nitrogen allocation to and within this network (Eastmond, 2006; Eastmond *et al.*, 1997; Higashi *et al.*, 2006; Li *et al.*, 2009, 2012; Schiltz *et al.*, 2004; Sulpice *et al.*, 2013). Understanding this process holistically is a major biological challenge.

To identify genes that impact plant composition, our strategy leveraged the model species Arabidopsis. Based on the postulate that the basic molecular genetic mechanisms for homeostasis are conserved across species (Li *et al.*, 2009), we selected Arabidopsis single-gene mutants that appeared morphologically like the wild type (WT) control, but differed in composition, and then determined the transcripts whose expression was impacted in the mutants. We anticipated identifying a combination of metabolic and regulatory genes by this approach. The *ss3* knockout mutant of Arabidopsis is high in starch, but has a normal morphological phenotype (Zhang *et al.*, 2005, 2008); among the genes whose expression is altered in *Atss3* mutants relative to WT plants is Qua-Quine Starch (*QQS*, locus At3g30720) (Li *et al.*, 2009). Reduction of *QQS* expression in transgenic *QQS* RNAi lines of Arabidopsis resulted in plants that were morphologically indistinguishable from control lines, but that expressed a 15%–30% increase in leaf starch content (Li *et al.*, 2009).

This study further defines how *QQS* functions in Arabidopsis, showing that altering expression of *QQS* in either over-expression or *QQS* RNAi lines results in shifts in leaf protein and starch content.

The *QQS* gene encodes a protein of only 59 amino acids whose homolog is not identifiable by primary sequence comparisons to any other sequenced species, not even the closely related *Brassica napus* (Li *et al.*, 2009) or *Arabidopsis lyrata*; as such, *QQS* is considered an orphan gene. Orphan genes (also referred to as species-specific genes, or, in prokaryotes, ORFans) can be defined as genes that encode proteins that are unique to a given species, having no identifiable sequence homologs in other species (Gollery *et al.*, 2006, 2007; Li *et al.*, 2009). The concept of orphan genes was first described by Fischer and Eisenberg in 1999 from studies of microbial genomes (Fischer and Eisenberg, 1999).

Although many have predicted that genes considered species specific would later turn out to be an artifact of sparse genome sequence, this has proved not to be the case (Arendsee et al., 2014; Gollery et al., 2006, 2007; Marsden et al., 2006; Neme and Tautz, 2013; Silveira et al., 2013; Tautz and Domazet-Loso, 2011). Orphan genes appear to be present in all species, and represent a significant fraction (approximately 0.5% to >8%) of analysed eukaryotic and prokaryotic genomes. The function of most orphan genes is obscure; however, they have been considered to be a determinant of species character (Gollery et al., 2006; Tautz and Domazet-Loso, 2011).

Orphan genes are thought to arise by multiple mechanisms (Carvunis et al., 2012; Donoghue et al., 2011; Tautz and Domazet-Loso, 2011; Wissler et al., 2013). Nongenic sequence can be defined as the sequence that is not a part of an organism's genes (Adler, 1992). In Saccharomyces cervesiae, nongenic sequences have been shown to be transcribed widely (Nagalakshmi et al., 2008) and some is also translated (Carvunis et al., 2012). Genes could arise de novo from such nongenic sequences via a noncoding or coding proto-gene that becomes more stabilized during evolution into an orphan gene (Carvunis et al., 2012). Orphan genes also arise via pre-existing genes, whose sequence can be highly modified by combinations of gene duplication, domain shuffling, shifting of location of translation frames and subsequent diversification (Carvunis et al., 2012; Ohno, 1987; Tautz and Domazet-Loso, 2011). The lack of even remote footprints of any A. thaliana genic sequence within QQS indicates it is likely an orphan gene that arose de novo (Silveira et al., 2013).

Another feature of orphans may be a general lack of co-expression with other genes. One way to consider this feature is to evaluate their representation in regulons. Regulons of eukaryotes can be defined as clusters of genes that have a prevailing pattern of co-expression across multiple/thousands of diverse conditions (Biehl et al., 2005; Feng et al., 2012; Mentzen et al., 2008); these have been detailed for the Arabidopsis transcriptome using a Markov chain cluster (MCL) approach (Mentzen and Wurtele, 2008). Only some of the genes of an organism can be classified into regulons; others do not have a prevailing pattern of co-expression with other genes (Feng et al., 2012; Mentzen and Wurtele, 2008). The Arabidopsis transcriptome can be partitioned into 872 regulons; about 49% of all Arabidopsis genes are members of these regulons (Mentzen and Wurtele, 2008). However, only 4.3% of orphans are members of regulons. Thus orphans are highly underrepresented among the co-expressed gene clusters.

We conjecture that some orphan genes might have arisen and stabilized because they confer a selective advantage by interacting with a previously existing protein. This previously existing protein could be relatively conserved, that is, present in many lineages. Because QQS influences composition, a process critical to plants in general, our working hypothesis is that QQS contains structural features that would interact with a conserved protein, such that introduction of the QQS gene into another species would influence its compositional traits.

We tested the hypothesis that QQS could function to regulate composition in another species by introducing the QQS gene into the major food crop, soybean. Here, we demonstrate the expression of the QQS gene in soybean increases leaf and seed protein and decreases leaf and seed carbohydrate. These results reveal that expression of the QQS gene increases carbon and nitrogen allocation to protein and that it can exert this function even in an ectopic species.

Results

Arabidopsis plants over-expressing and under-expressing QQS have altered carbon and nitrogen allocation to protein and starch

Our findings that down-regulation of QQS increases starch content in Arabidopsis (Li et al., 2009) might indicate that altered QQS impacts starch content without changing other aspects of leaf composition. Alternately, QQS might have a more general effect on carbon and nitrogen allocation, affecting protein and/or lipid content. To distinguish between these alternatives, we evaluated the protein and starch content of transgenic Arabidopsis plants that either over-expressed [under control of the constitutive cauliflower mosaic virus (CaMV) 35S promoter, see Figure S1] or suppressed (using QQS RNAi, Li et al., 2009) the accumulation of the QQS coding sequence. A total of ten independent QQS over-expression mutant lines and four independent QQS RNAi mutant lines were grown and evaluated using a randomized complete block design. The visual phenotype of these transgenic Arabidopsis lines appeared identical to the control lines throughout development, from seedlings to senescence (Figure 1a,b). However, leaf starch content in lines that over-expressed QQS (QQS-OE) was decreased by up to 23% (Figure S2). Conversely, QQS RNAi lines showed an increase in leaf starch content (Figures 1b and S2), consistent with our previous publication (Li et al., 2009). Leaf protein content was altered in each of the QQS-OE and QQS RNAi mutant lines that we tested; specifically, protein content was increased by about 3% in QQS-OE mutants and decreased by 3%–7% in QQS RNAi mutants (Figure 1c). These data indicate that QQS acts either directly or indirectly as a regulator of carbon and nitrogen metabolism and affects not only the accumulation of starch, but also protein.

The BLink algorithm (NCBI, http://www.ncbi.nlm.nih.gov/sutils/blink.cgi?mode=query) identifies 1155 orphan genes in Arabidopsis, based on the gene models described in Arabidopsis using the TAIR10 genome release (ftp://ftp.arabidopsis.org/home/tair/Genes/TAIR10_genome_release/), including 30 mitochondrial genes (as dated on September 23, 2013). In addition to these 1115 A. thaliana-specific genes, 839 genes can be identified using BLink that are unique to A. thaliana and A. lyrata but not identifiable outside of the Arabidopsis genus; these can be referred to as clade-specific genes or genus-specific genes, rather than the more restrictive term, orphans.

Orphans are typically shorter than typical genes (Amiri et al., 2003; Knowles and McLysaght, 2009; Wu et al., 2011). In A. thaliana, the median length of the predicted protein models from the 1155 orphan genes is 57 amino acid (range, 16–445 amino acid) while the median length of predicted protein models from all genes is 349 amino acid (range, 16–5393 amino acid) (Figure 2). When an alternate, more inclusive, algorithm for orphan genes is used, the predicted proteins have a similarly short length (Gollery et al., 2006).

Eukaryotic orphan genes generally tend to have greater tendency of being associated with transposable elements, relatively low expression that is organ specific, and less helical and sheet structure (as predicted computationally) (Arendsee et al., 2014; Donoghue et al., 2011; Wilson et al., 2007). QQS has only some of these characteristics. Like a typical orphan gene, it is a short peptide (Figure 2) with a GC content of 40%, slightly higher than the average for Arabidopsis genes (this average GC

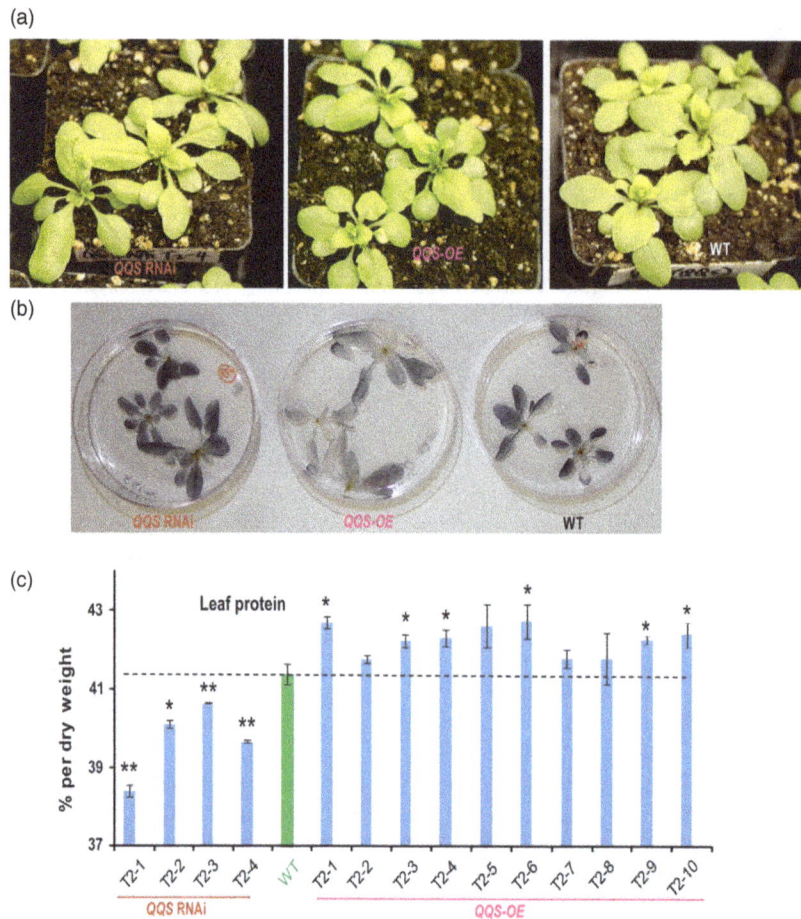

Figure 1 Phenotype and leaf composition of transgenic lines of Arabidopsis with down-regulation or over-expression of Qua-Quine Starch (*QQS*). Seedling shoots of *QQS* RNAi plants (Li *et al.*, 2009) derived from four independent transformation events, and *QQS-OE* plants derived from ten independent transformation events, together with wild type (WT) control plants, were sampled at the end of the light period. *QQS* RNAi plants are known to have increased starch (Li *et al.*, 2009). (a) All of these lines of transgenic *QQS* RNAi and *QQS-OE* plants are similar in morphology to WT plants throughout development. This phenotype (i.e. the *QQS-OE* plants were indistinguishable from WT in appearance) contrasts with the 35S:*QQS* transgenic phenotype of slower growth and rounded leaves that was described by Seo *et al.* (2011); possibly this is due to a difference in experimental conditions in which the plant were grown. (b) Starch staining shows increased starch in *QQS* RNAi mutants and decreased starch in *QQS-OE* mutants compared with WT. (c) Leaf protein is significantly decreased in *QQS* RNAi mutants and increased in *QQS-OE* mutants compared with WT. *QQS* mutants are each in the T2 generation, and each independent transformation events are designated by T2-independent transformation event #. All data in bar charts show mean ± SE (standard error), *n* = 3. Student's *t*-test was used to compare *QQS* RNAi and *QQS-OE* with WT, *$P < 0.05$; **$P < 0.01$.

content is 36% for Arabidopsis, Mishra *et al.*, 2009). *QQS* is embedded in a neighbourhood of chromosome 3 that is highly enriched in transposons (Figure S3) (Li *et al.*, 2009) and generally unusual. In the 3′ direction, *QQS* is 2.5 kb from At3g30725, a 'glutamine dumper-like' protein of unclear function; however, in the 5′ direction, *QQS* is a surprising 77.6 kb from the nearest gene, At3g30705, which is also an orphan gene and has no known function. The sequence immediately 5′ of the *QQS* start site contains multiple 5′ flanking siRNA repeated elements (Lister *et al.*, 2008; Silveira *et al.*, 2013).

Unlike what has been reported as general orphan character-istics, *QQS* expression in Arabidopsis ecotype Columbia (Col-0) is relatively not low and *QQS* is expressed in most plant organs; furthermore, its expression responds strongly to genetic and environmental perturbations (Li *et al.*, 2009). We evaluated qualitatively whether *QQS* expression is affected under two environmental conditions that are known to increase starch accumulation in Arabidopsis: high sucrose (Aloni *et al.*, 1997) and

low temperature (Espinoza *et al.*, 2010) using transgenic plants containing the *QQS* promoter driving the GUS coding sequence (Figure S4). *QQS* expression is decreased in plants grown in medium contain 5% sucrose compared with medium without added sucrose. *QQS* expression is also decreased in plants growing at 4 °C compared with 22 °C. These results further demonstrate the strong sensitivity of *QQS* expression to environ-mental perturbations.

Disorder prediction tools (http://www.disprot.org/predictors.php) indicate that QQS protein has a disordered N-terminal tail (approximately 20 first residues). The remainder of QQS protein is predicted to contain two α helices.

Soybean plants expressing *QQS* have increased leaf protein

To test the hypothesis that *QQS* could affect carbon and nitrogen allocation in an ectopic species, we determined whether expression of the *QQS* transgene would impact

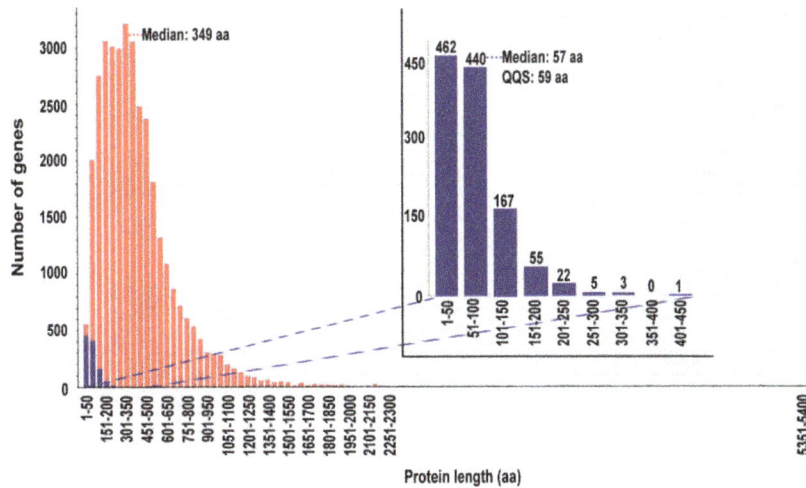

Figure 2 Distribution of orphan protein lengths in Arabidopsis. The median length of the predicted protein models from *Arabidopsis thaliana* orphan genes (blue bar) is smaller (57 amino acid) than the median length of all *A. thaliana* predicted protein models (orange bar) (349 amino acid) (range of orphan protein models: 16–445 amino acid; range of all protein models: 16–5393 amino acid). The *X*-axis represents the protein length (in amino acids); the *Y*-axis represents the number of genes. Orphan genes are predicted by NCBI BLink (http://www.ncbi.nlm.nih.gov/sutils/blink.cgi?mode=query). The insert provides a magnified view of the distribution of the sizes of orphan genes.

composition in the major crop plant, soybean (*Glycine max*). We chose soybean based on its evolutionary divergence from Arabidopsis and because of its importance as a direct or indirect source of dietary protein. Indeed, soy is the predominant source of protein for humans (Pathan and Sleper, 2008). In addition to its potential agronomic value, increasing the protein content in an already protein-rich crop could provide a stringent test of how *QQS* may affect compositional traits. Soybean (cultivar Williams 82) was transformed with the *QQS* coding sequence under the control of the constitutive CaMV 35S promoter (Figure 3). Transgenic soybean lines that expressed the *QQS* transgene were identified by selection for herbicide resistance conferred by the introduced vector, followed by real-time PCR analysis (Table S1).

T1-generation lines of soybean expressing the *QQS* gene survived from herbicide selection, from each of five independent transformation events (a total of 33 lines), together with Williams 82 controls, were evaluated after they were grown in growth chambers in a completely randomized design within each chamber. Visual examinations of the plants throughout development from seedling to senescence indicated that the morphology of the transgenic soybean plants expressing the *QQS* transcript was similar to the WT control plants (Figure 3a). However, in leaves of the transgenic *QQS*-expressing lines (*QQS-E*), starch content was reduced to levels as low as 20% of the WT control plants (Figure 3b). In contrast, leaf protein content was increased by up to 60% in the transgenic soybean plants (Figure 3c). The effect of *QQS* on the starch and protein accumulation in lines from the same transformation event may vary as the expression level of the transgene in different lines of the same transformation event may be different (Shou *et al.*, 2004). The plots of leaf starch and protein versus *QQS* transcript accumulation in Arabidopsis and in soybean indicate that for both species, when *QQS* RNA accumulation is elevated, protein accumulation is increased and starch accumulation is decreased. The relationship between *QQS* RNA level and starch and protein concentrations was not linear over the range of *QQS* RNA accumulation tested (Figure S5).

Soybean plants expressing *QQS* have increased seed protein

The increase in leaf protein associated with *QQS* gene expression in soybean led us to assess whether *QQS* had a broader impact, also affecting seed composition. As a preliminary indication of whether *QQS* expression alters seed protein content, T2 seeds from soybean plants that had been grown in growth chambers and survived from herbicide selection were screened by Nuclear Magnetic Resonance Spectrometry (NMR). These data (Figure S6) indicated a significant increase in seed protein in the *QQS*-expressing lines. To evaluate this finding in more detail, plants were propagated via single-seed descent. Progeny were grown in a completely randomized design in a greenhouse. The transgenic soybean plants expressing the *QQS* transcript survived from herbicide selection were indistinguishable in morphology and development from the WT control plants (Figure S7a). T3 seeds from the greenhouse-grown lines (seeds from the offspring of the same T1 plant were pooled and harvested together) were evaluated for composition by destructive chemical analysis. These seeds also showed a significant increase in seed protein in the *QQS*-expressing lines, compared with WT controls; the seed oil content of these lines was similar or slightly decreased (Figure S7b).

Based on the results of these analyses, we tested the effect of *QQS* expression in field-grown plants. Segregating seeds of independent transgenic lines were planted in a field in a randomized complete block design. Plants were monitored weekly during development. Plants were sprayed by herbicide and data on survival was monitored (Table S2). One line is possibly a homozygous line (16-6); in that line, all plants were herbicide resistant. WT nontransgenic siblings were identified by PCR analysis of DNA of leaf pieces from individual progeny of self-propagated T3 generation plants derived from three independent transformation events. Morphology and development of field-grown plants was indistinguishable among the populations of these WT-sibling lines and *QQS*-expressing lines (Figures 4a and S8a).

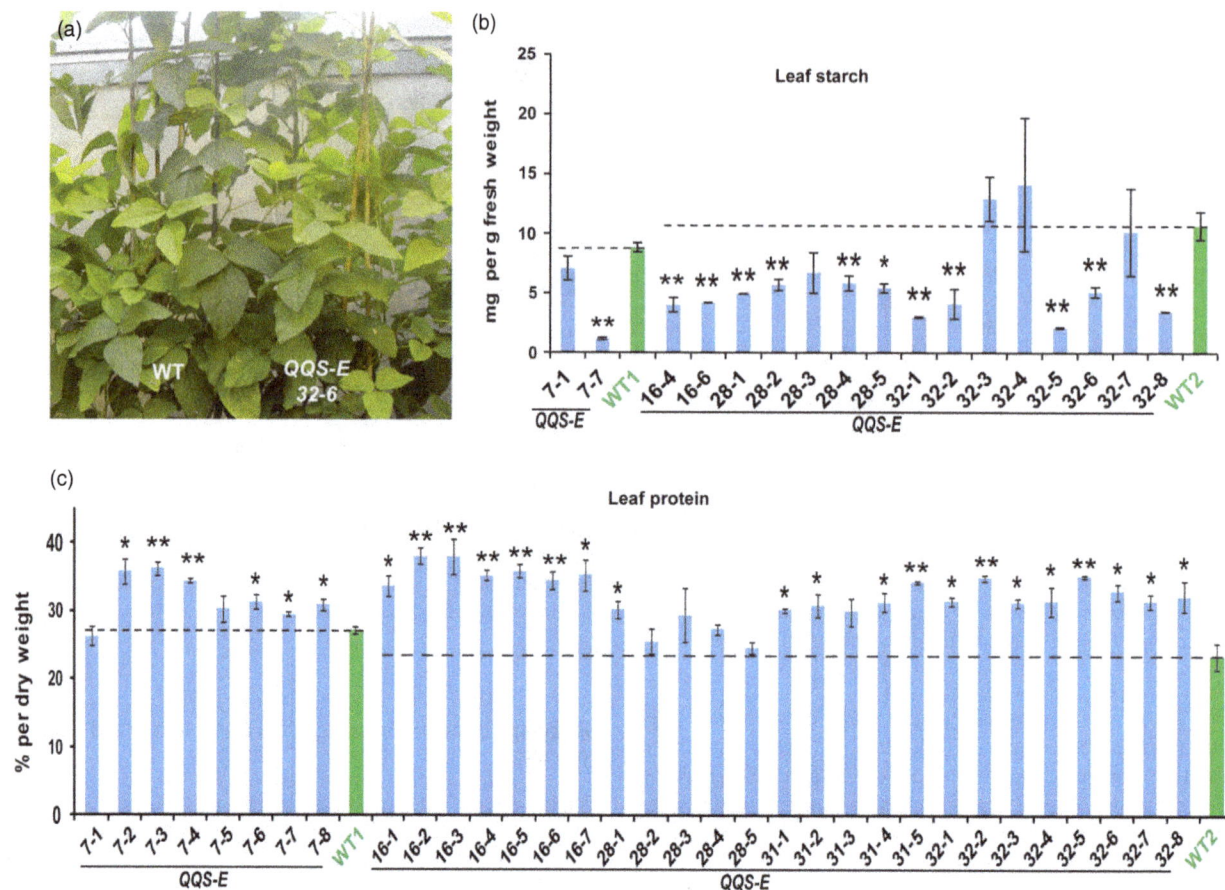

Figure 3 Characterization of leaves of growth chamber-grown transgenic soybean plants expressing Qua-Quine Starch (*QQS*). (a) Transgenic *QQS-E* plants (engineered in the Williams 82 background) are not visually distinguishable from Williams 82 control plants. (b) Leaf starch is decreased in *QQS-E* plants compared with wild type (WT) controls. (c) Leaf protein is increased in *QQS-E* plants compared with WT controls. WT1 and WT2, Williams 82 controls from two growth chambers; *QQS-E*, transgenic Williams 82 lines expressing the *QQS* coding sequence, selected by herbicide resistance. *QQS-E* mutants of T1 generation derived from 17 lines of four independent transformation events (for starch) and from 33 lines of five independent transformation events (for protein) are designated by: independent transformation event #–line #. All data in bar charts show mean ± SE, $n = 3$. Student's *t*-test was used to compare *QQS-E* and WT, *$P < 0.05$; **$P < 0.01$.

Seeds were harvested at maturity. Seed morphology, seed size and shape, seed weight per seed and per plant, and moisture content were similar among the populations of field-grown plants (Figures 4b and S8b,c). However, seed composition was significantly affected, as determined by independent methods: near infrared spectroscopy (NIRS, for eight lines, Table S3) and destructive chemical analyses (for four lines) to mature pooled seeds from herbicide-resistant mutant plants transformed with *QQS* compared to mature pooled seeds from WT plants. Protein content was from 10% to 18% higher in seeds of *QQS*-expressing lines as compared to those of WT-sibling controls (Figures 4c and S8c). The levels of several free amino acids were also increased (Table S4). The increase in the seed protein content did not affect the relative distribution of amino acids in hydrolysed proteins while the amino acid contents are significantly increased (Figure S8d and Table S5). Seed oil content in different lines ranged from a value similar to that of WT-sibling seeds to a 13% decrease (Figures 4c and S8c). The level of C16:0 is slightly decreased in the mutants (Table S6). There was no detectable change in levels of free fatty acids. Carbohydrate and fibre contents were decreased (Figure S8c). The total protein and oil content in seeds of *QQS*-expressing soybean was increased by

up to 6.5% as compared with that of the WT-siblings (Figures 4c and S8c). Thus, individual lines of soybean expressing the *QQS* transgene appeared morphologically indistinguishable but showed increases of 18.0%, 10.3%, 13.1%, 12.9% seed protein and 6.5%, 6.3%, 5.5% and 4.3 % seed protein + oil, compared with their sibling WTs. Segregation study of individual transgenic plants and their segregated siblings identified by PCR analysis of DNA of leaf pieces, and these individual plants' seed composition analysis by NIRS indicated that the high-protein trait was associated with *QQS* expression (Table S7).

Discussion

Orphan gene function in plant biology

Little is understood about the functional significance of the vast majority of orphan genes in any species. Indeed, to our knowledge, *QQS* is the only plant orphan gene that has been studied in any detail. In general, orphan genes have been considered to confer a species-specific function, for example, immunity to particular pathogens, self-recognition, or resistance/adaptation to an environmental stress (Gollery *et al.*, 2007; Khalturin *et al.*, 2009). Of the plant orphan genes that anything is known about

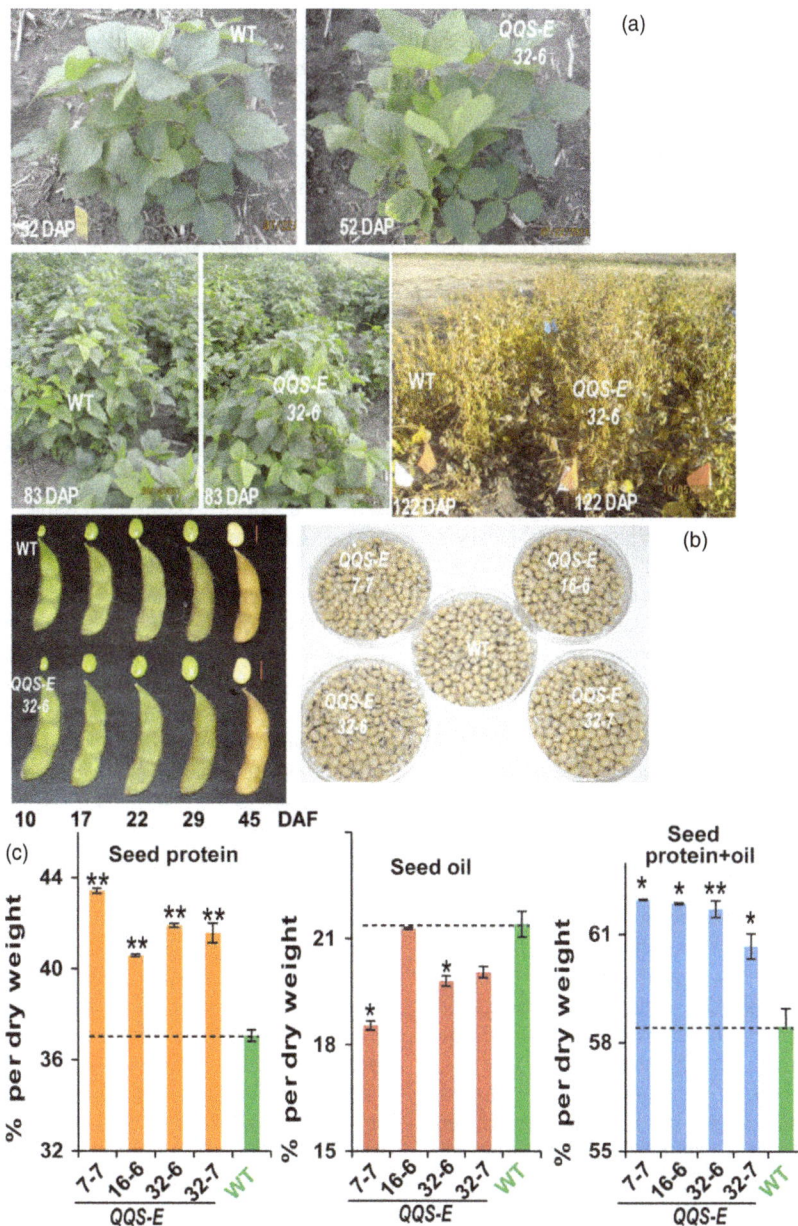

Figure 4 Characterization of seeds of field-grown transgenic soybean plants expressing Qua-Quine Starch (*QQS*). *QQS-E* plants derived from three independent transformation events (a total of four lines) and segregating wild type (WT)-sibling control plants were grown in a randomized block design in the field. Composition was determined in mature T4 seeds by chemical methods. Figure S8c shows near infrared spectroscopy analysis of different seed batches. (a) Transgenic *QQS-E* soybean plants are similar in morphology from WT-sibling plants throughout development and in visual phenotypes (Figure S8a). (b) Seed development and seed size are similar in *QQS-E* compared to WT-siblings. (c) Seed composition in *QQS-E* compared with that of the WT-sibling controls: seed protein content is increased, oil content is similar or decreased, and protein + oil content is increased. DAP, days after planting; DAF, days after flowering. WT, Williams 82 siblings identified by PCR from segregating populations of T3 plants (from transformation events 7 and 32); *QQS-E*, transgenic Williams 82 expressing the *QQS* coding sequence, selected by herbicide resistance. *QQS-E* mutants are designated by independent transformation event #–line #. All data in bar charts show mean $\pm$ SE, $n = 3$. Student's *t*-test was used to compare *QQS-E* and WT, *$P < 0.05$; **$P < 0.01$. Scale bar, 1 cm.

(other than *QQS*), most have been identified in mutant screens for genes that alter resistance to abiotic or biotic stresses (Gollery *et al.*, 2006, 2007; Luhua *et al.*, 2013). In the most comprehensive analysis of plant orphan genes to date, Mittler and colleagues (Luhua *et al.*, 2013) evaluated the responses to abiotic stresses of knockout mutants of 1007 Arabidopsis genes randomly selected from those annotated as of 'unknown function' (TAIR, 2005 gene model release). Of these genes annotated as of 'unknown

function', 12 were also orphan genes. Knockout mutants of nine of these 12 orphan genes conferred an altered response to one or more abiotic stresses (Luhua *et al.*, 2013).

Among the 839 clade-specific genes that are common to *A. thaliana* and *A. lyrata* are a small cluster of genes that, although divergent across the two species, are still recognizable by sequence; these genes have been shown to play a role in self-recognition (Takeuchi and Higashiyama, 2012). The genes encode cysteine-rich peptides (CRP810_1; AtLUREs) that are among the 300 defensin-like (*DEFL*) genes in Arabidopsis and function in self-recognition associated with pollen attraction (Takeuchi and Higashiyama, 2012). AtLURE1 (AT5G43285) from *A. thaliana*, when introduced into *Torenia fournieri*, enables *A. thaliana* pollen to be attracted to and penetrate *T. fournieri* ovules (Takeuchi and Higashiyama, 2012). Thus, with the possible exception of *QQS*, the few plant orphan or near-orphan (clade-specific) genes with functional information appear to play a role in recognition or defence-related processes (Gollery *et al.*, 2007; Kim *et al.*, 2009; Li *et al.*, 2009).

QQS and carbon and nitrogen allocation

QQS plays a role, direct or indirect, in regulating carbon and nitrogen allocation to starch (Li *et al.*, 2009) and protein, a process that would be expected to have considerable commonality among plant species. Although carbon and nitrogen allocation might be considered as very distinct from species-specific recognition, immune response, or defence, it actually is closely intertwined. Plants are 'planted' in one place, and therefore must respond with tremendous sensitivity to environmental cues. Global approaches to understand the processes of photosynthesis, nutrient supply and carbon and nitrogen allocation are revealing the intricate relationship among what some might consider distinct processes (Stitt *et al.*, 2010; Sulpice *et al.*, 2013; Thum *et al.*, 2008).

Among the most pronounced of the extremely varied pattern of *QQS* expression across changes in genotypes, environments and developments (Li *et al.*, 2009), *QQS* expression level is strongly changed in a variety of knockout mutants. For example, *QQS* expression is significantly higher (compared with WT controls) in mutants of genes as diverse as *PEN3*, a putative ATP binding cassette transporter that contributes to pathogen resistance (Stein *et al.*, 2006); starch synthase 3 (*SS3*) (Li *et al.*, 2009); *WIN1*, involved in regulating cuticular wax deposition (Kannangara *et al.*, 2007); and the brassinosteroid-induced *FER*, which functions in pollen tube-ovule interaction (Guo *et al.*, 2009). Plants expressing the *NahG* transgene, a bacterial gene that hydroxylates salicylic acid (SA) and reduces SA-mediated signalling (Takahashi *et al.*, 2004), have increased *QQS* expression (ArrayExpress experiment ID 'E-GEOD-5727', data submitted by Buchanan-Wollaston). SA plays a role in plant defense against pathogens (Lin *et al.*, 2013). Consistent with our direct demonstration that *QQS* impacts carbon and nitrogen allocation, *QQS* has been implicated in the ability of Arabidopsis to adjust to reduced carbon and energy environments, based on its altered expression in knockout lines of the EXORDIUM-LIKE1 (*EXL1*) gene (At1g35140) (Schroder *et al.*, 2011). In addition, the inverse relationship between *QQS* expression and starch accumulation across multiple environments fits our model that plants adjust *QQS* expression in response to stresses. Overall, the extremely variable expression pattern of *QQS*, combined with its lack of co-expression with other genes and association with compositional changes under changing environments (Li *et al.*, 2009), is consistent with the concept that the *QQS*-induced compositional changes in *A. thaliana* may aid in the metabolic adaptation of that species to its environment.

Ectopic function of QQS

Our demonstration that the introduction of the *QQS* gene into soybean results in a significant increase in protein accumulation and decrease in lipid accumulation in seeds indicate the potential of *QQS* as a molecular tool to increase the protein content of agronomic species. Comparison of composition in plants grown in growth chamber, greenhouse and field materials indicates that the general trend of high-protein content in *QQS-E* soybean holds for plants grown under these very different environments. Protein (and oil) content typically is extremely responsive to both genotype and environment (Arslanoglu *et al.*, 2011; Jing *et al.*, 2003; Singh *et al.*, 1993). Fertilizers containing nitrogen were applied to growth chamber- and greenhouse-grown soybeans, while no fertilizer was applied to field-grown soybeans. Therefore, the differences in overall composition (e.g. seed protein and oil) between the growth chamber-, greenhouse- and field-grown material, as well as the difference in composition between the *QQS-E* and control lines could be due to either genotype or environment, or a genotype and environment effect.

QQS expression in soybean causes greater increases in protein in soybean than in Arabidopsis. One possible explanation for the larger effect of *QQS* on soybean is that over time Arabidopsis has evolved mechanisms for homeostatic balances for the *QQS* gene; however, these mechanisms are not present in soybean. An alternate explanation is that the signalling mechanisms in soybean respond with different sensitivity than those in Arabidopsis.

Interestingly, leaves of *QQS-E* soybean lines with a ratio as low as 1.9 QQS RNA/18S rRNA by real-time PCR display the high-protein, low-starch trait. However, *QQS-E* 7-1 did not have high leaf protein, despite having significant *QQS* expression. Possible explanations are that Event 7 had multiple transgene insertion sites, and these multiple transgenes may have been retained in line 7-1, but not in 7-7. It may be that the insertion site of one of these transgenes interfered with some metabolic process, or that the expression of *QQS* was not stable in *QQS-E* 7-1. Indeed, there does not appear statistically significant linear regression of leaf starch and protein contents with leaf *QQS* expression (as determined by transcript level) in either Arabidopsis ($R^2 = 0.83$ for starch and 0.64 for protein) or soybean ($R^2 = 0.78$ for starch and 0.49 for protein); these calculations are complicated by the fact that the relationship between the composition traits and *QQS* transcript level is not necessarily linear, but complex, and have not yet been fully defined. This is perhaps not surprising given the very low level of QQS protein that accumulates in Arabidopsis even in mutant lines that highly express *QQS* (Li *et al.*, 2009). It is not unusual for regulatory proteins to have very low levels of expression (Nagaraj *et al.*, 2011). Thus, it may be that only a very small concentration of QQS saturates the QQS receptor, and any increases over this concentration do not affect protein and starch contents. Other possible explanations for a lack of strong linear correlation between levels of *QQS* transcript and composition are that QQS translational efficiency or stability or the effectiveness of the QQS protein to biochemically express its function or post-translational modification is limiting; also, a variety of post-translational regulatory mechanisms can come into play, as have been described for other transgenes (Lillo *et al.*, 2004; Vaucheret *et al.*, 2001). These considerations present interesting questions about the mechanism by which *QQS* acts in soybean; our current working hypothesis is that soybean and Arabidopsis have a common protein with which QQS interacts, and that QQS-interactor becomes saturated at low levels of *QQS* expression.

The experiments described do not distinguish whether the high-protein trait in the seeds is a maternal effect or a seed effect. The 35S promoter, which we are using in these studies, drives GUS expression in both leaf and in seed (Anderson and Botella, 2007; Wu *et al.*, 2010), which is consistent with either a seed or a maternal effect. Thus, it is possible that the QQS expression in the seed causes the high-protein trait (a seed effect). Alternately, a signalling molecule or a larger flux of organic carbon and nitrogen from the leaves might drive the high-protein trait in the soybean seeds- this would represent a maternal effect of QQS.

An increase in soybean protein of the magnitude reported in this study has societal relevance, as soybean provides a major source of global dietary vegetable protein (Pathan and Sleper, 2008; Wilson, 2008). Over 70 years of soybean breeding efforts have not been able to break the inverse relationship between seed protein content and oil content, or the inverse relationship between seed protein content and yield. However, the transgenic expression of QQS increases seed protein content in soybean grown under three diverse conditions (growth chamber, greenhouse and field) without detectably affecting plant or seed morphology or seed weight (these factors were determined under all three conditions). In seeds of field-grown plants, for which we made more detailed determinations, there were no significant differences in seed yield per plant, the relative composition of amino acids in the hydrolysed protein from seeds, the moisture content or the yield. We analysed free amino acids to determine whether the increase in protein we observed was associated with a very substantial increase in any free amino acid; this does not seem to be the case. Some free amino acid levels were altered in QQS-E soybean seeds. Glutamic acid and arginine are reported to help to avoid protein aggregating and precipitating (Golovanov et al., 2004); it is possible that these shifts in free glutamic acid and arginine might help to adapt to the increased protein content. Lysine and arginine are among the essential amino acid group according to their importance to nutrition and physiology values (Belitz et al., 2009). Thus increased free lysine and arginine in QQS-E seeds could potentially provide an increased value; however, the absolute level of free lysine and arginine is very low.

Taken together, our findings suggest that QQS could be introduced by breeding or transformation into an elite soybean variety with specific desirable agronomic traits to increase protein content. For example, QQS could be introduced into an Iowa soybean variety that is resistant to the soybean cyst nematode but relatively low in protein (http://www.cad.iastate.edu/gensoyrel.html), or in African soybean varieties highly resistant to rust, bacterial blight and leaf spot (http://www.iita.org/soybean-asset/-/asset_publisher/t3fl/content/better-soybean-varieties-offer-african-farmers-new-opportunities?redirect=%2Fsoybean#.U8ib1fldWSp).

Conclusions

Our data demonstrate that QQS expression alters plant composition. We show that expressing this gene in a plant species that has no QQS sequence homolog increases protein content of leaves and seeds, yet the morphology and development of the soybean expressing QQS cannot be distinguished from the WT sibling controls. Thus, the QQS-expressing mutant appears to preserve overall homeostasis while selectively effecting composition. QQS might be acting upstream of the process that controls carbon partitioning, or might be central to this process. These results also illustrate that orphan genes, although often poorly annotated and even ignored, may provide a valuable resource for new traits.

The evolutionary changes that resulted in the de novo origin (Silveira et al., 2013) of the QQS gene of A. thaliana must have been rapid and extensive as there is no gene homolog in even the closely related species, A. lyrata. Yet, soybean, a species that diverged from Arabidopsis approximately 100 million years ago (Hedges and Kumar, 2009), appears to contain a conserved receptor or mechanism that recognizes QQS and responds to its occurrence by conferring a compositional phenotype. This research reveals the fundamental capacity of a species-specific gene to act across species to impact the major metabolic function of carbon and nitrogen allocation.

Experimental procedures

Construction of QQS over-expression vector and Arabidopsis transformation and selection are provided in Appendix S1.

Soybean transformation, selection and nomenclature

Glycine max cultivar Williams 82 was transformed by Agrobacterium-mediated soybean transformation using half-seed explants (Paz et al., 2004). The transformation and selection of T1 plants were performed at the Plant Transformation Facility at Iowa State University (ISU) (http://www.agron.iastate.edu/ptf/index.aspx). Soybean plants from independent transformation events were selected based on herbicide resistance (segregated WTs were killed) and confirmed by real-time PCR analysis to identify transgenic lines that expressed the QQS transgene.

The progeny of each independent soybean transformation is referred to as an 'event' (a transformation event is considered independent if it is taken from an individual plate). Each plant germinated from one T1 seed is called a 'line' and the line designation continues throughout generations. So multiple lines stem from one independent transformation event; because each line is the result of sexual reproduction; these lines may not be genetically identical. A total of 33 lines from five independent transformation events were confirmed on the basis of BAR selection followed by PCR analysis for presence of the QQS gene.

Plant growth

This study used WT A. thaliana ecotype Columbia (Col-0), and transgenic lines derived from Col-0. Detailed information is provided in Appendix S1.

Detailed information about transgenic QQS-expressing (QQS-E) soybean grown in growth chambers (T1 generation) and in a greenhouse (T2 generation, selected on herbicide resistance and segregated WTs were killed), is provided in Appendix S1.

Transgenic QQS-E soybean (T3 generation) and WT (Williams 82) plants were planted at a randomized block design in the field at ISU Curtiss Farm in Ames, IA. One line (60 seeds) was planted in one row, with a total of three replicates in three rows. Each row was ten feet long and the rows were 2.5 feet apart. The criteria used for selecting the events-lines to study further in the field were: having sufficient seeds for NIRS analysis and planting in the field; having an increased leaf protein/decreased leaf starch trait; and the presence of the QQS transgene (as determined by PCR). The field conditions were harsh prior to germination, and there was considerable flooding; some seeds in flooded area were eaten by ground squirrels or other animals, and a number of lines were lost (including QQS-E 7-1). Eight mutant lines that survived germination were sprayed by herbicide, and numbers of herbicide-resistant and herbicide-sensitive progeny were counted. For yield trials, only seeds harvested from plants from the middle seven feet were used for yield estimate (grams of seed weight per plant). The seeds from different plants of the same line were pooled and used as different replicates for seed composition analysis by NIRS (all eight lines) and by chemical methods (four lines). Some plants were randomly marked for genotype screening by PCR to identify transgenic plants and their WT-siblings. Seeds from these plants were harvested per individual plant and were not pooled with seeds from other plants.

QQS determination by PCR is provided in Appendix S1.

Leaf composition analyses

For screening starch using I_2/KI staining in Arabidopsis, shoots were harvested at the end of the light period, and processed as described before (Li *et al.*, 2007). Detailed information is provided in Appendix S1.

Leaf starch and protein were determined at the end of the light period in Arabidopsis seedling shoots of 20 days after planting (DAP) grown in a growth chamber, and in soybean leaves that were newly, fully expanded, harvested from branches two to four from shoot apex from 58-DAP T1 plants (starch: four independent transformation events, protein: five). Three (for starch) or five (for protein) plants per replicate and three replicates from each independent T2 lines (Arabidopsis), and leaves from three positions that were used as three replicates in the T1 plants (soybean), were analysed. Detailed information about leaf starch and protein determination is provided in Appendix S1.

Soybean seed composition analyses

Information about NMR and NIRS screening is provided in Appendix S1.

Destructive chemical analyses were mostly conducted at Eurofins (Des Moines, IA). Methods were: protein content, AOCS Ba 4e-93 (American Oil Chemists' Society, 1997); oil content, AOCS Ac 3-44 (American Oil Chemists' Society, 1997); moisture content, AOCS Ac 2-41 (American Oil Chemists' Society, 1997); hydrolysed fatty acid profiling, AOCS Ce 2-66 and AOCS Ce 1-62 (American Oil Chemists' Society, 1991); free amino acid profiling, AOAC 999.13 modified (Fontaine *et al.*, 2000). About 30 g of seeds (approximately 170 seeds from nine plants, for protein content), 10 g (for oil content), 10 g (for moisture content), 10 g (for hydrolysed fatty acid profiling) and 20 g (for free amino acid profiling) per replicate were tested, respectively, with three biological replicates for each sample.

Hydrolysed amino acid composition analysis was conducted at the Experiment Station Chemical Laboratories, University of Missouri (http://www.aescl.missouri.edu/), using method AOAC 982.30 E (a,b,c) Ch. 45.3.05 (Association of Official Analytical Chemists (AOAC), 2006). About 30 g of seeds per replicate were tested, with three biological replicates for each sample.

Statistical analyses

For each experiment, plants were collected and analysed in a randomized complete block design or completely randomized design. Plant composition tests were conducted with a minimum of three biological tests. For all composition analyses, plant samples were assigned randomized numbers and provided to the analysis facilities for determination in a randomized order with no designator of genotype.

Data are presented as mean ± SE. Two sets of independent samples were compared using Student's *t*-test (two-tailed) with assumption of equal variances ($n = 3$). $P < 0.05$ was considered significant (*); $P < 0.01$ was considered very significant (**).

Bioinformatics analyses is provided in Appendix S1.

Acknowledgements

We are grateful to Walter Fehr for advice on soybean breeding, Zebulun Arendsee and Ruoran Li for orphan gene analyses, and Basil Nikolau and Jianming Yu for helpful suggestions on the manuscript. We thank Kan Wang and Diane Luth from the ISU Plant Transformation Facility for generating transgenic lines of soybean; Jianling Peng for the *QQS*-OE construct; Taner Sen and Vladimir Uversky for information on prediction of QQS protein structure and disorder; Kent Berns for field management; Wenguang Zheng, Sheng Huang, Marah Hoel, Xiaoran Shang, Alan Kading, Le Song, Hiwot Abebe, Ana Boehm and Sean Wefel for assistance with soybean growth and harvest; Charles Hurburgh and Glen Rippke from ISU Grain Quality Laboratory, for NIRS analysis; Dan Duvick for seed free fatty acid determinations; Jian Li for statistical consultation; Waitent Sow, Kevin Wenceslao, Dallas Jones, and ISU W. M. Keck Plant Metabolomics Laboratory for contributions to composition analysis; Eurofins (Des Moines, IA) for analysis of seed composition; ISU Soil and Plant Analysis Laboratory for leaf combustion analysis; Experiment Station Chemical Laboratories at University of Missouri for amino acid profiling; Pennington Caroline from UEA Consulting Limited for PCR analyses of *QQS* transcript level; and Sabry Elias, Oregon State University Seed Lab for NMR Spectrometry screening. This material is based in part upon work supported by National Science Foundation EEC-0813570 (to E.S.W.) and MCB-0951170 (to E.S.W. and L.L.); United Soybean Board award 2287 (to L.L.); ISU Research Foundation (to L.L.); and Center for Metabolic Biology (to E.S.W.). Any opinions, findings and conclusions or recommendations expressed in this material are those of the author(s) and do not necessarily reflect the views of the National Science Foundation.

Author contributions

L.L. and E.S.W. conceived the project. L.L. performed the experiments. L.L and E.S.W. co-supervised the project and contributed equally to data analysis and preparation of the paper.

References

Adler, R.G. (1992) Genome research: fulfilling the public's expectations for knowledge and commercialization. *Science*, **257**, 908–914.

Aloni, B., Karni, L., Zaidman, Z. and Schaffer, A.A. (1997) The relationship between sucrose supply, sucrose-cleaving enzymes and flower abortion in pepper. *Ann. Bot.* **79**, 601–605.

American Oil Chemists' Society (AOCS). (1991) *Official and Tentative Method of the American Oil Chemists' Society*, 4th edn. Champaign, IL: AOCS Press.

American Oil Chemists' Society (AOCS). (1997) *Official Methods and Recommended Practices of the American Oil Chemists' Society*, 5th edn. Champaign, IL: AOCS Press.

Amiri, H., Davids, W. and Andersson, S.G. (2003) Birth and death of orphan genes in Rickettsia. *Mol. Biol. Evol.* **20**, 1575–1587.

Anderson, D.J. and Botella, J.R. (2007) Expression analysis and subcellular localization of the *Arabidopsis thaliana* G-protein β-subunit AGB1. *Plant Cell Rep.* **26**, 1469–1480.

Arendsee, Z., Li, L. and Wurtele, E.S. (2014) Coming of age: the species-specific (orphan) genes of plants. *Trends Plant Sci.* In press.

Arslanoglu, F., Aytac, S. and Oner, E.K. (2011) Effect of genotype and environment interaction on oil and protein content of soybean (*Glycine max* (L.) Merrill) seed. *Afr. J. Biotechnol.* **10**, 18409–18417.

Association of Official Analytical Chemists (AOAC) (2006) *Official Methods of Analysis of the Association of Official Analytical Chemists*, Gaithersburg, MD: AOAC International.

Belitz, H.D., Grosch, W. and Schieberle, P. (2009) Amino acids, peptides, protein. In *Food Chemistry*, 4th revised and extended edn (Belitz, H.D., Grosch, W. and Schieberle, P., eds), pp. 8–34. Berlin: Springer.

Biehl, A., Richly, E., Noutsos, C., Salamini, F. and Leister, D. (2005) Analysis of 101 nuclear transcriptomes reveals 23 distinct regulons and their relationship to metabolism, chromosomal gene distribution and co-ordination of nuclear and plastid gene expression. *Gene*, **344**, 33–41.

Carvunis, A.R., Rolland, T., Wapinski, I., Calderwood, M.A., Yildirim, M.A., Simonis, N., Charloteaux, B., Hidalgo, C.A., Barbette, J., Santhanam, B., Brar, G.A., Weissman, J.S., Regev, A., Thierry-Mieg, N., Cusick, M.E. and Vidal, M. (2012) Proto-genes and de novo gene birth. *Nature*, **487**, 370–374.

Donoghue, M.T., Keshavaiah, C., Swamidatta, S.H. and Spillane, C. (2011) Evolutionary origins of Brassicaceae specific genes in *Arabidopsis thaliana*. *BMC Evol. Biol.* **11**, 47.

Eastmond, P.J. (2006) SUGAR-DEPENDENT1 encodes a patatin domain triacylglycerol lipase that initiates storage oil breakdown in germinating Arabidopsis seeds. *Plant Cell*, **18**, 665–675.

Eastmond, P.J., Dennis, D.T. and Rawsthorne, S. (1997) Evidence that a malate/inorganic phosphate exchange translocator imports carbon across the leucoplast envelope for fatty acid synthesis in developing castor seed endosperm. *Plant Physiol.* **114**, 851–856.

Espinoza, C., Degenkolbe, T., Caldana, C., Zuther, E., Leisse, A., Willmitzer, L., Hincha, D.K. and Hannah, M.A. (2010) Interaction with diurnal and circadian regulation results in dynamic metabolic and transcriptional changes during cold acclimation in Arabidopsis. *PLoS ONE*, **5**, e14101.

Feng, Y., Hurst, J., Almeida-De-Macedo, M., Chen, X., Li, L., Ransom, N. and Wurtele, E.S. (2012) Massive human co-expression network and its medical applications. *Chem. Biodivers.* **9**, 868–887.

Fischer, D. and Eisenberg, D. (1999) Finding families for genomic ORFans. *Bioinformatics*, **15**, 759–762.

Fontaine, J., Eudaimon, M., Fontaine, J. and Eudaimon, M. (2000) Liquid chromatographic determination of lysine, methionine, and threonine in pure amino acids (feed grade) and premixes: collaborative study. *J. AOAC Int.* **83**, 771–783.

Godfray, H.C., Beddington, J.R., Crute, I.R., Haddad, L., Lawrence, D., Muir, J.F., Pretty, J., Robinson, S., Thomas, S.M. and Toulmin, C. (2010) Food security: the challenge of feeding 9 billion people. *Science*, **327**, 812–818.

Gollery, M., Harper, J., Cushman, J., Mittler, T., Girke, T., Zhu, J.K., Bailey-Serres, J. and Mittler, R. (2006) What makes species unique? The contribution of proteins with obscure features. *Genome Biol.* **7**, R57.

Gollery, M., Harper, J., Cushman, J., Mittler, T. and Mittler, R. (2007) POFs: what we don't know can hurt us. *Trends Plant Sci.* **12**, 492–496.

Golovanov, A.P., Hautbergue, G.M., Wilson, S.A. and Lian, L.Y. (2004) A simple method for improving protein solubility and long-term stability. *J. Am. Chem. Soc.* **126**, 8933–8939.

Gomes, S.P., Nyengaard, J.R., Misawa, R., Girotti, P.A., Castelucci, P., Blazquez, F.H., de Melo, M.P. and Ribeiro, A.A. (2009) Atrophy and neuron loss: effects of a protein-deficient diet on sympathetic neurons. *J. Neurosci. Res.* **87**, 3568–3575.

Guo, H., Li, L., Ye, H., Yu, X., Algreen, A. and Yin, Y. (2009) Three related receptor-like kinases are required for optimal cell elongation in *Arabidopsis thaliana*. *Proc. Natl Acad. Sci. USA*, **106**, 7648–7653.

Hedges, S.B. and Kumar, S. (2009) *The Timetree of Life*. New York, NY: Oxford University Press.

Heitschmidt, R.K., Short, R.E. and Grings, E.E. (1996) Ecosystems, sustainability, and animal agriculture. *J. Anim. Sci.* **74**, 1395–1405.

Higashi, Y., Hirai, M.Y., Fujiwara, T., Naito, S., Noji, M. and Saito, K. (2006) Proteomic and transcriptomic analysis of Arabidopsis seeds: molecular evidence for successive processing of seed proteins and its implication in the stress response to sulfur nutrition. *Plant J.* **48**, 557–571.

Jing, Q., Jiang, D., Dai, T. and Cao, W. (2003) Effects of genotype and environment on wheat grain quality and protein components. *Chin. J. Appl. Ecol.* **14**, 1649–1653.

Kannangara, R., Branigan, C., Liu, Y., Penfield, T., Rao, V., Mouille, G., Hofte, H., Pauly, M., Riechmann, J.L. and Broun, P. (2007) The transcription factor WIN1/SHN1 regulates Cutin biosynthesis in *Arabidopsis thaliana*. *Plant Cell*, **19**, 1278–1294.

Khalturin, K., Hemmrich, G., Fraune, S., Augustin, R. and Bosch, T.C. (2009) More than just orphans: are taxonomically-restricted genes important in evolution? *Trends Genet.* **25**, 404–413.

Kim, M.J., Shin, R. and Schachtman, D.P. (2009) A nuclear factor regulates abscisic acid responses in Arabidopsis. *Plant Physiol.* **151**, 1433–1445.

Knowles, D.G. and McLysaght, A. (2009) Recent de novo origin of human protein-coding genes. *Genome Res.* **19**, 1752–1759.

Li, L., Ilarslan, H., James, M.G., Myers, A.M. and Wurtele, E.S. (2007) Genome wide co-expression among the starch debranching enzyme genes AtISA1, AtISA2, and AtISA3 in *Arabidopsis thaliana*. *J. Exp. Bot.* **58**, 3323–3342.

Li, L., Foster, C.M., Gan, Q., Nettleton, D., James, M.G., Myers, A.M. and Wurtele, E.S. (2009) Identification of the novel protein QQS as a component of the starch metabolic network in Arabidopsis leaves. *Plant J.* **58**, 485–498.

Li, Z., Gao, J., Benning, C. and Sharkey, T.D. (2012) Characterization of photosynthesis in Arabidopsis ER-to-plastid lipid trafficking mutants. *Photosynth. Res.* **112**, 49–61.

Lillo, C., Meyer, C., Lea, U.S., Provan, F. and Oltedal, S. (2004) Mechanism and importance of post-translational regulation of nitrate reductase. *J. Exp. Bot.* **55**, 1275–1282.

Lin, J., Mazarei, M., Zhao, N., Zhu, J.J., Zhuang, X., Liu, W., Pantalone, V.R., Arelli, P.R., Stewart Jr, C.N. and Chen, F. (2013) Overexpression of a soybean salicyclic acid methyltransferase confers resistance to soybean cyst nematode. *Plant Biotechnol. J.* **11**, 1135–1145.

Lister, R., O'Malley, R.C., Tonti-Filippini, J., Gregory, B.D., Berry, C.C., Millar, A.H. and Ecker, J.R. (2008) Highly integrated single-base resolution maps of the epigenome in Arabidopsis. *Cell*, **133**, 523–536.

Luhua, S., Hegie, A., Suzuki, N., Shulaev, E., Luo, X., Cenariu, D., Ma, V., Kao, S., Lim, J., Gunay, M.B., Oosumi, T., Lee, S.C., Harper, J., Cushman, J., Gollery, M., Girke, T., Bailey-Serres, J., Stevenson, R.A., Zhu, J.K. and Mittler, R. (2013) Linking genes of unknown function with abiotic stress responses by high-throughput phenotype screening. *Physiol. Plant.* **148**, 322–333.

Marsden, R.L., Lee, D., Maibaum, M., Yeats, C. and Orengo, C.A. (2006) Comprehensive genome analysis of 203 genomes provides structural genomics with new insights into protein family space. *Nucleic Acids Res.* **34**, 1066–1080.

Mentzen, W.I. and Wurtele, E.S. (2008) Regulon organization of Arabidopsis. *BMC Plant Biol.* **8**, 99.

Mentzen, W.I., Peng, J., Ransom, N., Nikolau, B.J. and Wurtele, E.S. (2008) Articulation of three core metabolic processes in Arabidopsis: fatty acid biosynthesis, leucine catabolism and starch metabolism. *BMC Plant Biol.* **8**, 76.

Mishra, A.K., Agarwal, S., Jain, C.K. and Rani, V. (2009) High GC content: critical parameter for predicting stress regulated miRNAs in *Arabidopsis thaliana*. *Bioinformation*, **4**, 151–154.

Muller, O. and Krawinkel, M. (2005) Malnutrition and health in developing countries. *CMAJ*, **173**, 279–286.

Nagalakshmi, U., Wang, Z., Waern, K., Shou, C., Raha, D., Gerstein, M. and Snyder, M. (2008) The transcriptional landscape of the yeast genome defined by RNA sequencing. *Science*, **320**, 1344–1349.

Nagaraj, N., Wisniewski, J.R., Geiger, T., Cox, J., Kircher, M., Kelso, J., Paabo, S. and Mann, M. (2011) Deep proteome and transcriptome mapping of a human cancer cell line. *Mol. Syst. Biol.* **7**, 548.

Neme, R. and Tautz, D. (2013) Phylogenetic patterns of emergence of new genes support a model of frequent de novo evolution. *BMC Genomics*, **14**, 117.

Ohno, S. (1987) Early genes that were oligomeric repeats generated a number of divergent domains on their own. *Proc. Natl Acad. Sci. USA*, **84**, 6486–6490.

Pathan, M.S. and Sleper, D.A. (2008) Advances in soybean breeding. In *Genetics and Genomics of Soybean* (Stacey, G., ed.), pp. 113–134. New York: Springer.

Paz, M.M., Shou, H., Guo, Z., Zhang, Z., Banerjee, A.K. and Wang, K. (2004) Assessment of conditions affecting Agrobacterium-mediated soybean transformation using the cotyledonary node explant. *Euphytica*, **136**, 167–179.

Pimentel, D. and Pimentel, M. (2003) Sustainability of meat-based and plant-based diets and the environment. *Am. J. Clin. Nutr.* **78**, 660S–663S.

Schiltz, S., Gallardo, K., Huart, M., Negroni, L., Sommerer, N. and Burstin, J. (2004) Proteome reference maps of vegetative tissues in pea. An investigation of nitrogen mobilization from leaves during seed filling. *Plant Physiol.* **135**, 2241–2260.

Schroder, F., Lisso, J. and Mussig, C. (2011) EXORDIUM-LIKE1 promotes growth during low carbon availability in Arabidopsis. *Plant Physiol.* **156**, 1620–1630.

Seo, P.J., Kim, M.J., Ryu, J.Y., Jeong, E.Y. and Park, C.M. (2011) Two splice variants of the IDD14 transcription factor competitively form nonfunctional heterodimers which may regulate starch metabolism. *Nat. Commun.* **2**, 303.

Shou, H., Frame, B.R., Whitham, S.A. and Wang, K. (2004) Assessment of transgenic maize events produced by particle bombardment or Agrobacterium-mediated transformation. *Mol. Breed.* **13**, 201–208.

Silveira, A.B., Trontin, C., Cortijo, S., Barau, J., Del Bem, L.E., Loudet, O., Colot, V. and Vincentz, M. (2013) Extensive natural epigenetic variation at a de novo originated gene. *PLoS Genet.* **9**, e1003437.

Singh, K.B., Bejiga, G. and Malhotra, R.S. (1993) Genotype–environment interactions for protein content in chickpea. *J. Sci. Food Agric.* **63**, 87–90.

Stein, M., Dittgen, J., Sanchez-Rodriguez, C., Hou, B.H., Molina, .A., Schulze-Lefert, P., Lipka, V. and Somerville, S. (2006) Arabidopsis PEN3/PDR8, an ATP binding cassette transporter, contributes to nonhost resistance to inappropriate pathogens that enter by direct penetration. *Plant Cell*, **18**, 731–746.

Stitt, M., Lunn, J. and Usadel, B. (2010) Arabidopsis and primary photosynthetic metabolism—more than the icing on the cake. *Plant J.* **61**, 1067–1091.

Sulpice, R., Flis, A., Ivakov, A.A., Apelt, F., Krohn, N., Encke, B., Abel, C., Feil, R., Lunn, J.E. and Stitt, M. (2013) Arabidopsis coordinates the diurnal regulation of carbon allocation and growth across a wide range of photoperiods. *Mol. Plant*, **7**, 137–155.

Takahashi, H., Kanayama, Y., Zheng, M.S., Kusano, T., Hase, S., Ikegami, M. and Shah, J. (2004) Antagonistic interactions between the sa and ja signaling pathways in Arabidopsis modulate expression of defense genes and gene-for-gene resistance to cucumber mosaic virus. *Plant Cell Physiol.* **45**, 803–809.

Takeuchi, H. and Higashiyama, T. (2012) A species-specific cluster of defensin-like genes encodes diffusible pollen tube attractants in Arabidopsis. *PLoS Biol.* **10**, e1001449.

Tautz, D. and Domazet-Loso, T. (2011) The evolutionary origin of orphan genes. *Nat. Rev. Genet.* **12**, 692–702.

Thum, K.E., Shin, M.J., Gutierrez, R.A., Mukherjee, I., Katari, M.S., Nero, D., Shasha, D. and Coruzzi, G.M. (2008) An integrated genetic, genomic and systems approach defines gene networks regulated by the interaction of light and carbon signaling pathways in Arabidopsis. *BMC Syst. Biol.* **2**, 31.

Vaucheret, H., Béclin, C. and Fagard, M. (2001) Post-transcriptional gene silencing in plants. *J. Cell Sci.* **114**, 3083–3091.

Victora, C.G., de Onis, M., Hallal, P.C., Blossner, M. and Shrimpton, R. (2010) Worldwide timing of growth faltering: revisiting implications for interventions. *Pediatrics*, **125**, e473–e480.

Wilson, R.F. (2008) Soybean: market driven research needs. In *Genetics and Genomics of Soybean* (Stacey, G., ed.), pp. 3–16. New York: Springer.

Wilson, G.A., Feil, E.J., Lilley, A.K. and Field, D. (2007) Large-scale comparative genomic ranking of taxonomically restricted genes (TRGs) in bacterial and archaeal genomes. *PLoS ONE*, **2**, e324.

Wissler, L., Gadau, J., Simola, D.F., Helmkampf, M. and Bornberg-Bauer, E. (2013) Mechanisms and dynamics of orphan gene emergence in insect genomes. *Genome Biol. Evol.* **5**, 439–455.

Wu, L., EL-Mezawy, A., Duong, M. and Shah, S. (2010) Two seed coat-specific promoters are functionally conserved between *Arabidopsis thaliana* and *Brassica napus*. *In Vitro Cell Dev. Biol. Plant*, **46**, 338–347.

Wu, D.D., Irwin, D.M. and Zhang, Y.P. (2011) De novo origin of human protein-coding genes. *PLoS Genet.* **7**, e1002379.

Young, V.R. and Pellett, P.L. (1994) Plant proteins in relation to human protein and amino acid nutrition. *Am. J. Clin. Nutr.* **59**, 1203S–1212S.

Zhang, X., Myers, A.M. and James, M.G. (2005) Mutations affecting starch synthase III in Arabidopsis alter leaf starch structure and increase the rate of starch synthesis. *Plant Physiol.* **138**, 663–674.

Zhang, X., Szydlowski, N., Delvalle, D., D'Hulst, C., James, M.G. and Myers, A.M. (2008) Overlapping functions of the starch synthases SSII and SSIII in amylopectin biosynthesis in Arabidopsis. *BMC Plant Biol.* **8**, 96.

9

Genetic enhancement of oil content in potato tuber (*Solanum tuberosum* L.) through an integrated metabolic engineering strategy

Qing Liu[1,*,†], Qigao Guo[1,2,†], Sehrish Akbar[1,3], Yao Zhi[1,4], Anna El Tahchy[1], Madeline Mitchell[1], Zhongyi Li[1], Pushkar Shrestha[1], Thomas Vanhercke[1], Jean-Philippe Ral[1], Guolu Liang[2], Ming-Bo Wang[1], Rosemary White[1], Philip Larkin[1], Surinder Singh[1] and James Petrie[1]

[1]Commonwealth Scientific and Industrial Research Organisation Agriculture, Black Mountain, ACT, Australia
[2]College of Horticulture & Landscape Architecture, Southwest University, Chongqing, China
[3]National University of Science and Technology (NUST) Islamabad, Islamabad, Pakistan
[4]State Key Laboratory of Agricultural Microbiology, Huazhong Agricultural University, Wuhan, China

*Correspondence

email qing.liu@csiro.au
[†]These authors contributed equally.

Keywords: triacylglycerol, potato, *Solanum tuberosum*, WRI1, DGAT1, oleosin.

Summary

Potato tuber is a high yielding food crop known for its high levels of starch accumulation but only negligible levels of triacylglycerol (TAG). In this study, we evaluated the potential for lipid production in potato tubers by simultaneously introducing three transgenes, including *WRINKLED 1 (WRI1)*, *DIACYLGLYCEROL ACYLTRANSFERASE 1 (DGAT1)* and *OLEOSIN* under the transcriptional control of tuber-specific (patatin) and constitutive (CaMV-35S) promoters. This coordinated metabolic engineering approach resulted in over a 100-fold increase in TAG accumulation to levels up to 3.3% of tuber dry weight (DW). Phospholipids and galactolipids were also found to be significantly increased in the potato tuber. The increase of lipids in these transgenic tubers was accompanied by a significant reduction in starch content and an increase in soluble sugars. Microscopic examination revealed that starch granules in the transgenic tubers had more irregular shapes and surface indentations when compared with the relatively smooth surfaces of wild-type starch granules. Ultrastructural examination of lipid droplets showed their close proximity to endoplasmic reticulum and mitochondria, which may indicate a dynamic interaction with these organelles during the processes of lipid biosynthesis and turnover. Increases in lipid levels were also observed in the transgenic potato leaves, likely due to the constitutive expression of *DGAT1* and incomplete tuber specificity of the patatin promoter. This study represents an important proof-of-concept demonstration of oil increase in tubers and provides a model system to further study carbon reallocation during development of nonphotosynthetic underground storage organs.

Introduction

The increasing food demand as the result of the growing world population will require substantial growth in future global vegetable oil supply. This has been exacerbated in recent years by the increasing recognition of vegetable oil as a renewable and potentially environmentally friendly alternative energy resource. Current global production of vegetable oil is dominated by just a few crop species including oil palm, soya bean, rapeseed and sunflower, and there is a clear need to create new oil production platforms to help meet growing demand.

Potato (*Solanum tuberosum* L.) is an important stolon tuber crop that has been regarded as the fourth most important staple food crop in the world. The world production of potatoes in 2013 was about 368 million tonnes (FAOSTAT, www.faostat.fao.orq). In contrast to oilseed species such as soya bean and rapeseed, potato tuber is rich in starch but very low in oil. Several studies have reported that it is possible to increase oil levels in seeds or leaf tissues by redirecting carbon allocation towards fatty acid biosynthesis and triacylglycerol (TAG) accumulation. This has been achieved by overexpressing the transcription factors and enzymes involved in lipid biosynthesis (Focks and Benning, 1998;

Santos Mendoza *et al.*, 2005; Vanhercke *et al.*, 2014; Vigeolas *et al.*, 2007; Zale *et al.*, 2016).

In plants, the fatty acid biosynthesis and TAG assembly processes are highly regulated, involving spatial separation of biosynthetic steps between different organelle compartments and the exquisite control of several biosynthetic steps by one or multiple biochemical mechanisms (Harwood, 2005; Weselake *et al.*, 2009). *De novo* fatty acid biosynthesis occurs mainly in the plastid where the biotin-containing enzyme acetyl-CoA carboxylase (ACCase) catalyses the first committed step by activating acetyl-CoA to malonyl-CoA. The malonyl group is then transferred from CoA to an acyl carrier protein (ACP) that serves as the carrier for the growing acyl chain, a step catalysed by a large dissociable multi-enzyme complex known as fatty acid synthase (FAS). The fatty acids thus synthesised are then exported to the endoplasmic reticulum (ER) where they become precursors for the production of both storage and membrane lipids. In the classical Kennedy pathway, *sn*-glycerol-3-phosphate acyltransferase (GPAT), lysophosphatidic acid acyltransferase (LPAAT) and diacylglycerol acyltransferase (DGAT) catalyse the sequential acylation reactions on a glycerol-3-phosphate backbone chain to produce TAG (Weselake *et al.*, 2009; Zou *et al.*, 1999). TAG synthesised within the ER membrane is then budded off as oil bodies or lipid droplets (LD) that have a single-layer phospholipid (PL) membrane with the polar head groups in contact with the cytosol and the nonpolar tails in contact with the internal neutral lipids (van Rooijen and Moloney, 1995; Yatsu and Jacks, 1972). Oleosins are small proteins that coat these LDs, helping to stabilise and protect TAG from cytosolic components and withstand the strains of both dehydration and rehydration in seeds (Siloto et al., 2006). The amount of oleosin proteins may also play a crucial role in determining the size of LDs (Siloto et al., 2006).

Many of the key enzymes involved in fatty acid biosynthesis and lipid assembly have now been isolated, and much effort has been spent in attempts to increase oil content via genetic modification. Overexpression of individual enzymes such as ACCase in Nicotiana tabacum leaves and potato tubers (Klaus et al., 2004; Madoka et al., 2002), or DGAT1 in N. tabacum leaves (Andrianov et al., 2010; Bouvier-Nave et al., 2000) resulted in only limited increases in TAG accumulation. Similarly, reduction in lipid turnover by reducing expression of TAG lipases and enzymes involved in β-oxidation resulted in only modest TAG accumulation increases (Fan et al., 2014; Slocombe et al., 2009). Alternative approaches have exploited transcription factors. WRINKLED1 (WRI1) is a transcription factor that was first identified in an Arabidopsis (Arabidopsis thaliana L.) mutant producing wrinkled seeds with lower oil content and higher content of sugars than wild type (WT), suggesting that WRI1 is involved in the regulation of carbon flux during seed development (Focks and Benning, 1998). Overexpression of WRI1 in wri1 Arabidopsis mutant or Brassica napus not only recovered the normal seed appearance but also increased oil content relative to WT Arabidopsis seeds (Cernac and Benning, 2004; Liu et al., 2010). Overexpression of WRI1 in maize (Zea mays L.) resulted in 30% higher seed oil content at the expense of starch accumulation (Shen et al., 2010). During the preparation process of this manuscript, it was reported that potato tuber overexpressing Arabidopsis WRI1 alone was able to raise TAG level up to 1% of tuber dry weight (DW) (Hofvander et al., 2016).

Recent work has also suggested that the coordinated regulation of the genes involved in transcriptional control of carbon metabolism and those involved in fatty acid and lipid biosynthesis is required for a maximum increase in TAG accumulation in transgenic plants. We have recently reported that the simultaneous overexpression of WRI1, DGAT1 and OLEOSIN genes was able to increase TAG content in tobacco leaf to 15% of its DW without severely impacting on plant development (Vanhercke et al., 2014). Each of these three genes plays a critical role in directing more carbon flux to fatty acid biosynthesis, TAG assembly and LD biogenesis or protection, respectively. The levels of TAG accumulation achieved by this approach have far exceeded the levels previously reported for engineering TAG yields in plant vegetative tissues, including the recent report in potato tuber expressing WRI1 alone (Hofvander et al., 2016). Such a finding has opened up new possibilities of using high biomass plants as alternative platforms for the production of high energy storage lipids.

The aim of this study was to extend our previous efforts in tobacco leaf to another high biomass plant tissue, with a view to apply our model to one of most extreme carbohydrate-based storage organ, namely potato tuber. We have examined the potential for lipid production and accumulation in potato tubers by introducing three genes: WRI1 and OLEOSIN under the transcriptional control of a tuber-specific promoter derived from the potato PATATIN gene, and DGAT1 driven by the constitutive CaMV-35S promoter. The synergistic functioning of these genes (Vanhercke et al., 2013) has resulted in significant increases in TAG and, to a lesser extent, PL and galactolipids (GL).

Results

Generation of transgenic potato plants containing transgenes for over-expression of WRI1, DGAT1 and OLEOSIN genes

The binary plasmid construct pJP3506 contained one gene cassette expressing Arabidopsis DGAT1 (atDGAT1) driven by the CaMV-35S promoter and two gene cassettes expressing Arabidopsis WRI1 (atWRI1) and sesame (Sesamum indicum L.) OLEOSIN (siOLEOSIN), respectively, under the transcriptional control of the tuber-specific patatin class I promoter B33 (Figure 1). The strong constitutive viral promoter CaMV-35S was selected for DGAT1 expression as this promoter had been successfully used to express the DGAT1 gene in tobacco, resulting in a modest TAG increase without a significant negative impact on plant growth (Bouvier-Nave et al., 2000; Vanhercke et al., 2013, 2014). The potato patatin promoter B33 was used as it is highly active in tuber but with low activity in other tissues (Rocha-Sosa et al., 1989). This promoter was selected in order to restrict the alteration of lipid biosynthesis to the tuber only, thereby minimising any potential negative impact on plant development.

Following the transformation of pJP3506 vector in potato (Solanum tuberosum cv Atlantic), 178 independent primary transgenic potato lines were selected on kanamycin-containing media and allowed to grow to maturity in potted soil under glasshouse condition.

Lipid analysis and molecular assessment of transgene expression in transgenic potato tubers

Assessment of lipid content was carried out in mature tubers harvested from both transgenic and wild-type (WT) plants that were grown alongside the transgenic plants. Total lipids were extracted and fractionated by thin-layer chromatography (TLC) for the analysis of TAG, PL and GL by gas chromatography (GC). In the primary screening of transgenic lines, 128 lines were found to contain TAG levels ranging between 0.05 and 4.70% of tuber DW, significantly increased in comparison with the 0.03% in WT. Eight lines with relatively high levels of TAG accumulation were selected to sprout and grow to maturity for further analysis. It is important to note that this was a vegetative reproduction of hemizygous transgenic lines and did not result in the transgenes being genetically fixed through homozygosity.

Among the eight potato transgenic lines, line #69 showed the highest TAG accumulation in tuber with an average level of 3.3% on a DW basis. This is an approximate 100-fold increase compared with WT (Figure 2a). The same line also accumulated the highest level of phosphotidylcholine (PC) at 0.27% which is 5.8-fold increase compared with WT; the highest monogalactosyldiacylglycerol (MGDG) at 0.06% which is 5.6-fold increase relative to WT; and the highest digalactosyldiacylglycerol (DGDG) at 0.86% which is 3.9-fold increase compared to WT (Figure 2b). The other selected seven lines also showed various levels of increases in TAG, PC, MGDG and DGDG as shown in Figure 2.

The enhanced lipid accumulation was accompanied by altered fatty acid composition in transgenic tubers. The TAG fraction showed significant reductions in the levels of saturated fatty acids and α-linolenic acid (ALA, $C18:3^{\Delta9,12,15}$) and significantly increased levels of monounsaturated fatty acids, mostly oleic acid ($C18:1^{\Delta9}$) and palmitoleic acid ($C16:1^{\Delta9}$), compared to WT tubers (Figure 3a). A similar trend was also observed in PC, MGDG and DGDG, with the increase in linolenic acid (LA, $C18:2^{\Delta9,12}$) being more evident than that in TAG (Figure 3b–d).

Cross sections of potato tubers stained with Sudan Red 7B showed enhanced accumulation of neutral lipids in the transgenic tubers in sharp contrast to WT (Figure 4a). It is visibly clear that that the accumulation of neutral lipids in transgenic line #68 is the highest among the three lines examined as indicated by its deepest red stain in tuber. The line #95 tuber showing a moderate increase in neutral lipids was stained in red colour ranging between line #68 and WT control line.

TAG accumulation in transgenic potato tuber (line #68) was visually evident as abundant LDs in tuber cells under confocal microscope following staining with Nile Red (Figure 4b), in clear contrast to WT tubers (Figure 4c). The bright-field image illustrated that LDs were mostly in close proximity to starch granules in tuber cells (Figure 4d,e).

Potato tubers of line #68 and WT were also compared by transmission electron microscopy (TEM). The LDs showed typically spherical and ovoid structures, in which the core of neutral lipids, believed to be mostly TAG, was surrounded by presumably a monolayer of PL showing uniform electron density (Figure 5). The sizes of LDs were ranged between 0.5 and 1.5 μm in diameter, which is consistent with the observation of the Nile Red-stained LDs under confocal microscope. LDs were often found in close proximity with the ER and mitochondria (Figure 5).

Transgene expression was not detected in WT tubers. The expression levels of the three transgenes in the developing tubers of the eight transgenic lines were assessed by real-time reverse transcription polymerase chain reaction (qRT-PCR) (Figure 6). In the transgenic lines, WRI1 expression was generally lower than that of DGAT1 and OLEOSIN, with the latter showing the highest expression of the three transgenes in all lines except lines #17 and #169. This may indicate that a moderate level of WRI1 expression is sufficient for the enhancement of TAG accumulation.

Lipid analysis and molecular assessment of transgene expression in transgenic potato leaves

Significant increases in TAG content, ranging from 2- to 12-fold, were also observed in fully expanded leaves at postanthesis stage in all the transgenic lines tested (Figure 7). The increases of PC, MGDG and DGDG were also evident, but to a much lesser extent compared to that of TAG. Interestingly, line #17 that showed the lowest TAG increase among all the eight transgenic lines showed the highest accumulation of GL (both MGDG and DGDG). Fatty acid composition of TAG in the transgenic leaves was also significantly altered (Figure 8a). ALA content was substantially reduced while the other fatty acids, including saturates, monounsaturates, LA and long-chain fatty acids (LCFA, including C20:0, C22:0, C24:0 and C20:1), were increased. Compared to WT, saturates in PC and GL were significantly reduced, rather than increased as in TAG (Figure 8b–d). The substantial increase in LCFA was not observed in PC and GL.

The expression levels of the three transgenes were also assessed with qRT-PCR in transgenic potato leaves (Figure 9).

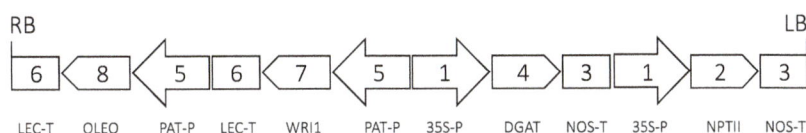

Figure 1 Schematic representation of the construct pJP3506 including the insertion region between the left and right borders. 1, cauliflower mosaic virus 35S promoter with duplicated enhancer region (35S-P); 2, neomycin phosphotransferase II (NPTII); 3, nopaline synthase terminator (NOS-T); 4, *Arabidopsis thaliana DGAT1* (DGAT); 5, potato patatin B33 promoter (PAT-P); 6, *Glycine max* lectin terminator (LEC-T); 7, *A. thaliana WRI1* (WRI1); 8, *Sesame indicum OLEOSIN* (OLEO).

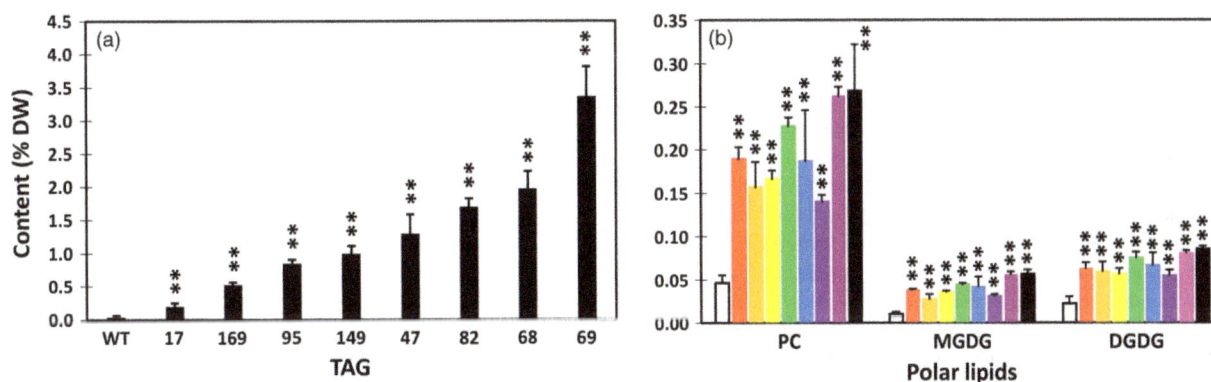

Figure 2 Triacylglycerol (TAG) (a), phosphotidylcholine (PC) and monogalactosyldiacylglycerol (MGDG), digalactosyldiacylglycerol (DGDG) (b) contents in the tubers of wild type (WT) (white bar) and eight selected transgenic potato lines, including #17 (red bar), #169 (orange bar), #95 (yellow bar), #149 (green bar), #47 (blue bar), #82 (indigo bar), #68 (violet bar) and #69 (black bar). Asterisks indicate statistically significant differences between transgenic line and WT using Student's t-test, with a significance threshold of 0.05 (*) and 0.01 (**). Error bars indicate standard deviations.

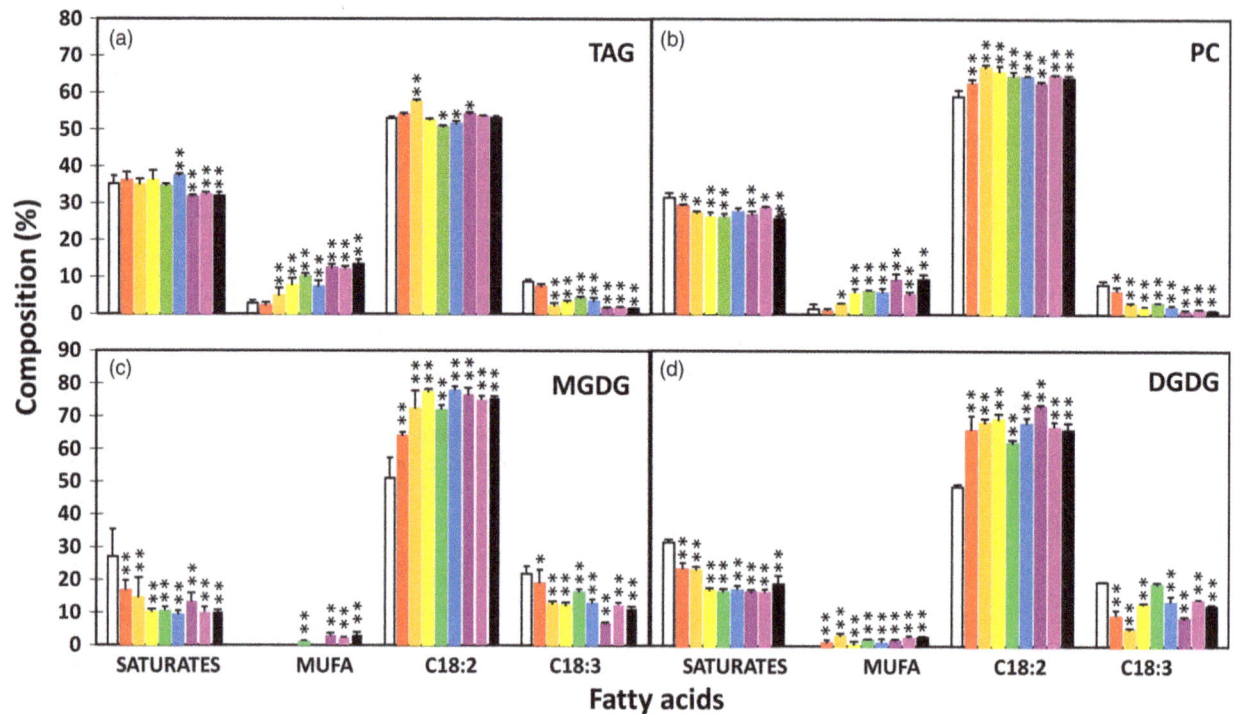

Figure 3 Fatty acid composition in triacylglycerol (TAG) (a), phosphotidylcholine (PC) (b) and monogalactosyldiacylglycerol (MGDG) (c) and digalactosyldiacylglycerol (DGDG) (d) in the tubers of wild type (WT) (white bar) and eight selected transgenic potato lines, including #17 (red bar), #169 (orange bar), #95 (yellow bar), #149 (green bar), #47 (blue bar), #82 (indigo bar), #68 (violet bar) and #69 (black bar). Asterisks indicate statistically significant differences between transgenic line and WT using Student's t-test, with a significance threshold of 0.05 (*) and 0.01 (**). Error bars indicate standard deviations.

Expression of *atDGAT1* driven by CaMV-35S promoter was evident in leaves of the eight transgenic lines, with relatively high expression levels shown by lines #68, #149 and #69, coincidental with their relatively higher levels of TAG accumulation in leaf among the eight transgenic lines. Despite the expressions of *atWRI1* and *siOLEOSIN* being transcriptionally controlled by the patatin B33 promoter, *atWRI1* expression was also clearly detectable in the leaves of the above-mentioned three high-lipid lines, while various levels of *siOLEOSIN* expression were shown in all the transgenic leaves. Similar to the expression in tubers, *siOLEOSIN* expression level was the highest among the three transgenes in transgenic potato leaf.

Analysis of sugar content, starch content and starch granule morphology of transgenic potato tuber

Significant increase in soluble sugar content and reduction in starch content were observed in transgenic lines compared to WT (Figure 10a). Such a trend in the change of carbohydrate accumulation was in a good correlation with the increase in TAG (Figure 10b). Scanning electron microscopy (SEM) revealed morphological differences of starch granules between WT and high-lipid transgenic tubers, which are represented by lines # 47 and #69 (Figure 11). In contrast to the large, smooth-surfaced ellipsoidal starch granules in WT (Figure 11a), the starch granules in the high-lipid lines had angular and irregular shapes with some granules showing indentations on the surface (Figure 11b,c). However, birefringence (a measure of the level of crystalline order in starch granules) did not appear to be affected in the high-lipid lines (Figure 11d–f) and all starch granules showed a distinct Maltese cross with similar brightness when examined by polarised light in both WT (Figure 11d) and transgenic lines (Figure 11e,f).

Discussion

Potato tubers are developed from underground stems. Like most other underground storage organs, potato tubers accumulate mostly starch that is used as energy reserves for sprouting and early plant establishment. Potato tubers also accumulate storage proteins, predominantly the glycoprotein patatin (Hoefgen and Willmitzer, 1990), but have very low levels of TAG. This is in sharp contrast to an oilseed where TAG is the major energy reserve supporting seed germination until the seedling is capable of photosynthesis.

In this study, we have demonstrated that TAG levels, and to a lesser extent PL and GL levels, can be raised in potato tubers by coordinated up-regulation of fatty acid production, TAG assembly and TAG packaging. This is an extension of the so-called Push, Pull and Protection strategy into an underground tuber crop, a concept developed in transgenic tobacco leaves where a massive increase in TAG content (up to 15% of DW) was observed by co-overexpressing *WRI1*, *DGAT1* and *OLEOSIN* genes (Vanhercke et al., 2014).

Yellow nutsedge (*Cyperus esculentus* L.) is perhaps the only plant known to accumulate high levels of dry matter (58%) and TAG (20%–36% DW) in tuber, which is comparable to a typical seed (Linssen et al., 1989; Turesson et al., 2010). But the utilisation of nutsedge tuber for vegetable oil production is not yet realistic because of its lack of domestication and environmental concerns. It is estimated that the transgenic potato with 25% dry matter and 3.3% TAG on DW basis could yield 393 kg vegetable oil per hectare in the USA where 47.7 tonnes potato fresh weight per hectare could be produced (FAOSTAT, http://faostat3.fao.org), which is comparable to current cotton seed oil yield.

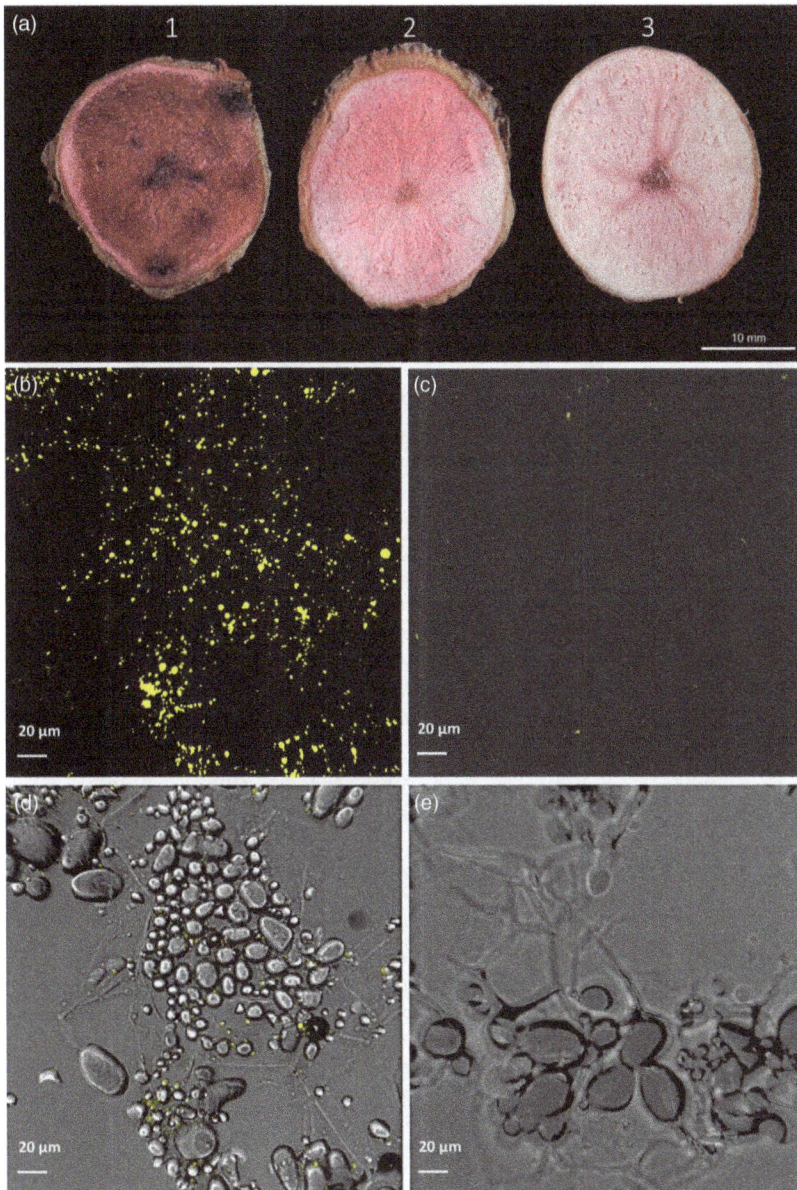

Figure 4 Visualisation of lipid accumulation in potato tubers. (a) Cross section of mature potato tubers showing lipid accumulation as stained with Sudan Red 7B. 1. Line #68; Transgenic potato line #68 showing high triacylglycerol (TAG) accumulation; 2. Line #95 showing moderate level of TAG accumulation; 3. Wild type (WT) showing low level TAG accumulation. (b) Confocal microscopy image of abundant lipid droplets (LDs) following staining with Nile Red in line #68. (c) Confocal microscopy image showing rare appearance of LDs following staining with Nile Red in WT. (d) Bright-field image of b. (e) Bright-field image of c.

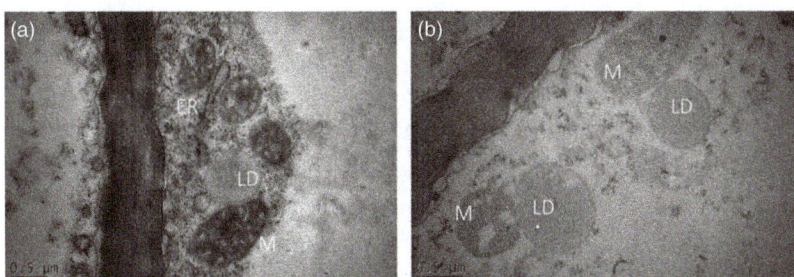

Figure 5 Representative TEM images of transgenic potato line #68, showing lipid droplets (LDs) and their close proximity to endoplasmic reticulum (ER) and mitochondria (M).

The present day knowledge of fatty acid biosynthesis and TAG accumulation in plants is mainly derived from studies on seeds in Arabidopsis, and other oilseed crop species, while much remains unknown in nonseed plant tissue, such as potato tuber. In model plants, it has been established that WRI1 is a positive regulator of genes encoding enzymes involved in late glycolysis and fatty acid biosynthesis (Cernac and Benning, 2004; Chapman and Ohlrogge, 2012). It has two APETALA2 domains by which it binds to the promoter of the target genes, thereby activating and enhancing their transcription levels. In vitro experiments have demonstrated that WRI1 binds directly to the AW box of several fatty acid synthesis genes, and the mutation of the AW box abolished the WRI1-mediated transcriptional activation in transient protoplast assays (Maeo et al., 2009). Expression studies

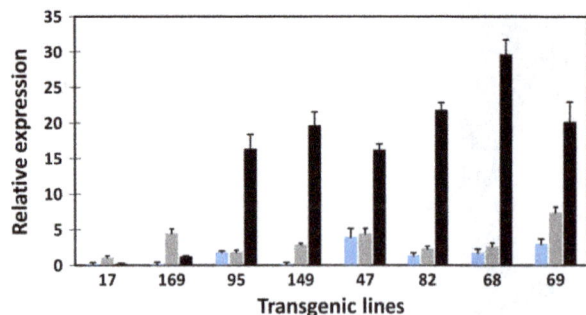

Figure 6 Real-time qRT-PCR analysis of transgene expression in developing transgenic potato tubers. *atWRI* (blue bar), *atDGAT1* (shade bar) and *siOLEOSIN* (black bar).

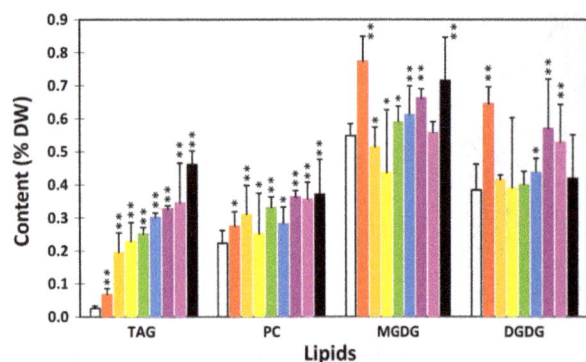

Figure 7 Triacylglycerol (TAG), phosphotidylcholine (PC) and monogalactosyldiacylglycerol (MGDG), digalactosyldiacylglycerol (DGDG) contents in the leaves of wild type (WT) (white bar) and eight selected transgenic potato lines, including #17 (red bar), #47 (orange bar), #82 (yellow bar), #95 (green bar), #169 (blue bar), #69 (indigo bar), #149 (violet bar) and #68 (black bar). Asterisks indicate statistically significant differences between transgenic line and WT using Student's *t*-test, with a significance threshold of 0.05 (*) and 0.01 (**). Error bars indicate standard deviations.

have shown that target genes of WRI1 may include those encoding the BCCP subunit of ACCase, ACP, enoyl-ACP reductase, β-ketoacyl-ACP reductase, plastidial pyruvate kinase, pyruvate dehydrogenase and FAD2 (Baud *et al.*, 2007; Ruuska *et al.*, 2002; To *et al.*, 2012).

In potato, over-expression of *ACCase* from Arabidopsis led to an increase in fatty acid biosynthesis and over fivefold increase in TAG content in tubers (Klaus *et al.*, 2004). The identification of BCCP subunit of ACCase as a target gene of *WRI1* up-regulation could be one of the key underlying mechanisms for TAG biosynthesis, as the result of *WRI1* over-expression (Sasaki and Nagano, 2004). In maize, the expression of either *WRI1* or its upstream regulator *LEC1* was able to increase the oil content in seeds, but *WRI1* is advantageous as it does not give rise to detrimental effects such as poor seed germination, and stunted growth as seen in *LEC1* over-expressing lines (Shen *et al.*, 2010). In addition to developing seed, a *WRI1* homologue from oil palm has been shown to be highly expressed in oil-accumulating mesocarp, 57-fold higher than the closely related date palm that does not accumulate oil (Bourgis *et al.*, 2011). Likely, as a result of the *WRI1* up-regulation, the expressions of genes involved in *de novo* fatty acid biosynthesis have been found elevated in oil

palm by 13-fold, compared to date palm, while the transcripts of the genes involved in TAG assembly were similar between oil and date palm mesocarps (Bourgis *et al.*, 2011). Subsequent experiments demonstrated that the oil palm *WRI1* homologue was able to rescue the Arabidopsis *wri1-1* (Ma *et al.*, 2013). This is regardless of the fact that the target gene *ACCase* in monocot species is highly divergent from that of dicot species. In the present study with potato tuber and leaf, the expression of *atWRI1* in transgenic potato tuber appeared to be moderate compared to *atDGAT1* and *siOLEOSIN,* which indicates a possible selection against those with high *WRI1* expression, and therefore, we have never recovered those tubers with very high level of *WRI1* expression. It is also possible that being a transcription factor gene, *WRI1* expression is normally at low levels compared to those encoding catalytic enzymes.

The expression of *DGAT1 and OLEOSIN* is unaffected by *WRI1* over-expression in either Arabidopsis or maize (Maeo *et al.*, 2009). Further, the peak expression of *DGAT* and *OLEOSIN* occurs much later compared to that of the *WRI1* during seed development (Kilaru *et al.*, 2015). It is therefore necessary to co-express *WRI1* together with *DGAT1* and *OLEOSIN* for an efficient enhancement of TAG production in transgenic plants, as we have demonstrated in tobacco leaf (Vanhercke *et al.*, 2014) and the potato tuber reported herein.

DGAT1 catalyses the final step in the Kennedy pathway for TAG biosynthesis, which is considered as a rate limiting step in oilseeds (Jako *et al.*, 2001; Lung and Weselake, 2006). The line #69 with the highest TAG accumulation also showed the highest *atDGAT1* expression in transgenic tubers; hence, the expression level of *atDGAT1* is a clear contributor to the high level TAG accumulation in potato. However, it is not clear how much of the TAG increase in potato tubers is attributable to the over-expression of *DGAT1* which is under the transcriptional control of the CaMV-35S promoter.

Both *atWRI1* and *siOLEOSIN* driven by patatin promoter have been found to be expressed in potato leaves, contrary to our initial anticipation of their tuber-specific expression. The enhancement of lipid accumulation in potato leaves is therefore unlike a sole effect of CaMV-35S driven *atDGAT1*. Perhaps at least partly due to some low level transcriptional activity of patatin promoter in leaf, the observed 12-fold increase in TAG is far less than what was previously observed in tobacco leaves (Vanhercke *et al.*, 2014). It has been previously reported that the expression of patatin is modulated by exogenous sucrose and DNA regions termed as sucrose responsive elements (SURE) elements have been identified in the proximal regions of the patatin promoter (Grierson *et al.*, 1994). The unexpected raise of soluble sugars in the high-lipid potato may have contributed to the leakiness of patatin promoter and led to TAG increase in potato leaf tissues.

DGAT1 uses DAG as a substrate that is also used by choline phosphotransferase to produce PC. In both potato tuber and leaf tissues, we have observed the concomitant increases of TAG and PC, rather than the increase in TAG at the expense of PC as shown in tobacco leaves (Vanhercke *et al.*, 2014). Clearly, the increase in PC is necessary in the potato cells with dramatically increased TAG to maintain its sufficient presence on the LD surface. However, it is not yet known what physiological impact the substantial increase in polar lipids, especially GL, might have in transgenic potato tubers.

Recently, plant oleosin and other lipid body-associated proteins have emerged as a new target for metabolic engineering of enhanced production in oil seeds as well as vegetative plant

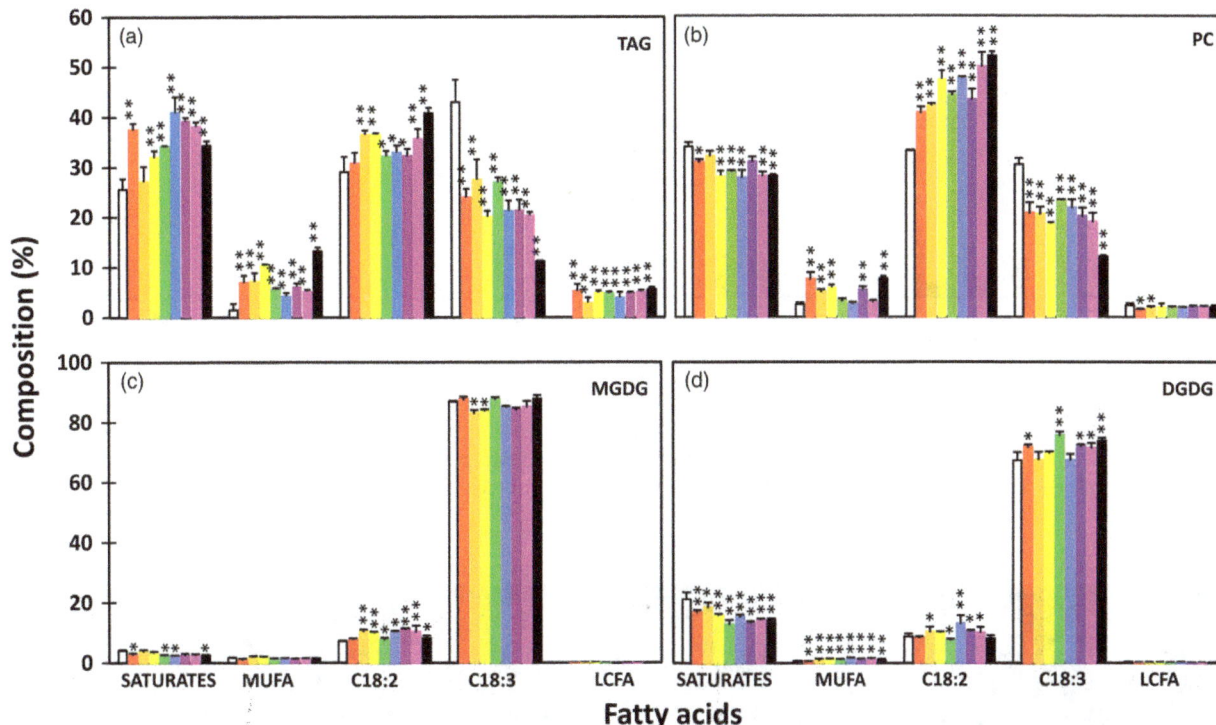

Figure 8 Fatty acid composition in triacylglycerol (TAG) (a), phosphotidylcholine (PC) (b) and monogalactosyldiacylglycerol (MGDG) (c) and digalactosyldiacylglycerol (DGDG) (d) in the leaves of wild type (WT) (white bar) and eight selected transgenic potato lines, including #17 (red bar), #47 (orange bar), #82 (yellow bar), #95 (green bar), #169 (blue bar), #69 (indigo bar), #149 (violet bar) and #68 (black bar). Asterisks indicate statistically significant differences between transgenic line and WT using Student's t-test, with a significance threshold of 0.05 (*) and 0.01 (**). Error bars indicate standard deviations.

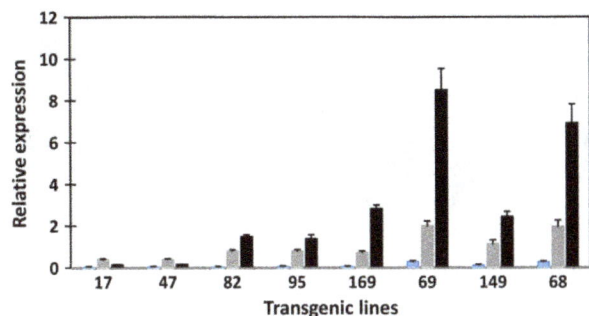

Figure 9 Real-time qRT-PCR analysis of transgene expression in transgenic potato leaves. *atWRI* (blue bar), *atDGAT1* (shade bar) and *siOLEOSIN* (black bar).

tissues. At least 17 differentially expressed *OLEOSIN* genes have been identified in Arabidopsis, suggesting that these genes are highly regulated even within a single plant species (Hsieh and Huang, 2004). The sesame *OLEOSIN* that was used in this study was previously demonstrated to enhance TAG accumulation in tobacco leaves (Vanhercke *et al.*, 2014; Winichayakul *et al.*, 2013). This is despite the fact that in Arabidopsis and maize the down-regulation of the *OLEOSIN* expression did not lead to a decrease in oil accumulation (Siloto *et al.*, 2006; Ting *et al.*, 1996). In this study, a high level expression of the introduced *OLEOSIN* gene was associated with the highest TAG accumulation in transgenic potato.

A previous study revealed that the *OLEOSIN*s are not highly expressed in nonseed oleaginous tissues such as the mesocarp of olive, oil palm and avocado (Kilaru *et al.*, 2015). Instead, lipid droplet-associated proteins (LDAP) have been identified in these tissues that were associated with LD formation (Horn *et al.*, 2013). It is therefore likely that oleosin may not be the optimal packaging protein in nonseed tissues to protect the accumulated oil from TAG lipase or other cytosolic enzyme activities. Additional research evaluating plant LDAP will be of particular interest in the further enhancement of TAG accumulation in potato tubers.

The observation of an increase in the number and size of LDs in transgenic potato indicated that the formation of LDs is clearly driven by the availability of TAG. A tight association of LDs with the ER has been observed in transgenic potato tubers, consistent with the hypothesis that LDs are formed on the surface of ER membrane (Thiam *et al.*, 2013). It can be envisaged that an expanded ER network dedicated to the lipid production was formed in the transgenic potato tuber cells to cope with the increased TAG production. Based on studies in yeast and mammalian cells, it has been proposed that large LDs can arise either by LD expansion or fusion (Martin and Parton, 2006). With the supply of excess fatty acids, LDs in yeast cells may grow up to 30-fold in volume within hours (Krahmer *et al.*, 2011). In this study, LD fusion was not observed despite examination of a large number of TEM specimens. We may therefore assume that during LD growth, TAGs are added to the droplet cores and PLs to the surfaces. The close association between the ER and LDs may facilitate the transfer of the newly synthesised TAG to the droplet core.

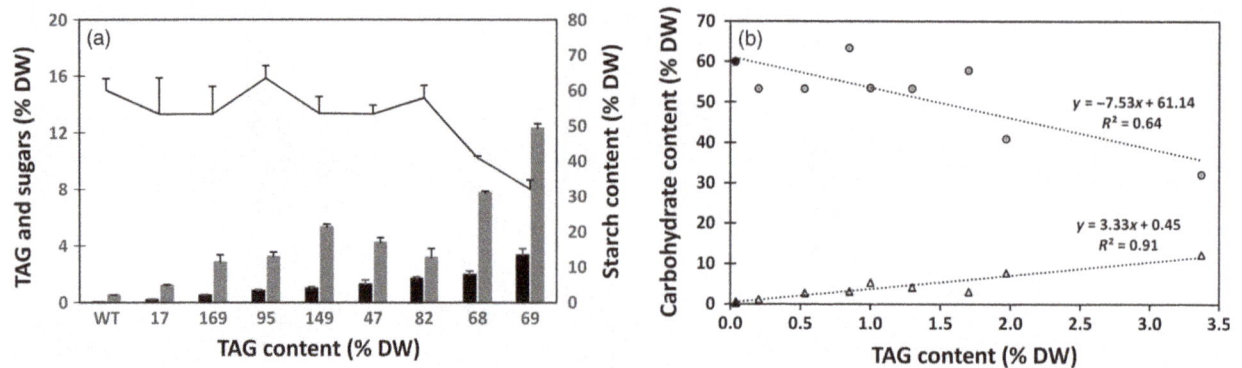

Figure 10 Carbohydrate accumulation in transgenic potato tubers. Contents of triacylglycerol (TAG) (black bar), soluble sugars (shade bar) and starch (line) in potato tubers (a). Relationship between TAG and carbohydrate content in wild type (WT) and transgenic potato lines (b): WT tuber (black symbols) has high starch (circles) and low soluble sugars (triangles) compared to transgenic potato lines (open symbols). Overall, the relationship between TAG and starch is negative ($y = -7.53x + 61.14$, $R^2 = 0.64$), while the relationship between TAG accumulation and soluble sugars is positive ($y = 3.33x + 0.45$, $R^2 = 0.91$).

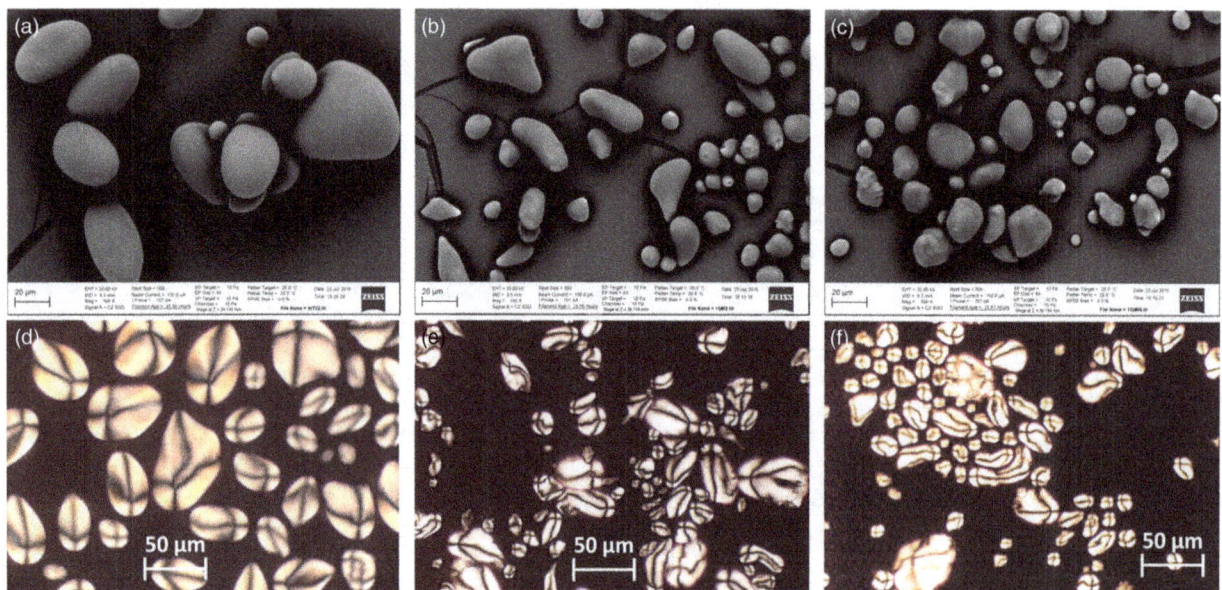

Figure 11 Starch granule morphology. SEM images of starch granules of wild type (WT) (a), transgenic lines #47 (b) and #69 (c). polarised light image of starch granules of WT (d), transgenic lines #47 (e) and #69 (f).

To determine whether the enhanced lipid accumulation has impacted on the other major carbon reserve in potato, we quantified the starch and soluble sugar levels in the tubers of a number of selected transgenic potato lines. It appears that the introduction of an engineered lipid biosynthesis pathway overexpressing *WRI1/DGAT1/OLEOSIN* triple gene cassettes may have negatively affected starch accumulation in most of the high-lipid lines. This is consistent with our previous observation with tobacco leaves where the up-regulation of fatty acid biosynthesis and TAG assembly pathways has resulted in substantial reduction of starch accumulation (Vanhercke *et al.*, 2014). In maize, the over-expression of a *WRI1* gene resulted in increased oil content in the grains by 10%–22%, with a concomitant 60% reduction in starch content (Shen *et al.*, 2010).

In our transgenic potato lines, consistent correlations were observed between TAG accumulation, starch reduction and soluble sugars accumulation, despite a few high-lipid lines where the starch content was not significantly reduced. Enhancement of

the lipid biosynthesis pathway in transgenic potato tubers likely had a detrimental effect on starch synthesis, leading to carbon accumulation in the form of soluble sugars. However, the relationship between starch and lipid accumulation in potato does not appear to be a simple redistribution of carbon but rather a complex mechanism that depends on the strategy used to reroute the carbon. For example, antisense down-regulation of the starch biosynthetic enzyme AGPase in potato tubers resulted in a reduction of starch and an increase in soluble sugars but not fatty acids or TAG levels (Klaus *et al.*, 2004). Although the underlying mechanism behind the decreased starch content in our study was not determined, it is likely the result of competition for carbon from the boosted lipid metabolic pathway, lowering the amount of metabolites available to be utilised for starch biosynthesis.

The hypothesis that competition for carbon decreases starch biosynthesis is consistent with the altered size and shape of starch granules in the transgenic potato tubers. Similar reductions in starch granule size, irregular granule shape and uneven surface morphology have been observed in several studies where the starch synthesis pathway was altered (Jobling et al., 2002; Regina et al., 2010; Shaik et al., 2016). However, in our case, preliminary investigation did not suggest any apparent change in starch birefringence or crystallinity. Further research is underway to investigate whether altered amylopectin/amylose composition is responsible for the observed changes in the size and morphology of starch granules in the transgenic potato tuber lines.

The fatty acid composition of polar lipids in potato tubers may reflect the biophysical features of the membrane structures of the cells and subcellular organelles. The degree of unsaturation of fatty acids in the constituent lipids of membranes is a major factor responsible for cell membrane fluidity at a given ambient temperature. We have observed a significant alteration of fatty acid composition in transgenic tubers, showing higher levels of oleic acid and LA, but lower levels of palmitic acid and ALA when compared to WT tubers. Our previous work with transgenic tobacco leaf showed a similar trend of alteration in fatty acid composition, but with more dramatic increase in oleic acid (Vanhercke et al., 2014). This could be as the result of WRI1 regulation on the expression of microsomal fatty acid desaturases as previously suggested by Ruuska et al. (2002). In transgenic tobacco leaf, the expression levels of FAD3 and FAD7 responsible for the ALA production in eukaryotic and prokaryotic pathways were found to be down-regulated, leading to the reduction of trienoic fatty acids (Vanhercke et al., 2014). WRI1 was also shown to transactivate the FATTY ACID ELONGASE1 gene (FAE1), and an AW box was also located in the promoter region of FAE1 (Maeo et al., 2009). This might partly explain our observation of a substantial increase in LCFAs in both tuber and leaf of the transgenic potato lines with increased TAG accumulation in this study. This is consistent with the characterisation of seed oil in Arabidopsis wri1 mutant that featured higher proportions of ALA and to a lesser extent, erucic acid (C22:1$^{\Delta 13}$) at the expense of oleic acid, LA and eicosenoic acid (C20:1$^{\Delta 11}$) (Cernac and Benning, 2004). It might also be possible that an enhanced DGAT activity may have selectively increased the flux of linoleoyl-DAG to TAG and deprived it of opportunity for further desaturation while associated with PC. In maize, a high-oil QTL with increased TAG content as the result of a single amino acid substitution in DGAT1 also showed significantly raised oleic acid contents at the expense of polyunsaturates (Zheng et al., 2008). The fatty acid profile of lipids has important implications for the functioning of biological membranes as well as for postharvest applications. Further research should be conducted on the plant physiology and adaptations to different growth temperatures by the transgenic potato with altered lipid profile. It could be envisaged that such an alteration in fatty acid composition is favourable for food applications, as the high ALA content in WT tubers leads to higher levels of autoxidation that is largely responsible for off-flavours and rancidity problems in processed potato food products. Galliard (1970, 1972) demonstrated that the presence of phospholipase, galactolipase and lipoxygenases could lead to rapid lipid peroxidation when tuber cells were broken. Together with the increase in saturated fatty acids, reduction in the level of ALA is anticipated to have marked effects on the improvement of oxidative stability of potato products. Further, the change of fatty acid composition is also favourable for biodiesel applications due to the improved oxidative stability, and possibly ignition quality and cold-temperature flow properties in potential biofuel applications (Rogalski and Carrer, 2011).

Concluding remarks

In this study, we have demonstrated as a proof of concept that potato tubers could be used as a potential target storage organ for TAG accumulation through a synergistic engineering approach, providing a basis for further optimisation of the 'Push', 'Pull' and 'Protect' strategy to further enhance TAG accumulation in a nonphotosynthetic underground sink organ. The moderately enhanced oil accumulation in potato tubers demonstrated herein should encourage further studies of TAG biosynthesis and turnover, and the intricate relationship between oil and starch in a classic starch-accumulating storage organ. Because the TAG content reaches more than 30% of DW in an anatomically similar nutsedge tuber, at least in theory it might be possible to further raise the TAG level in transgenic potato tubers. Increased knowledge of the lipid accumulation mechanism in transgenic potato would generate opportunities to adapt the large biomass tuber potato crop to meet the global food challenges of lipid production. There is also an opportunity to introduce novel fatty acids with health benefits or high industrial values to high-lipid tubers. Such 'dual purpose' crop strategies may have potential for direct use as a niche health food, animal feed or oleochemical feedstock.

Experimental procedures

Binary plasmid construct pJP3506

A binary plasmid construct, pJP3506, was derived from a pORE04-based binary expression vector, which contained NPTII gene driven by double enhancer CaMV 35S promoter (e35S) as the selectable marker and three gene expression cassettes, which are 35S::atDGAT1, B33::atWRI1 and B33::siOLEOSIN, respectively. Patatin B33 promoter derived from S. tuberosum was kindly provided by Dr Alisdair Fernie, Max Planck Institute of Molecular Plant Physiology in Germany. It is a truncated version with 185 bp deleted from the 5' end of the published GenBank sequence (GenBank accession number X14483). AtWRI1 is the codon-optimised coding region of A. thaliana WRI1 gene, and atDGAT1 referred to the WT A. thaliana DGAT1 gene and siOLEOSIN is an intron-interrupted S. indicum OLEOSIN gene as previously described in Vanhercke et al. (2014). A diagram showing the configuration of the arrangement of the transgene cassettes in pJP3506 is presented in Figure 1.

Potato transformation

In vitro seedlings of commercial potato cultivar, S. tuberosum cv. Atlantic were purchased from Toolangi Elite™, Healesville, Victoria, Australia. Stem internodes were excised into approximately 1-cm pieces in length under Agrobacterium tumefaciens strain AGL1 (OD$_{600}$ = 0.2) grown in a MS and LB mixture media (50 : 50, by volume). Following a brief blotting on sterile Whatman filter paper, the infected internodes were plated out onto MS media agar plates supplemented with 200 µg/L NAA and 2 mg/L BAP, and maintained at 24 °C. Following 2 days co-cultivation and further 8 days on fresh medium without selection, the explants were transferred onto MS medium supplemented with 2 mg/L BAP, 5 mg/L GA$_3$, 50 mg/L kanamycin (for transgenic selection) and 250 mg/L cefotaxime (to control Agrobacterium) until the emergence of green shoots which were excised and placed onto plain MS medium for root induction prior to transplanting into a 20-cm diameter pot containing potato potting mix (CSIRO). The plants were maintained at a greenhouse with 25/20 °C with extended light for 16-h photoperiod. The tubers harvested from the primary transgenic plants were used for initial screening, and selected eight lines, together with untransformed potato, were grown at the same condition for further analysis.

RNA extraction and transgene expression profiling in transgenic plants by qRT-PCR

Total RNAs from developing potato tubers at about 4 cm in diameter and fully expanded green leaves at the same stage were isolated using RNeasy Mini Kit (Qiagen, Hilden, Germany). Contaminating DNA was removed by digestion with Ambion TURBO RNA-free DNaseI (Thermo Fisher Scientific, Waltham, MA) according to the manufacturer's protocol. RNA concentrations were determined using a Nanodrop® spectrophotometer ND1000 (Thermo Fisher Scientific), and concentrations were standardised before analysis. To verify the RNA integrity, 1 µg of total RNA from each sample was visualised on an ethidium bromide-stained 1.5% agarose gel following electrophoresis.

The gene expression patterns were studied with qRT-PCR in triplicate using Platinum SYBR Green qPCR SuperMix (Bio-Rad, Hercules, CA) and run on ABI 7900HT Sequence Detection System. Each PCR contained 20 ng of total RNA template, 50 pmol each of the forward and reverse primers, 0.25 µL of reverse transcriptase, 5 µL One-step RT-PCR master mix reagents, increased to 10 µL total volumes with nuclease-free water. The primers for atWRI1 are sense: 5'-CTCCAACTACATCGACAGGC - 3'; antisense: 5'- GCAAGGTAGTAGAGGACGAAG-3'. The primers for atDGAT1 are sense: 5'-GGCGATTTTGGATTCTGCTGGC-3'; antisense: 5'-GGAACCAGAGAGAAGAAGTC-3'; and the primers for sesame OLEOSIN are sense: 5'-CAGCAGCAACAAA-CACGTG-3'; antisense: 5'-GAGAAGATCACCAGGAGAG-3'. A constitutively expressed S. tuberosum gene CYCLOPHILIN (stCYP) was used as the reference gene to normalise the relative quantities (GenBank accession number AF126551; Nicot et al., 2005). The oligo sequence of the primers for stCYP is sense: 5'-CTCTTCGCCGATACCACTC-3'; and antisense: 5'-CACACGGTG-GAAGGTTGAG-3'. The thermal cycling conditions were reverse transcription at 45 °C for 10 min, following deactivation by 95 °C for 2 min. This was followed by 40 cycles of 95 °C for 5 s and 60 °C for 50 s. The calculations were made using the comparative CT method ($2^{-\Delta\Delta Ct}$) as reported (Livak and Schmittgen, 2001).

Lipid analysis

Fresh tubers were harvested from potato plants at full maturity when the up-ground stems and leaves turned brown and dead. A thin slice was sampled from the middle of the tuber and freeze-dried for 72 h prior to lipid analysis. Full sized green potato leaf tissues were sampled for lipid analysis when the potato plant approach maturity and two-thirds of the leaves had turned brown. To extract the total lipids, the freeze-dried tuber tissues were first homogenised in chloroform/methanol (2 : 1, by volume) in an eppendorf tube containing a metallic ball using the TissueLyser (Qiagen) and mixed with one-third volume of 0.1 M KCl. Following centrifugation at 10 000 *g* for 5 min, the lower liquid phase containing lipids were collected and evaporated completely using N_2 flow before 3 µL of chloroform was added for each milligram of tuber DW. Lipid samples prepared as above were loaded on a TLC plate (10 cm × 20 cm, Silica gel 60; Merck, Darmstadt, Germany) and developed in hexane : diethyl ether : acetic acid (70 : 30 : 1, by volume). The TLC plate was

sprayed with primuline made in 80% acetone in water and visualised under UV. TAG and PLs were fractionated by cutting out the corresponding silica bands, and their respective fatty acid methyl esters (FAME) were produced by incubating corresponding silica bands in 1 N methanolic HCl (Supelco, Bellefonte, PA, USA) at 80 °C for 2 h together with triheptadecanoin (Nu-Chek PREP, Inc., Elysian, MN) as internal standard for lipid quantification. FAME were analysed by gas chromatography (7890A GC; Agilent Technologies, Santa Clara, CA) equipped with a 30 m BPX70 column (0.25 mm inner diameter, 0.25 mm film thickness; SGE, Austin, TX). Peaks were integrated with Agilent Technologies ChemStation software (Rev B.04.03).

Quantification of starch and soluble sugars

Fresh mature potato tubers were sampled for carbohydrate analysis when the up-ground stems and leaves had turned brown and dead. Triplicate samples (~10 mg) of ground freeze-dried potato tuber tissue were boiled four times in 80% ethanol. The supernatants (ethanol extract) were pooled for analysis of soluble sugars, while the pellet was retained for starch determination.

Aliquots of ethanol extract were boiled in anthrone reagent (2% anthrone in 70% H_2SO_4, by volume) for 10 min, and the absorbance at 630 nm was measured for the analysis of soluble sugar contents (Yemm and Willis, 1954). For starch determination, the pellet was resuspended in 350 µL 0.2 M NaOH, boiled for 30 min and neutralised using 3.5 µL glacial acetic acid. The starch content was determined relative to a control/blank aliquot using a Megazyme Total Starch Kit following the manufacturers' instruction (Megazyme International Ireland, Bray, Ireland). All spectrophotometric measurements were performed using a Thermo Multiskan® Spectrum plate reader (Thermo Fisher Scientific).

Observation of LDs and confocal fluorescence microscopy imaging

Thin slices of the freshly harvested immature potato tubers at about 4 cm in diameter were fixed in 4% paraformaldehyde, prior to embedding in CRYO-GEL™ embedding medium (Instrumedics Inc., Hackensack, NJ). The mounted samples were then frozen at −20 °C for 48 h prior to sectioning with a microtome. Glass microscope slides containing potato sections of 40 µm in thickness were stained with 1 µg/mL solution of Nile Red (Sigma-Aldrich, St. Louis, MO) and imaged using a Leica SP8 confocal scanning microscope (Leica Micrisystems, Wetzlar, Germany). The excitation wavelength was 488 nm, and the emission was recorded between 593 and 654 nm. LDs were imaged at 40× magnification and analysed using Leica LAS AF Lite software (www.leica-microsystems.com).

Isolation of starch and light and scanning electron microscopy imaging

Starch granules were isolated from mature potato tuber using a method adapted from Wischmann et al. (2007). Peeled potatoes were homogenised in 1% (w/v) sodium metabisulphite, and the resulting mixture was filtered through 200 µm nylon mesh. The slurry was steeped for at least 40 min to allow the starch to settle. The pelleted starch was repeatedly washed with and steeped in deionised water until the wash was clear. Purified starch was then washed three times with deionised water and dried overnight prior to observation of starch granules with light microscopy (Leica-DMR) using crossed polarised filters to reveal birefringence in the starch granules at a 400× magnification. Starch granule morphology was also examined with a scanning electron microscope (ZEISS EVO LS15, Carl Zeiss International, Oberkochen, Germany). Purified starches were sputter-coated with gold and scanned at 20 kV at room temperature at 500× magnification.

Transmission electron microscopy imaging

Freshly harvested immature potato tubers at about 4 cm in diameter were cut into approximately 2 × 1 mm pieces and directly fixed with 2.5% glutaraldehyde, 2% paraformaldehyde in 0.1 M phosphate buffer (pH 7.4), under vacuum for 24 h. The samples were then washed with 0.1 M phosphate buffer (pH 7.4) three times and for 10 min each. The secondary fixation with 1% osmium tetroxide was carried out at room temperature for 4 h. The fixed samples were rinsed with distilled water and dehydrated through an acetone series and embedded in Spurr's resin (Spurr, 1969) overnight at 70 °C. Subsequently, ultrathin sections (70–90 nm) were obtained with a Leica EM UC7 Ultra microtome. Sections were stained with uranyl acetate and lead citrate for 10 min each. Images were observed and recorded using the Hitachi 7100 TEM (Hitachi High Technologies America, Inc., Schaumburg, IL) at an accelerating voltage of 100 kv.

Statistical analysis

Significance of differences in the biochemical measurements of soluble sugars, starch and lipids was determined using ANOVA and student t-test. Differences were considered as statistically significant when $P < 0.05$ (represented by '*') and highly significant when $P < 0.01$ (represented by '**'), compared with the control group.

Acknowledgements

The authors wish to thank Dr Melanie Rug and Joanne Lee of the Centre for Advanced Microscopy, Australian National University for their excellent guidance and assistance with TEM work. QG and YZ wish to acknowledge financial support from Chinese Scholarship Council (CSC). SA wishes to acknowledge financial support from Pakistan Government. Excellent technical assistance from Lijun Tian, Jeni Prichard and Dr Dawar Hussain is gratefully acknowledged.

References

Andrianov, V., Borisjuk, N., Pogrebnyak, N., Brinker, A., Dixon, J., Spitsin, S., Flynn, J. et al. (2010) Tobacco as a production platform for biofuel: overexpression of Arabidopsis DGAT and LEC2 genes increases accumulation and shifts the composition of lipids in green biomass. Plant Biotechnol. J. **8**, 277–287.

Baud, S., Mendoza, M.S., To, A., Harscoet, E., Lepiniec, L. and Dubreucq, B. (2007) WRINKLED1 specifies the regulatory action of LEAFY COTYLEDON2 towards fatty acid metabolism during seed maturation in Arabidopsis. Plant J. **50**, 825–838.

Bourgis, F., Kilaru, A., Cao, X., Ngando-Ebongue, G.F., Drira, N., Ohlrogge, J.B. and Arondel, V. (2011) Comparative transcriptome and metabolite analysis of oil palm and date palm mesocarp that differ dramatically in carbon partitioning. Proc. Natl Acad. Sci. USA, **108**, 12527–12532.

Bouvier-Nave, P., Benveniste, P., Oelkers, P., Sturley, S.L. and Schaller, H. (2000) Expression in yeast and tobacco of plant cDNAs encoding acyl CoA:diacylglycerol acyltransferase. Eur. J. Biochem. **267**, 85–96.

Cernac, A. and Benning, C. (2004) WRINKLED1 encodes an AP2/EREB domain protein involved in the control of storage compound biosynthesis in Arabidopsis. Plant J. **40**, 575–585.

Chapman, K.D. and Ohlrogge, J.B. (2012) Compartmentation of triacylglycerol accumulation in plants. J. Biol. Chem. **287**, 2288–2294.

Fan, J., Yan, C., Roston, R., Shanklin, J. and Xu, C. (2014) Arabidopsis lipins, PDAT1 acyltransferase, and SDP1 triacylglycerol lipase synergistically direct fatty acids toward beta-oxidation, thereby maintaining membrane lipid homeostasis. Plant Cell, **26**, 4119–4134.

Focks, N. and Benning, C. (1998) wrinkled1: A novel, low-seed-oil mutant of Arabidopsis with a deficiency in the seed-specific regulation of carbohydrate metabolism. Plant Physiol. **118**, 91–101.

Galliard, T. (1970) The enzymic breakdown of lipids in potato tuber by phospholipid- and galactolipid-acyl hydrolase activities and by lipoxygenase. Phytochemistry, **9**, 1725.

Galliard, T. (1972) Fatty acid composition of immature potato tubers. Phytochemistry, **11**, 1899–1903.

Grierson, C., Du, J.S., de Torres Zabala, M., Beggs, K., Smith, C., Holdsworth, M. and Bevan, M. (1994) Separate cis sequences and trans factors direct metabolic and developmental regulation of a potato tuber storage protein gene. Plant J. **5**, 815–826.

Harwood, J.L. (2005) Fatty acid biosynthesis. In Plant Lipids: Biology, Utilization and Manipulation (Murphy, D.J., ed.), pp. 27–66. Oxford: Blackwell Publishing.

Hoefgen, R. and Willmitzer, L. (1990) Biochemical and genetic-analysis of different patatin isoforms expressed in various organs of potato (Solanum tuberosum). Plant Sci. **66**, 221–230.

Hofvander, P., Ischebeck, T., Turesson, H., Kushwaha, S.K., Feussner, I., Carlsson, A.S. and Andersson, M. (2016) Potato tuber expression of Arabidopsis WRINKLED1 increase triacylglycerol and membrane lipids while affecting central carbohydrate metabolism. Plant Biotechnol. J. doi: 10.1111/pbi.12550.

Horn, P.J., James, C.N., Gidda, S.K., Kilaru, A., Dyer, J.M., Mullen, R.T., Ohlrogge, J.B. et al. (2013) Identification of a new class of lipid droplet-associated proteins in plants. Plant Physiol. **162**, 1926–1936.

Hsieh, K. and Huang, A.H. (2004) Endoplasmic reticulum, oleosins, and oils in seeds and tapetum cells. Plant Physiol. **136**, 3427–3434.

Jako, C., Kumar, A., Wei, Y., Zou, J., Barton, D.L., Giblin, E.M., Covello, P.S. et al. (2001) Seed-specific over-expression of an Arabidopsis cDNA encoding a diacylglycerol acyltransferase enhances seed oil content and seed weight. Plant Physiol. **126**, 861–874.

Jobling, S.A., Westcott, R.J., Tayal, A., Jeffcoat, R. and Schwall, G.P. (2002) Production of a freeze-thaw-stable potato starch by antisense inhibition of three starch synthase genes. Nat. Biotechnol. **20**, 295–299.

Kilaru, A., Cao, X., Dabbs, P.B., Sung, H.-J., Rahman, M.M., Thrower, N., Zynda, G. et al. (2015) Oil biosynthesis in a basal angiosperm: transcriptome analysis of Persea Americana mesocarp. BMC Plant Biol. **15**, 1–19.

Klaus, D., Ohlrogge, J.B., Neuhaus, H.E. and Dormann, P. (2004) Increased fatty acid production in potato by engineering of acetyl-CoA carboxylase. Planta, **219**, 389–396.

Krahmer, N., Guo, Y., Wilfling, F., Hilger, M., Lingrell, S., Heger, K., Newman, H.W. et al. (2011) Phosphatidylcholine synthesis for lipid droplet expansion is mediated by localized activation of CTP:phosphocholine cytidylyltransferase. Cell Metab. **14**, 504–515.

Linssen, J.P.H., Cozijnsen, J.L. and Pilnik, W. (1989) Chufa (Cyperus esculentus): a new source of dietary fibre. J. Sci. Food Agric. **49**, 291–296.

Liu, J., Hua, W., Zhan, G., Wei, F., Wang, X., Liu, G. and Wang, H. (2010) Increasing seed mass and oil content in transgenic Arabidopsis by the overexpression of WRI1-like gene from Brassica napus. Plant Physiol. Biochem. **48**, 9–15.

Livak, K.J. and Schmittgen, T.D. (2001) Analysis of relative gene expression data using real-time quantitative PCR and the $(2^{-\Delta\Delta Ct})$ Method. Methods, **25**, 402–408.

Lung, S.C. and Weselake, R.J. (2006) Diacylglycerol acyltransferase: a key mediator of plant triacylglycerol synthesis. Lipids **41**, 1073–1088.

Ma, W., Kong, Q., Arondel, V., Kilaru, A., Bates, P.D., Thrower, N.A., Benning, C. et al. (2013) WRINKLED1, A ubiquitous regulator in oil accumulating tissues from Arabidopsis embryos to oil palm mesocarp. PLoS ONE, **8**, e68887.

Madoka, Y., Tomizawa, K.-I., Mizoi, J., Nishida, I., Nagano, Y. and Sasaki, Y. (2002) Chloroplast transformation with modified accD operon increases Acetyl-CoA Carboxylase and causes extension of leaf longevity and increase in seed yield in tobacco. Plant Cell Physiol. **43**, 1518–1525.

Maeo, K., Tokuda, T., Ayame, A., Mitsui, N., Kawai, T., Tsukagoshi, H., Ishiguro, S. *et al.* (2009) An AP2-type transcription factor, WRINKLED1, of *Arabidopsis thaliana* binds to the AW-box sequence conserved among proximal upstream regions of genes involved in fatty acid synthesis. *Plant J.* **60**, 476–487.

Martin, S. and Parton, R.G. (2006) Lipid droplets: a unified view of a dynamic organelle. *Nat. Rev. Mol. Cell Biol.* **7**, 373–378.

Nicot, N., Hausman, J.F., Hoffmann, L. and Evers, D. (2005) Housekeeping gene selection for real-time RT-PCR normalization in potato during biotic and abiotic stress. *J. Exp. Bot.* **56**, 2907–2914.

Regina, A., Kosar-Hashemi, B., Ling, S., Li, Z.Y., Rahman, S. and Morell, M. (2010) Control of starch branching in barley defined through differential RNAi suppression of starch branching enzyme IIa and IIb. *J. Exp. Bot.* **61**, 1469–1482.

Rocha-Sosa, M., Sonnewald, U., Frommer, W., Stratmann, M., Schell, J. and Willmitzer, L. (1989) Both developmental and metabolic signals activate the promoter of a class I patatin gene. *EMBO J.* **8**, 23–29.

Rogalski, M. and Carrer, H. (2011) Engineering plastid fatty acid biosynthesis to improve food quality and biofuel production in higher plants. *Plant Biotechnol. J.* **9**, 554–564.

van Rooijen, G.J. and Moloney, M.M. (1995) Structural requirements of oleosin domains for subcellular targeting to the oil body. *Plant Physiol.* **109**, 1353–1361.

Ruuska, S.A., Girke, T., Benning, C. and Ohlrogge, J.B. (2002) Contrapuntal networks of gene expression during Arabidopsis seed filling. *Plant Cell*, **14**, 1191–1206.

Santos Mendoza, M., Dubreucq, B., Miquel, M., Caboche, M. and Lepiniec, L. (2005) LEAFY COTYLEDON 2 activation is sufficient to trigger the accumulation of oil and seed specific mRNAs in Arabidopsis leaves. *FEBS Lett.* **579**, 4666–4670.

Sasaki, Y. and Nagano, Y. (2004) Plant acetyl-CoA carboxylase: structure, biosynthesis, regulation, and gene manipulation for plant breeding. *Biosci. Biotechnol. Biochem.* **68**, 1175–1184.

Shaik, S.S., Obata, T., Hebelstrup, K.H., Schwahn, K., Fernie, A.R., Mateiu, R.V. and Blennow, A. (2016) Starch granule re-structuring by starch branching enzyme and glucan water dikinase modulation affects caryopsis physiology and metabolism. *PLoS ONE*, **11**, e0149613.

Shen, B., Allen, W.B., Zheng, P., Li, C., Glassman, K., Ranch, J., Nubel, D. *et al.* (2010) Expression of *ZmLEC1* and *ZmWRI1* increases seed oil production in maize. *Plant Physiol.* **153**, 980–987.

Siloto, R.M., Findlay, K., Lopez-Villalobos, A., Yeung, E.C., Nykiforuk, C.L. and Moloney, M.M. (2006) The accumulation of oleosins determines the size of seed oil bodies in Arabidopsis. *Plant Cell*, **18**, 1961–1974.

Slocombe, S.P., Cornah, J., Pinfield-Wells, H., Soady, K., Zhang, Q., Gilday, A., Dyer, J.M. *et al.* (2009) Oil accumulation in leaves directed by modification of fatty acid breakdown and lipid synthesis pathways. *Plant Biotechnol. J.* **7**, 694–703.

Spurr, A.R. (1969) A low-viscosity epoxy resin embedding medium for electron microscopy. *J. Ultrastruct. Res.* **26**, 31–43.

Thiam, A.R., Antonny, B., Wang, J., Delacotte, J., Wilfling, F., Walther, T.C., Beck, R. *et al.* (2013) COPI buds 60-nm lipid droplets from reconstituted water-phospholipid-triacylglyceride interfaces, suggesting a tension clamp function. *Proc. Natl Acad. Sci. USA*, **110**, 13244–13249.

Ting, J.T., Lee, K., Ratnayake, C., Platt, K.A., Balsamo, R.A. and Huang, A.H. (1996) Oleosin genes in maize kernels having diverse oil contents are constitutively expressed independent of oil contents. Size and shape of intracellular oil bodies are determined by the oleosins/oils ratio. *Planta*, **199**, 158–165.

To, A., Joubès, J., Barthole, G., Lécureuil, A., Scagnelli, A., Jasinski, S., Lepiniec, L. *et al.* (2012) WRINKLED transcription factors orchestrate tissue-specific regulation of fatty acid biosynthesis in Arabidopsis. *Plant Cell*, **24**, 5007–5023.

Turesson, H., Marttila, S., Gustavsson, K.-E., Hofvander, P., Olsson, M.E., Bülow, L., Stymne, S. *et al.* (2010) Characterization of oil and starch accumulation in tubers of *Cyperus esculentus* var. sativus (Cyperaceae): a novel model system to study oil reserves in nonseed tissues. *Am. J. Bot.* **97**, 1884–1893.

Vanhercke, T., El Tahchy, A., Shrestha, P., Zhou, X.R., Singh, S.P. and Petrie, J.R. (2013) Synergistic effect of *WRI1* and *DGAT1* coexpression on triacylglycerol biosynthesis in plants. *FEBS Lett.* **587**, 364–369.

Vanhercke, T., El Tahchy, A., Liu, Q., Zhou, X.-R., Shrestha, P., Divi, U.K., Ral, J.-P. *et al.* (2014) Metabolic engineering of biomass for high energy density: oilseed-like triacylglycerol yields from plant leaves. *Plant Biotechnol. J.* **12**, 231–239.

Vigeolas, H., Waldeck, P., Zank, T. and Geigenberger, P. (2007) Increasing seed oil content in oil-seed rape (*Brassica napus* L.) by over-expression of a yeast glycerol-3-phosphate dehydrogenase under the control of a seed-specific promoter. *Plant Biotechnol. J.* **5**, 431–441.

Weselake, R.J., Taylor, D.C., Rahman, M.H., Shah, S., Laroche, A., McVetty, P.B. and Harwood, J.L. (2009) Increasing the flow of carbon into seed oil. *Biotechnol. Adv.* **27**, 866–878.

Winichayakul, S., Scott, R.W., Roldan, M., Hatier, J.-H.B., Livingston, S., Cookson, R., Curran, A.C. *et al.* (2013) *In vivo* packaging of triacylglycerols enhances Arabidopsis leaf biomass and energy density. *Plant Physiol.* **162**, 626–639.

Wischmann, B., Ahmt, T., Bandsholm, O., Blennow, A., Young, N., Jeppesen, L. and Thomsen, L. (2007) Testing properties of potato starch from different scales of isolations -A ring test. *J. Food Eng.* **79**, 970–978.

Yatsu, L.Y. and Jacks, T.J. (1972) Spherosome membranes: half unit-membranes. *Plant Physiol.* **49**, 937–943.

Yemm, E.W. and Willis, A.J. (1954) The estimation of carbohydrates in plant extracts by anthrone. *Biochem. J.* **57**, 508–514.

Zale, J., Jung, J.H., Kim, J.Y., Pathak, B., Karan, R., Liu, H., Chen, X. *et al.* (2016) Metabolic engineering of sugarcane to accumulate energy-dense triacylglycerols in vegetative biomass. *Plant Biotechnol. J.* **14**, 661–669.

Zheng, P., Allen, W.B., Roesler, K., Williams, M.E., Zhang, S., Li, J., Glassman, K. *et al.* (2008) A phenylalanine in DGAT is a key determinant of oil content and composition in maize. *Nat. Genet.* **40**, 367–372.

Zou, J., Wei, Y., Jako, C., Kumar, A., Selvaraj, G. and Taylor, D.C. (1999) The *Arabidopsis thaliana tag1* mutant has a mutation in a *diacylglycerol acyltransferase* gene. *Plant J.* **19**, 645–653.

10

Stochastic alternative splicing is prevalent in mungbean (*Vigna radiata*)

Dani Satyawan[1,2], Moon Young Kim[1,3] and Suk-Ha Lee[1,3,*]

[1]Department of Plant Science and Research Institute of Agriculture and Life Sciences, Seoul National University, Seoul, Korea
[2]Indonesian Center for Agricultural Biotechnology and Genetic Resources Research and Development, Bogor, Indonesia
[3]Plant Genomics and Breeding Institute, Seoul National University, Seoul, Korea

*Correspondence
email sukhalee@snu.ac.kr

Keywords: alternative splicing, mungbean (*Vigna radiata*), RNA sequencing, stochastic process, evolutionary conservation.

Summary

Alternative splicing (AS) can produce multiple mature mRNAs from the same primary transcript, thereby generating diverse proteins and phenotypes from the same gene. To assess the prevalence of AS in mungbean (*Vigna radiata*), we analysed whole-genome RNA sequencing data from root, leaf, flower and pod tissues and found that at least 37.9% of mungbean genes are subjected to AS. The number of AS transcripts exhibited a strong correlation with exon number and thus resembled a uniform probabilistic event rather than a specific regulatory function. The proportion of frameshift splicing was close to the expected frequency of random splicing. However, alternative donor and acceptor AS events tended to occur at multiples of three nucleotides (i.e. the codon length) from the main splice site. Genes with high exon number and expression level, which should have the most AS if splicing is purely stochastic, exhibited less AS, implying the existence of negative selection against excessive random AS. Functional AS is probably rare: a large proportion of AS isoforms exist at very low copy per cell on average or are expressed at much lower levels than default transcripts. Conserved AS was only detected in 629 genes (2.8% of all genes in the genome) when compared to *Vigna angularis*, and in 16 genes in more distant species like soya bean. These observations highlight the challenges of finding and cataloguing candidates for experimentally proven AS isoforms in a crop genome.

Introduction

Alternative splicing (AS) is the differential splicing of introns from pre-mRNA to yield several distinct mature mRNAs (isoforms) from a single gene. In general, four fundamental types of differential splicing can alter the coding region: intron retention, exon skipping, alternative donor and alternative acceptor (Breitbart *et al.*, 1987). In intron retention, intron sequences that are normally spliced out are retained in the mature mRNA, producing a longer transcript with extra coding sequences. By contrast, in exon skipping, some exon segments are spliced out from the final transcript to yield shorter mRNA molecules. The location of the splicing reaction can also change at only one of the splice sites; this situation is referred to as 'alternative donor' if the change occurs at the 5' end of the intron and 'alternative acceptor' if the change occurs at the 3' end of the intron.

The mature mRNAs produced by AS can harbour additional bases or lack some exon sequences, resulting in alteration of amino acid composition, physical characteristics or chemical function of the encoded proteins. Thus, AS can increase the number of protein types and phenotypes produced by a small number of genes. Inclusion of additional sequences and mis-sense splicing can also introduce premature stop codons into the transcripts, making them vulnerable to degradation by the non-sense-mediated decay (NMD) pathway (Neu-Yilik *et al.*, 2004) and decreasing the quantity of those particular transcripts in the cell. Several lines of evidence show that cells actually utilize this pathway to modulate and fine-tune the number of RNA molecules for a particular gene under certain conditions (Filichkin and Mockler, 2012; Kawashima *et al.*, 2014).

Consequently, AS could explain the complexity paradox, that is the observation that the genomes of certain complex organisms harbour a smaller number of coding regions than those of some simpler organisms (Graveley, 2001). Several experimentally proven AS isoforms produce multiple proteins with distinct characteristics and function from the same coding region (Inoue *et al.*, 1990; Lah *et al.*, 2014; Ullrich *et al.*, 1995), potentially explaining how a single gene could perform multiple functions in the cell. The advent of next-generation sequencing (NGS), which can generate large quantities of transcriptome data faster and more cheaply than previous methods, has aided in the identification of AS in many different organisms. Software and script packages, such as ASTALAVISTA (Foissac and Sammeth, 2007) and ASprofile (Florea *et al.*, 2013), have been developed to rapidly identify splicing variants by examining variations in exon–intron boundaries in genomewide alignment data generated using NGS. The results are quite surprising: in some cases, AS occurred in more than half of the annotated genes (Marquez *et al.*, 2012; Pan *et al.*, 2008; Shen *et al.*, 2014). If all AS produces functionally divergent proteins, then this process regulates the bulk of transcript generation and protein synthesis in the cell. Hence, proper annotation of the occurrence of AS in the genome is very important as a reference for functional genomic studies.

Several studies have attempted to catalogue the global occurrence of AS in the genomes of several plants, including soya bean (Shen *et al.*, 2014), *Arabidopsis* (Filichkin *et al.*, 2010; Marquez *et al.*, 2012) and maize (Thatcher *et al.*, 2014), by utilizing mRNA sequences obtained from different tissue types under diverse environmental conditions to capture as many transcript types as possible. Nevertheless, although those studies revealed that a large number of plant genes undergo AS, very little experimental evidence of functional AS proteins is available (Severing et al., 2009). Several groups have suggested that the scarcity of demonstrably functional AS isoforms could be due to the random nature of AS itself (Hon et al., 2013; Melamud and Moult, 2009a; Zhang et al., 2009), implying that most AS isoforms have no function because they are merely the by-products of erroneous splicing. Consequently, it is unlikely that all AS events are part of a distinct layer of gene regulation. That said, because AS isoforms that confer selective advantage could be retained by progeny with stronger AS signals for those isoforms, functional AS could still evolve and be retained by natural selection.

Because advantageous AS isoforms have a higher probability of being retained over the course of evolution, it should be possible to identify them in comparative studies of related species. Mungbean (Vigna radiata) and its close relatives in the Vigna genus, like adzuki bean (Vigna angularis), are good candidates for such studies. Their genome sequences have recently been published (Kang et al., 2014, 2015), enabling transcript alignment and facilitating identification of AS isoforms. They are also related to soya bean, whose genome is already well characterized, and for which comprehensive data regarding AS are available. Because mungbean and adzuki bean are widely planted for food consumption (annual plantation area of 6 million and 840 000 hectares, respectively), any practical applications that could be derived from genomic studies in these plants will have considerable economic impact (Nair et al., 2012; Rubatzky and Yamaguchi, 1997).

We performed global transcriptome analysis to identify and catalogue AS events that occur in mungbean. To infer the characteristics of AS regulation in this species, we tested for stochastic AS in the RNA population. To identify AS events with the strongest likelihood of being functional, for the purpose of subsequent in-depth studies, we investigated AS conservation in adzuki bean and soya bean. The resultant whole-genome annotation of AS isoforms represents a valuable contribution to the mungbean genome sequencing project.

Results

Characteristics of AS types in mungbean

The number of AS events (hereafter, AS number) of each type were detected in silico based on alignment of RNAseq data to the mungbean reference genome. Shotgun sequencing generated, on average, 38.6 million 100 bp reads per sample (Table S1), close to 10 times the size of the mungbean genome. The total length of annotated transcribed regions is 104 million bases; therefore, the sequence alignment produced roughly 37× sequencing coverage for all open reading frames. However, because most of the sequenced RNAs are derived from mature RNAs whose introns have been spliced out, the sequencing depth in exonic regions was 101× on average.

The number of AS events detected varied with the software pipeline: ASprofile annotated more AS events than ASTALAVISTA (Table 1). A closer inspection of the binary alignment (BAM) files using a genome browser revealed that the higher number of AS events detected by ASprofile was due to reporting of new exons not found in the mungbean genome annotation, as well as increased sensitivity in detecting rare splicing junctions. Depending on the AS types, ASTALAVISTA did not detect 85.3–90.1 per cent of AS events detected by ASprofile. However, ASprofile also

failed to detect 56.5–85 per cent of AS events detected by ASTALAVISTA. Neither pipeline is clearly superior to the other as they both missed AS events detected by the other pipeline, but ASprofile output was used for further analysis due to its increased sensitivity and better annotation system.

ASprofile estimated that 44.6% of mungbean genes are subjected to AS, whereas ASTALAVISTA estimated this proportion as 37.9%. Both figures are lower than the proportions reported for Arabidopsis (Marquez et al., 2012), maize (Thatcher et al., 2014), rice (Lu et al., 2010) and soya bean (Shen et al., 2014), but fairly similar to those of closely related legumes like Medicago and Phaseolus (Chamala et al., 2015).

The distribution of AS was generally similar across tissues (Figure 1), although tissue-specific AS isoforms were detected, and various tissues yielded different numbers of AS isoforms. Among all tissues, roots had the highest number of AS events, as well as the highest AS number per gene (Table 1). One potential reason for this is that roots express the largest number of tissue-specific genes, and these genes tend to be highly expressed (Table S2), potentially aiding in detection of AS isoforms in RNA from roots. About 2.3% of tissue-specific AS events were the consequence of tissue-specific gene expression, and their absence in other tissues is caused by the lack of expression of those genes; however, the remaining AS isoforms are tissue-specific even though the originating transcripts are expressed at significant levels in more than one tissue type. In many cases, we found that the absence of AS isoforms in other tissues was not merely caused by low expression levels and under-representation in RNA samples from these tissues.

The number of transcripts representing a particular isoform is difficult to quantify accurately without long-read sequence data, because some genes have multiple AS events and some isoforms may combine with others from the same gene to generate a distinct transcript structure. Because long-read RNAseq data were not available for this study, we simply assumed that such combinations were nonexistent and then estimated the quantity of each AS isoform based on the number of sequence fragments that aligned to the splice junction that underwent AS. Based on this assumption, a significant portion of detected AS isoforms (20.54%) had FPKM values lower than 1, that is their concentration is very low in an average cell. Moreover, a considerable proportion of AS isoforms (24.4%) were expressed at levels <10% of those of the more abundant constitutive splice forms.

Mungbean AS exhibits signs of stochastic splicing

The prevalence of AS isoforms with low concentration in our mungbean AS data raises the possibility that a significant number of mungbean AS could be the result of random errors with little effect on the protein composition of the cell. To determine whether stochastic splicing is prevalent in mungbean, we investigated the correlation between the presence of AS and several aspects of the plant's genomic features that may increase the probability of random splicing errors. We found that mean AS number was strongly correlated with the number of exons in a gene with a Pearson r-value of 0.879 and P-value of 4.09e-14 (Figure 2a). This is consistent with the random splicing error model: the higher the exon number, the larger the number of splicing junctions and the greater the chance of error associated with splicing of those junctions. However, an obvious consequence of probabilistic splicing error is that genes with a large number of introns will have a higher probability of accumulating useless splicing errors, which could be dangerous to the cell. To

Table 1 Number of AS isoform types in each tissue, based on isoform detection with ASTALAVISTA and ASprofile

Detection method	Tissue	Intron retention	Exon skipping	Alternative donor	Alternative acceptor	Affected genes
ASTALAVISTA	Flower	3620	659	1141	2101	4414
	Leaf	3512	647	1107	2060	4256
	Pod	4076	589	1093	2132	4541
	Root	5419	772	1320	2401	5429
ASprofile	Flower	4256	6494	3373	3059	8051
	Leaf	4173	6185	3278	2897	8152
	Pod	4582	6394	3455	3103	7931
	Root	5874	6724	3885	3336	7549

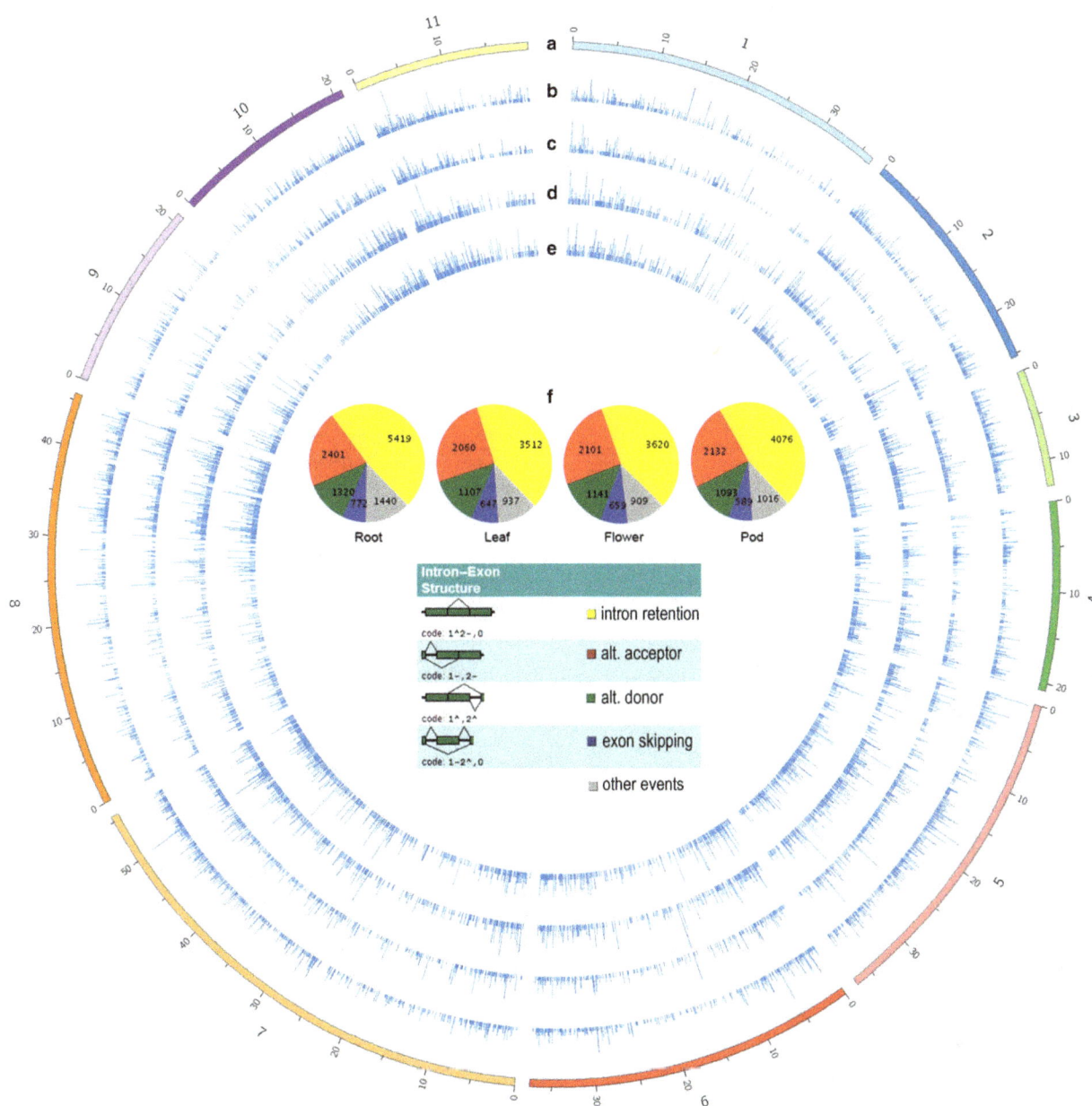

Figure 1 Types and chromosomal distribution of AS in four mungbean tissues. From outer to inner rings: (a) size of chromosome (in megabases); (b) histogram of AS number across chromosomes in root, (c) leaf, (d) flower and (e) pod tissues. (f) Proportions of each type of AS across the four tissues, as classified by ASTALAVISTA.

determine whether mungbean has evolved a mechanism to reduce the likelihood of such errors, we plotted the average number of AS per exon for genes with different exon numbers. The plot reveals a clear trend towards fewer AS events per exon as the number of exons increases ($r = -0.649$, $P = 7.755\text{e-}05$), although the pattern is less clear for genes containing more than 25 exons (Figure 3a).

Erroneous splicing is also more disadvantageous for highly expressed genes, because in such cases, it would create a large amount of mis-spliced mRNA, which in turn is more likely to be translated into a large quantity of nonfunctional protein. Consistent with this, we observed a trend towards fewer AS events per exon in highly expressed genes (Figure 3B), although the correlation was weak ($r = -0.057$ and P-value = 2.2e-16). A correlation plot of AS number vs expression level revealed a negative correlation ($r = -0.463$ and P-value = 0.001) between the two variables (Figure 2b). This observation contrasts with findings in other plants such as soya bean, in which highly expressed genes also usually have higher numbers of AS events (Shen *et al.*, 2014). One reason for this could be that, in mungbean, the average number of exons is lower among highly expressed genes (Figure S1), and exon number correlates more strongly to AS number than expression level. Moreover, the average number of AS was not higher in highly expressed genes than in genes carrying the same exon numbers expressed at lower levels (Figure S2).

Another important effect of AS on the final transcript sequence is the creation of frameshift mutations, which could significantly alter the amino acid composition downstream of the splice site.

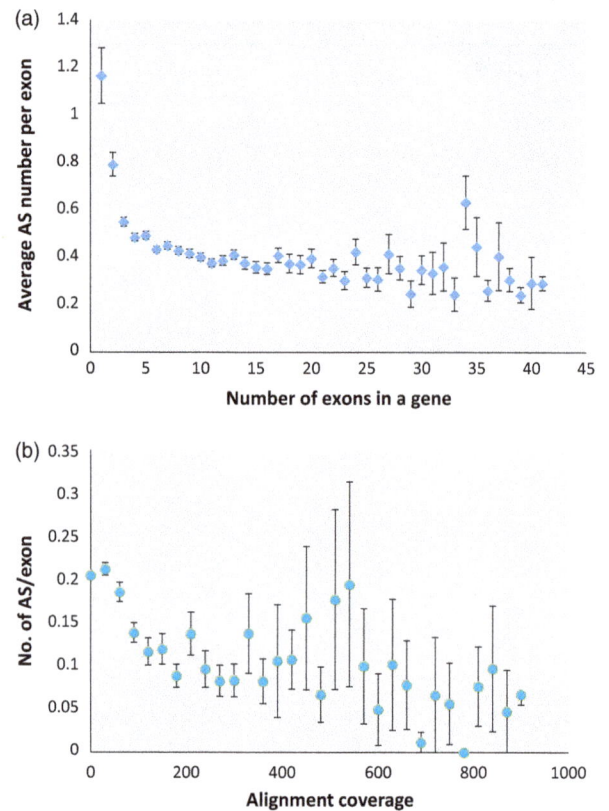

Figure 3 Comparison of the average number of AS events per exon, calculated by dividing the number of AS events in a gene with the number of exons in that gene, for genes containing different numbers of exons (a) and genes expressed at different levels (b), as determined by alignment coverage per base.

Splicing error can introduce or eliminate $n + 1$, $n + 2$ and $n + 3$ nucleotides to the mature mRNA, where n is a multiple of three nucleotides and only the $n + 3$ variant will preserve the downstream codons. Assuming that all three variants are equally likely to occur, random splicing error should introduce frameshift 67% of the time. Frameshift mutations have the highest probability of rendering the resulting protein nonfunctional; therefore, we were curious to see whether this phenomenon would be repressed in mungbean AS. Frameshift formation is close to 67% for AS events of the exon skipping and intron retention types (Table 2). However, the creation of frameshift was lower than expected in the alternative donor and alternative acceptor types of AS: plotting the number of AS events that occurred several bases from regular splicing sites revealed a preference for multiples of 3 (i.e. the length of a codon) in these types of AS (Figure 4). This could be partially explained by the common occurrence of NAGNAG motifs near the 3′ end of introns (Shi *et al.*, 2014). Because plant splice sites are normally located at AG bases at the 3′ ends of introns, such motifs would direct spliceosomes that miss their target to alternative targets located at distances that are integral multiples of codon lengths, thereby preventing the formation of frameshifted mRNA. Curiously, this effect was still visible at positions as far as 90 bases from the regular splice sites, distances at which NAGNAG motifs are unlikely to persist. Because frameshift mutation is very common in the more abundant intron retention and exon skipping types, it is unlikely

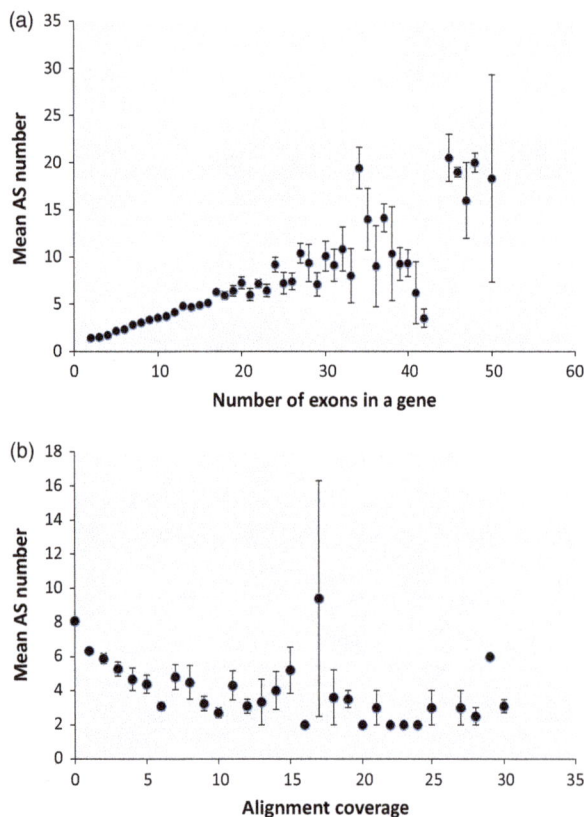

Figure 2 Correlation of mean AS number with number of exons in a gene (a) and gene expression level estimated from alignment coverage (b).

Table 2 Proportion of AS isoforms carrying frameshift codons in the mature mRNA, classified according to AS types

Type of AS	Root (%)	Leaf (%)	Pod (%)	Flower (%)
Alternative donor	34.3	33.9	32.6	29.9
Alternative acceptor	46.3	42.1	42.3	41.7
Exon skipping	72.0	72.5	71.2	72.2
Intron retention	61.6	63.2	62.9	62.6

that this is mostly caused by natural selection against frameshift splicing; therefore, other mechanisms could be responsible for these effects.

The role of sequence variation and the extent of AS conservation

The presence of motifs like NAGNAG at the 3' ends of introns raises the question of whether AS sites occur only at canonical splice sites or utilize other bases as well. We surveyed all splice junctions of the default and AS isoforms and categorized each AS site as high or low, using FPKM value of 10 as a threshold. At the 3' ends of introns, all isoforms utilize the AG splice site; by contrast, at the 5' end, a majority of splice sites occur at GU bases but a small fraction also occur at GC bases (Figure 5). There were no obvious sequence patterns around the two bases that could explain why some AS sites are more frequently spliced than others. Hence, although the pattern of AS appears random, it is still constrained by the availability of bases that can be used as splice sites. However, because the required motifs at each end are only two bases long, and the abundance of these dinucleotides in the genome is relatively high, it is not surprising that AS is so prevalent in many organisms.

Nevertheless, functional AS isoforms have been detected in the past (Inoue et al., 1990; Ullrich et al., 1995); hence, it is not unlikely that mungbean also harbours some functional isoforms in its transcriptome. We tried to identify candidate AS isoforms for more in-depth study of their function. However, given that a significant portion of AS in mungbean may not have any function at all, we paid extra attention to AS sites that are conserved in other species. Conservation among species does not necessarily imply function, but it at least indicates that the isoforms in question do not impose negative selection pressure on the plant over evolutionary timescales. To this end, we compared transcript sequences from mungbean to those from adzuki bean (*Vigna angularis*), a closely related species in the *Vigna* genus. BLAST analysis of exon sequences surrounding AS junctions identified

3600 AS sites with high sequence similarities in both species (Table S3), which is comparable to the results obtained by Chamala et al. (2015), who identified more than 5000 conserved AS between common bean and soya bean using a similar method. However, a closer examination also revealed that the exact splice sites are rarely conserved in both species. By applying the strict criteria that both splice sites must be located at the exact same position and the differences in nucleotide length between the two species must not introduce frameshift, we reduced the set of candidates to 629 genes, comprising 859 conserved AS events (Table S4).

Gene Ontology (GO) analysis of the genes carrying conserved AS identified 488 GO groups (Table S5) with significant enrichment for cellular components only (Figure S3), while other GO groups are not significantly different from background level (Figure 6). However, the number of conserved AS isoforms dwindles even further when the comparison is made between more distantly related species. A comparison of AS events between mungbean and soya bean (Glycine max) yielded only 16 conserved AS isoforms retained at the exact same base position in both species (Table S6). All but one AS junction was also conserved in adzuki bean, although two of them will create frameshift mutations in adzuki bean. Based on this observation, we conclude that AS events that confer selective advantages, and are thus retained over evolutionary timescales, are very rare. However, other groups have observed that when the conservation criteria are relaxed to ignore the exact splicing position and focus on exon sequence conservation among the same AS types, the number of conserved AS events increases considerably, and such events can be identified even in species outside the angiosperms (Chamala et al., 2015). While it is possible that such approach could identify conserved AS with similar function, it would be inadequate to identify possible inclusion of frameshift caused by differences in nucleotide length among species, which can create a very different protein if the isoforms are translated.

Discussion

Based on the findings in this study, we conclude that the noise hypothesis fits the pattern of AS events in mungbean; consequently, a large proportion of AS isoforms in mungbean probably have no function. However, the noisy splicing model does not exclude the possibility that useful and functional AS isoforms could emerge among the resultant nonfunctional isoforms. As suggested by our observations regarding genes with high exon numbers or expression levels, natural selection will act on genes that produce AS isoforms at concentrations that could be

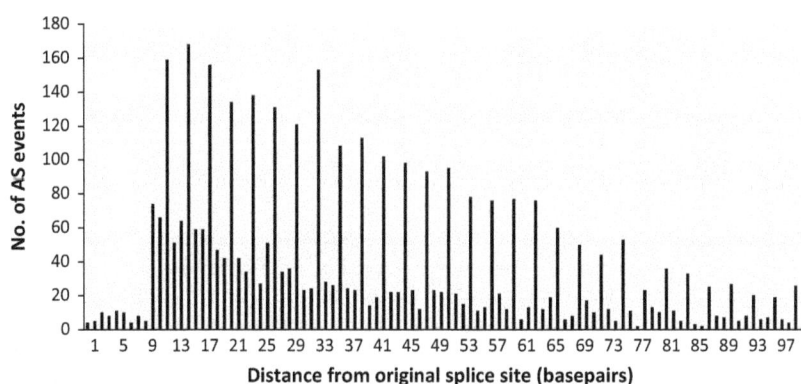

Figure 4 Number of AS isoforms, according to distance from the regular splice site, for alternative donor and alternative acceptor isoform types.

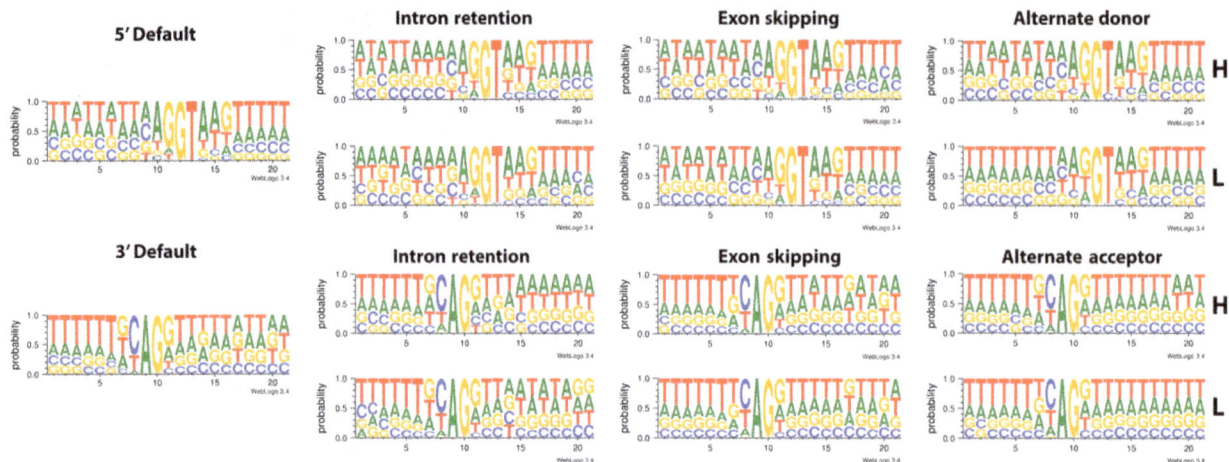

Figure 5 Proportion of DNA bases surrounding alternative splice types compared to the regular splice sites. AS isoforms with FPKM >10 were categorized as having high concentration (**H**), while those with FPKM <10 were grouped into the low concentration group (**L**).

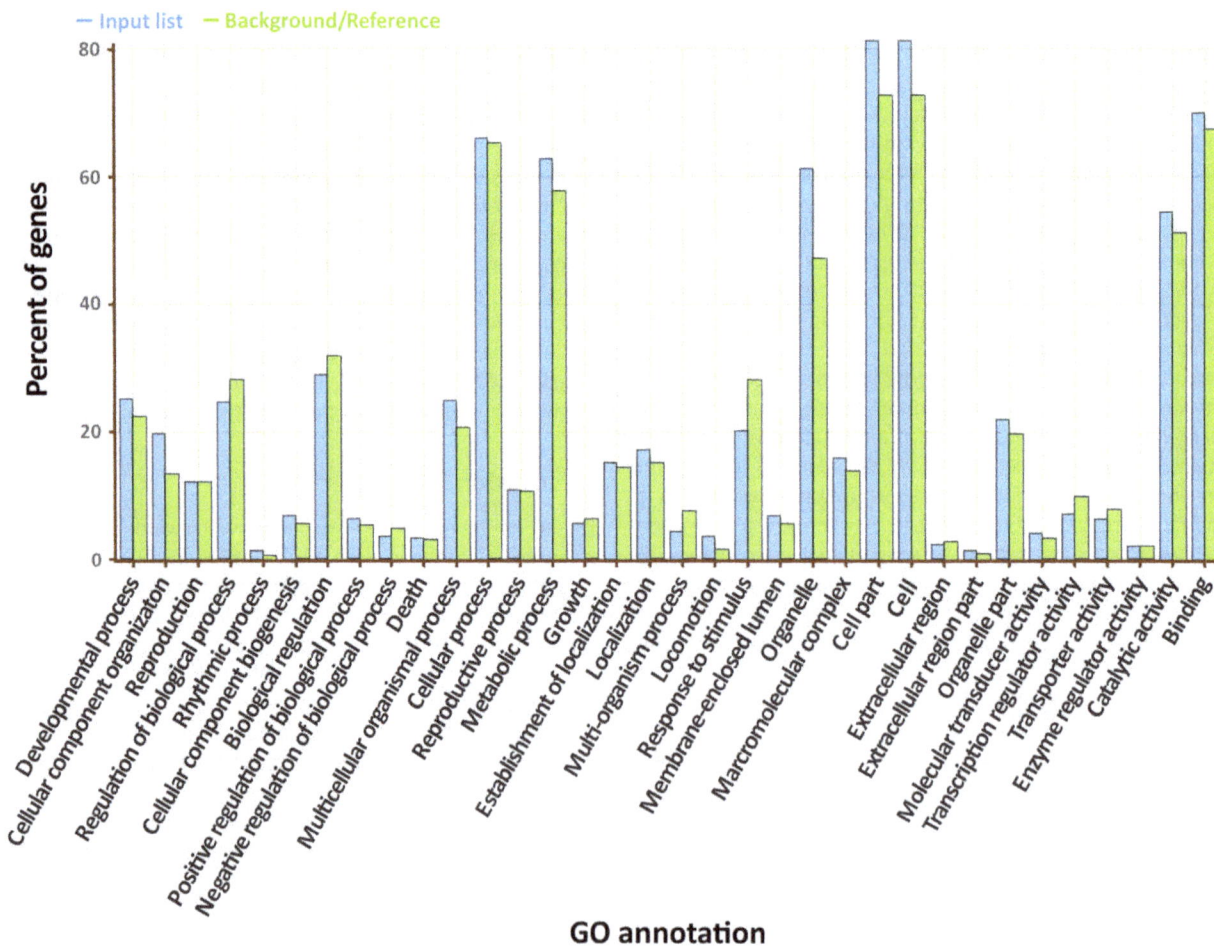

Figure 6 Gene ontology enrichment of genes with AS isoforms that are conserved in mungbean and adzuki bean.

dangerous to the cell, possibly by favouring the propagation of individuals with stronger splicing signals at the correct bases, or those that lack the bases that are used as AS sites. Similarly, AS isoforms that confer selective advantage will be retained or even strengthened, so that the spliceosome will regularly cut at the alternative site as well as the regular site. The presence of numerous low-abundance isoforms is probably not significantly harmful to mungbean cells, but it could be useful as a source of variants upon which natural selection can act.

The arguments in support of noisy splicing have been outlined by several groups (Hon *et al.*, 2013; Pickrell *et al.*, 2010; Zhang *et al.*, 2009). Melamud and Moult (2009b) noted that while some tissue-specific AS isoforms are conserved across species, these represent a relatively small fraction of AS events. A large proportion of AS isoforms also carry premature stop codons, which make them vulnerable to NMD. Even if these isoforms are translated, most of the alternative protein structures are predicted to be nonviable. The number of detected AS isoforms also tends to increase in genes with more introns or genes expressed at higher levels, in line with the view of AS as a probabilistic event. The more the introns to be processed, the greater the probability of noisy splicing. Similarly, higher levels of gene expression increase the likelihood of a splicing error among the pool of processed transcripts. This could explain why RNAseq data typically allow the detection of more AS isoforms, because the sequenced libraries usually have a very high sequencing depth. Thus, splicing errors that are not normally found in average cells become visible; this is compounded by the use of PCR during library preparation, which could amplify uncommon transcripts to a more easily detectable level.

Protein studies provide another line of evidence supporting noisy splicing. In human cells, the observed protein diversity revealed by high-throughput mass spectrophotometry is much lower than that predicted from AS studies of transcriptome data. Abascal et al. (2015) found that most human proteins exist as single dominant isoform and detected only 282 AS isoforms among 12 716 genes at the protein level; this is nowhere near the prediction of 95% based on transcriptome data (Pan et al., 2008). However, the absence of protein products of a given AS isoform may not necessarily mean that the isoform serves no function; in some cases, degradation of AS isoforms through the NMD pathway serves to regulate the concentration of transcripts in the cell (Drechsel et al., 2013; Fu et al., 2009). Nevertheless, this lack of representation at the protein level undermines the idea that AS increases the protein diversity generated by a given number of genes.

On the other hand, several compelling arguments support the notion that AS plays an important regulatory role in the cell. Barbazuk et al. (2008) presented several lines of evidence for the functional importance of AS: the predominance of AS in some gene families versus its absence in others; the existence of AS events that correlate with specific tissue and developmental cues; incorporation of AS products into ribosomes; and conservation of some AS events among distantly related species. Because noisy splicing has probably existed since the emergence of introns in eukaryotes, it is likely that a large number of useful isoforms have evolved from it, resulting in the phenomena detailed above. However, based on the observed low level of conservation among species, the contribution of AS to protein variation and regulation of gene expression does not appear to be significant.

The actual proportion of functional AS isoforms in mungbean obviously cannot be precisely determined without experimental tests. However, in general the number of AS isoforms is lower in plants than in animals (Kim et al., 2008). It will be interesting to speculate whether this is due to the different research focus in plants and animals, or instead to intrinsic differences in the splicing mechanisms in the two kingdoms. One issue that should be investigated is the effect of genome expansion (e.g. polyploidization) on AS. An organism whose genome cannot tolerate significant expansion will benefit greatly by the ability to increase the functionality of its existing genome via AS. Hence, to generate additional transcript diversity, it may be advantageous to maintain a less-specific splicing machinery. However, polyploid plants can easily obtain new gene variants by allowing duplicated

genes to evolve independently; this strategy is potentially safer because it does not disrupt the function of the original gene. In allotetraploid soya bean, duplicated genes undergo less AS (Shen et al., 2014); this is curious because in such cases, the penalty for incorrect splicing would be less severe because a backup copy is present elsewhere in the genome.

Because examples of functional AS in plants are rare, it would be prudent to assume that a significant portion of AS in plants has no distinct function; thus, more evidence is required to support the claim that there is extensive functional diversity generated by AS in plants. Thus, detected AS events should be treated like genetic marker data, which are useful to identify and catalogue, but strong experimental evidence such as QTL mapping and transgene expression are necessary to assert that a given sequence variation causes a particular phenotype. Additionally, approaches like comparative genomics, which are used to select candidate markers that are likeliest to alter a phenotype, could also be applied to finding AS isoforms that encode a novel function. We believe that our comparative AS detection data could be used as a starting point to perform more in-depth studies of the phenotypic diversity generated by AS.

Experimental procedures

Plant materials and RNA sequencing

RNA sequence data were obtained from Kang et al. (2014); in that study, a pure line mungbean plant from cultivar VC1973A (developed by AVRDC) was used as the source material for RNA extraction. The plants were planted in a greenhouse; following sowing, tissues were harvested from root after 2 weeks, leaf after 1 month, flower after 2 months and whole pods after 2.5 months.

Sequence alignment, transcript assembly and AS identification

The cleaned sequence data were aligned to the mungbean reference genome (Kang et al., 2014) using TopHat (Trapnell et al., 2009) with default settings. The resultant gapped alignment data in binary alignment format were then used as input for Cufflinks (Trapnell et al., 2012) under default settings to assemble the transcripts and identify splicing junctions from the alignment data. For AS detection and annotation, the assembled transcriptome files (in.gtf format) were submitted to ASTALA-VISTA (Foissac and Sammeth, 2007) web interface (http://genome.crg.es/astalavista/). AS events were also annotated with ASprofile (Florea et al., 2013), which also uses Cufflinks output as input data. The resulting AS annotations were checked at random by visual examination of the AS genome coordinates in the original binary alignment (.bam) files using Integrated Genome Viewer (Robinson et al., 2011).

Isoform quantitation

The FPKM (fragments per kilobase of transcript per million mapped reads) value for each AS isoform was provided by ASprofile, based on Cufflinks estimation, after assembly of each transcript. When the FPKM value was not available for a chromosomal segment of interest, transcript quantity was estimated based on alignment coverage on that segment, calculated using the coverageBed command in Bedtools (Quinlan and Hall, 2010). The number of aligned fragments was then multiplied by fragment length and divided by the number of bases in the segment of interest, to yield the average coverage per base value. Because each tissue had different sequence coverage, coverage per base was only compared within a tissue.

Statistical analysis

Basic arithmetical analysis, such as calculations of sums and means, was performed in Microsoft Excel. Calculation of descriptive statistics was performed in the R statistical package, using the 'describeBy' command in the psych library. Calculations of Pearson's correlations and the corresponding significance values were also carried out in R using the 'cor.test' command.

Sequence junction analysis

To visualize the presence of sequence conservation near splice sites, coordinates of splice sites along with 10 bases upstream and 10 bases downstream from those sites were input into Bedtools using the fastaFromBed command to obtain the DNA sequences between those coordinates. Sequences with FPKM value >10 were put in the high group, while the rest were put in the low group. The sequences were then used as input for weblogo (http://weblogo.threeplusone.com/) to visualize the proportion of bases commonly found surrounding the splice sites.

Comparative analysis

The sequence of 50 bp of exonic region surrounding AS sites and intron sequences from intron retention AS events were obtained using Bedtools from mungbean, adzuki bean (*Vigna angularis*) and soya bean (*Glycine max*). Splice sites were detected in adzuki bean and soya bean using ASprofile with the same settings as used for AS detection in mungbean. The RNA sequences used for AS detection were obtained from sequences provided by Kang *et al.* (2015) for adzuki bean and Shen *et al.* (2014) for soya bean. Sequences that share similarities were detected using local BLAST+ search (Camacho *et al.*, 2009), with mungbean sequences used as the local sequence database. The resulting matches were filtered using the following criteria: sequences at the splicing junctions must have exact match while sequences further away are allowed to have gaps and mismatches, intron sequences in intron retention type have at least 80% similarities, and the length difference of retained introns and skipped exons between two species must not exceed 30 nucleotides or introduce frameshift. Full sequences of proteins containing conserved AS were then identified, and their homologs in soya bean were identified using blastp in the BLAST+ package. Matching soya bean gene ID with the highest e-values were then submitted to agriGO (http://bioinfo.cau.edu.cn/agriGO/analysis.php) to obtain the gene ontology classification of those genes.

Acknowledgements

This work was supported by a grant from the Next Generation BioGreen 21 Programme (Code No. PJ01102601), Rural Development Administration, Republic of Korea.

The authors declare no conflict of interest.

References

Abascal, F., Ezkurdia, I., Rodriguez-Rivas, J., Rodriguez, J.M., del Pozo, A., Vazquez, J., Valencia, A. *et al.* (2015) Alternatively spliced homologous exons have ancient origins and are highly expressed at the protein level. *PLoS Comput. Biol.* **11**, e1004325.

Barbazuk, W.B., Fu, Y. and McGinnis, K.M. (2008) Genome-wide analyses of alternative splicing in plants: opportunities and challenges. *Genome Res.* **18**, 1381–1392.

Breitbart, R.E., Andreadis, A. and Nadal-Ginard, B. (1987) Alternative splicing: a ubiquitous mechanism for the generation of multiple protein isoforms from single genes. *Annu. Rev. Biochem.* **56**, 467–495.

Camacho, C., Coulouris, G., Avagyan, V., Ma, N., Papadopoulos, J., Bealer, K. and Madden, T.L. (2009) BLAST+: architecture and applications. *BMC Bioinformatics*, **10**, 1.

Chamala, S., Feng, G., Chavarro, C. and Barbazuk, W.B. (2015) Genome-wide identification of evolutionarily conserved alternative splicing events in flowering plants. *Frontiers Bioeng. Biotechnol.* **3**, 33.

Drechsel, G., Kahles, A., Kesarwani, A.K., Stauffer, E., Behr, J., Drewe, P., Rätsch, G. *et al.* (2013) Nonsense-mediated decay of alternative precursor mRNA splicing variants is a major determinant of the Arabidopsis steady state transcriptome. *Plant Cell*, **25**, 3726–3742.

Filichkin, S.A. and Mockler, T.C. (2012) Unproductive alternative splicing and nonsense mRNAs: a widespread phenomenon among plant circadian clock genes. *Biology Direct*, **7**, 20.

Filichkin, S.A., Priest, H.D., Givan, S.A., Shen, R., Bryant, D.W., Fox, S.E., Wong, W.K. *et al.* (2010) Genome-wide mapping of alternative splicing in Arabidopsis thaliana. *Genome Res.* **20**, 45–58.

Florea, L., Song, L. and Salzberg, S.L. (2013) Thousands of exon skipping events differentiate among splicing patterns in sixteen human tissues. *F1000Research*, **2**, 188.

Foissac, S. and Sammeth, M. (2007) ASTALAVISTA: dynamic and flexible analysis of alternative splicing events in custom gene datasets. *Nucleic Acids Res.* **35**, W297–W299.

Fu, Y., Bannach, O., Chen, H., Teune, J.H., Schmitz, A., Steger, G., Xiong, L. *et al.* (2009) Alternative splicing of anciently exonized 5S rRNA regulates plant transcription factor TFIIIA. *Genome Res.* **19**, 913–921.

Graveley, B.R. (2001) Alternative splicing: increasing diversity in the proteomic world. *Trends Genet.* **17**, 100–107.

Hon, C.C., Weber, C., Sismeiro, O., Proux, C., Koutero, M., Deloger, M., Das, S. *et al.* (2013) Quantification of stochastic noise of splicing and polyadenylation in Entamoeba histolytica. *Nucleic Acids Res.* **41**, 1936–1952.

Inoue, K., Hoshijima, K., Sakamoto, H. and Shimura, Y. (1990) Binding of the Drosophila sex-lethal gene product to the alternative splice site of transformer primary transcript. *Nature*, **344**, 461–463

Kang, Y.J., Kim, S.K., Kim, M.Y., Lestari, P., Kim, K.H., Ha, B.K., Jun, T.H. *et al.* (2014) Genome sequence of mungbean and insights into evolution within Vigna species. *Nat. Commun.* **5**, 5443.

Kang, Y.J., Satyawan, D., Shim, S., Lee, T., Lee, J., Hwang, W.J., Kim, S.K. *et al.* (2015) Draft genome sequence of adzuki bean, Vigna angularis. *Sci. Rep.* **5**, 8069.

Kawashima, T., Douglass, S., Gabunilas, J., Pellegrini, M. and Chanfreau, G.F. (2014) Widespread use of non-productive alternative splice sites in Saccharomyces cerevisiae. *PLoS Genet.* **10**, e1004249.

Kim, E., Goren, A. and Ast, G. (2008) Alternative splicing: current perspectives. *Bioessays*, **30**, 38–47.

Lah, G.J.-E., Li, J.S.S. and Millard, S.S. (2014) Cell-specific alternative splicing of Drosophila Dscam2 is crucial for proper neuronal wiring. *Neuron*, **83**, 1376–1388.

Lu, T., Lu, G., Fan, D., Zhu, C., Li, W., Zhao, Q., Feng, Q. *et al.* (2010) Function annotation of the rice transcriptome at single-nucleotide resolution by RNA-seq. *Genome Res.* **20**, 1238–1249.

Marquez, Y., Brown, J.W., Simpson, C., Barta, A. and Kalyna, M. (2012) Transcriptome survey reveals increased complexity of the alternative splicing landscape in Arabidopsis. *Genome Res.* **22**, 1184–1195.

Melamud, E. and Moult, J. (2009a) Stochastic noise in splicing machinery. *Nucleic Acids Res.* **37**, 4873–4886.

Melamud, E. and Moult, J. (2009b) Structural implication of splicing stochastics. *Nucleic Acids Res.* **37**, 4862–4872.

Nair, R., Schafleitner, R., Kenyon, L., Srinivasan, R., Easdown, W., Ebert, A. and Hanson, P. (2012) Genetic improvement of mungbean. *SABRAO J. Breed. Genet.* **44**, 177–190.

Neu-Yilik, G., Gehring, N.H., Hentze, M.W. and Kulozik, A.E. (2004) Nonsense-mediated mRNA decay: from vacuum cleaner to Swiss army knife. *Genome Biol.* **5**, 218.

Pan, Q., Shai, O., Lee, L.J., Frey, B.J. and Blencowe, B.J. (2008) Deep surveying of alternative splicing complexity in the human transcriptome by high-throughput sequencing. *Nat. Genet.* **40**, 1413–1415.

Pickrell, J.K., Pai, A.A., Gilad, Y. and Pritchard, J.K. (2010) Noisy splicing drives mRNA isoform diversity in human cells. *PLoS Genet.* **6**, e1001236.

Quinlan, A.R. and Hall, I.M. (2010) BEDTools: a flexible suite of utilities for comparing genomic features. *Bioinformatics*, **26**, 841–842.

Robinson, J.T., Thorvaldsdóttir, H., Winckler, W., Guttman, M., Lander, E.S., Getz, G. and Mesirov, J.P. (2011) Integrative genomics viewer. *Nat. Biotechnol.* **29**, 24–26.

Rubatzky, V.E. and Yamaguchi, M. (1997) Peas, beans, and other vegetable legumes. In *World Vegetables*, pp. 474–531. New York, Chapman Hall (ITP).

Severing, E.I., van Dijk, A.D., Stiekema, W.J. and van Ham, R.C. (2009) Comparative analysis indicates that alternative splicing in plants has a limited role in functional expansion of the proteome. *BMC Genom.* **10**, 154.

Shen, Y., Zhou, Z., Wang, Z., Li, W., Fang, C., Wu, M., Ma, Y. *et al.* (2014) Global dissection of alternative splicing in paleopolyploid soybean. *Plant Cell*, **26**, 996–1008.

Shi, Y., Sha, G. and Sun, X. (2014) Genome-wide study of NAGNAG alternative splicing in Arabidopsis. *Planta*, **239**, 127–138.

Thatcher, S.R., Zhou, W., Leonard, A., Wang, B.B., Beatty, M., Zastrow-Hayes, G., Zhao, X. *et al.* (2014) Genome-wide analysis of alternative splicing in Zea mays: landscape and genetic regulation. *Plant Cell*, **26**, 3472–3487.

Trapnell, C., Pachter, L. and Salzberg, S.L. (2009) TopHat: discovering splice junctions with RNA-Seq. *Bioinformatics*, **25**, 1105–1111.

Trapnell, C., Roberts, A., Goff, L., Pertea, G., Kim, D., Kelley, D.R., Pimentel, H. *et al.* (2012) Differential gene and transcript expression analysis of RNA-seq experiments with TopHat and Cufflinks. *Nat. Protoc.*, **7**, 562–578.

Ullrich, B., Ushkaryov, Y.A. and Südhof, T.C. (1995) Cartography of neurexins: more than 1000 isoforms generated by alternative splicing and expressed in distinct subsets of neurons. *Neuron*, **14**, 497–507.

Zhang, Z., Xin, D., Wang, P., Zhou, L., Hu, L., Kong, X. and Hurst, L.D. (2009) Noisy splicing, more than expression regulation, explains why some exons are subject to nonsense-mediated mRNA decay. *BMC Biol.* **7**, 23.

11

ARGOS8 variants generated by CRISPR-Cas9 improve maize grain yield under field drought stress conditions

Jinrui Shi*, Huirong Gao, Hongyu Wang, H. Renee Lafitte, Rayeann L. Archibald, Meizhu Yang, Salim M. Hakimi, Hua Mo and Jeffrey E. Habben

DuPont Pioneer, Johnston, IA, USA

*Correspondence

e-mail Jinrui.shi@Pioneer.com

Keywords: maize, ARGOS, CRISPR-Cas9, genome editing, drought tolerance, grain yield.

Summary

Maize *ARGOS8* is a negative regulator of ethylene responses. A previous study has shown that transgenic plants constitutively overexpressing *ARGOS8* have reduced ethylene sensitivity and improved grain yield under drought stress conditions. To explore the targeted use of *ARGOS8* native expression variation in drought-tolerant breeding, a diverse set of over 400 maize inbreds was examined for *ARGOS8* mRNA expression, but the expression levels in all lines were less than that created in the original *ARGOS8* transgenic events. We then employed a CRISPR-Cas-enabled advanced breeding technology to generate novel variants of *ARGOS8*. The native maize GOS2 promoter, which confers a moderate level of constitutive expression, was inserted into the 5′-untranslated region of the native *ARGOS8* gene or was used to replace the native promoter of *ARGOS8*. Precise genomic DNA modification at the *ARGOS8* locus was verified by PCR and sequencing. The *ARGOS8* variants had elevated levels of *ARGOS8* transcripts relative to the native allele and these transcripts were detectable in all the tissues tested, which was the expected results using the GOS2 promoter. A field study showed that compared to the WT, the *ARGOS8* variants increased grain yield by five bushels per acre under flowering stress conditions and had no yield loss under well-watered conditions. These results demonstrate the utility of the CRISPR-Cas9 system in generating novel allelic variation for breeding drought-tolerant crops.

Introduction

Developing more drought-tolerant crops in a sustainable manner is one means to meet the demand of an increasing human population that will require more food, feed and fuel. Improvement in drought tolerance of crops is ultimately measured by an increase in grain yield under water-limiting conditions. The physiological processes and metabolic networks underlying drought tolerance are complicated and often difficult to delineate. Nevertheless, the phytohormone ethylene is known to play an important role in regulating plant response to abiotic stress, including water deficits and high temperature (Hays *et al.*, 2007; Kawakami *et al.*, 2010, 2013). Field studies have shown that reducing ethylene biosynthesis by silencing *1-aminocyclopropane-1-carboxylic acid synthase6* in transgenic maize plants improves grain yield under drought stress conditions (Habben *et al.*, 2014). A higher yield also can be achieved by decreasing the sensitivity of maize to ethylene (Shi *et al.*, 2015). *ARGOS* genes are negative regulators of the ethylene response and modulate ethylene signal transduction, enhancing drought tolerance when overexpressed in transgenic maize plants (Guo *et al.*, 2014; Shi *et al.*, 2015).

In addition to a transgenic approach, natural genetic variation for traits that impact drought tolerance has also been used in maize breeding programmes to improve grain yield. By applying precision phenotyping and molecular markers as well as understanding the genetic architecture of quantitative traits, maize breeders developed hybrids (AQUAmax®) with increased grain yield under drought stress conditions (Cooper *et al.*, 2014; Gaffney *et al.*, 2015). The drought tolerance in these hybrids is governed by multiple genes which individually have small effects. Potentially, some of these key genes could be identified and

altered to generate new alleles to produce a larger effect, thus enhancing the breeding process. However, until recently, generating such allelic variation with physically or chemically induced mutagenesis was a random process, which made it difficult to produce intended DNA sequence changes at a target locus. In the past few years, efficient genome editing technologies have emerged, enabling rapid and precise manipulation of DNA sequences, and setting the stage for developing drought-tolerant germplasm by editing major genes in their natural chromosomal context.

Four genome editing tools, meganucleases, zinc-finger nucleases (ZFN), transcription activator-like effector nucleases (TALEN) and the clustered regularly interspaced short palindromic repeat (CRISPR)/CRISPR-associated nuclease protein (Cas) system, have provided targeted gene modification in plants (Čermák *et al.*, 2015; Gao *et al.*, 2010; Li *et al.*, 2012, 2013; Shukla *et al.*, 2009). Among these, the CRISPR-Cas9 system is easiest to implement and is highly efficient. The system consists of a Cas9 endonuclease derived from *Streptococcus pyogenes* and a chimeric single guide RNA that directs Cas9 to a target DNA sequence in the genome. CRISPR-Cas9 genome editing is accomplished by introducing a DNA double-strand break in the target locus via Cas9, followed by DNA repair through either the endogenous imprecise nonhomologous end-joining (NHEJ) or the high-fidelity homology-directed repair (HDR) pathways. NHEJ can induce small insertions or deletions at the repair junction while HDR stimulates precise sequence alterations, including programmed sequence correction as well as DNA fragment insertion and swap, when a DNA repair template is exogenously supplied. The system has been successfully tested in staple crops, such as maize, wheat, rice and soybean (Cai *et al.*, 2015; Du *et al.*, 2016; Jacobs *et al.*, 2015; Jiang *et al.*, 2013; Li *et al.*,

2015; Liang *et al.*, 2014; Shan *et al.*, 2015; Sun *et al.*, 2016; Svitashev *et al.*, 2015; Wang *et al.*, 2014; Zhang *et al.*, 2014; Zhou *et al.*, 2014, 2015).

In maize, endogenous *ARGOS8* mRNA expression is relatively low and spatially nonuniform. Previous field testing showed that constitutive overexpression of *ARGOS8* in transgenic plants increases grain yield under drought stress conditions without yield penalty in nonstress environments (Shi *et al.*, 2015). Aiming at creating novel *ARGOS8* variants which would confer beneficial traits for maize breeding, the genomic sequence of *ARGOS8* was edited using CRISPR-Cas-enabled advanced breeding technology to produce ubiquitous and elevated expression across multiple tissues and at different developmental stages. Here, we report the generation of maize lines carrying *ARGOS8* genome-edited variants and their hybrid yield performance in a field study. Our results demonstrate that modifying single native genes to change expression patterns can increase maize grain yield under drought stress conditions.

Results

Natural allelic variation of maize ARGOS8 and expression patterns

In wild-type (WT) inbreds PH184C (proprietary) and B73 (public), *ARGOS8* mRNA expression is very low in all the tissues tested, ranging from 3 to 25 transcripts per ten million (TPTM), as measured with RNA sequencing (Figure S1). The only exception is in kernels where expression was approximately 260 TPTM. For comparison, the transcripts of the ubiquitously and moderately expressed *GOS2* gene, the maize homolog of rice *GOS2* (de Pater *et al.*, 1992), are about 6000 TPTM in most tissues with the highest expression occurring in internodes (13 700 TPTM) and the lowest occurring in tassels (2500 TPTM). A survey of a diverse set of 419 proprietary and public inbred lines showed that the *ARGOS8* expression in leaves of 3-week-old seedlings only ranged from 0 to 20 TPTM (Figure S2), suggesting that the natural variation in expression levels among these inbreds was also low.

The protein encoded by the *ARGOS8* gene varies among inbred lines. The ARGOS8 protein in B73 has 118 amino acids (long version) while the protein from PH184C consists of 94 amino acids (short version). The difference in protein sequence is the

presence of an N-terminal extension of 24 amino acids in the long version. The extra coding sequence in the B73 allele is a result of a 7-bp duplication in the 5'-untranslated region (5'-UTR) which produces an in-frame ATG codon upstream of the original translation start codon. The 7-bp duplication may be a footprint left behind by a transposon excision event (Scott *et al.*, 1996). Like the long version from B73 (Shi *et al.*, 2015, 2016), the short version of *ARGOS8* also reduces ethylene responses when overexpressed, as demonstrated in Arabidopsis transgenic plants. In the ethylene triple response assay (Bleecker *et al.*, 1988), hypocotyls and roots of the etiolated 35S:*ARGOS8* Arabidopsis seedlings were longer than that in WT controls in the presence of the ethylene precursor aminocyclopropane-1-carboxylic acid (Figure 1). Although the short version of the ARGOS8 protein accumulated to a higher level than the B73 version in plants with a similar level of transcripts (data not shown), the native allelic variant still produces a very low level of the ARGOS8 protein in WT plants and it was not detectable by immunoblot analysis. This short allele, as well as the long B73 allele, was not able to confer drought-tolerant phenotypes without ectopic overexpression. Consequently, this observed functional native diversity in the *ARGOS8* gene is not enough for targeted drought breeding.

Novel ARGOS8 variants generated by the CRISPR-Cas9 system

To achieve a moderate level of constitutive expression of *ARGOS8*, the maize GOS2 promoter and the 5'-UTR with an intron, hereafter collectively referred to as GOS2 PRO, were either used to replace the native promoter of the *ARGOS8* gene or were inserted into the 5'-UTR of *ARGOS8* in the inbred PH184C. Because both the promoter swap and insertion require precise manipulation of genomic DNA, we employed an RNA-guided Cas9 endonuclease to generate DNA double-strand breaks in a site-specific manner, integrating the GOS2 PRO into the upstream region of *ARGOS8* via homology-directed DNA repair (Figure 2a). A DNA repair template and genome editing reagents were delivered into immature embryos by particle bombardment and plantlets were regenerated from embryogenic calli. The reagents include an *S. pyogenes Cas9* gene and a single guide RNA (sgRNA) gene, *CRISPR RNA1* (*CR1*), in the GOS2 PRO insertion or two sgRNAs (*CR2* and *CR3*) in the GOS2 PRO swap (Figure S3), as

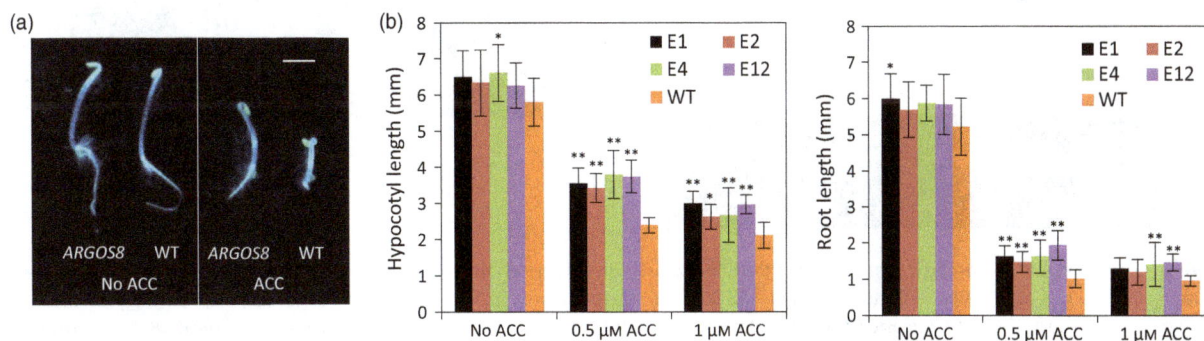

Figure 1 Maize *ARGOS8* reduces plant responses to ethylene when overexpressed in transgenic Arabidopsis plants. (a) Ethylene triple response of Arabidopsis *ARGOS8* transgenic plants (*ARGOS8*) and wild-type (WT) controls to 0.5 μM of the ethylene precursor aminocyclopropane-1-carboxylic acid (ACC). A short version of *ARGOS8* was overexpressed under control of the cauliflower mosaic virus 35S promoter (35S). Composite image of representative 3-day-old etiolated seedlings. Bar = 2 mm. (b) Hypocotyl and root lengths of etiolated Arabidopsis seedlings overexpressing the short version of *ARGOS8*. Four transgenic lines (E1, E2, E4 and E12) and wild-type (WT) controls were grown in the dark in the presence of indicated ACC concentrations for 3 days. Data are means ± SD, *n* = 15. Significant differences of the transgenic plants from the WT are denoted by asterisks (*$P < 0.05$, **$P < 0.01$, ANOVA, Tukey's HSD).

well as *phosphomannose isomerase* (*PMI*), *ovule development protein2* (*ODP2*) and *WUSCHEL* (*WUS*) for stimulation of transformation and seedling regeneration (Svitashev *et al.*, 2015). The DNA repair template consisted of the GOS2 PRO flanked by two DNA fragments of approximately 400-bp homologous to genomic sequences immediately adjacent to the Cas9 cleavage sites in the *ARGOS8* locus (Figure 2b). Of approximately 1000 immature embryos particle-bombarded for the promoter insertion and swap experiments, 194 and 334 shoots, respectively,

were regenerated on the selection medium (Table S1). To eliminate the shoots whose CRISPR RNA target sites (CTS) were not altered, a rapid screening was performed using a quantitative PCR (qPCR) assay (Table S2), which estimates the copy number of CTS. Shoots with no modification at CTS contained two copies of the wild-type CTS, shoots with CTS modification in one of the two sister chromosomes had one intact copy, while modification in both chromosomes would reduce the copy number to zero. With this screening, 190 and 172 regenerated shoots from the

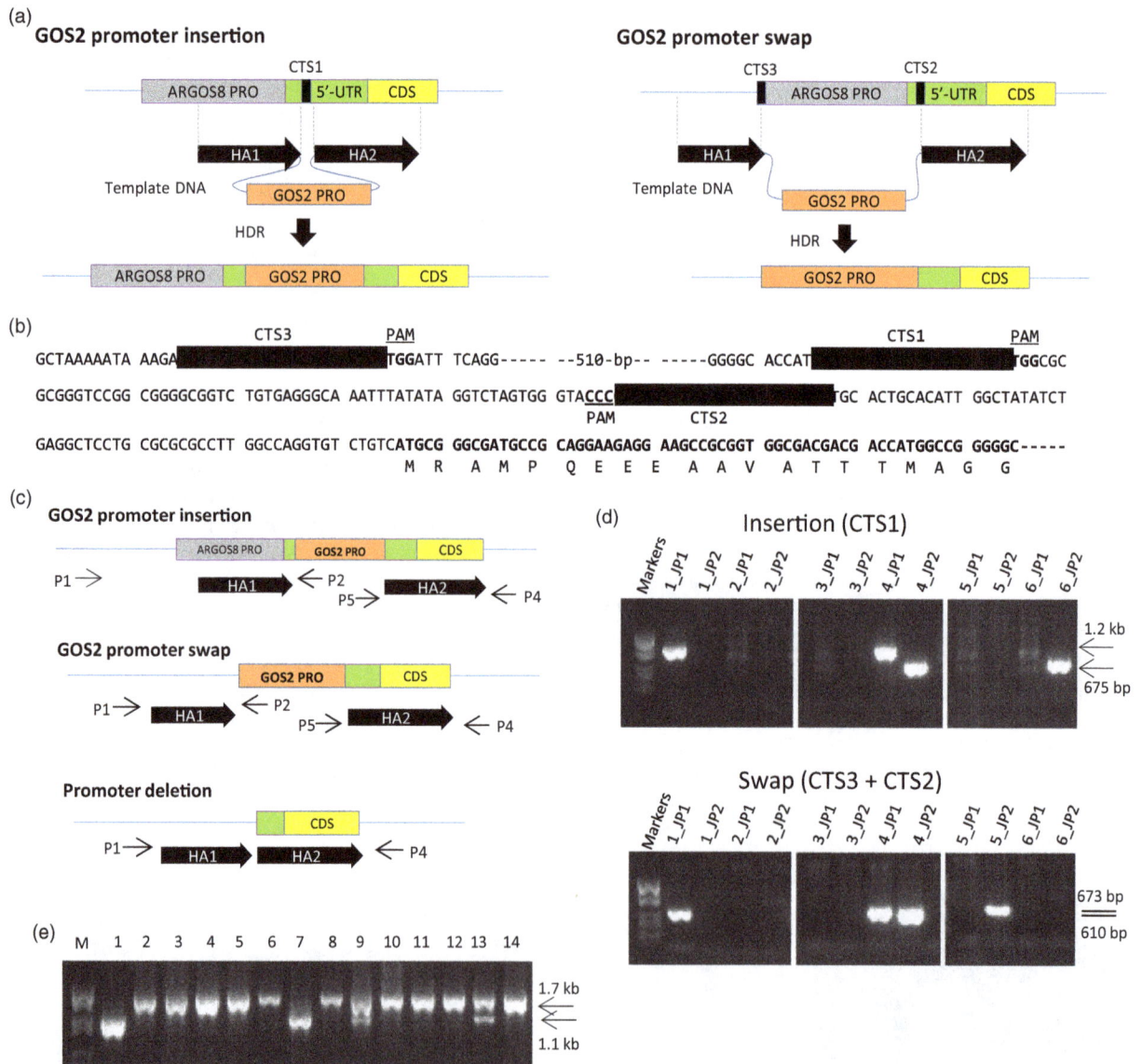

Figure 2 Editing the *ARGOS8* genomic sequence using the CRISPR/Cas9 system to generate variants with constitutive expression. (a) Schematic drawing illustrating the insertion of GOS2 PRO into the 5′-UTR of *ARGOS8* and the promoter swap. CTS, CRISPR-RNA target site; HA, homology arm; HDR, homology-directed repair; GOS2 PRO, maize GOS2 promoter and the 5′-UTR with an intron. (b) Genomic sequence of the *ARGOS8* 5′-UTR and the upstream region. The CRISPR-RNA target sites (CTS) are highlighted in red, and the protospacer adjacent motifs (PAM) are shown in blue font. The *ARGOS8* coding region is shown in bold font. (c) Diagram showing primers used in junction PCR for genotyping regenerated shoots and long PCR for amplifying and sequencing the entire modification region in homozygous plants. The relative position and direction of PCR primers (P) are indicated by arrows. P1 and P2 for the HR1 junction; P5 and P4 for the HR2 junction; P1 and P4 for the long PCR. (d) Junction PCR analysis of regenerated shoots. Agarose gel images are shown for representative regenerated shoots positive for one junction or two junctions and shoots negative in the junction PCR assay. JP1, HR1 junction PCR with the primer P1 and P2; JP2, HR2 junction PCR with P5 and P4. (e) PCR screening regenerated shoots for deletion in the *ARGOS8* locus. An agarose gel image is shown for PCR products amplified with the primer P1 and P4 in representative shoots (Lanes 1-14) generated from the *CRISPR RNA-3* and *RNA-1* transformation. M, DNA molecular weight markers.

insertion and swap experiments, respectively, were selected for genotyping with a junction PCR assay.

A pair of junction PCR assays was designed to detect GOS2 PRO inserts or swaps at CTS due to homologous recombination (Figure 2c). In the insertion experiments, five of the 190 shoots from the initial screening were found positive for one of the two junctions, and two shoots were positive for both junctions (Figure 2d and Table S1). These shoots were transferred to rooting media, and three plantlets were regenerated. Genotyping the T0 plants with the junction PCR assays revealed that one plant contained the GOS2 PRO insert in the ARGOS8 locus. The junction PCR products were sequenced, and expected sequences were confirmed (Figure S4). This line is referred to as ARGOS8-variant1 (ARGOS8-v1). For the GOS2 promoter swap, 23 of the 172 shoots obtained from the initial screening were positive for at least one junction. Among them, three were positive for both junctions (Figure 2d and Table S1). From these shoots, eight plantlets were regenerated. Of these T0 plants, two produced expected junction PCR products for the promoter swap in the ARGOS8 locus. Sequencing the PCR products confirmed correct sequences from both junctions (Figure S4). One of the lines is referred to as ARGOS8-variant2 (ARGOS8-v2). Genotyping also revealed that the ARGOS8-v1 and ARGOS-v2 were heterozygous.

F1 seeds of ARGOS8-v1 and ARGOS8-v2 were produced by crossing the T0 plants with WT PH184C plants. F1 plants were genotyped by PCR to select those that carry the ARGOS8-v1 or ARGOS8-v2, but were nulls for the genome editing reagents Cas9, sgRNA, PMI, ODP2 and WUS. To eliminate the plants containing random insertions of the DNA repair template, qPCR was performed to assess the copy number of the GOS2 PRO and ARGOS8. Selected clean F1 plants were backcrossed to produce BC1 seeds, or self-pollinated to obtain F2 seeds. Among the F2 segregants, homozygous plants were used to determine the sequence integrity of the newly created ARGOS8 variants. The entire genomic region was amplified using long PCR with primers P1 and P4, which were derived from genomic sequences further upstream and downstream of the homology arms used in the DNA repair templates (Figure 2c). Sequencing the long-PCR products confirmed that the ARGOS8-v1 and ARGOS8-v2 possess the expected DNA sequences (Figure 3a). The ARGOS8-v1 and ARGOS8-v2 segregated in a Mendelian fashion in BC1 and F2 populations (data not shown). Quantitative reverse-transcription PCR (qRT-PCR) analysis showed that the abundance of ARGOS8 transcripts in leaves of homozygous plants is approximately twice as much as in the heterozygotes for both lines (Figure 3b).

To obtain controls for analysing ARGOS8 gene expression in the ARGOS8 variants, the shoots regenerated from particle-bombarded immature embryos were screened for ARGOS8 promoter deletions in the promoter swap experiments using CR3 and CR1. Of 185 shoots screened, 30 produced a PCR product shorter than that expected for WT plants (Figure 2c and e), indicating a deletion between CTS3 and CTS1. Sequencing the PCR products from two T0 plants showed that both had an extra base pair (one line having T and the other A) at the junction of the nonhomologous end-joining (Figures 3a and S4). Similarly, approximately 13% (23 of 176) of the regenerated shoots were found contain deletion at the target sites in the swap experiments using CR3 and CR2 (data not shown). The deletion of the 550-bp genomic DNA fragment between CTS3 and CTS1 removed part of the ARGOS8 5'-UTR and the upstream promoter sequence (Figures 2b and 3a). One of the lines was referred to as ARGOS8-

variant3 (ARGOS8-v3). The ARGOS8 transcripts and proteins were undetectable in ARGOS8-v3 (Figure 4a and b). The line had normal growth and development, and no phenotypic defects were observed under normal growing conditions, indicating that the ARGOS8 gene is likely dispensable.

Expression patterns of ARGOS8 in genome-edited variants

The mRNA expression of ARGOS8 in the genome-edited variants was analysed in leaves, roots, silks and kernels using qRT-PCR. In the uppermost collared leaves of plants at the developmental stages V3, V6, V10 and V14, the ARGOS8 transcript level in ARGOS8-v1 and ARGOS8-v2 was significantly higher than that in WT plants with the highest expression found at V6 (Figure 4a). At the developmental stage R1, silks, roots and leaves all had higher levels of the ARGOS8 mRNA in the genome-edited plants relative to WT controls. In developing kernels 14 and 21 days after pollination (DAP), the ARGOS8 mRNA was also more abundant in the ARGOS8-v1 and ARGOS8-v2 than the WT. The ARGOS8 protein was detectable by immunoblot analysis in the developing kernels of the genome-edited variants, but not in the WT (Figure 4b). The two variants had a similar level of ARGOS8 mRNA expression in all the tissues tested (Figure 4b).

Improved grain yield under drought stress environments

The two genome-edited variants ARGOS8-v1 and ARGOS8-v2 were crossed with an inbred tester to create a hybrid for field evaluation. These variants were compared to a wild-type hybrid that had not undergone genome editing. Entries were evaluated across multiple environments at eight locations throughout the United States. At the end of the growing season, locations were grouped into three environmental types based on the occurrence of drought stress. Four locations had yields near or above 200 bushel per acre; these were classified as optimal locations (OPT) where water deficits were not a constraint. The remaining locations were grouped as either flowering stress (FS) or grain-filling stress (GFS), based on the EnClass location classification system (Loffler et al., 2005).

Significant differences among entries were observed for grain yield in the FS location group, with the ARGOS8-v1 and ARGOS8-v2 entries yielding approximately five bushel per acre more than the control (Table 1). In contrast, there was no significant difference in grain yield between the variants and WT in the GFS or OPT locations (Table 1). The GFS locations were characterized by limited soil moisture availability due to soil texture, and drought stress developed very quickly. This may have resulted in early cessation of grain filling in the ARGOS8-v1 and ARGOS8-v2 entries; grain moisture was significantly less than in the control for ARGOS8-v1 (Table S3). Plant height and ear height increased by a small (2.6 and 3.2 cm, respectively) but significant amount in the ARGOS8-v2 (Table S3) in the OPT locations. No differences were observed in thermal time to silk or to shed.

Discussion

Constitutive overexpression of ARGOS8 using a transgenic approach increases grain yield in maize under drought stress conditions (Shi et al., 2015). To explore the feasibility of recapitulating the ARGOS8 transgene effect using a conventional breeding approach, we determined the native expression levels of

(a)
ARGOS8-v1

GOS2 5'-UTR

-----CTCAA CAACCAAGTT TCCATGAGCG CTGGCGCGCG GGTCCGGCGG GGCGGTCTGT GAGGGCAAAT TTATATAGGT CTAGTGGGTA CCCGGCTACG

GATAGATATG ATGCTGCACT GCACATTGGC TATATCTGAG GCTCCTGCGC GCGCCTTGGC CAGGTGTCTG TCATGCGGGCGATGCCGCAGGAAGAGGA--
  M R A M P Q E E E

ARGOS8-v2

GOS2 5'-UTR

-----CTCAA CAACCAAGTT TCCATGGTAC GGATAGATAT GATGCTGCAC TGCACATTGG CTATATCTGA GGCTCCTGCG CGCGCCTTGG CCAGGTGTCT

GTCATGCGGG CGATGCCGCA GGAAGAGGAA ---
  M R A M P Q E E E

ARGOS8-v3

CTS3   CTS1

---AAATAAA GAGTTACTTC TCTAAGCACT CGCTGGCGCG CGGGTCCGGC GGGGCGGTCT GTGAGGGCAA ATTTATATAG GTCTAGTGGG TACCCGGCTA

CGGATAGATA TGATGCTGCA CTGCACATTG GCTATATCTG AGGCTCCTGC GCGCGCCTTG GCCAGGTGTC TGTCATGCGG GCGATGCCGC AGGAAGAG---
  M R A M P Q E E

(b)

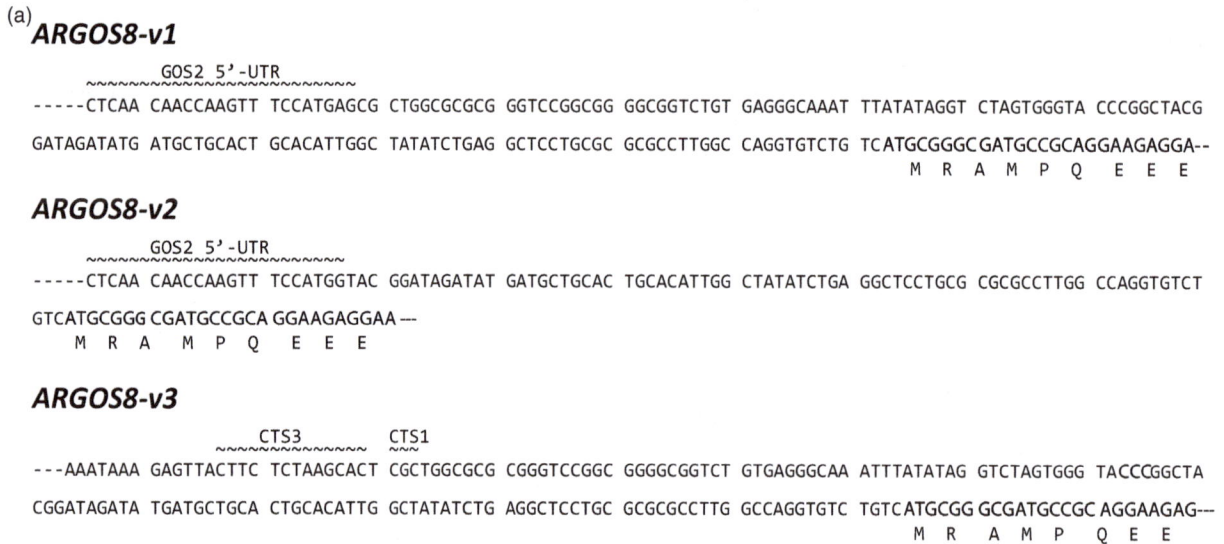

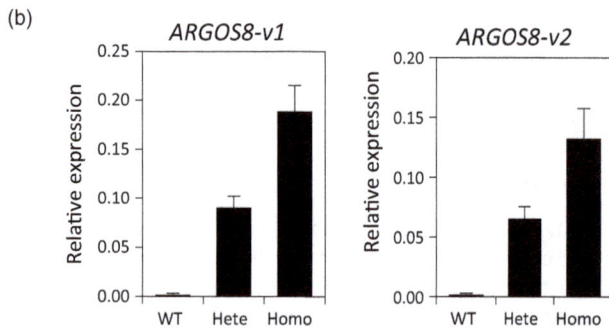

Figure 3 Maize genome-edited *ARGOS8* variants. (a) Genomic sequence upstream of the *ARGOS8* coding region in three genome-edited variants. The entire modification region in homozygous F2 plants was amplified using long PCR, and the PCR products were sequenced. Part of the GOS2 5'-UTR sequence (blue font) and the remaining 5'-UTR of *ARGOS8* as well as the 5'-terminus of *ARGOS8* coding sequence are shown. In the promoter deletion variant *ARGOS8-v3*, the remnant CTS3 and CTS1 sequences are highlighted. (b) Relative expression levels of *ARGOS8* in leaves as measured by qRT-PCR. Means ± SD are shown for F2 plants of 14-day-old *ARGOS8-v1* and 18-day-old *ARGOS8-v2*; *n* = 10–24. WT, wild-type; Hete, Heterozygote; Homo, homozygote.

Figure 4 Comparison of the *ARGOS8* expression in genome-edited variants and wild-type maize plants. (a) Relative expression of *ARGOS8* in a selection of maize tissues and stages. mRNA was quantified with qRT-PCR. Six individual plants were analysed for the genome-edited variants and two plants for WT controls. DAP, days after pollination. (b) ARGOS8 protein expression in developing kernels. Immature kernels (21 DAP) were analysed by immunoblotting using a monoclonal anti-ARGOS8 antibody.

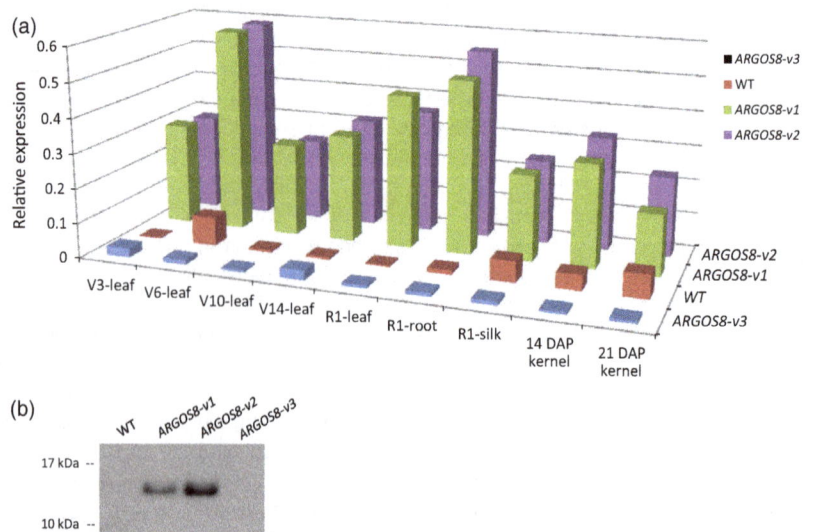

ARGOS8 in a set of public and proprietary maize inbred lines. None of the inbreds we examined had mRNA expression levels great enough to match those in the transgenic events. In addition, a naturally occurring variant of *ARGOS8* encoding a shorter protein also was not able to confer desired phenotypes without overexpression. Therefore, conventional breeding with this gene was not deemed worthwhile, and we elected to use a CRISPR-Cas enabled advanced breeding technology to generate

Table 1 Grain yield of *ARGOS8* genome-edited variants and wild type under flowering stress, grain-filling stress and optimal (well-watered) conditions.

	Flowering Stress	Grain-filling Stress	Optimal
	ton ha^{-1} (bushel acre^{-1})		
ARGOS8-v1	8.67 (138.0)*	7.47 (119.0)	13.13 (209.0)
ARGOS8-v2	8.67 (138.0)*	7.54 (120.0)	13.19 (210.0)
WT	8.34 (132.8)	7.72 (122.9)	13.01 (207.1)

Data are from two individual genome-edited variants (*ARGOS8-v1*, *ARGOS8-v2*) and wild type tested as one hybrid at eight locations in 2015. Predicted difference for each variant is compared with the wild type. All analyses were implemented using ASReml with output of the model presented as best linear unbiased predictions (see Experimental procedures).
*Predicted difference significant at $P < 0.1$.

new *ARGOS8* variants by changing the DNA sequence at the native *ARGOS8* locus. Replacement of the ARGOS8 promoter with a maize GOS2 promoter (GOS2 PRO), or insertion of a GOS2 PRO into the 5′-UTR of the *ARGOS8* gene, led to a change in the *ARGOS8* expression pattern from tissue preferred to ubiquitous, and from relatively low mRNA expression levels to significantly increased *ARGOS8* expression levels. Precise modification of the nucleotide sequence of *ARGOS8* at its native location in the genome was achieved, as determined by PCR assays of the entire region followed by sequencing. The *ARGOS8* variants were found to be stably inherited via analysis of over four generations. Field testing showed that the novel *ARGOS8* variants increased grain yield under drought stress conditions. These yield results are similar to previous results obtained from transgenic plants overexpressing *ARGOS8* (Shi *et al.*, 2015). These results demonstrate the utility of genome editing in creating novel allelic variation for enhancing crop drought tolerance.

The mutation rate at CRISPR-RNA target sites in the regenerated shoots ranged from 60% to 98%, similar to that reported in maize gene modification studies using stably transformed lines (Svitashev *et al.*, 2015). In the promoter swap experiments using two guide RNAs, we observed a frequency of approximately 16% (30 of 185) for DNA fragment deletion due to the nonhomologous end-joining. The homology-directed DNA swap at the *ARGOS8* locus occurred in approximately 1% (3 of 334) of the regenerated shoots. A comparable frequency (2 of 194) was found for insertion when one guide RNA was used. In a previous study, the insertion frequency at the maize *liguleless1* locus was 2.5%–4.1% (Svitashev *et al.*, 2015). CRISPR-RNA target sites, the surrounding genomic DNA sequences, insert sequences and the genotype of host plants as well as other factors may contribute to the difference in insertion frequencies. We also observed that nearly 60% (19 of 30; Table S1) of the regenerated shoots failed to produce T0 plants. Of the five two-junction-positive events identified in the shoot stages, only three T0 plants were recovered, indicating more genome-edited lines can be obtained by improving maize inbred transformation.

The maize *ARGOS8* gene is a negative regulator of ethylene responses. ARGOS8 proteins physically interact with the ethylene receptor signalling complex, modulating ethylene perception and the early stages of the ethylene signal transduction (Shi *et al.*, 2016). Transgenic maize and Arabidopsis plants overexpressing

ARGOS genes exhibit reduced sensitivity to ethylene (Rai *et al.*, 2015; Shi *et al.*, 2015), enhanced cell elongation and/or division resulting in taller plants, larger leaves, and longer ears in maize (Guo *et al.*, 2014; Shi *et al.*, 2015), as well as creating larger organs in other plant species (Feng *et al.*, 2011; Hu *et al.*, 2003; Kuluev *et al.*, 2011). Drought stress often reduces plant growth and can adversely affect development, leading to grain yield loss in crops. These stress-induced changes at the whole plant level are largely due to reduced cell number and/or size. Constitutively overexpressed ARGOS likely counteracts the effect of water deficiency by promoting cell expansion and/or division, mitigating the yield loss by enhancing plant growth under drought stress. Here, we show that maize variants of the *ARGOS8* gene generated by altering its regulatory elements can deliver a significant increase in grain yield under a flowering stress condition with no yield loss under an optimal condition, similar to that of *ARGOS8* transgenic plants (Shi *et al.*, 2015). However, this was not the case when variants were exposed to a grain-filling stress (Table 1). This result was not surprising given that much of the yield increase resulting from ectopic expression of *ARGOS8* under abiotic stress comes from an increase in kernel set (Shi *et al.*, 2015), which is primarily determined at flowering time.

Unlike transgenic *ARGOS8* plants, the maize inbred lines carrying the genome-edited variants contain no transformation selection markers or any nonmaize DNA. All the reagents (*i.e.* helper genes) used in maize DNA sequence editing, including *Cas9*, *sgRNA*, *PMI*, *ODP2* and *WUS* as well as plasmid backbones, were not required for the function of newly generated *ARGOS8* variants and were removed by backcrossing in the early stages of breeding. Instead of *Agrobacterium*-mediated transformation, particle bombardment was employed to deliver the genome editing reagents; thus, no plant pathogen was involved in the generation of these variants. The DNA repair template (GOS2 promoter flanked by homology arms) originated from maize genomic DNA; only the maize GOS2 promoter is site-specifically integrated into the *ARGOS8* locus via homologous recombination, leading to the designed modification of *ARGOS8* expression.

The *ARGOS8* editing process can be summarized as a two-step procedure: duplication of the GOS2 promoter and translocation to the *ARGOS8* locus. Both duplication and translocation of DNA fragments occur naturally in maize (Wang *et al.*, 2015; Zhang *et al.*, 2000). During its evolutionary history, the maize genome has undergone several rounds of whole-genome duplication. In addition, segmental duplication, which involves DNA fragments of different sizes ranging from a few base pairs up to many megabases which may or may not contain intact, functional genes, also play an important role in shaping the maize genome and increasing genetic diversity (Lai *et al.*, 2004; Schnable *et al.*, 2009). Similarly, this process occurs naturally in other plants and animals, including humans (Zhang *et al.*, 2005). Comparative genomic studies have shown that although segmental duplication drives the formation of clusters of closely related genes, duplicated sequences can translocate to different chromosomal locations, resulting in dispersal of paralogs throughout the genome (Freeling *et al.*, 2008; Lai *et al.*, 2004; Mendivil Ramos and Ferrier, 2012). For example, the plant disease resistance NBS-LRR genes are particularly prone to being transposed (Ameline-Torregrosa *et al.*, 2008; Baumgarten *et al.*, 2003; Leister *et al.*, 1998; Richly *et al.*, 2002), vividly attesting to segmental duplication and translocation occurring in plants. This natural rearrangement of DNA fragments occurs spontaneously at a low frequency in maize and has been exploited by breeders over the decades for maize improvement. Similarly, the CRISPR-Cas-enabled advanced breeding technology allows precise integration of duplicated genetic elements into a target locus and can enhance the grain yield of maize.

Since the adoption and widespread use of hybrid maize, natural allelic variations in a large number of genes, each with small effects, have improved drought tolerance, even though it has been suggested that the stress-tolerant alleles are present at relatively low frequencies in most elite breeding populations (Blum, 1988). With increasing knowledge on plant response to drought stress and molecular understanding of gene networks underlying the physiological processes that impact drought tolerance, monogenic drought tolerance in maize has become possible by a transgenic approach. Indeed, there are multiple examples of the validation of the efficacy of transgenes in elite hybrids under field conditions (Castiglioni et al., 2008; Guo et al., 2014; Habben et al., 2014; Leibman et al., 2014; Nuccio et al., 2015; Shi et al., 2015). With the advent of CRISPR-Cas enabled advanced technology, a new technique is now available to provide new sources of genetic variation for plant breeding. This genome editing study of *ARGOS8* demonstrates that single endogenous genes can be modified to create novel variants that have a significantly positive effect on a complex trait such as drought tolerance. In short, the generation and use of genome-edited variants is a seminal addition to the precision breeding toolbox that can enhance changes to the plant genome in a predictable manner.

Experimental procedures

Plasmids and reagents used for plant transformation

Plasmids were designed to insert a DNA fragment into the 5′-UTR of target genes using a single guide RNA (sgRNA), the Cas9 endonuclease gene derived from *Streptococcus pyogenes* and a DNA repair template. For the DNA fragment swap, two sgRNA were used. The sgRNA gene was adapted from Mali et al. (2013) and consists of a maize U6 polymerase III promoter (Svitashev et al., 2015), a CRISPR RNA (crRNA), a transactivating CRISPR RNA (tracrRNA) and a terminator (Figure S3). The Cas9 expression cassette contains the maize *UBIQUITIN1* promoter and potato protease inhibitor II terminator. The Cas9 sequence was maize codon optimized and added the potato ST-LS1 intron as well as the nuclear localization signals from the SV40 and the *Agrobacterium tumefaciens* Vir D2, as previously described (Svitashev et al., 2015), for appropriate expression and nuclear targeting in maize. The DNA repair template plasmid carries the maize GOS2 promoter (NCBI GenBank accession no. GQ184457; nucleotides 218 974–220 796 in reverse direction; Barbour et al., 2003) that was inserted between two homology arms each approximately 400 bp derived from the genomic sequence flanking the CRISPR-RNA target site (CTS; Figure 2a). The vector also contains a multiple cloning site of 61-bp immediately upstream of the GOS2 promoter. Constructs were assembled using chemically synthesized DNA fragments with standard DNA techniques (Sambrook et al., 1989). To facilitate delivery of the genome editing reagents into maize cells and regeneration of plants, three expression cassettes encoding transformation selection marker phosphomannose isomerase (PMI) as well as cell division and callus growth-promoting proteins ovule development protein2 (ODP2) and WUSCHEL (WUS) were constructed as previously described (Ananiev et al., 2009; Svitashev et al., 2015).

Maize transformation

Biolistic-mediated transformation of maize immature embryos was performed according to Svitashev et al. (2015). Briefly, gold particles, 0.6 μm in diameter, were washed with cold, 100% [v/v] ethanol and sterile distilled water. The DNA purified with QIAprep Spin Miniprep (Qiagen) was precipitated on the washed gold particles using a water-soluble cationic lipid TransIT-2020 (Mirus). Fifty microlitres of gold particles (water solution of 10 mg/mL) and 1 μL of TransIT-2020 water solution were added to the premixed DNA, mixed gently and incubated on ice for 10 min. DNA-coated gold particles were then centrifuged at 8 000 g for 1 min. The pellet was rinsed with 100 μL of 100% [v/v] ethanol and resuspended by a brief sonication. Immediately after sonication, DNA-coated gold particles were loaded onto the centre of a macrocarrier (10 μL of each) and allowed to air dry. Immature embryos 9–11 days after pollination were bombarded using a PDS-1000 Helium Gun (Bio-Rad) with a rupture pressure of 425 psi. Postbombardment culture, selection and plant regeneration were carried out as previously described (Gordon-Kamm et al., 2002). Regenerated shoots were sampled for initial qPCR screening and junction PCR. Leaf discs were taken from T0 seedlings 2 weeks post-transplanting for genotyping. F1 seeds were produced by crossing the T0 with wild-type plants. BC1 and F2 seeds were produced by backcrossing and self-pollination, respectively.

Arabidopsis transformation and ethylene response assay

The *35S::ARGOS8* construct was assembled and transgenic Arabidopsis plants generated as described (Shi et al., 2015). Transgenic lines were selected based on the expression of the fluorescent marker yellow fluorescence protein in T1 seeds. The *ARGOS8* transgene expression was confirmed by qRT-PCR. To determine the activity of *ARGOS8* in reducing plant response to ethylene, the ethylene triple response assay (Bleecker et al., 1988) was carried out in the presence of the ethylene precursor aminocyclopropane-1-carboxylic acid (ACC). Surface-sterilized and stratified seeds were germinated in the dark for 3 days in agar that contained one-half-strength Murashige and Skoog salts and 1% [w/v] sucrose supplemented with 0, 0.5, or 1 μM ACC. Hypocotyl and root lengths of the etiolated seedlings were measured by photographing the seedlings with a digital camera and using image analysis software ImageJ (National Institutes of Health).

DNA extraction, PCR genotyping and sequencing

DNA was extracted from regenerated shoots or leaf discs as described in Gao et al. (2010). PCR was performed using REDEtract-N-Amp PCR readyMix (Sigma) or Phusion High-Fidelity PCR Master Mix (NEB) according to the manufacturer's instructions. Quantitative PCR (qPCR) was carried out as described in Svitashev et al. (2015). The primer sequences used in PCR and qPCR are listed in Supporting Information Table S2. The PCR products were sequenced directly or cloned into the pCR2.1-TOPO vectors (Thermo Fisher Scientific) before sequencing.

RNA extraction, qRT-PCR and RNA sequencing

Total RNA was isolated with Qiagen RNA Isolation Kit (Qiagen). The DNaseI Enzyme Kit (Roche) was used to remove DNA from the RNA samples. Complementary DNA was synthesized from the total RNA using the High Capacity cDNA Reverse Transcription Kit

(Thermo Fisher Scientific). PCR amplifications were performed using the TaqMan probe-based detection system according to the manufacturer's instructions (Applied Biosystems). Primers and probes are shown in Table S2. Relative quantification values were determined using the difference in Ct from the target genes and the reference gene, maize UBIQUITIN5.

RNA sequencing (RNA-seq) was performed as described (Shi et al., 2015). In brief, total RNA was isolated from frozen maize tissues and used to prepare sequencing libraries using the TruSeq mRNA-Seq Kit (Illumina), and sequenced on the Illumina HiSeq 2000 system with Illumina TruSeq SBS v3 reagents. On average, 10 million sequences were generated for each sample. The resulting sequences were trimmed based on quality scores and mapped to the maize B73 reference genome sequence V2 and normalized to reads per kilobase of transcript per ten million mapped reads. The generated data matrix was visualized and analysed in GeneData Analyst software (Genedata AG, Basel, Switzerland).

Immunoblot analysis

To detect ARGOS8 proteins, extracts were prepared from immature kernels, proteins separated by SDS-PAGE, blotted to a nitrocellulose membrane and probed with a monoclonal anti-ARGOS8 antibody. The primary antibodies were detected with a HRP-conjugated goat anti-mouse secondary antibody and the Pierce SuperSignal® West Dura Extended Duration Substrate (Thermo Fisher Scientific).

Maize hybrid yield testing

To evaluate the genome-edited variants, field trials were conducted across multiple environments in small plots (approximately 4 m^2) with 2–4 replications at each of eight locations. Hybrid seed for these trials was generated by crossing the genome-edited ARGOS8 variants with an inbred tester. A wild-type hybrid containing the native ARGOS8 allele served as the comparator. The experimental variants and control were grown in field environments at research centres in Woodland, CA; Garden City, KS; Plainview, TX; York, NE; Marion, IA; Johnston, IA; and Princeton, IN in 2015. Some environments were managed to impose various levels of drought stress while others were managed for optimum yield/nonstress conditions. Fertilizer at each location was applied to achieve maximum yields. Weeds and pests were controlled according to local practices. Grain mass and grain moisture data were collected using a small plot combine. Grain yield was adjusted to a constant 15% moisture. Additional agronomic characteristics evaluated at selected locations were plant and ear height and thermal time to shed and silk.

The field experimental design was set up as a randomized complete block arrangement. Data analysis was by ASREML (VSN International Ltd), and the values reported are BLUPs (Best Linear Unbiased Predictions; Cullis et al., 1998; Gilmour et al., 2009). A mixed-model framework was used to perform the analysis. The model included replicate, row, column and heterogeneous residual variance with a separable autoregressive correlation for both row and column directions (AR1*AR1) within each location to reduce the impact of spatial variation in the field. In the analysis, the main effect of location type was considered a fixed effect. The main effect of entry and its interaction with location type were considered random effects. Statistical significance is reported with a P-value of 0.1 in a two-tailed test.

Acknowledgements

We thank Wally Marsh, Susan Wagner, Min Zeng, Megan Schroder, Joshua Young, Bruce Drummond, Mary Trimnell, Ben Weers, Jesse Ourada, Wang-Nan Hu, Dennis O'Neill, Andy Baumgarten, Norbert Brugiere, Gina Zastrow-Hayes, Mary Beatty, Jennifer Hanks, Matthew Hanson, Laura Church, Keith Allen and Jacque Hockenson for excellent assistance with this study. We acknowledge our colleagues at outlying breeding stations who conducted first-rate yield trials. We are grateful to Carl Simmons, Nathan Coles, Mark Cigan and Maria Fedorova for their critical review of this manuscript. We thank Mike Lassner and Mark Cigan for their support of this project. We also thank Tom Greene and Mark Cooper for their organizational leadership and helpful input.

References

Ameline-Torregrosa, C., Wang, B.B., O'Bleness, M.S., Deshpande, S., Zhu, H., Roe, B., Young, N.D. et al. (2008) Identification and characterization of nucleotide-binding site-leucine-rich repeat genes in the model plant Medicago truncatula. Plant Physiol. 146, 5–21.

Ananiev, E.V., Wu, C., Chamberlin, M.A., Svitashev, S., Schwartz, C., Gordon-Kamm, W. and Tingey, S. (2009) Artificial chromosome formation in maize (Zea mays L.). Chromosoma, 118, 157–177.

Barbour, E., Meyer, T.E. and Saad, M.E. (2003) Maize GOS2 Promoters. U.S. Patent No. 6504083 B1, Pioneer Hi-Bred International.

Baumgarten, A., Cannon, S., Spangler, R. and May, G. (2003) Genome-level evolution of resistance genes in Arabidopsis thaliana. Genetics, 165, 309–319.

Bleecker, A.B., Estelle, M.A., Somerville, C. and Kende, H. (1988) Insensitivity to ethylene conferred by a dominant mutation in Arabidopsis thaliana. Science, 241, 1086–1089.

Blum, A. (1988) Plant Breeding for Stress Environments. Boca Raton, FL: CRC Press. 223p.

Cai, Y., Chen, L., Liu, X., Sun, S., Wu, C., Jiang, B., Han, T. et al. (2015) CRISPR/Cas9-mediated genome editing in soybean hairy roots. PLoS ONE, 10, e0136064.

Castiglioni, P., Warner, D., Bensen, R.J., Anstrom, D.C., Harrison, J., Stoecker, M., Abad, M. et al. (2008) Bacterial RNA chaperones confer abiotic stress tolerance in plants and improved grain yield in maize under water-limited conditions. Plant Physiol. 147, 446–455.

Čermák, T., Baltes, N.J., Čegan, R., Zhang, Y. and Voytas, D.F. (2015) High-frequency, precise modification of the tomato genome. Genome Biol. 16, 232. doi:10.1186/s13059-015-0796-9.

Cooper, M., Gho, C., Leafgren, R., Tang, T. and Messina, C. (2014) Breeding drought-tolerant maize hybrids for the US corn-belt: discovery to product. J. Exp. Bot. 65, 6191–6204.

Cullis, B.R., Gogel, B.J., Verbyla, A.P. and Thompson, R. (1998) Spatial analysis of multi-environment early generation trials. Biometrics, 54, 1–18.

Du, H., Zeng, X., Zhao, M., Cui, X., Wang, Q., Yang, H., Cheng, H. et al. (2016) Efficient targeted mutagenesis in soybean by TALENs and CRISPR/Cas9. J. Biotechnol. 217, 90–97.

Feng, G., Qin, Z., Yan, J., Zhang, X. and Hu, Y. (2011) Arabidopsis ORGAN SIZE RELATED1 regulates organ growth and final organ size in orchestration with ARGOS and ARL. New Phytol. 191, 635–646.

Freeling, M., Lyons, E., Pedersen, B., Alam, M., Ming, R. and Lisch, D. (2008) Many or most genes in Arabidopsis transposed after the origin of the order Brassicales. Genome Res. 18, 1924–1937.

Gaffney, J., Schussler, J., Löffler, C., Cai, W., Paszkiewicz, S., Messina, C., Groeteke, J. et al. (2015) Industry scale evaluation of maize hybrids selected for increased yield in drought stress conditions of the U.S. Corn Belt. *Crop Sci.* **55**, 1608–1618.

Gao, H., Smith, J., Yang, M., Jones, S., Djukanovic, V., Nicholson, M.G., West, A. *et al.* (2010) Heritable targeted mutagenesis in maize using a designed endonuclease. *Plant J.* **61**, 176–187.

Gilmour, A.R., Gogel, B.J., Cullis, B.R. and Thompson, R. (2009) *ASReml User Guide.* Release 3.0. VSN International Ltd, Hemel Hempstead, HP1 1ES, UK

Gordon-Kamm, W., Dilkes, B.P., Lowe, K., Hoerster, G., Sun, X., Ross, M., Church, L. *et al.* (2002) Stimulation of the cell cycle and maize transformation by disruption of the plant retinoblastoma pathway. *Proc. Natl Acad. Sci. USA*, **99**, 11975–11980.

Guo, M., Rupe, M.A., Wei, J., Winkler, C., Goncalves-Butruille, M., Weers, B.P., Cerwick, S.F. *et al.* (2014) Maize ARGOS1 (ZAR1) transgenic alleles increase hybrid maize yield. *J. Exp. Bot.* **65**, 249–260.

Habben, J.E., Bao, X., Bate, N.J., DeBruin, J.L., Dolan, D., Hasegawa, D., Helentjaris, T.G. *et al.* (2014) Transgenic alteration of ethylene biosynthesis increases grain yield in maize under field drought-stress conditions. *Plant Biotechnol. J.* **12**, 685–693.

Hays, D.B., Do, J.H., Mason, R.E., Morgan, G. and Finlayson, S.A. (2007) Heat stress induced ethylene production in developing wheat grains induces kernel abortion and increased maturation in a susceptible cultivar. *Plant Sci.* **172**, 1113–1123.

Hu, Y., Xie, Q. and Chua, N.H. (2003) The Arabidopsis auxin-inducible gene ARGOS controls lateral organ size. *Plant Cell*, **15**, 1951–1961.

Jacobs, T.B., LaFayette, P.R., Schmitz, R.J. and Parrott, W.A. (2015) Targeted genome modifications in soybean with CRISPR/Cas9. *BMC Biotechnol.* **15**, 16.

Jiang, W., Zhou, H., Bi, H., Fromm, M., Yang, B. and Weeks, D.P. (2013) Demonstration of CRISPR/Cas9/sgRNA-mediated targeted gene modification in Arabidopsis, tobacco, sorghum and rice. *Nucleic Acids Res.* **41**, e188.

Kawakami, E.M., Oosterhuis, D.M. and Snider, J.L. (2010) Physiological effects of 1-methylcyclopropene on well-watered and water-stressed cotton plants. *J. Plant Growth Regul.* **29**, 280–288.

Kawakami, E.M., Oosterhuis, D.M. and Snider, J. (2013) High temperature and the ethylene antagonist 1-methylcyclopropene alter ethylene evolution patterns, antioxidant responses, and boll growth in Gossypium hirsutum. *Am. J. Plant Sci.* **4**, 1400–1408.

Kuluev, B.R., Knyazev, A.V., Iljassowa, A.A. and Chemeris, A.V. (2011) Constitutive expression of the ARGOS gene driven by dahlia mosaic virus promoter in tobacco plants. *Rus. J. Plant Physiol.* **58**, 507–515.

Lai, J., Ma, J., Swigonova, Z., Ramakrishna, W., Linton, E., Llaca, V., Tanyolac, B. *et al.* (2004) Gene loss and movement in the maize genome. *Genome Res.* **14**, 1924–1931.

Leibman, M., Shryock, J.J., Clements, M.J., Hall, M.A., Loida, P.J., McClerren, A.L., McKiness, Z.P. *et al.* (2014) Comparative analysis of maize (Zea mays) crop performance: natural variation, incremental improvements and economic impacts. *Plant Biotechnol. J.* **12**, 941–950.

Leister, D., Kurth, J., Laurie, D.A., Yano, M., Sasaki, T., Devos, K., Graner, A. *et al.* (1998) Rapid reorganization of resistance gene homologues in cereal genomes. *Proc. Natl Acad. Sci. USA*, **95**, 370–375.

Li, T., Liu, B., Spalding, M.H., Weeks, D.P. and Yang, B. (2012) High-efficiency TALEN-based gene editing produces disease-resistant rice. *Nat. Biotechnol.* **30**, 390–392.

Li, J.F., Norville, J.E., Aach, J., McCormack, M., Zhang, D., Bush, J., Church, G.M. *et al.* (2013) Multiplex and homologous recombination-mediated genome editing in Arabidopsis and *Nicotiana benthamiana* using guide RNA and Cas9. *Nat. Biotechnol.* **31**, 688–691.

Li, Z., Liu, Z.B., Xing, A., Moon, B.P., Koellhoffer, J.P., Huang, L., Ward, R.T. *et al.* (2015) Cas9-guide RNA directed genome editing in soybean. *Plant Physiol.* **169**, 960–970.

Liang, Z., Zhang, K., Chen, K. and Gao, C. (2014) Targeted mutagenesis in Zea mays using TALENs and the CRISPR/Cas system. *J. Genet. Genom.* **41**, 63–68.

Loffler, C.M., Wei, J., Fast, T., Gogerty, J., Langton, S., Bergman, M., Merrill, B. *et al.* (2005) Classification of maize environments using crop simulation and geographic information systems. *Crop Sci.* **45**, 1708–1716.

Mali, P., Yang, L., Esvelt, K.M., Aach, J., Guell, M., DiCarlo, J.E., Norville, J.E. *et al.* (2013) RNA-guided human genome engineering via Cas9. *Science*, **339**, 823–826.

Mendivil Ramos, O. and Ferrier, D.E. (2012) Mechanisms of gene duplication and translocation and progress towards understanding their relative contributions to animal genome evolution. *Int. J. Evol. Biol.* **2012**, 846421.

Nuccio, M.L., Wu, J., Mowers, R., Zhou, H.P., Meghji, M., Primavesi, L.F., Paul, M.J. *et al.* (2015) Expression of trehalose-6-phosphate phosphatase in maize ears improves yield in well-watered and drought conditions. *Nat. Biotechnol.* **33**, 862–869.

de Pater, B.S., van der Mark, F., Rueb, S., Katagiri, F., Chua, N.H., Schilperoort, R.A. and Hensgens, L.S. (1992) The promoter of the rice gene GOS2 is active in various different monocot tissues and bind rice nuclear factor ASF-1. *Plant J.* **2**, 837–844.

Rai, M.I., Wang, X., Thibault, D.M., Kim, H.J., Bombyk, M.M., Binder, B.M., Shakeel, S.N. *et al.* (2015) The ARGOS gene family functions in a negative feedback loop to desensitize plants to ethylene. *BMC Plant Biol.* **15**, 157. doi:10.1186/s12870-015-0554-x.

Richly, E., Kurth, J. and Leister, D. (2002) Mode of amplification and reorganization of resistance genes during recent Arabidopsis thaliana evolution. *Mol. Biol. Evol.* **19**, 76–84.

Sambrook, J., Fritsch, E.F. and Maniatis, T. (1989) *Molecular Cloning: A Laboratory Manual.* Cold Spring Harbor, NY: Cold Spring Harbor Laboratory.

Schnable, P.S., Ware, D., Fulton, R.S., Stein, J.C., Wei, F., Pasternak, S., Liang, C. *et al.* (2009) The B73 maize genome: complexity, diversity, and dynamics. *Science*, **326**, 1112–1115.

Scott, L., LaFoe, D. and Weil, C.F. (1996) Adjacent sequences influence DNA repair accompanying transposon excision in maize. *Genetics*, **142**, 237–246.

Shan, Q., Zhang, Y., Chen, K., Zhang, K. and Gao, C. (2015) Creation of fragrant rice by targeted knockout of the OsBADH2 gene using TALEN technology. *Plant Biotechnol. J.* **13**, 791–800.

Shi, J., Habben, J.E., Archibald, R.L., Drummond, B.J., Chamberlin, M.A., Williams, R.W., Lafitte, H.R. *et al.* (2015) Overexpression of ARGOS genes modifies plant sensitivity to ethylene, leading to improved drought tolerance in both Arabidopsis and maize. *Plant Physiol.* **169**, 266–282.

Shi, J., Drummond, B.J., Wang, H., Archibald, R.L. and Habben, J.E. (2016) Maize and Arabidopsis ARGOS proteins interact with ethylene receptor signaling complex, supporting a regulatory role for ARGOS in ethylene signal transduction. *Plant Physiol.*, **171**, 2783–2797.

Shukla, V.K., Doyon, Y., Miller, J.C., DeKelver, R.C., Moehle, E.A., Worden, S.E., Mitchell, J.C. *et al.* (2009) Precise genome modification in the crop species Zea mays using zinc-finger nucleases. *Nature*, **459**, 437–441.

Sun, Y., Zhang, X., Wu, C., He, Y., Ma, Y., Hou, H., Guo, X. *et al.* (2016) Engineering herbicide-resistant rice plants through CRISPR/Cas9-mediated homologous recombination of acetolactate synthase. *Mol. Plant*, **9**, 628–631.

Svitashev, S., Young, J.K., Schwartz, C., Gao, H., Falco, S.C. and Cigan, A.M. (2015) Targeted mutagenesis, precise gene editing and site-specific gene insertion in maize using Cas9 and guide RNA. *Plant Physiol.* **2**, 931–945.

Wang, Y., Cheng, X., Shan, Q., Zhang, Y., Liu, J., Gao, C. and Qiu, J.L. (2014) Simultaneous editing of three homoeoalleles in hexaploid bread wheat confers heritable resistance to powdery mildew. *Nat. Biotechnol.* **32**, 947–951.

Wang, D., Yu, C., Zuo, T., Zhang, J., Weber, D.F. and Peterson, T. (2015) Alternative transposition generates new chimeric genes and segmental duplications at the maize p1 locus. *Genetics*, **201**, 925–935.

Zhang, P., Chopra, S. and Peterson, T. (2000) A segmental gene duplication generated differentially expressed myb-homologous genes in maize. *Plant Cell*, **12**, 2311–2322.

Zhang, L., Lu, H.H., Chung, W.Y., Yang, J. and Li, W.H. (2005) Patterns of segmental duplication in the human genome. *Mol. Biol. Evol.* **22**, 135–141.

Zhang, H., Zhang, J., Wei, P., Zhang, B., Gou, F., Feng, Z., Mao, Y., *et al.* (2014) The CRISPR/Cas9 system produces specific and homozygous targeted gene editing in rice in one generation. *Plant Biotechnol. J.* **12**, 797–807.

Zhou, H., Liu, B., Weeks, D.P., Spalding, M.H. and Yang, B. (2014) Large chromosomal deletions and heritable small genetic changes induced by CRISPR/Cas9 in rice. *Nucleic Acids Res.* **42**, 10903–10914.

Zhou, J., Peng, Z., Long, J., Sosso, D., Liu, B., Eom, J.-S., Huang, S. *et al.* (2015) Gene targeting by the TAL effector PthXo2 reveals cryptic resistance gene for bacterial blight of rice. *Plant J.* **82**, 632–643.

12

BrpSPL9 (*Brassica rapa* ssp. *pekinensis SPL9*) controls the earliness of heading time in Chinese cabbage

Yali Wang[†], Feijie Wu[†], Jinjuan Bai and Yuke He*

National Key Laboratory of Plant Molecular Genetics, Shanghai Institute of Plant Physiology and Ecology, Shanghai Institutes for Biological Sciences, Chinese Academy of Sciences, Shanghai, China

*Correspondence
email ykhe@sibs.ac.cn
[†]These authors are contributed equally.

Keywords: *BrpSPL9*, Chinese cabbage, earliness, leafy head, miR156, phase transition.

Summary

The leafy heads of cabbage (*Brassica oleracea*), Chinese cabbage (*Brassica rapa* ssp. *pekinensis*), Brussels sprouts (*B. oleracea* ssp. *gemmifera*) and lettuce (*Lactuca sativa*) comprise extremely incurved leaves that are edible vegetable products. The heading time is important for high quality and yield of these crops. Here, we report that *BrpSPL9-2* (*B. rapa* ssp. *pekinensis SQUAMOSA PROMOTER BINDING-LIKE 9-2*), a target gene of microRNA brp-miR156, controls the heading time of Chinese cabbage. Quantitative measurements of leaf shapes, sizes, colour and curvature indicated that heading is a late adult phase of vegetative growth. During the vegetative period, miR156 levels gradually decreased from the seedling stage to the heading one, whereas *BrpSPL9-2* and *BrpSPL15-1* mRNAs increased progressively and reached the highest levels at the heading stage. Overexpression of a mutated miR156-resistant form of *BrpSPL9-2* caused the significant earliness of heading, concurrent with shortening of the seedling and rosette stages. By contrast, overexpression of miR156 delayed the folding time, concomitant with prolongation of the seedling and rosette stages. Morphological analysis reveals that the significant earliness of heading in the transgenic plants overexpressing *BrpSPL9-2* gene was produced because the juvenile phase was absent and the early adult phase shortened, whereas the significant delay of folding in the transgenic plants overexpressing *Brp-MIR156a* was due to prolongation of the juvenile and early adult phases. Thus, miR156 and *BrpSPL9* genes are potentially important for genetic improvement of earliness of Chinese cabbage and other crops.

Introduction

Chinese cabbage goes through a long period of vegetative growth to produce a leafy head. Traditionally, this period is divided into four stages: seedling, rosette, folding and heading, which are characterized by the morphological features of leaves (He *et al.*, 2000; Yu *et al.*, 2000). Primary leaves grow at the seedling stage; rosette leaves assist head formation by supplying photosynthates to head leaves; the folding leaves are curved inward to establish a posture for head formation and to provide a shade to the head; and the heading leaves around the shoot apexes, which are incurved enough to form a compact head and thus become the nutrient storage organ (He *et al.*, 2000; Opeña, 1984). Many factors such as temperature, light intensity, auxin concentration and ratio of carbohydrate to nitrogen may affect the formation of the leafy head (Ito and Kato, 1957). However, the genetic basis and molecular mechanism of the formation of leafy head are unclear.

Recent studies have begun to uncover the molecular mechanism of vegetative phase transition in Arabidopsis, maize and woody species (Huijser and Schmid, 2011; Wang *et al.*, 2011). In Arabidopsis, vegetative growth comprises juvenile and adult phases (Chuck *et al.*, 2007), which can be distinguished by leaf morphology. The transition from juvenile to adult phase can be observed in a variety of morphological traits, including leaf shape, occurrence of trichomes and wax (Chuck *et al.*, 2007; Huijser and Schmid, 2011; Wang *et al.*, 2011). miR156 and miR172 are the major genetic elements that control vegetative phase transition. miR156 is highly expressed early in shoot development and

decreases as the plant develops, while miR172 has inverse temporal expression (Wu *et al.*, 2009). SQUAMOSA PROMOTER BINDING-LIKE (SPL) transcription factors are regulated by miR156 and influence the transitions between these developmental phases (Schwab *et al.*, 2005). In Arabidopsis, there are 16 members of the *SPL* family. Overexpression of miR156, which down-regulates *SPL* genes, prolongs the juvenile phase and delays flowering. Constitutive overexpression of different *SPLs* accelerates the juvenile-to-adult phase change and promotes flowering (Wang *et al.*, 2009; Wu *et al.*, 2009). SPL9 directly activates the expression of miR172b to facilitate flowering by suppressing the APETALA2 (AP2) protein (Aukerman and Sakai, 2003; Wang *et al.*, 2009). miR156 and miR172, as well as their target genes, are conserved in rice, maize and trees (Chuck *et al.*, 2007; Lauter *et al.*, 2005; Wang *et al.*, 2011; Xie *et al.*, 2006).

Compared to Arabidopsis, which goes through seedling and rosette stages during the vegetative period, Chinese cabbage has two extra stages: folding and heading stages. Unfortunately, the vegetative phases and phase transition of Chinese cabbage have not been defined. Recently, a significant advance has been made in genome sequencing of Chinese cabbage (Wang *et al.*, 2011). The current genome structures of Chinese cabbage have been shaped by whole genome triplication followed by extensive diploidization (Cheng *et al.*, 2013; Fang *et al.*, 2012). This genome information enables us to test the knowledge gained in Arabidopsis for its applicability in the paleohexaploid crop. To exploit the molecular mechanism of head formation, we have clarified the differences in juvenile morphological traits in Brassica and Arabidopsis for the first time and studied the heading time by overexpression and underexpression of *BrpSPL9-2*. Our results indicate that *BrpSPL9-2* (*mBrpSPL9-2*) controls the timing of heading by shortening the seedling and rosette stages.

Results

Leaf incurvature reflects an additional adult phase in Chinese cabbage

Although the four stages of Chinese cabbage have been mentioned in several references (He *et al.*, 1994, 2000; Ito and Kato, 1957, 1958), the morphological identity of leaves at these four stages is not defined well. To demonstrate the morphological characteristics of Chinese cabbage leaves at different stages, we choose Bre, an inbred line of a round-headed cultivar. The seeds of Bre were sown in the growth room at 22 °C and/or transferred to the field for normal cultivation. The leaves of Bre plants showed distinctive shapes, sizes, colours and curvatures at different stages. In the growth room, the seedling leaves were small, elliptic, flat and blue-green, concomitant with the clear internodes (Figure 1a–e). The rosette leaves were with a circular arrangement at a similar height because of shortened internodes (Figure 1f). The folding leaves incurved and yellow-green (Figure 1g). The head leaves (in heads) were extremely incurved and white, with short petioles and leaf lobes (Figure 1h). In the field, the leaf shapes, sizes and curvatures between the stages were similar to those in the growth room. However, the duration of each stage was longer compared with that in the growth room.

Chinese cabbage and Arabidopsis plants belong to the Crucifer family and are close relatives. Both species go through seedling and rosette stages. Unlike Arabidopsis, Chinese cabbage must undergo two further steps, the folding stage and heading stage, to form edible leafy heads. In Arabidopsis plants, vegetative growth is divided into the juvenile phase (leaf 1 through 4) and the adult phase (leaf 5 and thereafter), mainly distinguished by the absence and the presence of abaxial trichomes, respectively (Clarke *et al.*, 1999; Telfer *et al.*, 1997). In the first four leaves of *Arabidopsis thaliana* Ecotype Columbia, corresponding to juvenile leaves, trichomes are absent on the abaxial side. By contrast, many trichomes were observed on abaxial side of the first two leaves of Bre plants (Figure 1a), indicating that the absence of abaxial trichome is not a morphological marker for the juvenile leaves of Chinese cabbage. The first four leaves of Bre plants were equivalent to the juvenile leaves of Arabidopsis and were uniformly small, blue-green, oblong and had long and narrow petioles (Figure 1a–d). Leaves 5–18 (5 through 18), corresponding to the adult phase of Arabidopsis, were distinct from the juvenile leaves in that they became much larger, light-green, round and had short petioles (Figure 1e). Leaves 19–21 were incurved and developed into the folding leaves (Figure 1f), and leaves 22–42 were extremely incurved (Figure 1h) and developed into the head leaves (inner leaves). We designated the folding leaves and head leaves as the late adult leaves because they were similar to each other in colour, shape, size and curvature, differing by the level of leaf incurvature. In this way, the late adult phase corresponded to the sum of the folding and heading stages.

The difference in leaf morphology between vegetative phases of Chinese cabbage was obvious. The juvenile leaves are small, with a big leaf base angle, high blade length-to-width ratio, serrate margins, simple reticulate venation and long petioles (Figure 1i). The early adult leaves are larger, with a smaller leaf base angle, lower blade length-to-width ratio, higher blade-to-

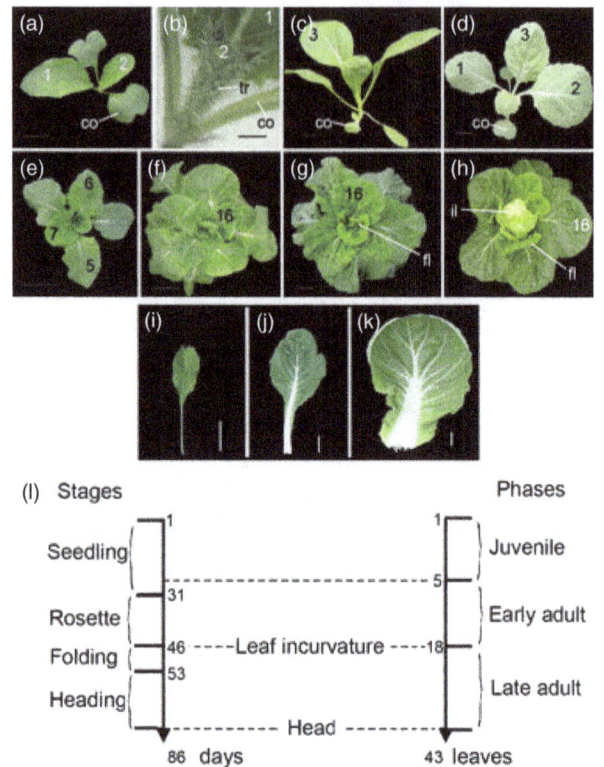

Figure 1 Relationship between vegetative phases and developmental stages of Chinese cabbage (Bre). (a–h) Bre plants at four stages. (a) Overview of a seedling with two juvenile leaves. (b) Side view of the seedling of (a) showing many abaxial trichomes on two juvenile leaves. (c) Side view of a seedling showing four juvenile leaves and elongated internodes. (d) Overview of a seedling with four juvenile leaves. (e) Seedling with early adult leaves (leaves 5–7). (f) Rosette with early adult leaves. (g) Folding plant with the first late adult leaf. (h) Heading plant with late adult leaves. (i–k) Leaves of juvenile (2nd) (i), early adult (10th) (b) and late adult (18th) (c) phases. (l) Diagram showing the relationship between vegetative phases and developmental stages of Chinese cabbage plants. Numbers refer to the positions of leaves as determined by their order of appearance. co, cotyledon; fl, folding leaf; il, inner leaf (head leaf); tr, trichomes. The plants observed were grown in the field as described in Experimental procedures. Scale bars: 2 cm in (a–d); 5 cm in (e–k).

petiole ratio and more complex venation than the juvenile leaves (Figure 1j). The late adult leaves have bulges, lobes and incurvature, short and wide petioles (Figure 1k).

To clarify the leaf identity of vegetative phases in Chinese cabbage, we measured leaf parameters quantitatively on the plants grown in the growth room. The density of the abaxial trichomes was the highest during the juvenile phase, gradually decreasing from the early juvenile to late adult phases (Table 1). The leaf shape index (the length-to-width ratio) during juvenile, early adult and late adult phases was 1.5, 1.2 and 1.2, respectively, revealing the transition of the oblong to round leaves from the juvenile to early adult leaves. The blade base angle was the largest in the juvenile leaves and gradually decreased. The blade margins of the juvenile leaves were less serrated, whereas the blade margins of the late adult leaves were severely serrated. The blade/petiole length ratio increased from the juvenile to the early adult phase and was highest during the late adult phase. By contrast, the petiole length-to-width ratio decreased from the juvenile to the early adult phase and was

Table 1 Morphological changes in leaves at different vegetative stages of Chinese cabbage. Seeds were sown in pots and grown at 22 °C in SIPPE Phytotron in a growth room

| Parameters | Leaves | | | | | |
| | Leaf 2 (Juvenile) | | Leaf 8 (Early adult) | | Leaf 18 (Late adult) | |
	WT	BS9-2	WT	BS9-2	WT	BS9-2
Abaxial trichomes	+++	++	++	++	+	++
Blade base angle (°)	143.33 ± 5.77	128.60 ± 4.41	128.33 ± 4.33	120.11 ± 4.85	124.50 ± 6.43	117.00 ± 9.19
Blade length (cm)	5.23 ± 0.68	7.32 ± 1.03	14.96 ± 0.98	14.55 ± 1.55	12.35 ± 2.09	8.51 ± 3.38
Blade width (cm)	3.56 ± 0.40	6.15 ± 1.37	12.52 ± 1.67	14.16 ± 1.21	10.05 ± 1.97	6.82 ± 2.93
Blade L/W ratio	1.46 ± 0.11	1.20 ± 0.12	1.20 ± 0.096	1.08 ± 0.08	1.24 ± 0.17	1.20 ± 0.16
B/P length ratio	1.47 ± 0.11	7.89 ± 3.15	4.49 ± 0.43	8.37 ± 3.55	5.32 ± 1.87	6.85 ± 1.78
Bulges	–	+	+	++	+++	+++
Hydathodes	14	28	22	32	30	42
Leaf Lobes	–	–	–	+	++	++
Petiole length (cm)	3.56 ± 0.60	1.29 ± 0.58	3.35 ± 0.37	1.91 ± 0.45	2.51 ± 0.85	1.90 ± 0.31
Petiole width	0.43 ± 0.12	0.68 ± 0.13	1.55 ± 0.35	1.66 ± 0.31	1.27 ± 0.30	1.80 ± 0.41
Petiole L/W ratio	8.47 ± 1.50	1.73 ± 0.70	2.37 ± 1.12	1.15 ± 0.20	2.14 ± 1.27	1.68 ± 0.59
Phyllotaxis	Op	Op	Sp	SP	Sp	Sp
Serration	+	++	++	+++	++	+++
Vein complexity	+	++	++	+++	+++	+++

For each measurement, more than 20 plants were evaluated. The data are the mean of 20 leaves, with standard deviation.

B/P, blade-to-petiole ratio; EA, early adult; EJ, early juvenile; LA, late adult; LJ, late juvenile. L/W, length-to-width ratio; Op, opposite, Sp, spiracle. –, no; +, a few; ++, medium; +++, many.

lowest during the late adult phase, indicating that the petioles of the late adult leaves were wide and short. The juvenile leaves of Bre plants had simple reticulate veins with 14–18 hydathodes, whereas the late adult leaves had a complex interconnected network of veins with 22–52 hydathodes. In addition, the late adult leaves were characterized by a number of bulges and lobes. These data revealed that the morphological identities of the three vegetative phases were distinct.

In the field, the lengths (number of leaves) of the juvenile phase, early adult and late adult phases were 4, 13 and 26, respectively (Figure 1l; Table 2). In the growth room, the lengths of the juvenile and early adult phases were 4 and 9 days, respectively.

MIR156 and *BrpSPL* genes in Chinese cabbage are differentially duplicated or triplicated

Several morphological markers for early juvenile leaves in Chinese cabbage were different from the ones of Arabidopsis; therefore, we wondered whether miR156 and its targeted genes controlled the heading time in Chinese cabbage. First, we searched for the genomic sequences of the *Brp-MIR156* and *BrpSPL* genes from our data on resequencing and transcriptome sequencing of Bre (Yu *et al.*, 2013), with the reference to the genome sequence of *B. rapa* ssp. *pekinensis* var. Chiifu-401-42 (http://brassicadb.org/brad/) (Wang *et al.*, 2011). Sequence alignment with the Arabidopsis miR156 precursor revealed 16 copies of *Brp-MIR156* (Figure S1a). These copies were classified into eight groups, according to the classification of their Arabidopsis homologs (Table S1). Within the *Brp-MIR156a* group of Chiifu-401-42, the genomic sequences of the two miRNA genes are highly conserved (99%).

According to the sequences of the *SPL* genes in Arabidopsis and their homologs in Chiifu-401-42, we searched for homologous genomic sequences and transcripts of *BrpSPL* genes from our genome resequencing and RNA-seq data of Bre (Yu *et al.*,

2013). The alignment of these sequences and their phylogenetic analysis indicated that most miR156-targeted *BrpSPLs* in Chinese cabbage were duplicated or triplicated (Figure S1b; Table S1).

miR156 and *BrpSPL* genes are inversely expressed in Chinese cabbage

During the vegetative period of Arabidopsis, miR156 is highly expressed early in shoot development and decreases as plant develops (Schwab *et al.*, 2005). To address whether leaf curvature was an extra adult phase, we analysed the expression levels of miR156 and miR156-targeted *BrpSPL* genes in the plants at different stages. We collected the shoot apices of Bre plants on days 10, 20, 30, 40, 50 and 60 after germination and performed small-RNA Northern blotting of miR156a and real-time PCR of *BrpSPL* genes. The level of miR156 expression was highest on day 10 and gradually decreased from day 10 to 60 (Figure 2a).

miR156-targeted *BrpSPL* genes were expressed differently in plants at various stages. The expression levels of *BrpSPL3-1*, *BrpSPL4-1*, *BrpSPL5-1*, *BrpSPL9-2* and *BrpSPL15-2* were higher on day 20 than on day 10 (Figure 2b). On day 40, expression levels of *BrpSPL9-2* and *BrpSPL15-1* were higher than on day 20, while those of *BrpSPL3-2* and *BrpSPL4-1* were lower. On day 60, expression levels of *BrpSPL9-2* and *BrpSPL15-1* were higher than on day 40. These results suggested that expression levels of *BrpSPL9-2* and *BrpSPL15-1* genes increased progressively over time, in an inverse relationship with the expression of miR156.

In situ hybridization showed that miR156 accumulated preferentially in shoot apical meristems, leaf primordia, leaf tips and vascular tissues of leaves (Figure 2c–f). *BrpSPL9-2* was preferentially expressed in shoot apical meristems, leaf primordia, leaf tips and vascular tissues of leaves (Figure 2g–j), which were almost the same locations as those of Brp-miR156. The preferential

Table 2 The folding time and heading time of the wild-type plants and the transgenic plants with *mBrpSPL9-2*. Seeds were sown in pots and grown at 22 °C in SIPPE Phytotron in a growth room. The seedlings were transferred to the field on 24 August 2012 on SIPPE Farm Station. For each measurement, more than 20 plants were evaluated

Parameters	In growth room			In the field	
	WT	BS9-1	B156a-1	WT	BS9-1
Days to folding	25 ± 1.8	15 ± 1.3*	37 ± 2.6*	46 ± 3.5	35 ± 2.4[†]
Days to heading	ND	ND	ND	55 ± 1.9	45 ± 2.8[†]
Days to maturity	ND	ND	ND	86 ± 7.1	74 ± 6.4
Leaves to folding	11 ± 0.8	7 ± 0.6*	17 ± 1.2*	18 ± 1.2	9 ± 0.6[†]
Leaves to heading	ND	ND	ND	23 ± 1.8	17 ± 2.2[†]
Leaves to maturity	ND	ND	ND	43 ± 3.6	37 ± 2.7
Number of juvenile leaves	4 ± 0.1	0*	7 ± 0.1*	4 ± 0.1	0[†]
Number of early adult leaves	9 ± 0.1	7 ± 0.1*	11 ± 0.2*	13 ± 0.2	11 ± 0.1[†]
Number of late adult leaves	ND	ND	ND	26 ± 1.4	28 ± 1.6

'Days to folding', 'days to heading' and 'days to maturity' represent numbers of days from the germination to the appearance of the first folding leaf, of the first head leaf and of the fully maturity of leafy head, respectively. 'Leaves to folding', 'leaves to heading' and 'leaves to maturity' represent total numbers of leaves produced from germination to the appearance of the first folding leaf, of the first head leaf and of the fully maturity of leafy head, respectively. Leaves longer than 1 cm were recorded and accounted. The data are the mean of 20 plants, with standard deviation. ND, not detected.

* and [†] indicate the significant differences (Student's *t*-test) between the transgenic lines and the wild type in the growth room and in the field, respectively.

expression of *BrpSPL15-1* was the same as that of *BrpSPL9-2* except that it was not obvious in vascular tissues (Figure 2j–m). At the seedling stage, miR156 was mainly expressed in shoot apical meristems, whereas *BrpSPL9-2* and *BrpSPL15-1* were preferentially expressed in developing leaves.

The gradual decrease in miR156 accumulation and gradual increase in *BrpSPL* expression during vegetative growth suggested that the time and length of the juvenile and early adult phases are regulated by miR156 and *BrpSPL* genes.

Overexpression of *BrpSPL9* causes early heading of Chinese cabbage

To determine whether miR156-targeted genes affect the heading time of Chinese cabbage, we chose *BrpSPL9-2* gene for genetic transformation, because the Arabidopsis homolog of *BrpSPL9-2* plays a central role in the transition from the juvenile to adult phase and in the vegetative phase change among *AtSPL* genes (Fornara and Coupland, 2009). We first constructed *mBrpSPL9-2* (mutated miR156-resistant version of *BrpSPL9-2*; Figure S2) under the control of CaMV 35S promoter. We transferred the construct to Chinese cabbage using the vernalization–infiltration method (Bai *et al.*, 2013). Two transgenic plants were obtained. *BrpSPL9-2* overexpression was detected in T2 plants, but *BrpSPL9-2* expression decreased remarkably in T3 plants. We hypothesized that expression of *mBrpSPL9* under the control of the CaMV 35S promoter was unstable in Chinese cabbage.

To overcome the problem of low and unstable expression of the transgene, we replaced the 35S promoter with the AA6 promoter (PCT/NL2006/050314) in the binary vectors (Figure 3a). The AA6 promoter was originally isolated from tomato plants. Seven hygromycin-resistant seedlings were selected from more than 80 000 seeds on medium containing hygromycin, meaning that the transformation frequency was about 0.1%. The three transformants, termed BS9-1, BS9-3 and BS9-4, were self-fertilized for three generations. We performed PCR to identify these three lines using the primers specific for AA6. The sequences of the PCR products showed that all three lines were transgenic because the AA6 promoter was detected in the

genomes (Figure 3b). *BrpSPL9-2* was up-regulated eightfold in BS9-1, and nearly threefold in BS9-4 line, compared with that of the wild type (Figure 3c). In addition, brp-miR172, whose homolog in Arabidopsis is directly downstream of *SPL9*, was up-regulated in BS9-1 plants (Figure 3d), meaning that *BrpSPL9-2* regulates the miR172-directed pathway in Chinese cabbage as does *AtSPL9* in Arabidopsis.

In the growth room, the folding time of the transgenic lines BS9-1 and BS9-4 was much earlier than that of the wild type (Figure 3e). For BS9-1 plants, the days to folding and the leaves to folding were fewer 10 and 4, respectively, than those of the wild type (Table 2). These data revealed that the folding time of BS9-1 was significantly earlier than that of the wild type in terms of number of days and leaves. In the other words, the seedling and rosette stages were much shorter than those of the wild type. To determine the reasons for early folding and shorter seedling and rosette stages in the transgenic plants expressing *mBrpSPL9-2*, we measured the length of the juvenile and early adult phases. The juvenile leaves of BS9-1 were absent and the number of the early adult leaves decreased compared with that of the wild-type ones. This reveals that the earliness of folding in the transgenic plants was due to the absence of the juvenile phase and the shortening of the early adult phase.

The colour of the BS9-1 leaves was dark-green, in contrast with the light-green colour of the wild-type leaves (Figure 3e). The chlorophyll contents in BS9-1 plants were nearly twice as high as those of the wild type (Figure S3), meaning that the activity of *BrpSPL9* affected the synthesis of chlorophyll. The first four leaves of BS9-1 plants were approximately round, in contrast with the elliptical shape of the wild type (Figure 3f), whereas their petiole length, blade base angle and serration were comparable to those of early adult leaves on the wild-type plants (Table 2). In the growth room, the folding plants continued to grow, but failed to form the compact head.

In the field, the folding time of BS9-1 plants was much earlier than that of the wild type as in the growth room. The heading time (the day when the first head leaves appeared and reached 1 cm in length) for BS9-1 plants was significantly earlier (12 days)

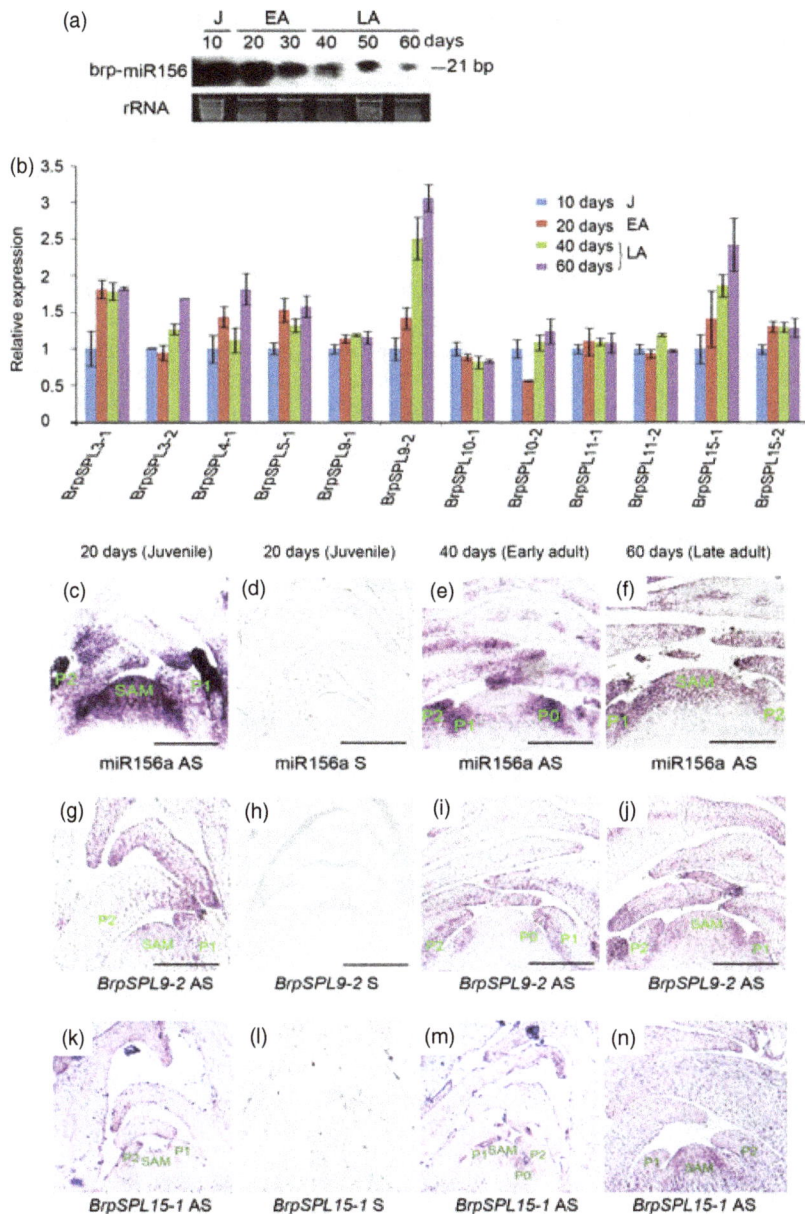

Figure 2 Temporal and spatial expression of brp-miR156 and *BrpSPL* genes in Chinese cabbage (Bre) grown in the field. (a) Northern blotting showing the changes in expression of brp-miR156 in shoot apices on different days after germination. (b) Real-time PCR showing relative expression of *BrpSPL* genes in shoot apices on different days after germination. Error bars indicate the standard deviation. (c–n) *in situ* hybridization showing expression of brp-miR156 (c–f), *BrpSPL9-2* (g–j) and *BrpSPL15-1* (k–n) in the shoot apices on different days after germination. AS, antisense probe; S, sense probe; EA, early adult; J, juvenile; LA, late adult; SAM, shoot apical meristem; P0, P1, P2 and P3, the four stages of leaf primordia. Scale bars: 200 μm in (c–n).

than that of the wild type (Table 2), thus confirming the earliness of the transgenic line BS9-1 in the growth room. Comparison of phase transition indicated that the juvenile phase for BS9-1 plants was absent, and the juvenile and early adult leaves were fewer six than the wild type, consistent with the results observed in the growth room. The maturity of BS9-1 heads was earlier than that of the wild type. But the heading stage of BS9-1 was longer than that of the wild type, thus affecting the net output of head maturity. On the average, the harvesting time for BS9-1 was earlier 8 days than that of the wild type. These results indicated that overexpression of *BrpSPL9-2* caused the earliness of heading by shortening the juvenile and early adult phases rather than the late adult phase.

The first four leaves of BS9-1 plants were not distinct from the early adult, in contrast with the juvenile leaves of the wild type (Figure 3g,j). When the wild-type plants were at the rosette stage, with early adult leaves, BS9-1 plants were at the folding stage, with late adult leaves (Figure 3h,k). In appearance, the

shape and size of the leafy heads of BS9-1 plants were equivalent to those of the wild type (Figure 3i,l).

To link the earliness of heading to the development of individual leaves, we measured various leaf parameters on plants in growth room. In the wild type, the blade base angles for the early juvenile phase were about 140°, and the ratios of blade length to width, blade to petiole length and petiole length to width were 1.46, 1.47 and 8.47, respectively (Table 1). For the juvenile leaves of BS9-1 plants, the blade base angle and blade length-to-width ratio were almost equal to those of the early adult phase of the wild type; however, blade-to-petiole length was higher and petiole length-to-width ratio was lower than those of the early adult phase of the wild type. These data revealed that the juvenile phase of BS9-1 plants was replaced by the early adult phase.

BS9-1 plants exhibited the early head formation; therefore, we wondered whether *mBrpSPL9-2* affected other head traits. The measurement of head parameters indicated that head diameter,

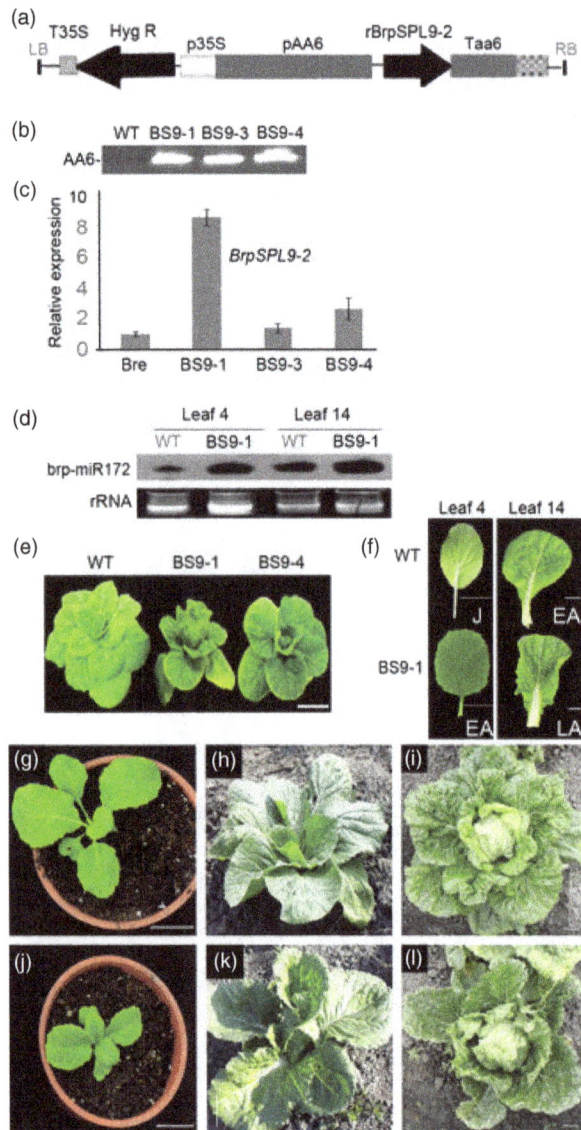

Overexpression of *Brp-MIR156a1* delays the heading time

According to the earliness of heading in BS9-1 plants, we inferred that underexpression of *BrpSPL9* would delay heading time. To address this question, we transferred *Brp-MIR156a1* into Bre (Figure 4a). In total, five independent lines with *pAA6::Brp-MIR156a1* were obtained. We identified three transgenic lines by detecting a fragment of the AA6 promoter using PCR (Figure 4b). In the two transgenic lines, B156a-1 and B156a-2, brp-miR156 was highly accumulated (Figure 4c) and *BrpSPL9-1*, *BrpSPL9-2* and *BrpSPL15-1* were underexpressed (Figure 4d).

Overexpression of miR156 delayed the time of leaf folding (Figure 4e). In growth room, the heading time for B156a-1 plants was 12 days later than that of the wild type, and the seedling and rosette stages were significantly longer than those of the wild type. Comparison of phase transition indicated that the length of the juvenile and early adult phases for B156a-1 plants was significantly longer than that of the wild type. B156a-1 plants had five fewer juvenile and early adult leaves than the wild type (Table 2). Therefore, silencing of miR156-targeted *BrpSPL* genes prolonged the seedling and rosette stages by lengthening the juvenile and early adult phases.

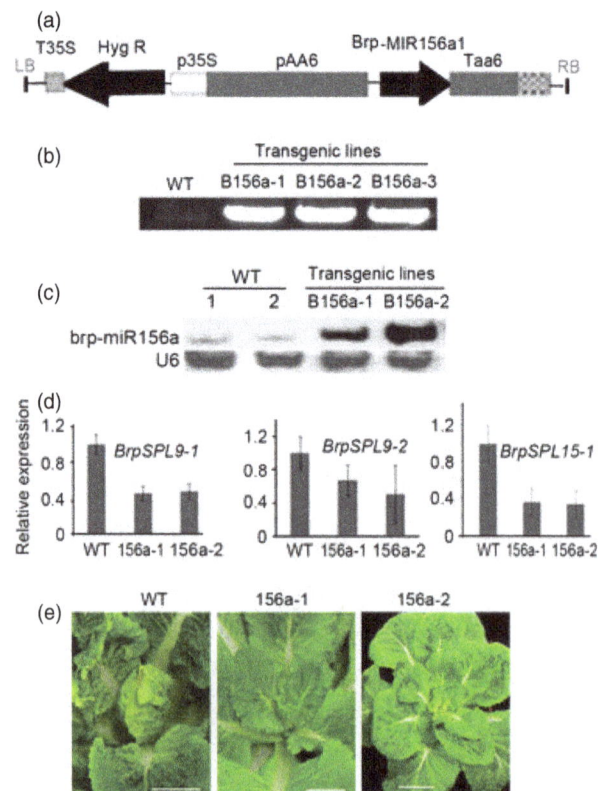

Figure 3 Leaf and shoot morphology of Bre plants transgenic for a mutated miR156-resistant version of *BrpSPL9-2* (*mBrpSPL9-2*). (a) The *mBrpSPL9-2* construct under the control of the AA6 promoter. (b) Detection of the transgenic lines by PCR with the primers spanning AA6 promoter and *mBrpSPL9-2* fragment. (c) Real-time PCR showing the relative expression of *BrpSPL9-2* in the shoot apices of the transgenic lines with *mBrpSPL9-2* at the heading stage. Error bars indicate the standard deviation. (d) Northern blotting showing the expression of brp-miR172 in shoot apices of BS9-1 plants at the stages of leaf 4 and leaf 14. (e) Transgenic lines BS9-1 and BS9-4 with earliness of leaf folding. (f) Leaves of line BS9-1 showing the absence of juvenile leaves. (g–l) Wild-type plants at seedling (g), rosette (h) and heading (i) stages. BS9-2 plants at the stages corresponding to seedling (j), rosette (k) and heading (l) stages of the wild-type plants. p35S, CaMV 35S promoter; Hyg R, hygromycin-resistant gene; pAA6, AA6 promoter; LB, left border of T-DNA; RB, right border of T-DNA; T35S, CaMV35S terminator; Taa6, AA6 terminator. Scale bars: 2 cm in (e–f); 5 cm in (g–k).

Figure 4 Leaf and shoot morphology of the Bre plants transgenic for *Brp-MIR156a1*. (a) The *Brp-MIR156a1* construct under the control of the AA6 promoter. (b) Detection of the transgenic lines by PCR with primers specific for the AA6 promoter. (c) Northern blotting showing the expression of brp-miR156a in shoot apices of 60-day-old plants. (d) Real-time PCR showing the relative expression of *BrpSPL9-1*, *BrpSPL9-2* and *BrpSPL15-1* in shoot apices at the late adult stage. Error bars indicate the standard deviation. (e) The wild-type plants and the transgenic lines B156a-1 and B156a-2. p35S, CaMV 35S promoter; Hyg R, hygromycin-resistant gene; pAA6, AA6 promoter; LB, left border of T-DNA; RB, right border of T-DNA; T35S, CaMV 35S terminator; Taa6, AA6 terminator. Scale bar: 2 cm in (e).

height and weight of BS9-1 plants were almost the same as the wild type (Table S2). The difference in head shape index (head height/diameter ratio) between them was not significant. We concluded that *BrpSPL9-2* regulates earliness of Chinese cabbage without obvious impact on the other head traits.

Discussion

Morphological markers for the juvenile leaves of Chinese cabbage are different from those of Arabidopsis

In Arabidopsis, trichomes on the abaxial surface of the leaves are often used as a morphological marker for the phase transition (Clarke et al., 1999). Leaf shape, marginal serration, hydathodes and veins are also regarded as the morphological markers. Recently, blade base angle has been introduced as a new marker (Li et al., 2012). Among these morphological markers, the absence of abaxial trichomes, round shape and small blade base angle are the three morphological markers for the juvenile phase in Arabidopsis. However, this is not the case in Chinese cabbage. The juvenile leaves of Chinese cabbage have many abaxial trichomes, long elliptical shape and a large blade base angle. Instead, the late adult leaves of Chinese cabbage have a round shape and small blade base angle. This implies that some of the morphological markers for juvenile leaves in Chinese cabbage are the same as those of the adult leaves in Arabidopsis. Abaxial trichomes are seen on the juvenile and adult leaves, even though their density on adult leaves is higher than that on juvenile leaves. The difference in abaxial trichome density between Chinese cabbage and Arabidopsis supports the hypothesis that abaxial trichome production is regulated by both the level of a trichome inducer and the competence of the abaxial epidermis to respond to this inducer (Telfer et al., 1997).

Incurvature, bulges, lobes and short and wide petioles are the other morphological markers for the late adult leaves of Chinese cabbage. The juvenile leaves of Chinese cabbage are downwardly curved, whereas the early adult leaves are basically flat and the late adult leaves are incurved. Bulges and lobes are not obvious on the juvenile leaves and early adult leaves, but are obvious on the late adult leaves. Petioles of the adult leaves are very short and wide relative to the blades, corresponding to the juvenile leaves in Arabidopsis.

In the plants transgenic for mBrpSPL9-2, the first four leaves were round, with a small blade base angle and low ratios of blade to petiole length and petiole length to width, compared with the ones of the early adult leaves. This indicates that BrpSPL9 represses the morphological markers for the juvenile leaves, but promotes those of the early adult leaves. The early adult phase of the transgenic plants overexpressing BrpSPL9-2 is shorter than that of the wild type; therefore, we hypothesized that BrpSPL9 accelerates the appearance of the morphological markers for the late adult leaves. It would be interesting to define the genetic relationship between these morphological markers and phase transition in different plant species.

BrpSPL9 controls the earliness of heading by shortening juvenile and early adult phases

The cause for early heading in the transgenic plants is mBrpSPL9-2. Overexpression of BrpSPL9-2 shows the strong effects on phase transition in the two aspects: the juvenile phase is absent when the first four juvenile leaves are replaced by early adult leaves, and the early adult phase is shortened, compared with the wild type. The repression of SPL9 in the juvenile phase has been reported in Arabidopsis. Overexpression of SPL9 and its closely related paralog SPL15 not only shortens the juvenile phase and produces an early flowering phenotype but also affects changes in the cell size and cell number typical of adult leaves (Usami et al., 2009). In the transgenic plants of Chinese cabbage with

mBrpSPL9-2, the shortening of juvenile and adult phases is consistent with those of Arabidopsis. However, the prolonged length of the late adult phase in the transgenic Chinese cabbage plants is not reported in Arabidopsis. Anyway, the net outcome of BrpSPL9 overexpression is that the transition from the early adult phase to late adult phase occurs much earlier. Therefore, the reasons for earliness of heading in BS9-1 are the short length of the juvenile and early adult phases and the decreased number of juvenile and early leaves.

The absence of juvenile phase combined with shortening of the early adult phase is equivalent to the shortened period of seedling and rosette stages. In the transgenic plants overexpressing BrpSPL9-2, seedling and rosette stages are much shorter and fewer leaves are produced at these stages, compared with the wild type. By contrast, the duration of the heading stage becomes longer and more leaves are produced at heading stage, compared with the wild type. This implies that after heading time, the transgenic plants overexpressing BrpSPL9-2 require the longer time to form the compact head than the wild type. Nevertheless, the heading time of the transgenic plants is so early that the resulting head remains matured much earlier than that of the wild type. We suggest that BrpSPL9 genes cause the earliness of maturity by repressing the juvenile and early adult phases.

miR156 and BrpSPL genes are potentially important for genetic improvement of earliness of crops

Leafy heads supply mineral nutrients, crude fibre and vitamins in the human diet. Heading time is one of the most important traits that affect the yield and quality of these vegetables and hence is the focus of many vegetable breeders. The net output of the transgenic plants overexpressing BrpSPL9-2 is the earliness of heading. This is reflected by the decreased number of days for the seedling and rosette stages and the number of leaves for these stages. Cultivars of Chinese cabbage, cabbage, lettuce and Brussels sprouts show wide variations in the numbers of days and leaves necessary for the initiation of leafy heads. Breeding of the cultivars with early heading is usually performed by the selection of germplasm with the traits of shorter seedling and rosette stages and/or of fewer seedling and rosette leaves. Interestingly, these two traits are combined in our transgenic lines with mBrpSPL9-2. While the seedling and rosette stages are shortened in the transgenic plants, the heading stage prolongs slightly. Long heading stage usually leads to high yield of Chinese cabbage. This is especially beneficial for those cultivars with low yield. Importantly, the other head traits such as head weight, head size and head shape in the transgenic plants did not change significantly. Therefore, BrpSPL9-2 is potentially important for the genetic improvement of earliness in Chinese cabbage and other related crops without impairing the other agronomic traits.

In the transgenic plants overexpressing Brp-MIR156a, the folding time is delayed. This is not helpful for earliness of Chinese cabbage in general, but may be beneficial for some cultivars with late heading that are harvested in the second year. Thus, manipulation of BrpSPL9 level will optimize the earliness of Chinese cabbage.

Chinese cabbage and cabbage are the biennial plants. Vernalization during the winter is necessary for flowering of these crops in the second year. In the production of Chinese cabbage and cabbage, however, many cultivars undergo vernalization and start bolting before head maturity (BBHM), which usually causes the huge losses in yield and quality. In our transgenic plants with *mBrpSPL9-2*, the gradual increase in *BrpSPL9-2* expression and miR172 accumulation is correlated. Earliness of the leafy head is coupled with early flowering (data not shown), apparently because SPL9 directly activates the expression of miR172b to facilitate flowering by suppressing the APETALA2 (AP2) protein (Aukerman and Sakai, 2003; Wang *et al.*, 2009), whereas miR172 promotes the transition from vegetative growth to flowering (Huijser and Schmid, 2011; Willmann and Poethig, 2011). For breeding of Chinese cabbage and cabbage, the breeders have tried a lot to uncouple head earliness from early flowering, which is very difficult in practice. Recently, Bergonzi *et al.* (2013) and Zhou *et al.* (2013) reported simultaneously that miR156 plays a role in the age-dependent response to winter temperature in perennial flowering of *Arabis alpina* and *Cardamine flexuosa*, in which the miR156 and miR172 levels are uncoupled. The APETALA2 transcription factor, a target of microRNA miR172, prevents flowering before vernalization (Bergonzi *et al.*, 2013). We expect that genetic manipulation of Chinese cabbage and cabbage for the new cultivars with head earliness and late flowering will be possible.

Experimental procedures

Plant materials and growth conditions

An inbred line of the cultivar Bre was used in this study. The seeds of the wild-type and the transgenic plants were surface-sterilized in 70% ethanol for 1 min, followed by 0.1% $HgCl_2$ for 10 min, and then washed four times in sterile distilled water and plated on top of solid Murashige and Skoog medium containing 500 mg/L carbenicillin and incubated at 22 °C for 14 days. The seedlings were transplanted to peat soil in plastic pots and moved in a growth room of SIPPE Phytotron. In this growth chamber, the plants were grown at 22 °C with 16 h of light per day under a light source of warm white fluorescent tubes (colour code 990), an irradiance of 150 $\mu mol/m^2/s$ and a light intensity on the plant canopy of 75 $\mu mol/m^2/s$. The relative humidity was 65%–70%, and the air velocity was approximately 0.9 m/s. All of the seedlings were grouped randomly and grown under identical conditions for 7 weeks. For cultivation in the field, the 4-week-old seedlings were acclimated for 3 days and transferred to the field in SIPPE Farm Station on 24 August 2012.

Observation and measurement of heading time

More than 20 individual plants for each treatment were observed and measured. Leaves (longer than 1 cm) were recorded and accounted during plant growth. The folding time, heading time and maturity time were represented by the days or leaves to folding, heading and maturity, respectively, which are the numbers of the days lasted or the leaves produced from the germination to the appearance of the first folding leaf, the first head leaf and the fully maturity (final head size and weight), respectively. Lengths of juvenile, early adult and late adult phases were the numbers of leaves produced during each phase. The diameter of head was measured in the middle of the head. Head height was the distance between the bottom and top. Head weight indicated fresh weight per head. For leaf shape analysis, fully expanded leaves were harvested and flattened on a blank cardboard for scanning with digital scanner. The measurement of the blade base angle was according to Liu's method (Li *et al.*, 2012; Liu *et al.*, 2010).

Genetic transformation

The fragments of *BrpSPL9-2* coding sequence were cloned by PCR with cDNA from Bre seedlings as the template using KOD TOYOBO DNA polymerase (TOYOBO, Osaka, Japan). Polymerase. *mBrpSPL9-2* was made by two rounds of mutagenic PCR (Cadwell and Joyce, 1994). Fragments of *mBrpSPL9-2* were inserted into pCAMBIA1300 binary vectors and placed under the control of the AA6 promoter. The binary constructs were delivered into *Agrobacterium tumefaciens* strain GV3101 (pMP90RK) using a freeze–thaw method (Weigel and Glazebrook, 2006). The Bre plants were transformed using vernalization–infiltration method (Bai *et al.*, 2013). This method is based on vacuum infiltration method (Bechtold *et al.*, 1993). Briefly, a pot of germinated seeds were put in growth room and incubated at 4 °C for 25 days of vernalization treatment. The seedlings in the pots were transferred to growth room at 22 °C and grew for 2 weeks. Then, the plants with small flower buds at the early bolting stage were placed upside down in vacuum desiccator that contained infiltration medium and the engineered Agrobacterium (Bechtold *et al.*, 1993) for vacuum infiltration. The Agrobacterium-infected plants were transferred to dark growth room and incubated for 2 days at 22 °C and then transferred to growth room. These plants were pollinated by hand with the pollens from the Bre plants without vacuum infiltration. The pots were placed back to the growth room. The seeds were harvested and screened by germination on agar medium containing hygromycin. The hygromycin-resistant seedlings were transplanted in growth room and grown at 22 °C with 16-h light. The transgenic plants of T1 generation were self-fertilized by hand pollination, and then, T2 and T3 seedlings were identified for the insertion of exogenous genes and analysed for the segregation of population.

Expression analyses

Total RNA was extracted from vegetative shoot apices with TRIzol (Invitrogen, Carlsbad, CA) and treated with DNase I (TaKaRa, Otsu, Japan) to remove DNA contamination. Four micrograms of RNA was used for cDNA synthesis with oligo (dT) primer and reverse transcriptase (TaKaRa). Real-time PCR was performed with SYBR-Green PCR Mastermix (Bio-Rad, Richmond, CA) using specific pairs of primers (Table S3). The comparative threshold cycle (Ct) method was used to determine relative transcript levels in the real-time PCR (MyiQ2 Two-color Real-time PCR Detection System; Bio-Rad). Relative amounts of the genes examined are normalized to *BrpACTIN3* of Chinese cabbage. Two biological replicates and three technical replicates were performed. Error bars indicate standard deviation.

In situ hybridization

Sections (7 µm thick) of shoot apices from both the wild-type and mutant plants were prepared following pretreatment and hybridization methods described previously (Jackson, 1991). Hybridization probes corresponding to coding sequences were defined as follows: a coding sequence of *BrpSPL9-2*; a *BrpSPL15-1*-specific probe located at 282–767 bp. Digoxigenin-labelled probes were prepared by *in vitro* transcription (Roche, Basel, Switzerland) according to the manufacturer's protocol. Locked nucleic acid (LNA)-modified probes of miR156a were synthesized and labelled with digoxigenin (DIG) at the 3'-end by TaKaRa.

miRNA isolation and Northern blot analysis

A volume containing 40–60 µg RNA was separated on 19% polyacrylamide denaturing gels. The RNA was transferred to a Hybond membrane (Amersham Biosciences, GE Healthcare, Uppsala, Sweden) for 4 h at 150 mA. After cross-linking by UV irradiation for 3 min, the Hybond membrane was hybridized with biotin-labelled DNA probes complementary to the predicted miRNA sequences at 42 °C overnight. The membrane was washed at 42 °C twice with 2 × SSC and 0.1% SDS, followed by two higher stringency washes of 0.1 × SSC and 0.1% SDS at 42 °C. Subsequently, the membrane was incubated with a stabilized streptavidin–horseradish peroxidase conjugate (Thermo Scientific, Waltham, MA) in nucleic acid detection blocking buffer and then washed four times with 1 × washing buffer. After washing with substrate equilibration buffer and adding stable peroxide solution and enhancer solution, the blots were imaged using an FLA-5000 Phosphor-imager (FujiFilm, Tokyo, Japan). To confirm uniform loading, the rRNA or the hybrid U6 was used as a control. The DNA probes used for small-RNA Northern blotting were synthesized and biotin-labelled using a 3'-end DNA labelling method (Invitrogen).

Acknowledgements

This work is supported by National Basic Research Program of China (Grant No. 2012CB113903) and Natural Science Foundation of China (Grant No. 30730053 and 31070696).

Author contributions

Y.H. designed the research. Y.W. and F.W. performed the research and analysed the data. J.B. contributed to genetic transformation of Chinese cabbage. Y.W. and F.W. wrote the article.

References

Aukerman, M.J. and Sakai, H. (2003) Regulation of flowering time and floral organ identity by a microRNA and its *APETALA2*-like target genes. *Plant Cell*, **15**, 2730–2741.

Bai, J., Wu, F., Mao, Y. and He, Y. (2013) *In planta* transformation of *Brassica rapa* and *B. napus* via vernalization-infiltration methods. *Protoc. Exch.* doi:10.1038/protex.2013.067.

Bechtold, N., Ellis, J. and Pelletier, G. (1993) *In Planta* Agrobacteria mediated gene transfer by infiltration of adult *Arabidopsis thaliana* plants. *C. R. Acad. Sci. Paris Life Sci.* **316**, 1194–1199.

Bergonzi, S., Albani, M.C., Ver Loren van Themaat, E., Nordström, K.J., Wang, R., Schneeberger, K., Moerland, P.D. and Coupland, G. (2013) Mechanisms of age-dependent response to winter temperature in perennial flowering of *Arabis alpina*. *Science*, **340**, 1094–1097.

Cadwell, R.C. and Joyce, G.F. (1994) Mutagenic PCR. *PCR Methods Appl.* **3**, S136–S140.

Cheng, F., Mandáková, T., Wu, J., Xie, Q., Lysak, M.A. and Wang, X. (2013) Deciphering the diploid ancestral genome of the mesohexaploid *Brassica rapa*. *Plant Cell*, **25**, 1541–1554.

Chuck, G., Cigan, A.M., Saeteurn, K. and Hake, S. (2007) The heterochronic maize mutant Corngrass1 results from overexpression of a tandem microRNA. *Nat. Genet.* **39**, 544–549.

Clarke, J.H., Tack, D., Findlay, K., Van Montagu, M. and Van Lijsebettens, M. (1999) The *SERRATE* locus controls the formation of the early juvenile leaves and phase length in Arabidopsis. *Plant J.* **20**, 493–501.

Fang, L., Cheng, F., Wu, J. and Wang, X. (2012) The impact of Genome Triplication on Tandem Gene Evolution in *Brassica rapa*. *Front. Plant Sci.* **3**, 261.

Fornara, F. and Coupland, G. (2009) Plant phase transitions make a SPLash. *Cell*, **138**, 625–627.

He, Y.K., Wang, J.Y., Gong, Z.H., Wei, Z.M. and Xu, Z.H. (1994) Root development initiated by exogenous auxin genes in Brassica crops. *Plant Physiol. Biochem.* **32**, 492–500.

He, Y.K., Xue, W.X., Sun, Y.D., Yu, X.H. and Liu, P.L. (2000) Leafy head formation of the progenies of transgenic plants of Chinese cabbage with exogenous auxin genes. *Cell Res.* **10**, 151–160.

Huijser, P. and Schmid, M. (2011) The control of developmental phase transitions in plants. *Development*, **138**, 4117–4129.

Ito, H. and Kato, T. (1957) Studies on the head formation of Chinese cabbage: histological and physiological studies of head formation. *J. Jpn. Soc. Hortic. Sci.* **26**, 154–204.

Ito, H. and Kato, T. (1958) Studies on the head formation of Chinese cabbage. II. Relation between auxin and head formation (in Japanese). *Agric. Hortic.* **26**, 771.

Jackson, D. (1991) In situ hybridization in plants. In *Molecular Plant Pathology, A Practical Approach* (Gurr, S.J., McPherson, M. and Bowles, D.J., eds), pp. 163–174. Oxford, England: Oxford University Press.

Lauter, N., Kampani, A., Carlson, S., Goebel, M. and Moose, S.P. (2005) microRNA172 down-regulates glossy15 to promote vegetative phase change in maize. *Proc. Natl Acad. Sci. USA*, **102**, 9412–9417.

Li, S., Yang, X., Wu, F. and He, Y. (2012) HYL1 controls the miR156-mediated juvenile phase of vegetative growth. *J. Exp. Bot.* **63**, 2787–2798.

Liu, Z., Jia, L., Mao, Y. and He, Y. (2010) Classification and quantification of leaf curvature. *J. Exp. Bot.* **61**, 2757–2767.

Opeña, R.T., Kuo, C.G. and Yoon, J.Y. (1988) Breeding and Seed Production of Chinese Cabbage in Tropics and Subtropics. Technical Bulletin No. 17. AVRDC, Shanhua, Tainan. pp. 13–15, 92pp.

Schwab, R., Palatnik, J.F., Riester, M., Schommer, C., Schmid, M. and Weigel, D. (2005) Specific effects of MicroRNAs on the plant transcriptome. *Dev. Cell*, **8**, 517–527.

Telfer, A., Bollman, K.M. and Poethig, R.S. (1997) Phase change and the regulation of trichome distribution in Arabidopsis thaliana. *Development*, **124**, 645–654.

Usami, T., Horiguchi, G., Yano, S. and Tsukaya, H. (2009) The more and smaller cells mutants of Arabidopsis thaliana identify novel roles for *SQUAMOSA PROMOTER BINDING PROTEIN-LIKE* genes in the control of heteroblasty. *Development*, **136**, 955–964.

Wang, J.W., Czech, B. and Weigel, D. (2009) miR156-regulated SPL transcription factors define an endogenous flowering pathway in Arabidopsis thaliana. *Cell*, **138**, 738–749.

Wang, J.W., Park, M.Y., Wang, L.J., Koo, Y.J., Chen, X.Y., Weigel, D. and Poethig, R.S. (2011) MiRNA control of vegetative phase change in trees. *PLoS Genet.* **7**, e1002012.

Weigel, D. and Glazebrook, J. (2006) In planta transformation of Arabidopsis. *CSH Protoc.* doi: 10.1101/pdb.prot4668.

Willmann, M.R. and Poethig, R.S. (2011) The effect of the floral repressor FLC on the timing and progression of vegetative phase change in Arabidopsis. *Development*, **138**, 677–685.

Wu, G., Park, M.Y., Conway, S.R., Wang, J.W., Weigel, D. and Poethig, R.S. (2009) The sequential action of miR156 and miR172 regulates developmental timing in Arabidopsis. *Cell*, **138**, 750–759.

Xie, K.B., Wu, C.Q. and Xiong, L.Z. (2006) Genomic organization, differential expression, and interaction of SQUAMOSA promoter-binding-like transcription factors and microRNA156 in rice. *Plant Physiol.* **142**, 280–293.

Yu, X.H., Peng, J.S., Feng, X.Z., Yang, S.X., Zheng, Z.R., Tang, X.R. and He, Y.K. (2000) Cloning and structural and expressional characterization of *BcpLH* gene preferentially expressed in folding leaf of Chinese cabbage. *Sci. China, C Life Sci.* **43**, 321–329.

Yu, X., Wang, H., Zhong, W., Bai, J. and He, Y. (2013) QTL mapping of leafy heads by genome resequencing in the RIL population of *Brassica rapa*. *PLoS ONE*, **8**, e76059.

Zhou, C.M., Zhang, T.Q., Wang, X., Yu, S., Lian, H., Tang, H., Feng, Z.Y., Zozomova-Lihová, J. and Wang, J.W. (2013) Molecular basis of age-dependent vernalization in *Cardamine flexuosa*. *Science*, **340**, 1097–1100.

13

Unlocking the secondary gene-pool of barley with next-generation sequencing

Neele Wendler[1], Martin Mascher[1], Christiane Nöh[2], Axel Himmelbach[1], Uwe Scholz[1], Brigitte Ruge-Wehling[2] and Nils Stein[1,*]

[1]Leibniz Institute of Plant Genetics and Crop Plant Research (IPK), Gatersleben, Germany

[2]Julius Kühn-Institut (JKI), Institute for Breeding Research on Agricultural Crops, Groß Lüsewitz, Germany

*Correspondence

email stein@ipk-gatersleben.de

Accession numbers: Accession numbers for sequence data generated in this study are ERP004444 (GBS reads) and ERP004445 (exome capture reads).

Keywords: *Hordeum bulbosum*, crop wild relatives, introgression line, genotyping-by-sequencing, exome capture, next-generation sequencing.

Summary

Crop wild relatives (CWR) provide an important source of allelic diversity for any given crop plant species for counteracting the erosion of genetic diversity caused by domestication and elite breeding bottlenecks. *Hordeum bulbosum* L. is representing the secondary gene pool of the genus *Hordeum*. It has been used as a source of genetic introgressions for improving elite barley germplasm (*Hordeum vulgare* L.). However, genetic introgressions from *H. bulbosum* have yet not been broadly applied, due to a lack of suitable molecular tools for locating, characterizing, and decreasing by recombination and marker-assisted backcrossing the size of introgressed segments. We applied next-generation sequencing (NGS) based strategies for unlocking genetic diversity of three diploid introgression lines of cultivated barley containing chromosomal segments of its close relative *H. bulbosum*. Firstly, exome capture-based (re)-sequencing revealed large numbers of single nucleotide polymorphisms (SNPs) enabling the precise allocation of *H. bulbosum* introgressions. This SNP resource was further exploited by designing a custom multiplex SNP genotyping assay. Secondly, two-enzyme-based genotyping-by-sequencing (GBS) was employed to allocate the introgressed *H. bulbosum* segments and to genotype a mapping population. Both methods provided fast and reliable detection and mapping of the introgressed segments and enabled the identification of recombinant plants. Thus, the utilization of *H. bulbosum* as a resource of natural genetic diversity in barley crop improvement will be greatly facilitated by these tools in the future.

Introduction

Genetic diversity of crop plants has narrowed constantly throughout the process of domestication and breeding. To ensure sustainability or even increase crop yields under changing environmental conditions, permanent breeding for crop improvement is a prerequisite. Crop wild relatives (CWR) are a valued source of allelic diversity for broadening the genetic basis of breeding germplasm (Hajjar and Hodgkin, 2007; Tanksley and McCouch, 1997). Traits were transferred from CWR in at least 13 crops (Hajjar and Hodgkin, 2007) - among them the four world's most important cereals; wheat (*Triticum aestivum* L.), rice (*Oryza sativa* L.), maize (*Zea mays* L.), and barley (*Hordeum vulgare* L.). It was calculated that the benefit from crop-wild introgressions used or imported in North America accounted for more than 340 million US dollar per year (Prescott-Allen and Prescott-Allen, 1986). As limitations due to crossing barriers may be overcome (i.e. Akano *et al.*, 2002; Mallikarjuna, 1999) there is an increasing need for advanced molecular tools that facilitate the efficient utilization of CWR in (pre)-breeding programs.

In barley, three gene pools may be considered as a source of new advantageous alleles in breeding (von Bothmer *et al.*, 1995). The primary gene pool of barley includes domesticated barley (*H. vulgare* L. ssp. v*ulgare;* in the following *Hv*) and wild barley (*H. vulgare* L. *ssp. spontaneum;* in the following *Hs*). *Hs* has been repeatedly used to improve *Hv* elite cultivars (Kalladan *et al.*, 2013; Lakew *et al.*, 2013; Nevo, 1992). *H. bulbosum* L. (in the following *Hb*) is the only member of the secondary gene pool of

barley. Crossing barriers between *Hb* and *Hv* can be overcome by modifying environmental conditions (Pickering, 1984), use of specific genotypes and biotechnological tools such as embryo rescue (Kasha and Sadasiva, 1971; Pickering, 1983). Thirty wild *Hordeum* species constitute the tertiary gene pool of barley; however, strong crossability barriers and low chromosome pairing in hybrids with *Hv* have interfered with their use in classical barley breeding (von Bothmer *et al.*, 1995). Interspecies hybrids between wild *Hordeum chilense* and domesticated tetraploid durum wheat (*Triticum durum*) were used to form a novel crop species *Tritordeum* with novel nutritional properties (Martin *et al.*, 1999) and *H. chilense* introgressions allowed to introduce abiotic stress tolerances into wheat (Forster *et al.*, 1990).

Hb has been recognized and was used as an important source of advantageous traits for *Hv*, especially with respect to pathogen resistance or tolerance (Pickering *et al.*, 1987; Ruge *et al.*, 2003; Szigat and Szigat, 1991; Walther *et al.*, 2000; Xu and Snape, 1988). The progeny of *Hv/Hb* hybrids were surveyed to identify *Hv* plants carrying preferentially a single *Hb* introgression with a gene of interest (in the following *Hv^b*). Advantageous traits of *Hb* can often only be transferred at the cost of negative linkage drag that is introduced with the introgressed *Hb* segments—causing a strong penalty on breeding programs. Thus, in order to enable efficient utilization of introgressed *Hb*-loci for barley breeding, marker-assisted reduction of the size of introgressed *Hb* segments is needed. The availability of molecular markers is no longer a limiting factor for the differentiation of barley cultivars (Close *et al.*, 2009; Comadran *et al.*, 2012; Mascher *et al.*, 2013c;

Poland et al., 2012; Sato et al., 2009; Stein et al., 2007); however, markers that differentiate between Hv and Hb alleles are rare. The paucity of Hb specific molecular markers has so far been one of the main bottlenecks for the incorporation of Hb traits into elite barley germplasm.

The development of molecular markers is greatly facilitated by access to genomic or expressed sequence information of the genotypes of interest (Davey et al., 2011). Even though next-generation sequencing (NGS) costs are continuously decreasing, sequencing of the whole barley genome still remains very costly due to its size (>5 Gbp) and high repetitive DNA content. Targeted sequencing of a sample with reduced complexity would simultaneously decrease sequencing costs and minimize redundancy.

One approach to complexity reduction is a targeted sequence capture, which is used to enrich for defined sequences prior to NGS (Asan et al., 2011; Sulonen et al., 2011). Recently, a barley exome capture assay, targeting ~62 Mb of exon sequence was developed and used to capture genomic sequence information of several barley cultivars and barley CWR (Mascher et al., 2013b). A second approach, genotyping-by-sequencing (GBS), combines complexity reduction, multiplexing of samples, and the use of NGS methods for genotyping of whole mapping populations (Elshire et al., 2011; Poland et al., 2012).

Here, we utilized both, exome capture and GBS, to survey sequence polymorphisms in three independent Hv^b lines. The sequence information obtained by exome sequencing for one introgression on chromosome 2HL was used to design a marker assay to genotype a population segregating for the introgressed segment on chromosome 2HL. Both approaches proved to be highly efficient and are readily applicable for evaluation and characterization of CWR introgressions in other crop species.

Results

Identification of Hv/Hb SNPs and the Hb introgression intervals by exome capture

Three Hv^b lines and the respective donor lines were (re)-sequenced after exome capture, in order to survey sequence polymorphisms between Hv and Hb and to identify and locate the Hb introgression segment in the Hv^b lines. Although all three Hv^b lines were previously characterized, the working hypothesis of the present study was that the new approaches should principally allow the identification of the introgressions de novo at much higher precision than by previous attempts.

For the identification and comparison of sequence polymorphisms between Hv, Hb and Hv^b, sequencing reads of all samples were mapped to the barley reference. This allowed comparing sequence reads of each Hv^b line and the respective Hv and Hb progenitor genotypes based on their position within the reference sequence. Nearly, 90% and 70% of total Hv and Hb reads could be mapped to the reference, respectively (Table 1). Sequence or SNP targets between samples can only be compared if they have a given minimum read coverage across all samples. At a minimum 10-fold coverage, the three analysed Hv^b lines shared more than 45 Mb of sequence with both respective donor lines (data not shown).

To localize the Hb introgression segments in the three Hv^b lines, we filtered for (only homozygous) single-nucleotide-polymorphisms (SNPs) between Hv^b and the respective Hv donor lines —assuming that sequences of these two genotypes should differ nearly exclusively in the region of the Hb introgression segment.

Table 1 Exome capture and GBS mapping results

Method	Sample	% mapped	% mapped to EC[*,‡]	% mapped to HC[†,‡]	% mapped to LC[†,‡]
EC	A42	69	85	63	17
EC	Borwina	87	76	46	22
EC	Vogelsanger G.	88	76	47	22
EC	Hv^b 2HL	89	79	49	23
EC	Hv^b 5HL	80	63	38	18
EC	Hv^b 6HS	89	75	46	22
GBS	A42	37	50	43	13
GBS	Borwina	82	34	29	14

*EC: Exome capture target region.
[†]HC/LC: High/low confidence genes of barley (IBSC et al., 2012).
[‡]The percentage of reads mapped to EC, HC, and LC regions is relative of mapped reads.

To further validate the identified polymorphisms, Hv/Hv^b SNPs were only included if they had the same genotype calls in Hv^b (only homozygous) and its respective Hb donor lines (heterozygous and homozygous calls allowed). Filtered SNPs that fit all these criteria were considered as Hb-specific. The frequency of Hb-specific SNPs was then computed in non-overlapping 5 Mb bins along the physical map of barley (IBSC et al., 2012), assuming that SNP frequency should be highest within the introgressed Hb segment (Figure 1). This assumption was based on previous findings (Mascher et al., 2013b), that SNP frequency between the barley reference 'Morex' and Hb was at least 5 times higher than between the barley reference and cultivars of Hv.

Hb-specific SNP frequency revealed clear peaks (>600 to >1000 SNPs per 5 Mb bin) at 2HL, 5HL, and 6HS for the Hv^b lines holding previously allocated Hb introgressions at 2HL, 5HL, and 6HS, respectively (Figure 1). SNPs that matched our criteria for Hb-specific SNPs could also be indentified on other chromosomes (or chromosome arms) mainly towards the telomeres of the three Hv^b lines, albeit at much lower frequencies (<200 SNPs per 5 Mb bin). Instead of representing SNPs truly originating from a Hb introgression, these SNPs likely represent false positives, i.e. sequencing errors, incorrect read mapping, or SNPs between the (re)-sequenced Hv genotype and the original Hv donor plant that was used to generate the initial hybrid.

In case of the 2HL Hv^b genotype two introgressed Hb segments, separated by approximately 22 CentiMorgans (cM) (23 Mb) of Hv DNA, could be determined. Based on the barley reference (IBSC et al., 2012), the proximal introgressed segment was located at approximately 110–114 cM while the distal introgression was placed approximately at 136–149 cM in the iSelect genetic map (Comadran et al., 2012), and corresponded to 5 and 16 Mb of physical sequence, respectively. Within these two Hb segments, more than 4000 Hb-specific SNPs were detected. The two other Hv^b introgressions on chromosomes 5HL and 6HS each comprised a single Hb introgression at approximately 167–169 cM (8 Mb) of chromosome 5HL and 0–10 cM (10 Mb) of chromosome 6HS, respectively. In both cases, more than 900 Hb-specific SNPs were assigned to the introgressed segments.

96-plex SNP Genotyping of a 2HL Hv^b mapping population

Out of 4000 SNPs from exome capture (re)-sequencing, nearly 200 fulfilled the 'Golden Gate assay' criteria, and a custom assay

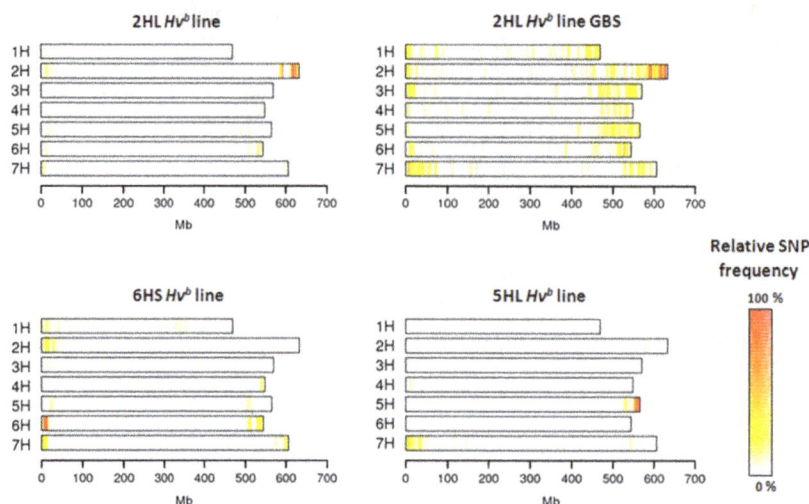

Figure 1 *Hb*-specific SNP frequency distribution in *Hv^b* lines. *Hb*-specific SNPs were discovered by exome capture-based (re)-sequencing of three independent *Hb* introgression lines, as well as by Genotyping-by-Sequencing (GBS) in case of 2HL *Hv^b* (upper right panel). The SNP frequency was visualized as a heat map of 5 Mb bins along the physical length of the seven barley chromosomes. The positions of contigs were taken from the barley reference (IBSC *et al.*, 2012). Relative SNP frequency was visualized in 1/1000 colour steps from white (0% SNP frequency) to red (100% SNP frequency) using heat.colors with the R statistical environment (R Core Team, 2012).

targeting 96 independent SNP loci was designed. The aim of this assay was to verify the exome capture SNP information and to identify recombination events in the segment introgressed from *Hb*. SNP markers were selected considering a more or less equidistant genetic distribution across the two introgressed segments according to the physical/genetic map of the barley reference. To analyse the detected pattern of the two (separated) introgressed segments in the 2HL *Hv^b* line, six SNPs were selected to be located proximal to the proximal *Hb* introgression and nine SNPs to be placed in between the two *Hb* introgressions.

The assay was used to genotype a 2HL *Hv^b* F_7 mapping population comprising 88 individuals, including three replicates of each donor (*Hv* and *Hb*) genotype. Genotyping resulted in 90% successful SNP markers (88 successful SNP markers), which gave clear and consistent genotype calls (max. 16% missing data) and all successful markers were polymorphic between the donor lines as predicted from the assay design. The 50% GC score of a SNP— a metric to indicate the reliability for the genotype calls (0 = failed, 1 = excellent) revealed that 88% of the loci targeted by the assay (84 SNPs) produced a 50% GC score of 0.5 or higher (Figure S1). Genotyping revealed six recombinant plants and proved the presence of the two detected *Hb* introgression segments as well as the presence of the *Hv* fragment that was identified between those segments. The proximal *Hb* introgression as well as the putative intercalary *Hv* segment was not segregating in the analysed mapping population; thus, orientation of these segments to each other could not be resolved. However, the high success rate and the high quality of the obtained genotype calls revealed the strong reliability of the sequences obtained by exome capture.

Genotyping-by-sequencing of the 2HL *Hv^b* introgression

Genotyping-by-sequencing (GBS) was applied as an alternative strategy for generating new molecular markers for *Hb* introgressions targeting the same individuals that were genotyped by the Golden Gate assay. Similar to the analysis of the exome capture

data, GBS sequence reads were mapped to the barley reference for sequence variant detection. More than 80% of sequence reads of *Hv* but <40% of the *Hb* reads mapped to the barley reference (Table 1), respectively. A shared total of 2 Mb of sequence was represented by a minimum 10-fold sequence read coverage in 95% of all samples of the 2HL *Hv^b* mapping population, enabling high confidence SNP detection (Figure 2). High quality GBS SNPs between the *Hv* and *Hb* donor genotypes were selected based on genotype call quality, percentage of missing data, sequence read coverage and segregation ratio of parental alleles among the population.

In total, 57 *Hv/Hb* SNP markers could be scored on the basis of GBS data. Some SNP markers clustered with tight physical linkage of only several bp distances, thus, were redundant for the purpose of detecting recombination events. In such cases, only SNP markers lying on different reference contigs were selected as independent markers. Overall, 33 non-redundant GBS markers were used for mapping and identification of six recombination events in the introgression interval.

The analysed 2HL *Hv^b* mapping population was only segregating for the distal *Hb* introgression. Due to the filtering of GBS markers with a 1 : 2 : 1 segregation pattern, no markers were detected for the proximal *Hb* introgression; however, *Hb*-specific GBS SNP frequency indicated the detection of the proximal *Hb* introgression to the same region as identified by exome capture (Figure 1). Thus, GBS is an efficient method to detect and to precisely define even small *Hb* introgressions in an *Hv* background, similarly to exome capture.

Genetic map of a 2HL *Hv^b* population and collinearity with barley

For construction of the genetic maps, a GBS marker that was showing three additional recombinants was excluded, as it was referring to the same barley reference contig with two other GBS markers, which did not support these recombination events. Furthermore, one GBS and one Golden Gate marker were (re)-sequenced using Sanger sequencing in order to revise two

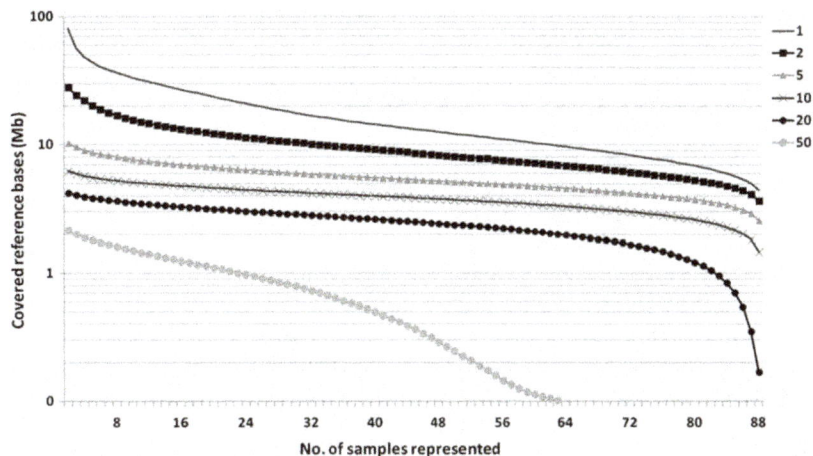

Figure 2 Read coverage across 88 GBS samples of the 2HL Hv^b mapping population. The cumulative size of sequence intervals on the barley reference, covered by at least x sequence reads (x = 1, 2, 5, 10, 20 or 50) that is shared in different numbers of GBS-sequenced samples of the 2HL Hv^b mapping population is visualized.

unexpected genotypes in one individual, respectively. Subsequently, genotyping results were corrected accordingly for map construction. Genetic maps were constructed only for the distal segment of the 2HL Hv^b line, since the proximal Hb introgression segment was not segregating.

When genotyping the 2HL Hv^b mapping population, GBS as well as the Golden Gate assay markers each revealed six recombinant plants. Two of these recombinants were differing between the data sets. GBS detected one recombination event towards the centromere, which was not recognized by GBS. The Golden Gate assay, however, could identify an additional distal recombination event, which in turn was not documented by the Golden Gate assay. Thus, both marker sets were spanning the same genomic regions with slightly differing extensions on either side and each of the maps covered a region that was comprised of the same 24 anchored BAC contigs of the physical map of barley (Ariyadasa et al., 2014) with an estimated cumulative length of >17 Mb (Table S1). Both maps were spanning a distance of 3.4 cM and a consensus map interval of 4.5 cM, respectively. Recombination frequency was greatly reduced as compared to the published Hv genetic map of the same genomic region (Mascher

et al., 2013a), where the same genetic interval spanned more than 12 cM. When Golden Gate and GBS marker order was compared to the predicted marker position in the barley genomic reference (Mascher et al., 2013a), two and three markers showed a rearranged order, respectively (Figure 3). However, further analysis is needed to determine if this finding is real or attributable to the limited genetic resolution of the barley reference genome.

Discussion

We demonstrated two efficient strategies for generating large numbers of molecular markers for the genetic resource of Hv^b lines. Our analysis took advantage of the recently published physical and genetic framework of barley (IBSC et al., 2012; Mascher et al., 2013b) that served both as a reference for computational read mapping and for the chromosomal localization of variants that were discovered.

Both methods allowed to precisely define the locations and sizes of the Hb introgressions and to identify recombinant plants. Exome capture based (re)-sequencing as well as GBS of the 2HL Hv^b line revealed an Hb introgression located distally at

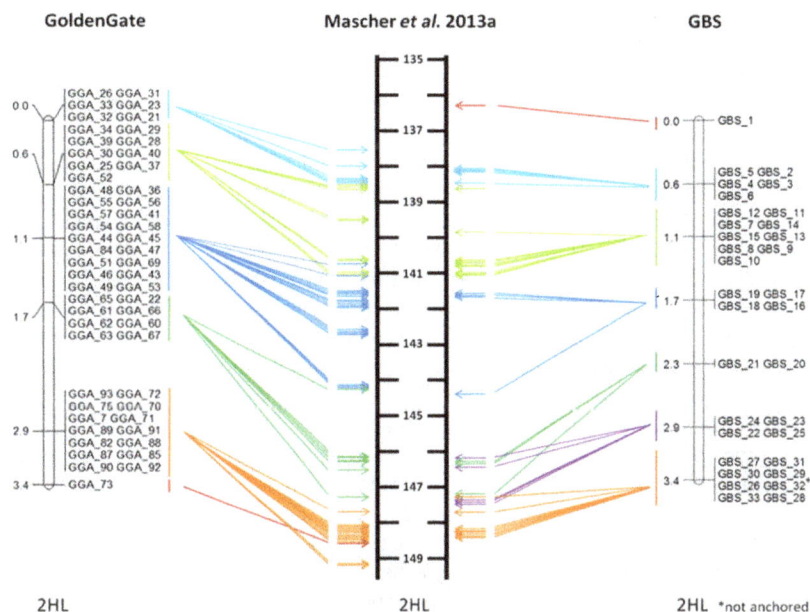

Figure 3 Comparison of marker order. The genetic maps derived by genotyping of the 2HL Hv^b mapping population using either GBS or the Golden Gate assay were compared to the genetic reference map of Hv (Mascher et al., 2013a). Distances are given in centiMorgans (cM) and arrows indicate marker positions on the reference map.

chromosome 2HL. These findings are in line with previous results by Ruge-Wehling *et al.* (2006) who genotyped this Hv^b line using RFLP, STS, and SSR markers. In this study, we identified an additional proximal, smaller *Hb* introgression at 110–114 cM, presumably 22 cM proximal to the larger *Hb* introgression at 136–149 cM on 2HL. Interestingly, this smaller and more proximal *Hb* segment was not detected earlier (Ruge-Wehling *et al.*, 2006), most likely due to a lack of markers. This underpins the strength of both NGS-based marker development strategies applied in the present work. Moreover, the *Hb* segments of two additional Hv^b lines could be identified in the expected chromosomal regions by means of exome capture, corroborating the sensitivity, reproducibility, and reliability of the approach. It is noteworthy that exome capture generates a substantial amount of additional information, such as a large number of SNP calls (i.e. 900–4000 SNPs for the *Hb* segments in the 3 Hv^b lines). Furthermore, the complete sequence of candidate genes may be contained in the exome capture data. In contrast, GBS does not come with this additional information; instead, it would require either deeper sequencing or new GBS libraries using different restriction enzymes, to generate further independent SNP (marker) information.

The power of both methods for producing SNP information was mainly limited by the efficiency of mapping *Hb* sequence reads to the barley reference (IBSC *et al.*, 2012). This limitation was more pronounced in the GBS approach compared to exome capture (re)-sequencing. Both methods allowed mapping of more than 80% of *Hv* sequence reads, while <40–70% of *Hb* sequence reads could be mapped for GBS and exome capture, respectively. Principally, it should be possible to increase mapping rates of *Hb* reads by decreasing stringency of the mapping parameters. Default parameters that were used in this study allow a maximum of 4% sequence polymorphism for a read to be mapped. This permits theoretically 1 sequence polymorphism to appear every 25 bp. As sequence divergence between *Hv* and *Hb* was found to be 1 SNP per 32.4 bp in introns and 1 SNP per 69.6 bp in exons (Johnston *et al.*, 2009), the default parameters are already beyond polymorphism rates. Therefore, decreasing stringencies seems to be unfavourable and would probably only increase miss-mapping of reads within repetitive DNA rather than increasing efficiency. In general, robust mapping of *Hb* reads to the *Hv* reference sequence was mainly restricted to gene regions (HighConfidence genes, LowConfidence genes and exome capture targets, Table 1), as genes appear to be highly conserved between both species. Thus, overall higher success of *Hb* sequence read mapping in exome capture can be attributed to the stronger enrichment for genic regions if compared to GBS.

Exome capture-enriched (re)-sequencing and GBS significantly shorten the time-consuming steps that have been involved to locate and map introgressions from *Hb*. In brief, the rough localization of an *Hb* introgression has typically been carried out using a combination of genomic *in situ* hybridization and fluorescent *in situ* hybridization steps (Pickering *et al.*, 2000). Subsequently, molecular markers had to be found and distributed across the *Hb* introgression. The development of new markers without access to genomic or expressed sequence information of the genotypes of interest was generally time consuming, and transferability of existing non-genic *Hv* markers to Hv^b was limited due to the high degree of sequence divergence between *Hv* and *Hb*. For instance, barley sequence tagged site (STS) markers were found to be unreliable in Hv^b lines. They either failed completely on any *Hb* genotype or they preferentially amplified *Hv* alleles in

heterozygotes (Pickering and Johnston, 2005). Johnston *et al.* (2009) developed EST-based markers using the HarvEST database (Wanamaker *et al.*, 2006) to map Hv^b lines. This revealed 92 markers of an original set of 216 amplicons, giving a success rate of <50%. When the 1536-SNP barley BOPA1 set (Close *et al.*, 2009) was applied to localize introgressions of *H. vulgare* ssp. *spontaneum* (wild barley, primary gene pool) in *Hv*, only ~55% of all markers gave useful and polymorphic genotypes (Schmalenbach *et al.* 2011). Considering this low success rate and the high diversity between *Hv* and *Hb* (secondary gene pool), such primer-based, pre-designed marker assays do not seem promising to analyse Hv^b lines. Mascher *et al.* (2013b) reported efficient SNP calling from exome capture data obtained for *Hb* accessions, but transferring existing *Hv*/*Hb* markers or SNP information to new Hv^b lines is difficult, owing to the highly polymorphic nature of the *Hb* genepool.

In contrast, in this study, GBS as well as exome capture based approaches proved useful to precisely delineate *Hb* introgressions without the need for initial screenings. A large set of markers/SNP information and a precise allocation of introgressed segments both were obtained in the same experiment.

By applying exome capture and GBS a dense map of the 2HL Hv^b introgression could be generated. This map, however, was substantially condensed (nearly 4-fold reduction in interspecific recombination frequency of *Hv* and *Hb*) compared to the corresponding region of the barley reference map (Mascher *et al.*, 2013a). Such suppression of recombination frequency was previously observed for introgressed segments from *Hb* (Johnston *et al.*, 2009, 2013; Ruge *et al.*, 2003; Ruge-Wehling *et al.*, 2006), which magnifies the need for inexpensive and easy to handle molecular markers. Large numbers of markers will not compensate for the lack of recombination. However, the possibility to allocate them to the physical/genetic and sequence framework of barley (Ariyadasa *et al.*, 2014; IBSC *et al.*, 2012; Mascher *et al.*, 2013a) helps to more robustly estimate physical distances represented by recombination events which will also provide an estimate of the involved genes of the respective genomic segments.

The number of suitable markers was reported to be one of the major impediment when using CWR (Hajjar and Hodgkin, 2007) in plant breeding. The results of this study strongly suggest that GBS as well as exome capture are two promising tools that will be applicable to other crop-wild introgression systems, as well.

Accuracy and ease-of-use are not the only considerations when selecting a genotyping strategy targeting a specific research aim. Among other factors, the cost in terms of both money and time required by different approaches often play a pivotal role in choosing a method. In this study, the two approaches were compared according to their cost and time requirements (Figure 4 and Table 2). Both methods can be considered as being cost efficient. A caveat for the multiplex SNP assay is that the manufacturer set a minimum order size of 480 genotyping reactions, making this approach only efficient when 480 or more samples are analysed (Figure 4). Furthermore, costs of the initial exome capture experiment for marker discovery are fixed, favouring a larger amount of subsequently analysed samples. In contrast, per-sample costs of GBS do not change as the number of samples analysed increases. Both strategies rely on NGS and may thus become even cheaper in the near future as sequencing costs are expected to continue to decrease. However, it needs to be considered that to apply both methods, sufficient computational power and bioinformatics knowledge is needed to analyse

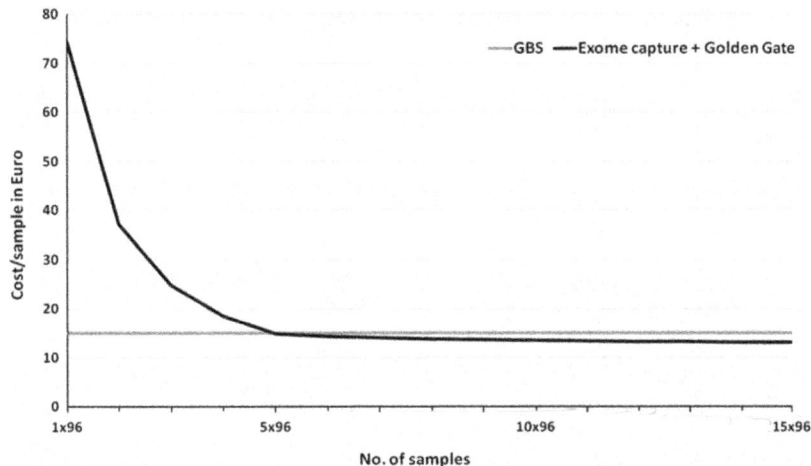

Figure 4 Cost comparison. Per-sample costs for sequencing and genotyping by GBS or exome capture/Golden Gate are compared across different numbers of analysed samples (96-plex format). Costs were calculated based on the experiences during the present study. However, costs may vary considerably due to different manufacturers, reagents, scales, and laboratories. Labour costs have not been considered.

Table 2 Time flowchart of exome capture (EC), Golden Gate assay (GGA) and GBS experiments

	Method		
Exp. step	EC (days)	GGA/96 samples (days)	GBS/96 samples (days)
Library preparation	14*		6
Sequencing	11		7
Analysis	7		7
GGA design		20	
GGA manufacturing		(30–60)**	
GGA genotyping		3	
GGA analysis		1	
Total	56	20	

*Assay assumed to be already in place ** not included.

these fairly large amounts of NGS data. The latter might be a disadvantage for the presented methods as some laboratories or institutes might be lacking the required infrastructure, a limitation, however, that might be easily circumvented by establishing scientific collaborations. GBS was found to be the more flexible approach compared to exome capture/multiplex SNP assay if considering the time required carrying out the experiments (Table 2). This is mainly due to the initial exome capture experiment as well as the time-consuming steps of SNP marker selection for the SNP assay and its manufacturing by the company. However, once the SNP assay is in place, genotyping of 96 samples takes only 2–3 days.

In conclusion, both GBS and exome capture-enriched (re)-sequencing in combination with a multiplexed marker assay are powerful and convenient methods for the precise localization and genotyping of the *Hb* introgression segments in *Hv^b* lines. This will ultimately pave the way to incorporating advantageous *Hb* traits into elite *Hv* cultivars. Furthermore, we suggest that both methods are highly promising to be successfully implemented in other crop-wild introgression systems as well. An attractive immediate application of the two approaches outlined here

would be the characterization of a larger set of *Hv^b* lines (i.e. Johnston *et al.*, 2009). Such an experiment may further validate putative introgressions and exactly delimit introgression intervals. Analogous to the vast collection of nearly-isogenic barley mutant lines (Druka *et al.*, 2011), a comprehensive, well-characterized catalogue of *Hv^b* lines would be an excellent community resource and may pave the way for the wide-spread utilization of *Hv^b* lines in applied breeding and basic research.

Experimental procedures

Plant material and DNA extraction

The plant material of the 2HL introgression (*Hv^b*) line originated from a single diploid resistant F_2 recombinant plant, which had been obtained from an interspecific tetraploid *Hordeum vulgare* L. (*Hv*) cv. 'Borwina' × *Hordeum bulbosum* L. (*Hb*, accession 'A42') hybrid (Szigat and Szigat, 1991) as described previously (Ruge-Wehling *et al.*, 2006). The resulting F_7 population that was subject to GBS and the Golden Gate assay, segregated for the introgressed *Hb* segment, which was mapped distally to the long arm of chromosome 2H (Ruge-Wehling *et al.*, 2006). To obtain this population, a single plant heterozygous for the *Hb* introgression was self-pollinated. The 2HL *Hv^b* plant targeted in the exome capture, was a single plant of the ancestral F_6 population, homozygous for the *Hb* introgression. Furthermore, a diploid plant of cultivar 'Borwina' and a tetraploid clonal progeny plant of the 'A42' donor line were used for exome capture and were included in the GBS experiments to capture information on donor genotype specific DNA polymorphisms.

Two additional *Hv^b* plants known to carry *Hb* introgressions on barley chromosomes 6HS (Ruge *et al.*, 2003) and 5HL (unpublished results), respectively, were analysed by exome capture. The 6HS *Hv^b* genotype (F_5) originated from the same hybridization event that gave rise to the above mentioned 2HL *Hv^b* line. The 5HL *Hv^b* genotype (F_2) was kindly provided by Dr. U. Walther (BAZ, Aschersleben). It was developed by crossing a tetraploid *Hv* (cv. 'Vogelsanger Gold') with an unknown tetraploid *Hb* parent. Both *Hv^b* plants (5HL and 6HS) selected for exome capture (re)-sequencing were confirmed before to be homozygous for the *Hb* introgression interval using RFLP, STS, and SSR markers. A diploid plant of cultivar 'Vogelsanger Gold' was (re)-sequenced by exome capture to replace the original donor genotype of the initial hybrid for subsequent variant detection. As the *Hb* donor line of the 5HL introgression was unknown, the 'A42' *Hb* accession was used as a surrogate for the original *Hb* parent.

DNA extraction was performed as described previously (Stein *et al.*, 2001).

Exome capture and sequencing

Construction of exome capture libraries, and sequencing followed previously established procedures (Mascher *et al.*, 2013b). In brief, Illumina TruSeq Paired End libraries (Illumina Part # 15026486) were prepared according to the manufacturer using DNA Adapter Indexes 4, 5, 6 and 12 (Illumina, Inc., San Diego, CA). Genomic DNA (1 µg) was fragmented (size range: 200–300 bp) using Covaris microTUBES and the Covaris S220 Instrument (175W ultrasonic power, 10% duty factor, 200 cycles per burst, 100 s treatment time, Covaris, Inc., Woburn, MA). Adapter ligated DNA products were isolated (size range: 320–420 bp) by excision from an SYBR-Gold stained agarose gel (Life Technologies GmbH, Invitrogen, Darmstadt, Germany) and purified using AmPure XP beads (Beckman Coulter GmbH, Krefeld, Germany). Correctly ligated DNA fragments were enriched using a pre-capture LM-PCR reaction (ligation-mediated PCR) and purified as described (Haun *et al.*, 2011; Mascher *et al.*, 2013b). DNA was quantified with a Qubit 2.0 fluorometer (Life Technologies GmbH, Invitrogen) and analysed using an Agilent 2100 Bioanalyser (Agilent Technology, Santa Clara, Part# 5067-1506) on a DNA 7500 chip (between 250 and 500 bp).

Equimolar amounts of up to four amplified libraries containing different indices were pooled for hybridization. The library capture, liquid array processing, quantification, and sequencing on the Illumina HiSeq2000 device were as described (Mascher *et al.*, 2013a). Equimolar amounts of up to five pooled libraries, each containing different mixes of indices, were pooled for sequencing.

Selection of SNPs for the Golden Gate assay

For the 2HL *Hv^b* line, a Veracode Golden Gate assay (Illumina, Inc.) of 96 SNP markers was designed based on SNP information gained by exome captured (re)-sequencing of the 2HL *Hv^b* genotype and the donor genotypes 'Borwina' (*Hv*) and 'A42' (*Hb*), respectively. To improve the selection of high quality SNPs and to avoid false-positive SNP calls, single nucleotide variants between *Hv* and *Hb* with samtools SNP call score <200 and which were located within 200 bp of the end of a reference sequence contig were excluded from further evaluation. Furthermore, SNPs were only selected if they were supported by a minimum of fivefold sequence read coverage in the SNP itself, plus the flanking sequence (50–60 bp) of the (re)-sequenced *Hv^b* as well as the two donor genotypes.

The selected SNPs along with 50–60 bp flanking sequence to either side of the SNP position and harbouring at most one additional SNP between in each of the flanking sides were provided to the company Illumina Inc., for a pre-assessment using the Array Design Tool (ADT) (Illumina Inc.). ADT provides a feasibility rank score for each SNP ranging from 0 to 1, giving a benchmark about how likely a particular SNP will convert into a successful marker assay. According to recommendations by Illumina Inc., only SNPs with preliminary design evaluation rank scores of 0.6 or higher were selected for the custom assay design (Table S2).

Illumina Golden Gate assay-based genotyping

Genotyping was carried out in a 96-well, 96 SNP format (Fan *et al.*, 2003; Shen *et al.*, 2005) using the Illumina BeadXpress Reader Array platform (Illumina Inc.) according to manufacturer's instructions. Raw data processing and genotype calling were performed using the GenomeStudio software (Illumina Inc.). As recommended by Illumina, the minimum threshold for the 'GenCall' score (GenomeStudio, Illumina Inc.) was set to 0.25. In addition, all generated SNP clusters were visually inspected and manually re-clustered, using the GenomeStudio software (Illumina Inc.), if SNP clusters appeared to be incorrect.

GenCall (GC) scores and 50% GC scores were calculated with the GenomeStudio software (Illumina Inc.). The GC score is a quality value between 0 and 1 (0 = failed, 1 = excellent) to indicate the reliability for each genotype call, while the 50% GC score of a SNP represents the 50th percentile rank for all GenCall scores for that SNP. For visualization the obtained 50% GC score decimals were rounded to the nearest tenth.

GBS library construction and sequencing

Genomic DNA (200 ng) was cleaved over night at 37 °C using 14 units each of the restriction enzymes PstI-HF (CTGCAG, NEB Inc., Ipswich, UK) and MspI (CCGG, NEB Inc.), respectively. The reactions contained 1× NEB buffer 4 and 100 ng/µL BSA (NEB Inc.) in a total volume of 20 µL. After restriction digest, samples were incubated at 65 °C for 20 min to inactivate the restriction enzymes and were subsequently used for adapter ligation without purification. Illumina sequencing libraries were constructed as described with some modifications (Meyer and Kirchner, 2010). Briefly, for the reaction clean-up steps (solid phase reversible immobilization, SPRI) AMPure XP beads (Beckman Coulter GmbH, Krefeld, Germany) were replaced by MagNa beads (Thermo Scientific, Inc. Waltham, MA) for cost reduction (Rohland and Reich, 2012). The initial blunt-end repair step was omitted, as the adapter mix (P57_GBS) was provided with overhangs complementary to the ends obtained from the restriction digest. The P57_GBS adapter mix was prepared by annealing oligonucleotides P5_GBS_PstI (IS1_GBS_PstI: A*C*A*C*TCTTTCCCTACACGACGCTCTTCCGATCT*T*G*C*A and IS3_adapter.P5 + P7: A*G*A*T*CGGAA*G*A*G*C) and oligonucleotides P7_GBS_MspI (I2_adapter.P7: G*T*G*A*CTG-GAGTTCAGACGTGTGCTCTTCCG*A*T*C*T and IS3_GBS_MspI: C*G*A*G*ATCGGAA*G*A*G*C; '*' indicates a PTO bond) as described for P5 and P7, respectively (Meyer and Kirchner, 2010). Adapter ligation and adapter fill-in were performed as published (Meyer and Kirchner, 2010). Following adapter fill-in, the DNA was purified and eluted in 20 µL EB (10 mM Tris-HCl, pH 8.0). For indexing PCR, 8 µL of the eluate was used as template DNA. The indexing PCR was performed in 50 µL volume with a final concentration of 1× Phusion HF buffer, 200 µM each dNTP, 200 nM primer IS4_indPCR.P5 (Meyer and Kirchner, 2010), 200 nM indexing primer (Table S3) and 0.02 U/µL Phusion Hot Start Flex (NEB Inc., Ipswich, UK). After initial incubation at 98 °C for 30 s, the amplification was performed for 16 cycles (98 °C for 10 s, 60 °C for 30 s, 72 °C for 5 s) followed by a final extension (72 °C, 10 min). Products of adapter ligation, indexing PCR and primer positions are shown in Figure S2. The products were purified using SPRI and eluted in 25 µL EB. DNA was then quantified using the Quant-iT PicoGreen dsDNA assay kit (Life Technologies GmbH) and a Synergy HT microplate reader (BioTek, Bad Friedrichshall, Germany). Subsequently, the indexed samples were pooled in equimolar ratios. 500 ng pooled DNA was size-fractionated electrophoretically using a 2% agarose gel (Life Technologies GmbH, Invitrogen) and SYBR Gold (Life Technologies GmbH, Invitrogen) staining. The DNA (size between 150 and 600 bp) was recovered by excision and purified using a MinElute Spin column according to the manufacturer's instructions (Qiagen, Hilden, Germany). The GBS library was eluted from the column in 20 μL EB and analysed electrophoretically with an Agilent 2100 Bioanalyser (Agilent Technology, Santa Clara) using the Agilent High Sensitivity DNA kit (Part# 5067-4626).

Finally, the library was quantified using qPCR essentially as described previously (Mascher et al., 2013b). The concentration was determined based on a standard curve and the average size of the GBS library. The sample was diluted to 10 nM for cluster formation on an Illumina cBot (Illumina, Inc.). In this study, 200 μL custom sequencing primer (Read1_GBS_PstI: CTTTCCCTACAC GACGCTCTTCCGATCTTGCAG; 500 nM in HT1 buffer from Illumina Inc.) were provided per lane for processing on the cBot (Illumina Inc., program: SR_Amp_Lin_Block_TubeStripHyb.v.8.0). Cluster formation and 1 × 100 bp single-end sequencing-by-synthesis using Illumina's HiSeq2000 instrument were performed according to protocols provided by the manufacturer (Illumina Inc.).

NGS read mapping and SNP calling

Sequence data were de-multiplexed using the CASAVA pipeline 1.8 (Illumina, Inc.). Exome capture read data for 'Borwina', 'Vogelsanger Gold' and the Hb accession 'A42' were previously published (Mascher et al., 2013b) (accession number PRJEB1810) and were analysed together with sequence data generated in the present study.

Adapter trimming of GBS sequence reads was performed with cutadapt (Martin, 2011) and reads shorter than 30 bp after adapter removal were discarded. Trimmed GBS and exome capture reads were mapped against the barley reference (IBSC et al., 2012) with BWA version 0.6.2 (Li and Durbin, 2009) (commands aln and samse, GBS reads or sampe, exome capture reads). The BWA command aln was called with the parameter '-q 15' for quality trimming, otherwise default parameters were used. Duplicate reads were removed from the exome sequencing mapping files with samtools rmdup. The read mapping files of three GBS replicates of 'A42' and 'Borwina' were merged using samtools merge (Li et al., 2009) before analysis. SNP calling was performed with the SAMtools pipeline (version 0.1.18) using the command samtools aln and bcftools with default parameters (Li et al., 2009). The additional parameter '-D' was used for samtools mpileup to obtain per-sample read depth.

For GBS genotyping, genotype calls were filtered with a custom AWK script and the R statistical environment (R Core Team, 2012), selecting only SNPs matching the following criteria: (i) read depth at least 10 (5, 20) and genotype quality at least 10 (5, 20) for homozygous and heterozygous calls, (ii) at most 5% missing genotype calls (genotype calls with insufficient coverage or quality were set to missing) (iii) a 1 : 2 : 1 segregation ratio in the segregating population.

Exome capture SNP calls were discarded if their SAMtools SNP quality score was below 40, the sequence read coverage was smaller than fivefold or they were positioned on a reference assembly contig shorter than 2000 bp. SNP frequency as determined from GBS and from exome capture data at a minimum 10-fold coverage was visualized along the integrated physical/genetic map of barley (IBSC et al., 2012) using R (R Core

Team, 2012). GBS SNP frequency was analysed for minimum SNP call scores of 200. Read depth statistics were performed using SAMtools depth (Li et al., 2009). Only properly paired non-duplicated reads were considered for coverage calculation in the exome sequencing data. The overlap with different genomic intervals such as gene models (IBSC et al., 2012) and exome capture targets (Mascher et al., 2013b) was analysed using bedtools intersect (Quinlan and Hall, 2010) and samtools flagstat (Li et al., 2009).

Primer design and Sanger sequencing

PCR primers were designed using the online software Batch primer 3 (Untergasser et al., 2012) and were obtained from Eurofins (Ebersberg, Germany) (Table S4). PCR amplification of genomic DNA was performed in a final volume of 20 μL, containing 20 ng genomic DNA, 0.1 μM forward and reverse primers, 0.5 U Hot-star polymerase (Qiagen), 1× PCR reaction buffer and 0.1 mM dNTPs (Fermentas, Fisher scientific, Schwerte, Germany). PCR amplification was performed with 15 min at 95 °C for initial denaturation, followed by 10 cycles of 30 s at 95 °C, 30 s annealing (60–55 °C, −0.5 °C per cycle), 1 min at 72 °C for extension, and 35 cycles of 30 s at 95 °C, 30 s at 55 °C for extension, 1 min at 72 °C, and a final extension for 7 min at 72 °C.

Sanger cycle-sequencing of PCR products was performed following the manufacturer's instructions. In brief, PCR products were purified using the NucleoFast 96 PCR Kit (Macherey-Nagel, Dueren, Germany). The concentration of purified amplicons was subsequently analysed on an agarose gel (Life Technologies GmbH, Invitrogen). Normalized amplicons (10 ng per 100 bp of template fragment) were used for cycle-sequencing (BigDye®Terminator v3.1, Applied Biosystems, Darmstadt, Germany) by using ABI-3730xl technology. Sequence analysis was performed using Sequencher 4.7 (Gene Codes, Ann Arbor).

Map construction and comparison to barley

Genetic maps based on GBS and/or Golden Gate markers were constructed with JoinMap v4.0 (Van Ooijen, 2006) using the Kosambi function. The resulting marker order of each map was compared to the genetic order by Mascher et al. (2013a).

Acknowledgements

We gratefully acknowledge the excellent technical support by Ines Walde, Manuela Knauft and Sandra Driesslein (IPK). We thank Doreen Stengel for raw data submission. The work was financially supported as part of the collaborative project 'TRANS-BULB' by a grant (0315966) from the German Federal Ministry of Education and Research (BMBF) to BRW and NS.

References

Akano, A.O., Dixon, A.G.O., Mba, C., Barrera, E. and Fregene, M. (2002) Genetic mapping of a dominant gene conferring resistance to cassava mosaic disease. Theor. Appl. Genet. 105, 521–525.

Ariyadasa, R., Mascher, M., Nussbaumer, T., Schulte, D., Frenkel, Z., Poursarebani, N., Zhou, R.N., Steuernagel, B., Gundlach, H., Taudien, S., Felder, M., Platzer, M., Himmelbach, A., Schmutzer, T., Hedley, P.E., Muehlbauer, G.J., Scholz, U., Korol, A., Mayer, K.F.X., Waugh, R., Langridge, P., Graner, A. and Stein, N. (2014) A sequence-ready physical map of barley anchored genetically by two million single-nucleotide polymorphisms. Plant Physiol. 164, 412–423.

Asan, Xu.Y., Jiang, H., Tyler-Smith, C., Xue, Y.L., Jiang, T., Wang, J.W., Wu, M.Z., Liu, X., Tian, G., Wang, J., Wang, J., Yang, H.M. and Zhang, X.Q.

(2011) Comprehensive comparison of three commercial human whole-exome capture platforms. *Genome Biol.* **12**, R95.

von Bothmer, R., Jacobsen, N., Baden, C., Jorgensen, R.B. and Linde-Laursen, I. (1995) *An Ecogeographical Study of the Genus Hordeum.* Rome: International Plant Genetic Resources Institute (IPGRI).

Close, T.J., Bhat, P.R., Lonardi, S., Wu, Y.H., Rostoks, N., Ramsay, L., Druka, A., Stein, N., Svensson, J.T., Wanamaker, S., Bozdag, S., Roose, M.L., Moscou, M.J., Chao, S.A.M., Varshney, R.K., Szucs, P., Sato, K., Hayes, P.M., Matthews, D.E., Kleinhofs, A., Muehlbauer, G.J., DeYoung, J., Marshall, D.F., Madishetty, K., Fenton, R.D., Condamine, P., Graner, A. and Waugh, R. (2009) Development and implementation of high-throughput SNP genotyping in barley. *BMC Genomics*, **10**, 582.

Comadran, J., Kilian, B., Russell, J., Ramsay, L., Stein, N., Ganal, M., Shaw, P., Bayer, M., Thomas, W., Marshall, D., Hedley, P., Tondelli, A., Pecchioni, N., Francia, E., Korzun, V., Walther, A. and Waugh, R. (2012) Natural variation in a homolog of Antirrhinum CENTRORADIALIS contributed to spring growth habit and environmental adaptation in cultivated barley. *Nat. Genet.* **44**, 1388–1392.

Davey, J.W., Hohenlohe, P.A., Etter, P.D., Boone, J.Q., Catchen, J.M. and Blaxter, M.L. (2011) Genome-wide genetic marker discovery and genotyping using next-generation sequencing. *Nat. Rev. Genet.* **12**, 499–510.

Druka, A., Franckowiak, J., Lundqvist, U., Bonar, N., Alexander, J., Houston, K., Radovic, S., Shahinnia, F., Vendramin, V., Morgante, M., Stein, N. and Waugh, R. (2011) Genetic dissection of barley morphology and development. *Plant Physiol.* **155**, 617–627.

Elshire, R.J., Glaubitz, J.C., Sun, Q., Poland, J.A., Kawamoto, K., Buckler, E.S. and Mitchell, S.E. (2011) A robust, simple genotyping-by-sequencing (GBS) approach for high diversity species. *PLoS ONE*, **6**, e19379.

Fan, J.B., Oliphant, A., Shen, R., Kermani, B.G., Garcia, F., Gunderson, K.L., Hansen, M., Steemers, F., Butler, S.L., Deloukas, P., Galver, L., Hunt, S., McBride, C., Bibikova, M., Rubano, T., Chen, J., Wickham, E., Doucet, D., Chang, W., Campbell, D., Zhang, B., Kruglyak, S., Bentley, D., Haas, J., Rigault, P., Zhou, L., Stuelpnagel, J. and Chee, M.S. (2003) Highly parallel SNP genotyping. *Cold Spring Harb. Symp. Quant. Biol.* **68**, 69–78.

Forster, B.P., Phillips, M.S., Miller, T.E., Baird, E. and Powell, W. (1990) Chromosome location of genes-controlling tolerance to salt (Nacl) and vigor in *Hordeum-vulgare* and *Hordeum-chilense. Heredity*, **65**, 99–107.

Hajjar, R. and Hodgkin, T. (2007) The use of wild relatives in crop improvement: a survey of developments over the last 20 years. *Euphytica*, **156**, 1–13.

Haun, W.J., Hyten, D.L., Xu, W.W., Gerhardt, D.J., Albert, T.J., Richmond, T., Jeddeloh, J.A., Jia, G.F., Springer, N.M., Vance, C.P. and Stupar, R.M. (2011) The composition and origins of genomic variation among individuals of the soybean reference cultivar Williams 82. *Plant Physiol.* **155**, 645–655.

IBSC, Mayer, K.F.X., Waugh, R., Langridge, P., Close, T.J., Wise, R.P., Graner, A., Matsumoto, T., Sato, K., Schulman, A., Muehlbauer, G.J., Stein, N., Ariyadasa, R., Schulte, D., Poursarebani, N., Zhou, R.N., Steuernagel, B., Mascher, M., Scholz, U., Shi, B.J., Langridge, P., Madishetty, K., Svensson, J.T., Bhat, P., Moscou, M., Resnik, J., Close, T.J., Muehlbauer, G.J., Hedley, P., Liu, H., Morris, J., Waugh, R., Frenkel, Z., Korol, A., Berges, H., Graner, A., Stein, N., Steuernagel, B., Taudien, S., Groth, M., Felder, M., Platzer, M., Brown, J.W.S., Schulman, A., Platzer, M., Fincher, G.B., Muehlbauer, G.J., Sato, K., Taudien, S., Sampath, D., Swarbreck, D., Scalabrin, S., Zuccolo, A., Vendramin, V., Morgante, M., Mayer, K.F.X., Schulman, A. and Conso, I.B.G.S. (2012) A physical, genetic and functional sequence assembly of the barley genome. *Nature*, **491**, 711–716.

Johnston, P.A., Timmerman-Vaughan, G.M., Farnden, K.J.F. and Pickering, R. (2009) Marker development and characterisation of *Hordeum bulbosum* introgression lines: a resource for barley improvement. *Theor. Appl. Genet.* **118**, 1429–1437.

Johnston, P.A., Niks, R.E., Meiyalaghan, V., Blanchet, E. and Pickering, R. (2013) Rph22: mapping of a novel leaf rust resistance gene introgressed from the non-host *Hordeum bulbosum* L. into cultivated barley (*Hordeum vulgare* L.). *Theor. Appl. Genet.* **126**, 1613–1625.

Kalladan, R., Worch, S., Rolletschek, H., Harshavardhan, V.T., Kuntze, L., Seiler, C., Sreenivasulu, N. and Roder, M.S. (2013) Identification of quantitative trait loci contributing to yield and seed quality parameters under terminal drought in barley advanced backcross lines. *Mol. Breed.* **32**, 71–90.

Kasha, K.J. and Sadasiva, R.S. (1971) Genome relationship between *Hordeum-Vulgare* L and *H-Bulbosum* L. *Chromosoma*, **35**, 264–287.

Lakew, B., Henry, R.J., Ceccarelli, S., Grando, S., Eglinton, J. and Baum, M. (2013) Genetic analysis and phenotypic associations for drought tolerance in *Hordeum spontaneum* introgression lines using SSR and SNP markers. *Euphytica*, **189**, 9–29.

Li, H. and Durbin, R. (2009) Fast and accurate short read alignment with Burrows-Wheeler transform. *Bioinformatics*, **25**, 1754–1760.

Li, H., Handsaker, B., Wysoker, A., Fennell, T., Ruan, J., Homer, N., Marth, G., Abecasis, G., Durbin, R. and Proc, G.P.D. (2009) The sequence alignment/map format and SAMtools. *Bioinformatics*, **25**, 2078–2079.

Mallikarjuna, N. (1999) Ovule and embryo culture to obtain hybrids from interspecific incompatible pollinations in chickpea. *Euphytica*, **110**, 1–6.

Martin, M. (2011) Cutadapt removes adapter sequences from high-throughput sequencing reads. *EMBnet. J.* **17**, 10–12.

Martin, A., Alvarez, J.B., Martin, L.M., Barro, F. and Ballesteros, J. (1999) The development of tritordeum: a novel cereal for food processing. *J. Cereal Sci.* **30**, 85–95.

Mascher, M., Muehlbauer, G.J., Rokhsar, D.S., Chapman, J., Schmutz, J., Barry, K., Munoz-Amatriain, M., Close, T.J., Wise, R.P., Schulman, A.H., Himmelbach, A., Mayer, K.F., Scholz, U., Poland, J.A., Stein, N. and Waugh, R. (2013a) Anchoring and ordering NGS contig assemblies by population sequencing (POPSEQ). *Plant J.* **76**, 718–727.

Mascher, M., Richmond, T.A., Gerhardt, D.J., Himmelbach, A., Clissold, L., Sampath, D., Ayling, S., Steuernagel, B., Pfeifer, M., D'Ascenzo, M., Akhunov, E.D., Hedley, P.E., Gonzales, A.M., Morrell, P.L., Kilian, B., Blattner, F.R., Scholz, U., Mayer, K.F., Flavell, A.J., Muehlbauer, G.J., Waugh, R., Jeddeloh, J.A. and Stein, N. (2013b) Barley whole exome capture: a tool for genomic research in the genus Hordeum and beyond. *Plant J.* **76**, 494–505.

Mascher, M., Shuangye, W., Amand, P.S., Stein, N. and Poland, J. (2013c) Application of genotyping-by-sequencing on semiconductor sequencing platforms: a comparison of genetic and reference-based marker ordering in barley. *PLoS ONE*, **8**, e76925.

Meyer, M. and Kirchner, M. (2010) Illumina Sequencing Library Preparation for Highly Multiplexed Target Capture and Sequencing. *Cold Spring Harb. Protoc.* doi: 10.1101/pdb.prot5448.

Nevo, E. (1992) Origin, evolution, population genetics and resources for breeding of wild barley, *Hordeum spontaneum*, in the fertile crescent. In: *Genetics, Biochemistry, Molecular Biology and Biotechnology* (Shewry, P.R., ed.), pp. 19–43. Oxford: C.A.B. International, The Alden Press.

Pickering, R.A. (1983) The location of a gene for incompatibility between *Hordeum-Vulgare*-L and *H-Bulbosum* L. *Heredity*, **51**, 455–459.

Pickering, R.A. (1984) The influence of genotype and environment on chromosome elimination in crosses between *Hordeum-Vulgare*-L × *Hordeum-Bulbosum* L. *Plant Sci. Lett.* **34**, 153–164.

Pickering, R.A. and Johnston, P.A. (2005) Recent progress in barley improvement using wild species of Hordeum. *Cytogenet. Genome Res.* **109**, 344–349.

Pickering, R.A., Rennie, W.F. and Cromey, M.G. (1987) Disease resistant material available from the wide hybridization programme at DSIR. *Barley Newsl.* **31**, 248–250.

Pickering, R.A., Malyshev, S., Kunzel, G., Johnston, P.A., Korzun, V., Menke, M. and Schubert, I. (2000) Locating introgressions of *Hordeum bulbosum* chromatin within the *H-vulgare* genome. *Theor. Appl. Genet.* **100**, 27–31.

Poland, J.A., Brown, P.J., Sorrells, M.E. and Jannink, J.L. (2012) Development of high-density genetic maps for barley and wheat using a novel two-enzyme genotyping-by-sequencing approach. *PLoS ONE*, **7**, e32253.

Prescott-Allen, C. and Prescott-Allen, R. (1986) *The First Resource: Wild Species in the North American Economy.* New Haven: Yale University.

Quinlan, A.R. and Hall, I.M. (2010) BEDTools: a flexible suite of utilities for comparing genomic features. *Bioinformatics*, **26**, 841–842.

R Core Team (2012) *Language and Environment for Statistical Computing.* Austria: R Foundation for Statistical Computing. ISBN3-900051-900007-900050.

Rohland, N. and Reich, D. (2012) Cost-effective, high-throughput DNA sequencing libraries for multiplexed target capture. *Genome Res.* **22**, 939–946.

Ruge, B., Linz, A., Pickering, R., Proeseler, G., Greif, P. and Wehling, P. (2003) Mapping of Rym14(Hb), a gene introgressed from *Hordeum bulbosum* and conferring resistance to BaMMV and BaYMV in barley. *Theor. Appl. Genet.* **107**, 965–971.

Ruge-Wehling, B., Linz, A., Habekuss, A. and Wehling, P. (2006) Mapping of Rym16(Hb), the second soil-borne virus-resistance gene introgressed from *Hordeum bulbosum. Theor. Appl. Genet.* **113**, 867–873.

Sato, K., Nankaku, N. and Takeda, K. (2009) A high-density transcript linkage map of barley derived from a single population. *Heredity*, **103**, 110–117.

Schmalenbach, I., March, T.J., Bringezu, T., Waugh, R. and Pillen, K. (2011) High-Resolution Genotyping of Wild Barley Introgression Lines and Fine-Mapping of the Threshability Locus thresh-1 Using the Illumina GoldenGate Assay. *G3-Genes Genomes Genetics*, **1**, 187–196.

Shen, R., Fan, J.B., Campbell, D., Chang, W.H., Chen, J., Doucet, D., Yeakley, J., Bibikova, M., Garcia, E.W., McBride, C., Steemers, F., Garcia, F., Kermani, B.G., Gunderson, K. and Oliphant, A. (2005) High-throughput SNP genotyping on universal bead arrays. *Mutat. Res.* **573**, 70–82.

Stein, N., Herren, G. and Keller, B. (2001) A new DNA extraction method for high-throughput marker analysis in a large-genome species such as *Triticum aestivum. Plant Breed.* **120**, 354–356.

Stein, N., Prasad, M., Scholz, U., Thiel, T., Zhang, H.N., Wolf, M., Kota, R., Varshney, R.K., Perovic, D., Grosse, I. and Graner, A. (2007) A 1,000-loci transcript map of the barley genome: new anchoring points for integrative grass genomics. *Theor. Appl. Genet.* **114**, 823–839.

Sulonen, A.M., Ellonen, P., Almusa, H., Lepisto, M., Eldfors, S., Hannula, S., Miettinen, T., Tyynismaa, H., Salo, P., Heckman, C., Joensuu, H., Raivio, T., Suomalainen, A. and Saarela, J. (2011) Comparison of solution-based exome capture methods for next generation sequencing. *Genome Biol.* **12**, R94.

Szigat, G. and Szigat, G. (1991) Amphidiploid hybrids between *Hordeum vulgare* and *H. bulbosum*—basis for the development of new initial material for winter barley breeding. In *Vortr Pflanzenzüchtg* (Winkel, A., ed.), pp. 34–39. Wageningen: EUCARPIA.

Tanksley, S.D. and McCouch, S.R. (1997) Seed banks and molecular maps: unlocking genetic potential from the wild. *Science*, **277**, 1063–1066.

Untergasser, A., Cutcutache, I., Koressaar, T., Ye, J., Faircloth, B.C., Remm, M. and Rozen, S.G. (2012) Primer3-new capabilities and interfaces. *Nucleic Acids Res.* **40**, e115.

Van Ooijen, J.W. (2006) *JoinMap® 4, Software for the Calculation of Genetic Linkage Maps in Experimental Populations.* Wageningen, Netherlands: Kyazma BV.

Walther, U., Rapke, H., Proeseler, G. and Szigat, G. (2000) *Hordeum bulbosum* - a new source of disease resistance - transfer of resistance to leaf rust and mosaic viruses from *H-bulbosum* into winter barley. *Plant Breed.* **119**, 215–218.

Wanamaker, S., Close, T., Roose, M. and Lyon, M. (2006) *HarvEST: Barley EST database version 1.44.*

Xu, J. and Snape, J.W. (1988) The cytology of hybrids between *Hordeum vulgare* and *H. bulbosum* revisited. *Genome Biol.* **30**, 486–494.

14

A *myo*-inositol-1-phosphate synthase gene, *IbMIPS1*, enhances salt and drought tolerance and stem nematode resistance in transgenic sweet potato

Hong Zhai[†], Feibing Wang[†], Zengzhi Si, Jinxi Huo, Lei Xing, Yanyan An, Shaozhen He and Qingchang Liu*

Beijing Key Laboratory of Crop Genetic Improvement/Laboratory of Crop Heterosis and Utilization, Ministry of Education, China Agricultural University, Beijing, China

Summary

*Correspondence
email liuqc@cau.edu.cn
[†]These authors contributed equally to this work.

Myo-inositol-1-phosphate synthase (MIPS) is a key rate limiting enzyme in *myo*-inositol biosynthesis. The *MIPS* gene has been shown to improve tolerance to abiotic stresses in several plant species. However, its role in resistance to biotic stresses has not been reported. In this study, we found that expression of the sweet potato *IbMIPS1* gene was induced by NaCl, polyethylene glycol (PEG), abscisic acid (ABA) and stem nematodes. Its overexpression significantly enhanced stem nematode resistance as well as salt and drought tolerance in transgenic sweet potato under field conditions. Transcriptome and real-time quantitative PCR analyses showed that overexpression of *IbMIPS1* up-regulated the genes involved in inositol biosynthesis, phosphatidylinositol (PI) and ABA signalling pathways, stress responses, photosynthesis and ROS-scavenging system under salt, drought and stem nematode stresses. Inositol, inositol-1,4,5-trisphosphate (IP_3), phosphatidic acid (PA), Ca^{2+}, ABA, K^+, proline and trehalose content was significantly increased, whereas malonaldehyde (MDA), Na^+ and H_2O_2 content was significantly decreased in the transgenic plants under salt and drought stresses. After stem nematode infection, the significant increase of inositol, IP_3, PA, Ca^{2+}, ABA, callose and lignin content and significant reduction of MDA content were found, and a rapid increase of H_2O_2 levels was observed, peaked at 1 to 2 days and thereafter declined in the transgenic plants. This study indicates that the *IbMIPS1* gene has the potential to be used to improve the resistance to biotic and abiotic stresses in plants.

Keywords: *IbMIPS1*, salt and drought tolerance, stem nematode resistance, sweet potato.

Introduction

Salinity and drought reduce the productivity of crops worldwide (Munns and Tester, 2008). The development of crops with elevated levels of salt and drought tolerance is therefore highly desirable. Sweet potato, *Ipomoea batatas* (L.) Lam., is the seventh most important food crop in the world (FAO, 2013), and its yield is often reduced by salt and drought stresses. Especially, sweet potato as source of bio-energy will mainly be planted on marginal land, and it is therefore important to improve its salt and drought tolerance for maintaining productivity. Stem nematode (*Ditylenchus destructor* Thorne) is a serious disease limiting the production of many crops such as sweet potato, potato, peanut and onion. This disease usually decreases sweet potato yield by 20%–50%, and even no yield in the field if seriously infected by stem nematodes (Xie *et al.*, 2004). The breeding of stem nematode-resistant varieties has become one of the most important objectives in sweet potato.

Genetic engineering has been shown to have the potential for improving the resistance to biotic and abiotic stresses in sweet potato. The *OCI* gene from rice increased stem nematode resistance in transgenic sweet potato (Gao *et al.*, 2011a,b). The *BADH* gene from spinach improved tolerance to salt, MV-mediated oxidative and low temperature stresses in transgenic sweet potato (Fan *et al.*, 2012). Overexpression of the genes, *IbOr*, *IbLCY-ε*, *IbNFU1*, *IbP5CR*, *IbMas* and *IbSIMT1*, from sweet potato enhanced salt tolerance of this crop (Kim *et al.*, 2013a,b; Liu *et al.*, 2014a,b,c, 2015; Wang *et al.*, 2013).

Myo-inositol-1-phosphate synthase (MIPS, EC 5.5.1.4) is a key rate limiting enzyme of *myo*-inositol biosynthesis which catalyses the reaction from glucose-6-phosphate (G-6-P) to *myo*-inositol-1-phosphate (Ins1P), which is subsequently dephosphorylated by *myo*-inositol monophosphatase (MIPP) to form free inositol (Abreu and Aragão, 2007). In plants, inositol is a precursor for many inositol containing compounds and is implicated in various physiological and biochemical processes such as growth regulation, cell membrane biogenesis, hormonal regulation, programmed cell death, stress signalling, plant immunity (Kaur *et al.*, 2013; Tan *et al.*, 2013). The genes encoding MIPS have been isolated from several plant species, such as *Arabidopsis* (Johnson, 1994), rice (Yoshida *et al.*, 1999), maize (Larson and Raboy, 1999), soya bean (Hegeman *et al.*, 2001), smooth cordgrass (Joshi *et al.*, 2013) and *Medicago falcate* (Tan *et al.*, 2013). Overexpression of the *MIPS* gene enhanced tolerance to abiotic stresses including salt, dehydration and chilling in *Arabidopsis*, rice, tobacco and *Brassica juncea* (Das-Chatterjee *et al.*, 2006; Goswami *et al.*, 2014; Joshi *et al.*, 2013; Kaur *et al.*, 2013; Tan *et al.*, 2013). Expression of *MIPS* was increased in resistant soya bean and decreased in susceptible soya bean after *Fusarium solani* infection (Iqbal *et al.*, 2002). However, its role in resistance to biotic stresses has not been reported.

In our previous study, the *IbMIPS1* gene was isolated from sweet potato cv. Nongda 603 and its expression was strongly induced after stem nematode infection, indicating this gene may play an important role in responses against nematode infection (Zhai and Liu, 2009), but its function has not been characterized.

In this study, we developed the *IbMIPS1*-overexpressing sweet potato plants and found that *IbMIPS1* significantly enhanced salt and drought tolerance and stem nematode resistance of the transgenic plants.

Results

Expression of *IbMIPS1* is induced by NaCl, PEG, ABA and stem nematodes

Real-time quantitative PCR (qRT-PCR) analysis showed that the expression of *IbMIPS1* in sweet potato cv. Nongda 603 was significantly induced by NaCl, polyethylene glycol (PEG)6000, abscisic acid (ABA) and stem nematode stresses (Figure S1). An increase in the *IbMIPS1* transcript was observed after 2 h of exposure to 200 mM NaCl, peaked at 6 h with the 9.04-fold higher expression level than that of untreated control and then declined. For 30% PEG6000 stress, the expression of *IbMIPS1* was induced to the highest level (2.68-folds) at 12 h, followed by a decrease. The expression of *IbMIPS1* reached highest level (3.39-folds) at 6 h of 100 μM ABA treatment. The highest level of *IbMIPS1* expression up to 52.79-folds was observed at 4 day of stem nematode infection (Figure S1).

Production of the *IbMIPS1*-overexpressing sweet potato plants

A total of 1250 cell aggregates of sweet potato cv. Lizixiang (Figure S2a) cocultivated with *Agrobacterium tumefaciens* formed 315 hygromycin (Hyg)-resistant calluses on Murashige and Skoog (MS) medium with 2.0 mg/L 2,4-dichlorophenoxyacetic acid (2,4-D), 100 mg/L carbenicillin (Carb) and 25 mg/L Hyg (Figure S2b). These calluses produced 418 putatively transgenic plants, named L1, L2, L3,…, L418, respectively, on MS medium with 1.0 mg/L ABA, 100 mg/L Carb and 25 mg/L Hyg (Figure S2c,d). GUS and PCR analyses showed that 342 of them were transgenic plants (Figure S2e–h).

Overexpression of *IbMIPS1* enhances salt tolerance

The randomly sampled 30 transgenic sweet potato plants were cultured on MS medium with 0 or 86 mM NaCl for 4 weeks. The growth and rooting of all plants were normal without NaCl (Table S1). There were no significant differences in proline and malondialdehyde (MDA) content and superoxide dismutase (SOD) activity between the transgenic plants and wild type (WT) (Table S1). And at 86 mM NaCl, the transgenic plants exhibited vigorous growth and good rooting in contrast to the poor-growing WT (Figure 1a; Table S1). This observation showed that the transgenic plants, especially L264, L316, L164, L328 and L258, had higher salt tolerance than WT. qRT-PCR analysis indicated that there was positive correlationship between expression level of *IbMIPS1* and salt tolerance of transgenic plants (Figure S3). Further analysis demonstrated that proline content and SOD activity were significantly higher, while MDA content was significantly lower in the 18 transgenic plants, especially L264, L316, L164, L328 and L258, than in WT (Table S1).

The 18 transgenic plants and WT showed 100% survival after they were transferred to the soil in a greenhouse and a field (Figure S2i,j). No morphological variations were observed. The cuttings of these 18 transgenic plants and WT were cultured for 4 weeks in the Hoagland solution containing 0 and 86 mM NaCl, respectively. The growth and rooting of all cuttings were normal without NaCl (Table S2). And at 86 mM NaCl, the 5 transgenic plants (L264, L316, L164, L328 and L258) formed obvious new

leaves and roots; the 4 transgenic plants survived, but failed to form new leaves; the 9 transgenic plants and WT gradually turned brown to death (Figure 1b–d; Table S2). This observation was in agreement with that of *in vitro* assay, in which L264, L316, L164, L328 and L258 exhibited significantly higher salt tolerance compared to the other transgenic plants and WT.

The 5 salt-tolerant transgenic plants and WT were grown in a transplanting box and irrigated with a 200 mL of 200 mM NaCl solution once every 2 days. After 2 weeks, the transgenic plants maintained significantly higher photosynthetic rate, stomatal conductance, transpiration rate and chlorophyll relative content, which were increased by 34%–77%, 31%–68%, 40%–78% and 34%–62%, respectively, compared to WT (Figure S4). After 4 weeks, the transgenic plants showed good growth and increased physical size and their fresh weight (FW) and dry weight (DW) were increased by 180%–322% and 28%–102%, respectively, while WT died (Figure 1e–h). The 5 salt-tolerant transgenic plants and WT were further planted in a saline field, and it was found that the transgenic plants showed good growth and produced storage roots, while WT died (Figure 1i–k).

Overexpression of *IbMIPS1* increases drought tolerance

For evaluating drought tolerance of the 5 salt-tolerant transgenic plants, the cuttings of the transgenic plants and WT were cultured for 2 weeks in the Hoagland solution containing 30% PEG6000, and then recovered for 2 weeks in Hoagland solution without PEG6000. The transgenic plants recovered quickly and formed obvious new leaves and roots, while WT gradually turned brown to death (Figure 2a–c; Table S3). FW and DW of the transgenic plants were increased by 261%–333% and 42%–74%, respectively, compared to WT (Figure 2d).

The 5 transgenic plants and WT were grown in a transplanting box for evaluation of drought tolerance. There were no significant differences in SOD activity and proline, relative water, soluble sugar and MDA content between the transgenic plants and WT without drought stress (Figure S5). After 4 weeks of drought stress, the transgenic plants showed better growth and their photosynthetic rate, stomatal conductance, transpiration rate and chlorophyll relative content were increased by 86%–131%, 37%–57%, 56%–93% and 32%–49%, respectively, compared to WT (Figure S4). SOD activity and proline, relative water and soluble sugar content were significantly increased, whereas MDA content was significantly decreased in the transgenic plants compared to WT (Figure S5). After 6 weeks of drought stress, the transgenic plants showed good growth and increased physical size and their FW and DW were increased by 43%–225% and 18%–59%, respectively, while WT died (Figure 2e–h). These tolerant plants were further evaluated in the drought stress facility. Following drought stress, WT gradually turned brown to death, while the transgenic plants showed good growth and produced storage roots (Figure 2i–l).

Overexpression of *IbMIPS1* improves stem nematode resistance

The 18 transgenic plants which exhibited higher salt tolerance based on *in vitro* assay were identified for their stem nematode resistance by stem nematode inoculation test. The results showed that 12 transgenic plants were resistant, 3 middle resistant and 3 susceptible to stem nematodes (Figure 3a; Table S4). Storage roots of WT and 3 susceptible plants (L140, L248 and L44) were

Figure 1 Responses of the IbMIPS1-overexpressing sweet potato plants under salt stress. (a) The growth and rooting of transgenic plants and WT cultured for 4 weeks on MS medium with 86 mM NaCl. (b) Transgenic plants and WT before treatment in Hoagland solution. (c) and (d) Phenotypes of transgenic plants and WT incubated for 4 weeks in Hoagland solution with 86 mM NaCl. (e) Transgenic plants and WT before treatment in a transplanting box. (f), (g) and (h) Phenotypes, FW and DW of transgenic plants and WT grown in a transplanting box under 200 mM NaCl stress. (i), (j) and (k) Performance and yield of salt-tolerant transgenic plants and WT planted in a saline field. * and ** indicate a significant difference from that of WT at $P < 0.05$ and <0.01, respectively, by Student's t-test.

Figure 2 Responses of the salt-tolerant transgenic sweet potato plants under drought stress. (a) Transgenic plants and WT before treatment in Hoagland solution. (b), (c) and (d) Phenotypes, FW and DW of transgenic plants and WT incubated for 2 weeks in Hoagland solution with 30% PEG6000, then recovered for 2 weeks in Hoagland solution. (e) Transgenic plants and WT before treatment in a transplanting box. (f), (g) and (h) Phenotypes, FW and DW of transgenic plants and WT grown in a transplanting box after 6 weeks of drought stress. (i) Transgenic plants and WT before treatment in the drought stress facility. (j), (k) and (l) Performance and yield of drought-tolerant transgenic plants and WT planted in the drought stress facility. * and ** indicate a significant difference from that of WT at $P < 0.05$ and <0.01, respectively, by Student's t-test.

rotten almost completely; the rotten areas of 3 middle resistant plants (L42, L90 and L234) reached 16–26 mm from the hole; the inoculated nematodes did not spread in storage roots of 4 resistant plants (L264, L316, L164 and L328) except for the nematodes spread to 2–8 mm from the role in the remaining 8 resistant plants (L258, L220, L238, L142, L178, L210, L100 and L2).

qRT-PCR analysis indicated that there was positive correlationship between expression level of *IbMIPS1* and stem nematode resistance of transgenic plants (Figure S6). The 12 resistant plants showed significantly higher *IbMIPS1* expression level and IbMIPS1 activity than middle resistant, susceptible and WT plants after stem nematode infection. Further analysis demonstrated that SOD, glutathione peroxidase (GPX), catalase (CAT), ascorbate peroxidase (APX) and peroxidase (POD) activities were significantly higher, while MDA content was significantly lower in the resistant plants than in WT after stem nematode infection (Table S5). Without infection, the resistant plants and WT showed no significant differences in these physiological parameters (Table S5).

The resistant plants were planted in a field infected by stem nematodes, and their stem nematode resistance was further evaluated. The results were in agreement with those of stem nematode inoculation test (Figure 3b).

Overexpression of *IbMIPS1* up-regulates the related genes

Transcriptome analysis of L264, tolerant to salt and drought and resistant to nematodes, was conducted using RNA-Seq technol-ogy. A total of 536 genes (54%) were up-regulated (more than twofold changes at $P < 0.05$) and 458 genes (46%) were down-regulated (more than twofold changes at $P < 0.05$) in L264 compared to WT under 86 mm NaCl stress. Of the 536 up-regulated genes, 39% were unknown genes; 9% were involved in responses to stresses/defences; others were related to metab-olism, protein fate, signal transduction, transcription regulation, cell structure, energy, etc. (Figure S7a). A total of 804 genes (42%) were up-regulated and 1120 genes (58%) was down-regulated in L264 compared to WT after stem nematode infection. Of the 804 up-regulated genes, 25% were unknown genes; 7% were involved in responses to diseases/defences; others were related to metabolism, protein fate, signal transduc-tion, transcription regulation, cell growth/division, etc. (Fig-ure S7b).

Further analysis showed that the genes related to inositol biosynthesis, phosphatidylinositol (PI) and ABA signalling path-ways, stress responses and ROS-scavenging were up-regulated under salt and nematode stresses (Table 1). In addition, the expression level of *psbA* and *PRK* genes encoding D1 protein and phosphoribulokinase (PRKase), respectively, was increased by 1.70-folds and 1.65-folds, respectively, in L264 compared to WT under salt stress. These results indicated that *IbMIPS1* might be involved in multiple processes of expression regulation.

To verify the transcriptome data, qRT-PCR was carried out using the 5 resistant plants (L264, L316, L164, L328 and L258) and WT under salt, drought and stem nematode stresses. The results were in agreement with those of transcriptome analysis (Figure S8).

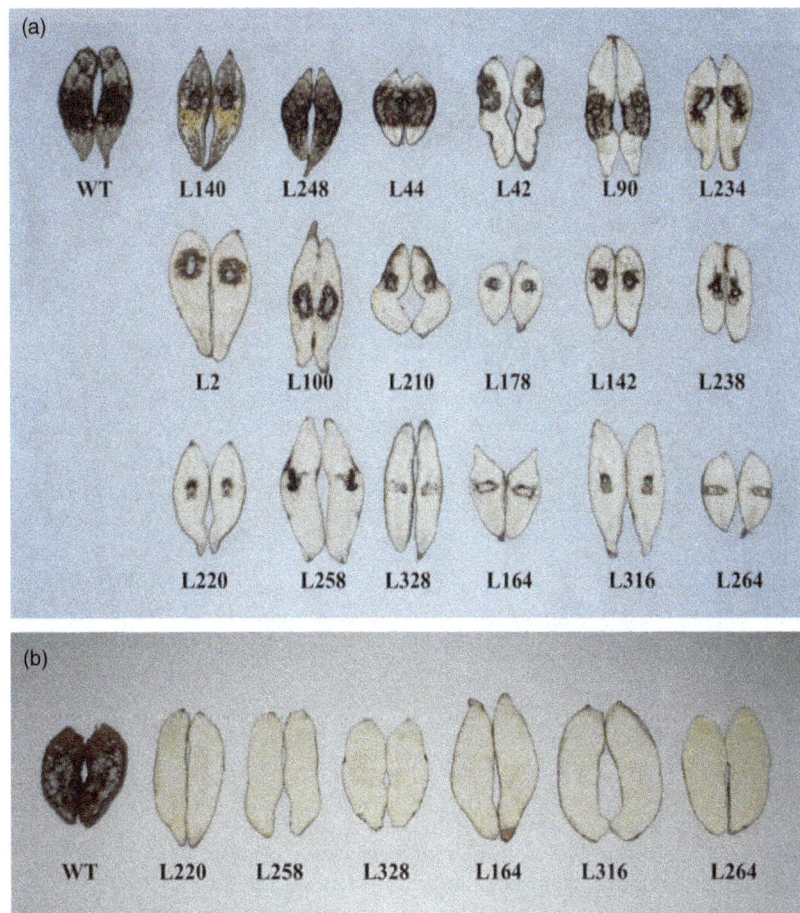

Figure 3 Assay for stem nematode resistance of the IbMIPS1-overexpressing sweet potato plants. (a) Storage roots of the transgenic plants and WT displaying different reactions to stem nematodes after 4 weeks of inoculation. L140, L248 and L44, susceptible; L42, L90 and L234, middle resistant; L2, L100, L210, L178, L142, L238, L220, L258, L238, L164, L316 and L264, resistant. (b) Storage roots of the transgenic plants and WT displaying different reactions to stem nematode in the field infected by stem nematodes.

Table 1 Stress-related genes up-regulated in L264 under salt or stem nematode stress

Gene	Fold increase		Stress response[‡]
	Salt stress	Stem nematode stress	
MIPS, *myo*-inositol-1-phosphate synthase	13.02*	12.08	A, Di, Dr, S
MIPP, *myo*-inositol monophosphatase	3.07	4.10	
PIS, phosphatidylinositol synthase	2.55	7.12	A, Dr, S
PI4K, phosphatidylinositol-4-kinase	3.50	11.33	
PIP5K, phosphatidylinositol-4-phosphate 5-kinase	2.10	6.43	
PLC, phospholipase C	4.28	4.76	A
DGK, diacylglycerol kinase	2.86	9.31	Di, Dr, S
PLD, phospholipase D	3.03	8.16	A
ZEP, zeaxanthin epoxidase	2.18	2.50	A, Dr, S
NCED, 9-cis-epoxycarotenoid dioxygenase	9.21	8.85	Dr, S
ABA2, xanthoxin dehydrogenase	2.63	2.86	Dr, S
TPS, trehalose-6-phosphate synthase	2.75	–[†]	A, Dr, S
TPP, trehalose-6-phosphate phosphatase	4.78	–	A, Dr, S
P5CS, pyrroline-5-carboxylate synthase	3.12	–	A, Dr, S
P5CR, pyrroline-5-carboxylate reductase	4.75	–	A, Dr, S
NHX, Na^+/H^+ antiporter cation exchanger	2.01	–	A, S
LEA, late embryogenesis abundant protein	4.06	–	A, Dr, S
CAS, callose synthase	–	12.73	A, Di
HRGP, hydroxyproline-rich glycoprotein	–	4.95	Di
LTP, lipid transfer protein	–	2.80	Di
LRP, leucine-rich repeat protein	–	10.68	Di
CPI, cysteine proteinase inhibitor	–	99.43	Di
PI, proteinase inhibitor	–	12.91	Di
AMP, anti-microbial protein	–	11.46	Di
SOD, superoxide dismutase	3.19	3.29	Di, Dr, S
GPX, glutathione peroxidase	3.23	3.74	Di, Dr, S
CAT, catalase	3.66	3.47	Di, Dr, S
APX, ascorbate peroxidase	2.66	6.82	Di, Dr, S
POD, peroxidase	2.93	9.35	Di, Dr, S

*Numbers in boldface indicate up-regulation by more than twofold changes ($P < 0.05$).

[†]No obvious change.

[‡]Genes responsive to ABA (A), disease (Di), drought (Dr) and salt (S) stress are based on transcriptome analysis.

In addition, Southern blot analysis indicated that the transgenic plants displayed different integration patterns and the increased copy number of integrated *IbMIPS1* gene varied from 1 to 3, but there was no clear relationship between expression levels of related genes and copy number of integrated *IbMIPS1* gene (Figure S9). Clear relationship between salt and drought tolerance and stem nematode resistance and the copy number was not also found (Figure S9).

Overexpression of *IbMIPS1* alters the content of the related components

Inositol, inositol-1,4,5-triphosphate (IP_3), phosphatidic acid (PA), Ca^{2+}, ABA, trehalose and K^+ content and K^+/Na^+ values were significantly increased, while H_2O_2 and Na^+ content was significantly decreased in the resistant plants compared to WT under salt stress (Figure S10). Under drought stress, inositol, IP_3, PA, Ca^{2+}, ABA and trehalose content were significantly increased, while H_2O_2 was significantly less accumulated in the resistant plants compared to WT (Figure S10).

After stem nematode infection, inositol, PA, Ca^{2+}, ABA, callose and lignin content were significantly increased in the resistant plants compared to WT (Figure S11). The continuous increase of IP_3 content in the resistant plants was found after infection (Figure S12). H_2O_2 levels exhibited a rapid increase, peaked at 1 to 2 days and then declined (Figure S12). Sand column assay showed that the percentage of movement of stem nematodes was significantly decreased under H_2O_2 treatment, indicating that H_2O_2 could reduce the activity of nematodes, but the percentage had no obvious change under different concentrations of IP_3 (Figure S13).

Discussion

Overexpression of *IbMIPS1* enhances salt and drought tolerance and stem nematode resistance

The *MIPS* gene has been shown to enhance tolerance to abiotic stresses including salt, dehydration and chilling in several plant species (Das-Chatterjee et al., 2006; Goswami et al., 2014; Joshi et al., 2013; Kaur et al., 2013; Tan et al., 2013). However, its role in resistance to biotic stresses is still unclear. This study has indicated that the expression of *IbMIPS1* in sweet potato was strongly induced by NaCl, PEG, ABA and stem nematodes (Figure S1). Its overexpression significantly enhanced stem nematode resistance as well as salt and drought tolerance in the transgenic sweet potato (Figures 1, 2 and 3).

Overexpression of *IbMIPS1* up-regulates the genes involved in PI and ABA signalling pathways

Myo-inositol-1-phosphate synthase is a key rate limiting enzyme of inositol biosynthesis (Abreu and Aragão, 2007). Inositol generates the second messengers diacylglycerol (DAG) and IP_3 in PI signalling pathway (Valluru and Ende, 2011). DAG is rapidly converted to PA by DGK (Liu et al., 2013). PA can also be directly formed by the action of PLD (Testerink and Munnik, 2011). PA was involved in ABA signal transduction (Katagiri et al., 2005; Liu et al., 2013; Testerink and Munnik, 2011). In addition, the increase of IP_3 levels can stimulate the release of Ca^{2+} from internal stores under various stresses (Nie et al., 2014; Zhu, 2002). Ca^{2+} is known to mediate early ABA signalling (Kudla et al., 2010; Zhu, 2002).

Overexpression of *MIPS* in *B. juncea*, tobacco, *Arabidopsis* and rice enhanced the tolerance to salt, dehydration and chilling due to the increased production of inositol (Das-Chatterjee et al., 2006; Goswami et al., 2014; Kaur et al., 2013; Majee et al., 2004; Tan et al., 2013). In this study, inositol, IP_3, PA, Ca^{2+} and ABA levels were significantly increased and the genes (*PIS*, *PI4K*, *PIP5K*, *PLC*, *DGK*, *PLD*, *ZEP*, *NCED* and *ABA2*) involved in PI and ABA signalling pathways were systematically up-regulated in the *IbMIPS1*-overexpressing sweet potato plants under salt, drought and nematode stresses (Figures S8, S10 and S11). Therefore, it is thought that overexpression of *IbMIPS1* enhances the resistance to salt, drought and nematodes due to the up-regulation of genes involved in PI and ABA signalling pathways, which increase the production of signalling molecules and further the expression of resistance-responsive genes (Figure 4).

Overexpression of *IbMIPS1* up-regulates the resistance-responsive genes

It has been reported that ABA regulates the expression of abiotic stress-responsive genes (*P5CS, P5CR, TPS, TPP, NHX* and *LEA*) and biotic stress-responsive genes (*CAS, POD, HRGP, LTP, LRP1, CPI, PI* and *AMP*) in several plant species (Abraham *et al.*, 2003; Asselbergh *et al.*, 2008; Dalal *et al.*, 2009; Gao *et al.*, 2009, 2013; Hildmann *et al.*, 1992; Iturriaga *et al.*, 2009; Jung *et al.*, 2004; Lee and Hwang, 2009; Muñoz-Mayor *et al.*, 2012; Pramanik and Imai, 2005; Schluepmann and Paulb, 2009; Song *et al.*, 2011; Sripinyowanich *et al.*, 2013; Sun *et al.*, 2014; Tseng *et al.*, 2013; Yokoi *et al.*, 2002; Zhu, 2002). The application of ABA has been shown to enhance the resistance to *Pythium irregulare* in *Arabidopsis* (Adie *et al.*, 2007), necrosis virus in tobacco (Iriti and Faoro, 2008) and *Alternaria solani* in tomato (Song *et al.*, 2011). In the present study, the resistance-responsive genes mentioned above were significantly up-regulated in the *IbMIPS1*-overexpressing sweet potato plants under salt, drought and nematodes (Figure S8). Our results support that high ABA levels up-regulate these resistance-responsive genes, which lead to the enhanced resistance to salt, drought and nematodes in the *IbMIPS1*-overexpressing sweet potato plants (Figure 4).

In addition, it is reported that Ca^{2+} can directly induce the expression of the resistance-responsive genes, resulting in the improvement of resistance (Kanchiswamy *et al.*, 2014; Zhu, 2002). In *Arabidopsis*, Ca^{2+} was described as a necessary but not sufficient component in mediating the molecular events associated with hyperosmotic and salt induction of *P5CS* (Knight *et al.*, 1997). The increase of Ca^{2+} levels elevated P5CR activity under water stress in maize and wheat (Nayyar, 2003). Ca^{2+}

addition was required for CAS activity (Aidemark *et al.*, 2009). *StLTPa7* was induced by Ca^{2+} in potato (Gao *et al.*, 2009). Thus, the up-regulation of the resistance-responsive genes in the *IbMIPS1*-overexpressing sweet potato plants might be also due to the increase of Ca^{2+} (Figures S8, Figure 4).

Overexpression of *IbMIPS1* alters the resistance-associated components

In plants, the up-regulation of *P5CS, P5CR, TPS, TPP* and *NHX* has been found to increase proline and trehalose content and K^+/Na^+ ratio, which result in the enhanced salt and drought tolerance (Krasensky and Jonak, 2012; Li *et al.*, 2011; Liu *et al.*, 2014a; Qiu, 2012; Yue *et al.*, 2012). We also found that the *IbMIPS1*-overexpressing sweet potato plants had significantly higher proline and trehalose content and K^+/Na^+ ratio compared to WT under salt and drought stresses (Table S1; Figures S5 and S10), indicating the marked improvement of their salt and drought tolerance. Furthermore, the *IbMIPS1*-overexpressing sweet potato plants exhibited the increased photosynthesis capacity (Figure S4). Also, *psbA* and *PRK* genes were up-regulated in the transgenic plants (Figure S8). The increased photosynthesis is thought to be due to more accumulation of proline, which might provide protection against photoinhibition in the transgenic plants under salt and drought stresses (Liu *et al.*, 2014a; Zhang *et al.*, 2012). In addition, the systematic up-regulation of ROS-scavenging genes was found in the transgenic plants under salt and drought stresses (Figure S8). Our results support that more proline accumulation stimulates ROS-scavenging system, which leads to the improved salt and drought tolerance in the *IbMIPS1*-overexpressing sweet potato plants (Liu *et al.*, 2014a,b).

H_2O_2 plays an important role in plant defence mechanisms (Desikan *et al.*, 2005; Mai *et al.*, 2013; Zebelo and Maffei, 2015). Orozco-Cardenas and Ryan (1999) found that during herbivore attack, H_2O_2 rapidly increased, maximized at 4–6 h, then declined in the wounded leaves of tomato and thought that the elevated H_2O_2 levels could potentiate the plants' defence responses against invading pathogens. In our study, H_2O_2 levels

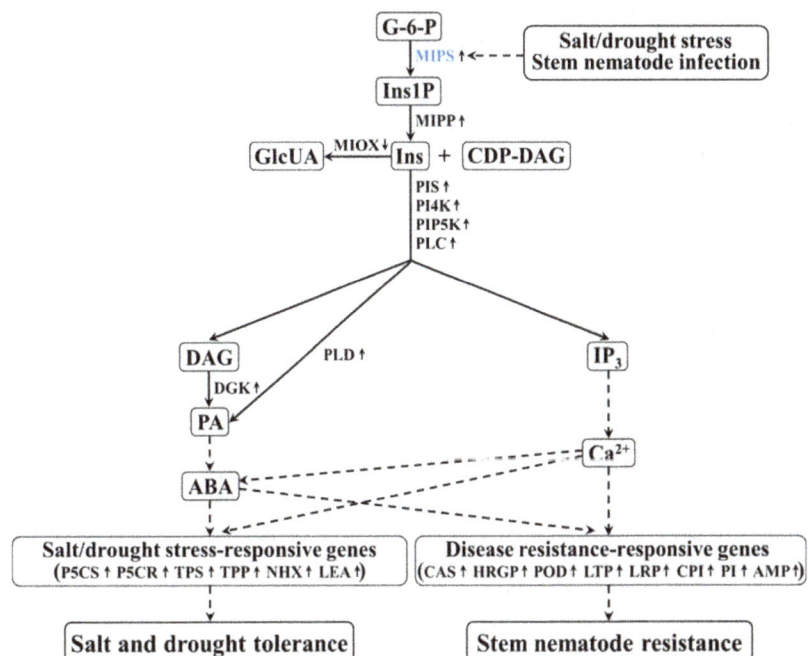

Figure 4 Diagram showing the regulation of inositol biosynthesis, PI signalling pathway, ABA signalling pathway and stress responses in the IbMIPS1-overexpressing sweet potato plants. Biosynthesis pathways are shown with solid arrows and regulatory interactions are shown with broken arrows. ↑ and ↓ indicate up-regulation and down-regulation of genes coding these enzymes (proteins), respectively.

exhibited a rapid increase, peaked at 1 to 2 days and then declined (Figure S12). The early, strong generation of H_2O_2 might inhibit the invasion of stem nematodes in the *IbMIPS1*-overexpressing sweet potato plants. Sand column assay provided further evidence that H_2O_2 treatment can reduce the activity of stem nematodes (Figure S13). The early, strong generation of H_2O_2 has been also shown to prevent aphid attack in pea (Mai *et al.*, 2013). Therefore, it is proposed that excessive H_2O_2 generated at the early invasion stage is an important factor in resistance against stem nematodes in the *IbMIPS1*-overexpressing sweet potato plants.

The deposition of callose and lignin plays an essential role in the defence response to invading pathogens (Miedes *et al.*, 2014; Naumann *et al.*, 2013). The up-regulation of *CAS* and *POD* has been demonstrated to enhance callose and lignin synthesis in several plant species (Hong *et al.*, 2001; Kim *et al.*, 2008; Lagrimini, 1991; Wally and Punja, 2010). In the present study, the *IbMIPS1*-overexpressing sweet potato plants exhibited significantly higher callose and lignin levels compared to WT after stem nematode infection, indicating measurable improvement of stem nematode resistance (Figure S11).

In conclusion, we found novel functions of the *IbMIPS1* gene in the resistance to plant diseases. Overexpression of *IbMIPS1* significantly enhanced stem nematode resistance as well as salt and drought tolerance in the transgenic sweet potato. These findings suggest that the *IbMIPS1* gene may be used to improve biotic and abiotic resistance in sweet potato and other plants.

Experimental procedures

Plant materials

Sweet potato cvs. Nongda 603 and Lizixiang were used in this study. Nongda 603 was resistant to stem nematodes and was employed for the cloning (Zhai and Liu, 2009) and expression analysis of *IbMIPS1*. Lizixiang was sensitive to salt and drought and susceptible to stem nematodes and was used to characterize the function of *IbMIPS1*.

Expression analysis of *IbMIPS1*

The 4-week-old *in vitro* grown plants of Nongda 603 were submerged in 1/2 MS medium containing 200 mM NaCl, 30% PEG6000 and 100 µM ABA, respectively, and sampled at 0, 2, 4, 6, 12, 24 and 48 h after treatment to analyse expression of *IbMIPS1*. The storage roots of Nongda 603 were inoculated with stem nematodes according to the method of Gao *et al.* (2011a) and sampled at 0 h, 6 h, 1, 2, 4, 6, 8 and 10 days after treatment to analyse expression of *IbMIPS1*. Total RNA was isolated using RNAprep Pure Plant Kit (Tiangen Biotech, Beijing, China) and first-strand cDNA was prepared by Quantscript Reverse Transcriptase Kit (Tiangen Biotech). qRT-PCR was performed as described by Liu *et al.* (2014a). The primers used for PCR were listed in Table S6.

Production of transgenic plants

Agrobacterium tumefaciens strain EHA105 harbouring a binary vector, plasmid pCAMBIA1301, was used in this study. This vector contained *IbMIPS1* gene under the control of CaMV 35S promoter and NOS terminator of the expression box and *gusA* and *hptll*genes driven by a CaMV 35S promoter, respectively. Transgenic plants were produced according to the method of Yu *et al.* (2007). Histochemical GUS assay was conducted as described by Jefferson *et al.* (1987). Genomic DNA was extracted using EasyPure Plant Genomic DNA Kit (Transgen Biotech, Beijing, China). PCR amplifications were performed with an initial denaturation at 95 °C for 5 min, followed by 35 cycles at 94 °C for 30 s, 55 °C for 30 s, 72 °C for 90 s and final extension at 72 °C for 5 min. The specific primers for *IbMIPS1* were listed in Table S6.

Assay for salt tolerance

In vitro assay for salt tolerance was conducted using MS medium with 86 mM NaCl at 27 ± 1 °C under 13 h of cool-white fluorescent light at 54 µM/m^2/s. Three plants were treated for each line. The growth and rooting ability were continuously observed for 4 weeks. Proline and MDA content and SOD activity were analysed as described by Gao *et al.* (2011c). The transgenic plants and WT were transferred to soils in a greenhouse and in a field. The cuttings about 25 cm in length from 6-week-old plants grown in a field were cultured in the Hoagland solution (Hoagland and Arnon, 1950) with 0 and 86 mM NaCl, respectively. Hoagland solution was renewed once every 3 days. Three cuttings were treated for each line. The growth and rooting ability were continuously observed for 4 weeks. The cuttings of the 6-week-old transgenic plants and WT grown in a field were planted in a transplanting box containing a mixture of soil, vermiculite and humus (1 : 1 : 1, v/v/v) in a greenhouse. Evaluation for salt tolerance and measurements of photosynthetic rate, stomatal conductance and transpiration rate and relative chlorophyll content were conducted according to the method of Liu *et al.* (2014a).

The salt-tolerant transgenic plants and WT were planted in a saline field (EC$_{1 : 5}$ = 1517–1803 µS/cm, annual rainfall 634.7 mm, 118°53′44″N, 37°88′38″E), Dongying Experimental Station, Shandong, China. The field trial was arranged in a randomized complete block design (RCBD) with three replications per experiment. Each plot contained 10 plants spaced at 25 cm within rows and 80 cm between rows. All of the experiments were conducted under natural field conditions. Approximately 120 days after planting, the plants were harvested and FW of storage roots were measured.

Assay for drought tolerance

The 25-cm-long cuttings of the 6-week-old salt-tolerant transgenic plants and WT grown in a field were cultured in the Hoagland solution with 30% PEG6000 for 2 weeks, then returned to normal growing conditions for 2 weeks. Hoagland solution was renewed once every 3 days. Three cuttings were treated for each line. The growth and rooting ability were continuously observed for 4 weeks, and FW was measured immediately. The plants were then dried for 24 h in an oven at 80 °C and weighed (DW). The cuttings of the 6-week-old transgenic plants and WT grown in a field were planted in a transplanting box. All plants were irrigated sufficiently with half-Hoagland solution for 10 days until the cuttings formed new leaves and then were subjected to drought stress without water for 6 weeks. After 4 weeks, photosynthetic rate, stomatal conductance, transpiration rate, relative chlorophyll, proline and MDA content and SOD activity in the leaves of the transgenic plants and WT were measured according to the method of Liu *et al.* (2014a). The relative water and soluble sugar content were analysed as described by Yang *et al.* (2009). After 6 weeks, FW and DW were measured.

The drought-tolerant transgenic plants and WT were planted in a drought stress facility of Shangzhuang Experimental Station, Beijing, China. Three plants per line were planted with a space at

25 cm within rows and 80 cm between rows. At 30 days following planting, plants were subjected to drought stress for 90 days. The plants were harvested and FW of storage roots were measured.

Assay for stem nematode resistance

Transgenic plants were identified for stem nematode resistance by nematode inoculation test based on the method of Gao et al. (2011a). After 4 weeks of inoculation, IbMIPS1 activity was analysed as described by Barnett et al. (1970) and Bradford (1976). MDA content and SOD, GPX, CAT, APX and POD activities were measured according to the methods of Gao et al. (2011c), He et al. (2009), Wilson et al. (1989), Yang et al. (2009), Nakano and Asada (1981) and Luriea et al. (1997), respectively.

The resistant transgenic plants were planted in a field infected by stem nematodes at a density of 300 nematodes per 100 g soil and their stem nematode resistance was evaluated for 2 years according to the method of Gao et al. (2011a).

Transcriptome analysis

Total RNA was extracted from transgenic plants and WT cultured for 4 weeks on MS medium with 86 mM NaCl and their storage roots infected by stem nematodes for 4 weeks by Trizol method (Invitrogen, Carlsbad, CA) and 10 μg RNA was used for Illumina RNA-seq. Transcriptome sequencing, de novo transcriptome assembly and evaluation were performed by Beijing SinoGenoMax (Beijing, China). The differentially expressed transcripts (≥200 bp) with more than twofold changes in the transgenic plants compared to WT were submitted for homology and annotation searches using Blast2GO software v2.4.4 (Conesa et al., 2005). For BLASTX against the NR database, the threshold was set to E-value $(P) < 10^{-5}$. GO classification was achieved using WEGO software (Ye et al., 2006). Kyoto Encyclopedia of Genes and Genomes (KEGG) (Kanehisa et al., 2004) pathways were retrieved from KEGG web server (http://www.genome.jp/kegg/).

Expression analysis of the related genes

Leaves of pot-grown transgenic and WT plants treated for 2 weeks with 200 mM NaCl or for 4 weeks by drought stress and the storage roots infected by stem nematodes for 4 weeks were used to analyse expression of the genes related to inositol biosynthesis, PI and ABA signalling pathways, stress responses, photosynthesis and ROS-scavenging as described above. The primers for PCR were listed in Table S6.

Southern blot analysis

Southern blot analysis was conducted as described by Liu et al. (2014a). Coding sequence of the 578 bp IbMIPS1 was used as probe (Table S6). The labelling of probe, prehybridization, hybridization and detection were performed using DIG High Prime DNA Labeling and Detection Starter Kit II (Roche, Basel, Switzerland).

Measurements of the related components

The content of inositol, IP$_3$, Ca^{2+}, ABA, trehalose, H$_2$O$_2$, K$^+$ and Na$^+$ in the leaves of pot-grown transgenic and WT plants treated for 2 weeks with 200 mM NaCl or for 4 weeks by drought stress was measured according to the methods of Bieleski and Redgwell (1977), Xiong et al. (2001), Baisakh et al. (2012), Gao et al. (2011c), Jiang et al. (2014), Alexieva et al. (2001) and Storey

(1995), respectively. PA was extracted following the protocol of Welti et al. (2002), and its content was measured with Plant PA Test Kit (Uscn Life Science Inc., China) (Uscn Life Science Inc., Wuhan, China).

The content of inositol, PA, Ca^{2+} and ABA in the storage roots of transgenic and WT plants infected by stem nematodes for 4 weeks was measured as described above. Callose and lignin content was measured according to the methods of Köhle et al. (1985) and Syros et al. (2004), respectively. Changes of IP$_3$ and H$_2$O$_2$ content in the storage roots after different times of infection were analysed as described by Xiong et al. (2001) and Alexieva et al. (2001), respectively. Sand column assay to assess movement of stem nematodes was based on the method of Twomey et al. (2002). The 300 stem nematodes were treated for each plastic tube with sand (low iron, 40–100 μm particle size). The sand was soaked with 2.5 mL of 0, 50, 100, 200 and 300 μM IP$_3$ or 0, 50, 100, 200 and 300 mM H$_2$O$_2$. The number of stem nematodes which migrated through sand columns into dishes was counted after 24 h at 25 °C.

Statistical analysis

The experiments were repeated three times and the data presented as the mean ± SE were analysed by Student's t-test in a two-tailed analysis. A value of $P < 0.05$ or <0.01 was considered to be statistically significant.

Acknowledgements

We thank Dr. Daniel Q. Tong, University of Maryland, USA, for English improvement. This work was supported by National Natural Science Foundation of China (31271777) and China Agriculture Research System (CARS-11).

References

Abraham, E., Rigo, G., Szekely, G., Nagy, R., Koncz, C. and Szabados, L. (2003) Light-dependent induction of proline biosynthesis by abscisic acid and salt stress is inhibited by brassinosteroid in Arabidopsis. Plant Mol. Biol. **51**, 363–372.

Abreu, E.F.M. and Aragão, F.J. (2007) Isolation and characterization of a myo-inositol-1-phosphate synthase gene from yellow passion fruit (Passiflora edulis f. flavicarpa) expressed during seed development and environmental stress. Ann. Bot. **99**, 285–292.

Adie, B.A.T., Pérez-Pérez, J., Pérez-Pérez, M.M., Godoy, M., Sánchez-Serrano, J.J., Schmelz, E.A. and Solano, R. (2007) ABA is an essential signal for plant resistance to pathogens affecting JA biosynthesis and the activation of defenses in Arabidopsis. Plant Cell, **19**, 1665–1681.

Aidemark, M., Andersson, C.J., Rasmusson, A.G. and Widell, S. (2009) Regulation of callose synthase activity in situ in alamethicin-permeabilized Arabidopsis and tobacco suspension cells. BMC Plant Biol. **12**, 9–27.

Alexieva, V., Sergiev, I., Mapelli, S. and Karanov, E. (2001) The effect of drought and ultraviolet radiation on growth and stress markers in pea and wheat. Plant, Cell Environ. **24**, 1337–1344.

Asselbergh, B., Achuo, A.E., Hofte, M. and Van Gijsegem, F. (2008) Abscisic acid deficiency leads to rapid activation of tomato defence responses upon infection with Erwinia chrysanthemi. Mol. Plant Pathol. **9**, 11–24.

Baisakh, N., RamanaRao, M.V., Rajasekaran, K., Subudhi, P., Janda, J., Galbraith, D., Vanier, C. and Pereira, A. (2012) Enhanced salt stress tolerance of rice plants expressing a vacuolar H$^+$-ATPase subunit c1 (SaVHAc1) gene from the halophyte grass Spartina alterniflora Löisel. Plant Biotechnol. J. **10**, 453–464.

Barnett, J.E.G., Brice, R.E. and Corina, D.L. (1970) A colorimetric determination of inositol monophosphates as an assay for D-glucose 6-phosphate-1L-myoinositol 1-phosphate cyclase. Biochem. J. **119**, 183–186.

Bieleski, R.L. and Redgwell, R.J. (1977) Synthesis of sorbitol in apricot leaves. *Aust. J. Plant Physiol.* **4**, 1–10.

Bradford, M.M. (1976) A rapid and sensitive method for the quantitation of microgram quantities of protein utilizing the principle of protein-dye binding. *Anal. Biochem.* **72**, 248–254.

Conesa, A., Gotz, S., Garcia-Gomez, J.M., Terol, J., Talon, M. and Robles, M. (2005) Blast2GO: a universal tool for annotation, visualization and analysis in functional genomics research. *Bioinformatics*, **21**, 3674–3676.

Dalal, M., Tayal, D., Chinnusamy, V. and Bansal, K.C. (2009) Abiotic stress and ABA-inducible Group 4 *LEA* from *Brassica napus* plays a key role in salt and drought tolerance. *J. Biotechnol.* **139**, 137–145.

Das-Chatterjee, A., Goswami, L., Maitra, S., Dastidar, K.G., Ray, S. and Majumder, A.L. (2006) Introgression of a novel salt-tolerant L-*myo*-inositol 1-phosphate synthase from *Porteresia coarctata* (Roxb.) Tateoka (*PcINO1*) confers salt tolerance to evolutionary diverse organisms. *FEBS Lett.* **580**, 3980–3988.

Desikan, R., Hancock, J. and Neil, S. (2005) Reactive oxygen species as signalling molecules. In *Antioxidants and Reactive Oxygen Species in Plants* (Smirnoff, N., ed.), pp. 169–196. Oxford: Blackwell Publishing Ltd.

Fan, W.J., Zhang, M., Zhang, H.X. and Zhang, P. (2012) Improved tolerance to various abiotic stresses in transgenic sweet potato (*Ipomoea batatas*) expressing spinach betaine aldehyde dehydrogenase. *PLoS One*, **7**, e37344.

FAO. (2013) www.fao.org/giews/english/fo/index.htm.

Gao, G., Jin, L.P., Xie, K.Y. and Qu, D.Y. (2009) The potato *StLTPa7* gene displays a complex Ca^{2+}-associated pattern of expression during the early stage of potato–Ralstonia solanacearum interaction. *Mol. Plant Pathol.* **10**, 15–27.

Gao, S., Yu, B., Yuan, L., Zhai, H., He, S.Z. and Liu, Q.C. (2011a) Production of transgenic sweetpotato plants resistant to stem nematodes using oryzacystatin-I gene. *Sci. Hortic.* **128**, 408–414.

Gao, S., Yu, B., Zhai, H., He, S.Z. and Liu, Q.C. (2011b) Enhanced stem nematode resistance of transgenic sweetpotato plants expressing oryzacystatin-I gene. *Agric. Sci. China*, **10**, 519–525.

Gao, S., Yuan, L., Zhai, H., Liu, C.L., He, S.Z. and Liu, Q.C. (2011c) Transgenic sweetpotato plants expressing an *LOS5* gene are tolerant to salt stress. *Plant Cell, Tissue Organ Cult.* **107**, 205–213.

Gao, W.D., Bai, S., Li, Q.M., Gao, C.Q., Liu, G.F., Li, G.D. and Tan, F.L. (2013) Overexpression of *TaLEA* gene from *Tamarix androssowii* improves salt and drought tolerance in transgenic poplar (*Populus simonii* × *P. nigra*). *PLoS One*, **8**, e67462.

Goswami, L., Sengupta, S., Mukherjee, S., Ray, S., Mukherjee, R. and Majumder, A.L. (2014) Targeted expression of L-*myo*-inositol 1-phosphate synthase from *Porteresia coarctata* (Roxb.) Tateoka confers multiple stress tolerance in transgenic crop plants. *J. Plant Biochem. Biotechnol.* **23**, 316–330.

He, S.Z., Han, Y.F., Wang, Y.P., Zhai, H. and Liu, Q.C. (2009) In vitro selection and identification of sweetpotato (*Ipomoea batatas* (L.) Lam.) plants tolerant to NaCl. *Plant Cell, Tissue Organ Cult.* **96**, 69–74.

Hegeman, C.E., Good, L.L. and Grabau, E.A. (2001) Expression of D-*myo*-inositol-3-phosphate synthase in soybean. Implications for phytic acid biosynthesis. *Plant Physiol.* **125**, 1941–1948.

Hildmann, T., Ebneth, M., Peña-Cortés, H., Sánchez-Serrano, J.J., Willmitzer, L. and Prat, S. (1992) General roles of abscisic and jasmonic acids in gene activation as a result of mechanical wounding. *Plant Cell*, **4**, 1157–1170.

Hoagland, D.R. and Arnon, D.I. (1950) The water-culture method for growing plants without soil. *Calif. Agric. Exp. Stn. Circ.* **347**, 1–39.

Hong, Z.L., Delauney, A.J. and Verma, D.P.S. (2001) A cell plate-specific callose synthase and its interaction with phragmoplastin. *Plant Cell*, **13**, 755–768.

Iqbal, M.J., Afzal, A.J., Yaegashi, S., Ruben, E., Triwitayakorn, K., Njiti, V.N., Ahsan, R., Wood, A.J. and Lightfoot, D.A. (2002) A pyramid of loci for partial resistance to *Fusarium solani* f. sp. *glycines* maintains *myo*-inositol-1-phosphate synthase expression in soybean roots. *Theor. Appl. Genet.* **105**, 1115–1123.

Iriti, M. and Faoro, F. (2008) Abscisic acid is involved in chitosan-induced resistance to tobacco necrosis virus (TNV). *Plant Physiol. Biochem.* **46**, 1106–1111.

Iturriaga, G., Suárez, R. and Nova-Franco, B. (2009) Trehalose metabolism: from osmoprotection to signaling. *Int. J. Mol. Sci.* **10**, 3793–3810.

Jefferson, R.A., Kavanagh, T.A. and Bevan, M.W. (1987) GUS fusion: β-glucuronidase as a sensitive and versatile gene fusion marker in higher plants. *EMBO J.* **6**, 3901–3907.

Jiang, T., Zhai, H., Wang, F.B., Zhou, H.N., Si, Z.Z., He, S.Z. and Liu, Q.C. (2014) Cloning and characterization of a salt tolerance-associated gene encoding trehalose-6-phosphate synthase in sweetpotato. *J. Integr. Agric.* **13**, 1651–1661.

Johnson, M.D. (1994) The *Arabidopsis thaliana myo*-inositol 1-phosphate synthase (EC 5.5.1.4). *Plant Physiol.* **105**, 1023–1024.

Joshi, R., Ramanarao, M.V. and Baisakh, N. (2013) *Arabidopsis* plants constitutively overexpressing a *myo*-inositol 1-phosphate synthase gene (*SaINO1*) from the halophyte smooth cordgrass exhibits enhanced level of tolerance to salt stress. *Plant Physiol. Biochem.* **65**, 61–66.

Jung, E.H., Jung, H.W., Lee, S.C., Han, S.W., Heu, S. and Hwang, B.K. (2004) Identification of a novel pathogen-induced gene encoding a leucine-rich repeat protein expressed in phloem cells of *Capsicum annuum. Biochim. Biophys. Acta*, **1676**, 211–222.

Kanchiswamy, C.N., Malnoy, M., Occhipinti, A. and Maffei, M.E. (2014) Calcium imaging perspectives in plants. *Int. J. Mol. Sci.* **15**, 3842–3859.

Kanehisa, M., Goto, S., Kawashima, S., Okuno, Y. and Hattori, M. (2004) The KEGG resource for deciphering the genome. *Nucleic Acids Res.* **32**, 277–280.

Katagiri, T., Ishiyama, K., Kato, T., Tabata, S., Kobayashi, M. and Shinozaki, K. (2005) An important role of phosphatidic acid in ABA signaling during germination in *Arabidopsis thaliana. Plant J.* **43**, 107–117.

Kaur, H., Verma, P., Petla, B.P., Rao, V., Saxena, S.C. and Majee, M. (2013) Ectopic expression of the ABA-inducible dehydration-responsive chickpea L-*myo*-inositol 1-phosphate synthase 2 (*CaMIPS2*) in *Arabidopsis* enhances tolerance to salinity and dehydration stress. *Planta*, **237**, 321–335.

Kim, S.H., Ahn, Y.O., Ahn, M.J., Jeong, J.C., Lee, H.S. and Kwak, S.S. (2013a) Cloning and characterization of an orange gene that increases carotenoid accumulation and salt stress tolerance in transgenic sweetpotato cultures. *Plant Physiol. Biochem.* **70**, 445–454.

Kim, S.H., Kim, Y.H., Ahn, Y.O., Ahn, M.J., Jeong, J.C., Lee, H.S. and Kwak, S.S. (2013b) Downregulation of the lycopene ε-cyclase gene increases carotenoid synthesis via the β-branch-specific pathway and enhances salt-stress tolerance in sweetpotato calli. *Physiol. Plant.* **147**, 432–442.

Kim, Y.H., Kim, C.Y., Song, W.K., Park, D.S., Kwon, S.Y., Lee, H.S., Bang, J.W. and Kwak, S.S. (2008) Overexpression of sweetpotato *swpa4* peroxidase results in increased hydrogen peroxide production and enhances stress tolerance in tobacco. *Planta*, **227**, 867–881.

Knight, H., Trewavas, A.J. and Knight, M.R. (1997) Calcium signaling in *Arabidopsis thaliana* responding to drought and salinity. *Plant J.* **12**, 1067–1078.

Köhle, H., Jeblick, W., Poten, F., Blaschek, W. and Kauss, H. (1985) Chitosan-elicited callose synthesis in soybean as a Ca^{2+}-dependent process. *Plant Physiol.* **77**, 544–551.

Krasensky, J. and Jonak, C. (2012) Drought, salt, and temperature stress-induced metabolic rearrangements and regulatory networks. *J. Exp. Bot.* **63**, 1593–1608.

Kudla, J., Batistic, O. and Hashimoto, K. (2010) Calcium signals: the lead currency of plant information processing. *Plant Cell*, **22**, 541–563.

Lagrimini, L.M. (1991) Wound-induced deposition of polyphenols in transgenic plants overexpressing peroxidase. *Plant Physiol.* **96**, 577–583.

Larson, S.R. and Raboy, V. (1999) Linkage mapping of maize and barley *myo*-inositol 1-phosphate synthase DNA sequences: correspondence with a low phytic acid mutation. *Theor. Appl. Genet.* **99**, 27–36.

Lee, S.C. and Hwang, B.K. (2009) Functional roles of the pepper antimicrobial protein gene, *CaAMP1*, in abscisic acid signaling, and salt and drought tolerance in *Arabidopsis. Planta*, **229**, 383–391.

Li, H.W., Zang, B.S., Deng, X.W. and Wang, X.P. (2011) Overexpression of the trehalose-6-phosphate synthase gene *OsTPS1* enhances abiotic stress tolerance in rice. *Planta*, **234**, 1007–1018.

Liu, D.G., He, S.Z., Song, X.J., Zhai, H., Liu, N., Zhang, D.D., Ren, Z.T. and Liu, Q.C. (2015) *IbSIMT1*, a novel salt-induced methyltransferase gene from *Ipomoea batatas*, is involved in salt tolerance. *Plant Cell, Tissue Organ Cult.* **120**, 701–715.

Liu, D.G., He, S.Z., Zhai, H., Wang, L.J., Zhao, Y., Wang, B., Li, R.J. and Liu, Q.C. (2014a) Overexpression of *IbP5CR* enhances salt tolerance in transgenic sweetpotato. *Plant Cell, Tissue Organ Cult.* **117**, 1–16.

Liu, D.G., Wang, L.J., Liu, C.L., Song, X.J., He, S.Z., Zhai, H. and Liu, Q.C. (2014b) An *Ipomoea batatas* iron-sulfur cluster scaffold protein gene, *IbNFU1*, is involved in salt tolerance. *PLoS One*, **9**, e93935.

Liu, D.G., Wang, L.J., Zhai, H., Song, X.J., He, S.Z. and Liu, Q.C. (2014c) A novel α/β-hydrolase gene *IbMas* enhances salt tolerance in transgenic sweetpotato. *PLoS One*, **9**, e115128.

Liu, X.X., Zhai, S.M., Zhao, Y.J., Sun, B.C., Liu, C., Yang, A.F. and Zhang, J.R. (2013) Overexpression of the phosphatidylinositol synthase gene (*ZmPIS*) conferring drought stress tolerance by altering membrane lipid composition and increasing ABA synthesis in maize. *Plant, Cell Environ.* **36**, 1037–1055.

Luriea, S., Fallika, E., Handrosa, A. and Shapirab, R. (1997) The possible involvement of peroxidase in resistance to Botrytis cinerea in heat treated tomato fruit. *Physiol. Mol. Plant Pathol.* **50**, 141–149.

Mai, V.C., Bednarski, W., Borowiak-Sobkowiak, B., Wilkaniec, B., Samardakiewicz, S. and Morkunas, I. (2013) Oxidative stress in pea seedling leaves in response to *Acyrthosiphon pisum* infestation. *Phytochemistry*, **93**, 49–62.

Majee, M., Maitra, S., Dastidar, K.G., Pattnaik, S., Chatterjee, A., Hait, N.C., Das, K.P. and Majumder, A.L. (2004) A novel salt-tolerant L-*myo*-inositol-1-phosphate synthase from *Porteresia coarctata* (Roxb.) Tateoka, a halophytic wild rice: molecular cloning, bacterial overexpression, characterization, and functional introgression into tobacco-conferring salt tolerance phenotype. *J. Biol. Chem.* **279**, 28539–28552.

Miedes, E., Vanholme, R., Boerjan, W. and Molina, A. (2014) The role of the secondary cell wall in plant resistance to pathogens. *Front. Plant Sci.* **5**, a358.

Munns, R. and Tester, M. (2008) Mechanisms of salinity tolerance. *Annu. Rev. Plant Biol.* **59**, 651–681.

Muñoz-Mayor, A., Pineda, B., Garcia-Abellán, J.O., Antón, T., Garcia-sogo, B., Sanchez-Bel, P., Flores, F.B., Atarés, A., Angosto, T., Pintor-Toro, J.A., Moreno, V. and Bolarin, M.C. (2012) Overexpression of dehydrin *tas14* gene improves the osmotic stress imposed by drought and salinity in tomato. *J. Plant Physiol.* **169**, 459–468.

Nakano, Y. and Asada, K. (1981) Hydrogen peroxide is scavenged by ascorbate-specific peroxidase in spinach chloroplasts. *Plant Cell Physiol.* **22**, 867–880.

Naumann, M., Somerville, S.C. and Voigt, C.A. (2013) Differences in early callose deposition during adapted and non-adapted powdery mildew infection of resistant *Arabidopsis* lines. *Plant Signal. Behav.* **8**, e24408.

Nayyar, H. (2003) Accumulation of osmolytes and osmotic adjustment in waterstress wheat (*Triticum aestivum*) and maize (*Zea mays*) as affected by calcium and its antagonists. *Environ. Exp. Bot.* **50**, 253–264.

Nie, Y., Huang, F., Dong, S., Li, L., Gao, P., Zhao, H., Wang, Y. and Han, S. (2014) Identification of inositol 1,4,5-trisphosphate-binding proteins by heparin-agarose affinity purification and LTQ ORBITRAP MS in *Oryza sativa*. *Proteomics*, **14**, 2335–2338.

Orozco-Cardenas, M. and Ryan, C.A. (1999) Hydrogen peroxide is generated systemically in plant leaves by wounding and systemin via the octadecanoid pathway. *Proc. Natl Acad. Sci. USA*, **96**, 6553–6557.

Pramanik, M.H.R. and Imai, R. (2005) Functional identification of a trehalose 6-phosphate phosphatase gene that is involved in transient induction of trehalose biosynthesis during chilling stress in rice. *Plant Mol. Biol.* **58**, 751–762.

Qiu, Q.S. (2012) Plant and yeast NHX antiporters: roles in membrane trafficking. *J. Integr. Plant Biol.* **54**, 66–72.

Schluepmann, H. and Paulb, M. (2009) Trehalose metabolites in *Arabidopsis*-elusive, active and central. *Arabidopsis Book*, **14**, e0122.

Song, W.W., Ma, X.R., Tan, H. and Zhou, J.Y. (2011) Abscisic acid enhances resistance to *Alternaria solani* in tomato seedlings. *Plant Physiol. Biochem.* **49**, 693–700.

Sripinyowanich, S., Klomsakul, P., Boonburapong, B., Bangyeekhun, T., Asami, T., Gu, H., Buaboocha, T. and Chadchawan, S. (2013) Exogenous ABA induces salt tolerance in indica rice (*Oryza sativa* L.): the role of *OsP5CS1* and *OsP5CR* gene expression during salt stress. *Environ. Exp. Bot.* **86**, 94–105.

Storey, R. (1995) Salt tolerance, ion relations and the effect of root medium on the response of citrus to salinity. *Aust. J. Plant Physiol.* **22**, 101–114.

Sun, X.L., Yang, S.S., Sun, M.Z., Wang, S.T., Ding, X.D., Zhu, D., Ji, W., Cai, H., Zhao, C.Y., Wang, X.D. and Zhu, Y.M. (2014) A novel *Glycine soja* cysteine proteinase inhibitor GsCPI14, interacting with the calcium/calmodulin-binding receptor-like kinase GsCBRLK, regulated plant tolerance to alkali stress. *Plant Mol. Biol.* **85**, 33–48.

Syros, T., Yupsanis, T., Zafiriadis, H. and Economou, A. (2004) Activity and isoforms of peroxidases, lignin and anatomy, during adventitious rooting in cuttings of *Ebenus cretica* L. *J. Plant Physiol.* **161**, 66–77.

Tan, J.L., Wang, C.Y., Xiang, B., Han, R.H. and Guo, Z.F. (2013) Hydrogen peroxide and nitric oxide mediated cold- and dehydration-induced *myo*-inositol phosphate synthase that confers multiple resistances to abiotic stresses. *Plant, Cell Environ.* **36**, 288–299.

Testerink, C. and Munnik, T. (2011) Molecular, cellular, and physiological responses to phosphatidic acid formation in plants. *J. Exp. Bot.* **62**, 2349–2361.

Tseng, I.C., Hong, C.Y., Yu, S.M. and Ho, T.H. (2013) Abscisic acid- and stress-induced highly proline-rich glycoproteins regulate root growth in rice. *Plant Physiol.* **163**, 118–134.

Twomey, U., Rolfe, R.N., Warrior, P. and Perry, R.N. (2002) Effects of the biological nematicide, DiTera®, on movement and sensory responses of second stage juveniles of *Globodera rostochiensis*, and stylet activity of *G. rostochiensis* and fourth stage juveniles of *Ditylenchus dipsaci*. *Nematology*, **4**, 909–915.

Valluru, R. and Ende, W.V. (2011) *Myo*-inositol and beyond-emerging networks under stress. *Plant Sci.* **181**, 387–400.

Wally, O. and Punja, Z.K. (2010) Enhanced disease resistance in transgenic carrot (*Daucus carota* L.) plants over-expressing a rice cationic peroxidase. *Planta*, **232**, 1229–1239.

Wang, L.J., He, S.Z., Zhai, H., Liu, D.G., Wang, Y.N. and Liu, Q.C. (2013) Molecular cloning and fanctional characterization of a salt tolerance-associated gene *IbNFU1* from sweetpotato. *J. Integr. Agric.* **12**, 27–35.

Welti, R., Li, W., Li, M., Sang, Y., Biesiada, H., Zhou, H.E., Rajashekar, C.B., Williams, T.D. and Wang, X. (2002) Profiling membrane lipids in plant stress responses. Role of phospholipase Dα in freezing-induced lipid changes in *Arabidopsis*. *J. Biol. Chem.* **277**, 31994–32002.

Wilson, S.R., Zucker, P.A., Huang, R.R.C. and Spector, A. (1989) Development of synthetic compounds with glutathione peroxidase activity. *J. Am. Chem. Soc.* **111**, 5936–5939.

Xie, Y.Z., Yin, Q.H., Dai, Q.W. and Qiu, R.L. (2004) Inheritance and breeding for resistance to sweetpotato nematodes. *J. Plant Genet. Resour.* **5**, 393–396.

Xiong, L., Lee, B.H., Ishitani, M., Lee, H., Zhang, C. and Zhu, J.K. (2001) *FIERY1* encoding an inositol polyphosphate 1-phosphatase is a negative regulator of abscisic acid and stress signaling in *Arabidopsis*. *Genes Dev.* **15**, 1971–1984.

Yang, Y.F., Guan, S.K., Zhai, H., He, S.Z. and Liu, Q.C. (2009) Development and evaluation of a storage root-bearing sweetpotato somatic hybrid between *Ipomoea batatas* (L.) Lam. and *I. triloba* L. *Plant Cell, Tissue Organ Cult.* **99**, 83–89.

Ye, J., Fang, L., Zheng, H.K., Zhang, Y., Chen, J., Zhang, Z.J., Wang, J., Li, S.T., Li, R.Q., Bolund, L. and Wang, J. (2006) WEGO: a web tool for plotting GO annotations. *Nucleic Acids Res.* **34**, W293–W297.

Yokoi, S., Quintero, F.J., Cubero, B., Ruiz, M.T., Bressan, R.A., Hasegawa, P.M. and Pardo, J.M. (2002) Differential expression and function of *Arabidopsis thaliana* NHX Na+/H+ antiporters in the salt stress response. *Plant J.* **30**, 529–539.

Yoshida, K.T., Wada, T., Koyama, H., Mizobuchi-Fukuoka, R. and Naito, S. (1999) Temporal and spatial patterns of accumulation of the transcript of *myo*-inositol-1-phosphate synthase and phytin-containing particles during seed development in rice. *Plant Physiol.* **119**, 65–72.

Yu, B., Zhai, H., Wang, Y.P., Zang, N., He, S.Z. and Liu, Q.C. (2007) Efficient *Agrobacterium tumefaciens*-mediated transformation using embryogenic suspension cultures in sweetpotato *Ipomoea batatas* (L.) Lam. *Plant Cell, Tissue Organ Cult.* **90**, 265–273.

Yue, Y., Zhang, M.C., Zhang, J.C., Duan, L.S. and Li, Z.H. (2012) *SOS1* gene overexpression increased salt tolerance in transgenic tobacco by maintaining a higher K+/Na+ ratio. *J. Plant Physiol.* **169**, 255–261.

Zebelo, S.A. and Maffei, M.E. (2015) Role of early signalling events in plant-insect interactions. *J. Exp. Bot.* **66**, 435–448.

15

Overexpression of a novel peanut NBS-LRR gene *AhRRS5* enhances disease resistance to *Ralstonia solanacearum* in tobacco

Chong Zhang[1,2,†], Hua Chen[1,2,†], Tiecheng Cai[1,2], Ye Deng[1,2], Ruirong Zhuang[2], Ning Zhang[2], Yuanhuan Zeng[2], Yixiong Zheng[2,3], Ronghua Tang[4], Ronglong Pan[5] and Weijian Zhuang[1,2,*]

[1]*College of Plant Protection, Fujian Agriculture and Forestry University, Fuzhou, China*

[2]*Fujian Key Laboratory of Crop Molecular and Cell Biology, Fujian Agriculture and Forestry University, Fuzhou, Fujian, China*

[3]*College of Agronomy, Zhongkai Agriculture and Engineering College, Guangzhou, Guangdong, China*

[4]*Cash Crops Research Institute, Guangxi Academy of Agricultural Sciences, Nanning, China*

[5]*Department of Life Science and Institute of Bioinformatics and Structural Biology, College of Life Science, National Tsing Hua University, Hsinchu, Taiwan*

*Correspondence

email weijianz@fafu.edu.cn
[†]Equal contribution as co-first authors.

Keywords: *Arachis hypogaea*,
resistance gene, bacterial wilt, signal
transduction, NPR1, tobacco.

Summary

Bacterial wilt caused by *Ralstonia solanacearum* is a ruinous soilborne disease affecting more than 450 plant species. Efficient control methods for this disease remain unavailable to date. This study characterized a novel nucleotide-binding site-leucine-rich repeat resistance gene *AhRRS5* from peanut, which was up-regulated in both resistant and susceptible peanut cultivars in response to *R. solanacearum*. The product of *AhRRS5* was localized in the nucleus. Furthermore, treatment with phytohormones such as salicylic acid (SA), abscisic acid (ABA), methyl jasmonate (MeJA) and ethephon (ET) increased the transcript level of *AhRRS5* with diverse responses between resistant and susceptible peanuts. Abiotic stresses such as drought and cold conditions also changed *AhRRS5* expression. Moreover, transient overexpression induced hypersensitive response in *Nicotiana benthamiana*. Overexpression of *AhRRS5* significantly enhanced the resistance of heterogeneous tobacco to *R. solanacearum*, with diverse resistance levels in different transgenic lines. Several defence-responsive marker genes in hypersensitive response, including SA, JA and ET signals, were considerably up-regulated in the transgenic lines as compared with the wild type inoculated with *R. solanacearum*. Nonexpressor of pathogenesis-related gene 1 (*NPR1*) and non-race-specific disease resistance 1 were also up-regulated in response to the pathogen. These results indicate that *AhRRS5* participates in the defence response to *R. solanacearum* through the crosstalk of multiple signalling pathways and the involvement of *NPR1* and R gene signals for its resistance. This study may guide the resistance enhancement of peanut and other economic crops to bacterial wilt disease.

Introduction

Bacterial wilt caused by *Ralstonia solanacearum* is a destructive soilborne bacterial disease in plants, including peanut (Arachis hypogaea L.), worldwide (Wicker *et al.*, 2007). This disease is the key limiting factor for the production yield and quality of peanut, an important oil and food crop in China and the world (Yu *et al.*, 2011). *R. solanacearum* infects more than 450 plant species, including many important crops, such as peanut, tomato, tobacco, potato, pepper, soybean and rape. However, effective techniques to control this disease remain unavailable to date (Gururani *et al.*, 2012; Yu *et al.*, 2011). The employment of resistant cultivars has been the most efficient strategy to control this disease, but the enhancement has not been conducted successfully in crops thus far (Bhatnagar-Mathur *et al.*, 2015; Keneni *et al.*, 2012; Reddy, 2016; Sunkara *et al.*, 2014). A recent report has indicated that stable resistant varieties of peanut have been bred to overcome the incidence of serious bacterial wilt in large areas effectively. This report implies that peanut might contain resistant gene resources that are potentially important in controlling this disease. However, few resistant varieties of peanut have been developed in high yield and quality so far (Sunkara *et al.*, 2014). Therefore, elucidating the molecular

mechanism underlying the resistance of crops to bacterial wilt is urgently required to breed ideal varieties.

Plants have developed a complete defence mechanism against the infection of pathogens, such as bacteria, viruses, fungi and insects during evolution (Henry *et al.*, 2013; Jones and Dangl, 2006; Thomma *et al.*, 2011; Zvereva and Pooggin, 2012). Several pathogens are killed by the first defence system, whereas some are suppressed by the plant innate immune (PTI) system (Jones and Dangl, 2006; Zhang and Zhou, 2010). Notwithstanding, various successful pathogens deploy effectors for pathogen virulence. Many effectors can interfere with PTI to some extent as effector-triggered susceptibility (Jones and Dangl, 2006). A given effector is 'specifically recognized' by plant NB-LRR proteins (R genes) during effector-triggered immunity (ETI) (Jones and Dangl, 2006). In general, R gene-triggered resistance is associated with a rapid defence response termed hypersensitive response (HR) (Dangl *et al.*, 1996; Greenberg, 1997; Keen, 1990; Thomma *et al.*, 2011). HR brings a localized cell and tissue death at the infection site following a series of downstream defence responses (Baker *et al.*, 1997; Lamb *et al.*, 1989; Ryals *et al.*, 1996; Zvereva and Pooggin, 2012).

NBS-LRR genes are classified into two subfamilies, namely TIR-NBS-LRR and non-TIR-NBS-LRR, on the basis of the motifs located in the N-terminal region (Liu et al., 2007). The former subfamily contains a Drosophila Toll/mammalian interleukin-1 receptor (TIR) domain, whereas the latter subfamily consists of a coiled coil (CC)/leucine zip motif (Van Ooijen et al., 2008). Thus far, more than 70 disease resistance genes have been cloned and characterized in monocots and dicots (Liu et al., 2007). Most of these genes are NBS-LRR genes obtained using map-based cloning and transposon tagging methods in crops (Hulbert et al., 2001; McDowell and Woffenden, 2003; Meyers et al., 2005; Takken and Joosten, 2000).

R gene products can directly or indirectly recognize pathogen effector proteins (avirulence protein) and induce resistance (Cesari et al., 2013; Flor, 1971; Sohn et al., 2014). Furthermore, some NB-LRR proteins act downstream of R protein activation. The tobacco 'N-required gene 1' and tomato 'NB-LRR protein required for HR-associated cell death 1' (NRC1) (both as CC-NB-LRR proteins) are required for TIR-NB-LRR protein N-mediated resistance to tobacco mosaic virus and receptor-like protein Cf-4-mediated resistance to tomato leaf mould, respectively (Gabriëls et al., 2007; Peart et al., 2005). The CC-NB-LRR activated disease resistance 1 family of proteins in Arabidopsis is required for salicylic acid (SA)-dependent ETI (Bonardi et al., 2011). The downy mildew resistance locus RPP2 in Arabidopsis Col-0 comprises two closely linked NB-LRR genes, RPP2A and RPP2B, for resistance (Sinapidou et al., 2004). The rice Pia locus for blast (Magnaporthe) resistance includes two divergently transcribed CC-NB-LRR genes, RGA4 and RGA5, for resistance (Cesari et al., 2013).

Quantitative trait loci (QTL) controlling resistance to bacterial wilt have been identified in several crops, such as tomato (Carmeille et al., 2006; Danesh et al., 1994; Mangin et al., 1999; Thoquet et al., 1996; Wang et al., 2000), eggplant (Lebeau et al., 2013) and tobacco (Qian et al., 2012), as well as in model plants, such as Arabidopsis thaliana (Godiard et al., 2003) and Medicago truncatula (Ben et al., 2013). However, only two resistance genes have been identified thus far: the A. thaliana ERECTA gene involved in polygenic resistance and the A. thaliana RRS1-R gene involved in monogenic resistance. RRS1-R is a typical TIR-NB-LRR resistance gene generated through map-based cloning in Arabidopsis (Deslandes et al., 2002). RRS1-R contains a WRKY transcription factor domain at the C-terminus to activate downstream gene expression and a nuclear localization signal (NLS) at its N-terminus (Deslandes et al., 2002). PopP2 is the corresponding avirulence gene of RRS1-R. It was recognized and recruited with the LRR domain of RRS1-R and trafficked to the nucleus through NLS. ERECTA, a quantitative resistance locus for bacterial wilt, encodes a leucine-rich repeat receptor-like kinase. ERECTA-controlled resistance is triggered by disease defence response through the phosphorylation of extracellular kinase-regulated downstream genes (Godiard et al., 2003). However, resistance genes to bacterial wilt have yet to be cloned in crops other than Arabidopsis, thereby hindering genetic enhancement towards the disease. In addition, the molecular mechanism and details in the signalling pathway of R gene resistance to R. solanacearum have yet to be elucidated.

In this study, the up-regulated NBS-LRR resistant gene AhRRS5 was screened from peanut through microarray analysis. This gene was induced by R. solanacearum containing the typically conserved motifs of an NBS-LRR gene. AhRRS5 was localized in the nucleus and could be up-regulated relatively higher in the resistant than susceptible peanut cultivars against bacterial wilt. This gene responded differently to phytohormones, such as

salicylic acid (SA), abscisic acid (ABA), methyl jasmonate (JA) and ethephon (ET), among distinct resistance varieties. The transient overexpression of AhRRS5 induced HR responses in Nicotiana benthamiana, whereas the overexpression of this gene in Nicotiana tabacum significantly enhanced the resistance of peanut to R. solanacearum. The underlying mechanism presumably involved the significant up-regulation of several representative stress-responsive and resistance marker genes. We concluded that AhRRS5 indirectly participates in the defence response to R. solanacearum in plants through multiple signalling regulatory networks.

Results

Cloning and phylogenetic analysis of AhRRS5

The 5' and 3' unknown cDNA sequences of AhRRS5 were cloned by rapid amplification of cDNA ends (RACE) on the basis of the known fragment. The full-length cDNA sequence of AhRRS5 was isolated from the total RNA of peanut leaf through reverse transcription polymerase chain reaction (RT-PCR), and the genomic DNA sequence of AhRRS5 was cloned from the genomic DNA of peanut through PCR. The full-length cDNA contained a 3157-bp open reading frame encoding a polypeptide of 943 amino acids, an 88-bp 5' untranslated terminal region (5' UTR), and a 138-bp 3' UTR. The genomic DNA sequence of AhRRS5 was 3662-bp, including a 535-bp intron. The entire sequence of the AhRRS5 protein has 76% identity with an NBS-LRR resistance protein, RPM1-like, in Glycine max (Figure 1; Data S1 and Data S2). A comparison of the AhRRS5 amino acid sequence with the R gene of a known function demonstrates that it most closely resembles RXO1 (33% identity and 53% positive) from Zea mays, which confers resistance to X. o. pv. Oryzicola containing avrRxo1, and RPM1 (32% identity and 53% positive) from A. thaliana, resisting Pseudomonas syringae pv. maculicola 1 containing AvrBand, AvrRpm1, and Pid3 (33% identity and 53% positive) from Rice and resisting Magnaporthe oryzae (Data S2). The former two were resistant to bacterial pathogens.

Sequence analysis showed that the deduced AhRRS5 protein contained conserved NBS motifs, such as P-loop (MGGVGKT), GLPL (GLPLALK), kinase-2 (LLVLDDVVW), kinase-3a (GSRVLVTTR) and RNBS-C (YEVxxLSDEEAWELFCKxAF) motif (Bertioli et al., 2003; Zheng et al., 2012), and 4 LRR-conserved domains (LxxLxxLxxLxLxxC/A-xx) (Leah McHale et al., 2012) (Figure 1; Data S1). On the basis of the conserved domains at the N-terminus of the deduced NBS-LRR genes, the AhRRS5 gene had the typical structure of non-TIR-NBS-LRR genes (Wan et al., 2012), with RNBS-A-non-TIR (FnLxAWVCvSQxF) domains (Figure 1).

The phylogenetic analysis of 29 types of NBS-LRR resistance proteins from GenBank together with AhRRS5 generated two clades coarsely (Figure 2; Data S3). The topology of the phylogenetic analysis showed that the NBS-LRR-type resistance proteins can be divided into two types, namely TIR-NBS-LRR and non-TIR-NBS-LRR, and that the non-TIR-NBS-LRR-type resistance proteins can be subdivided into two classes, namely NBS-LRR and CC-NBS-LRR. AhRRS5 is a NBS-LRR-type resistance protein that is similar to NBS-LRR resistance proteins, such as RPM1 (XP_006587620.1|) from Glycine max, RPP8 (GenBank: XP_003612691.1) from M. truncatula, RXO1 (GenBank: AAX31149.1) from Zea mays, RPM1 (GenBank: AGC12590) from A. thaliana and Pi9 (GenBank: ABB88855.1) from Oryza sativa. These similarities indicate that these resistance genes share

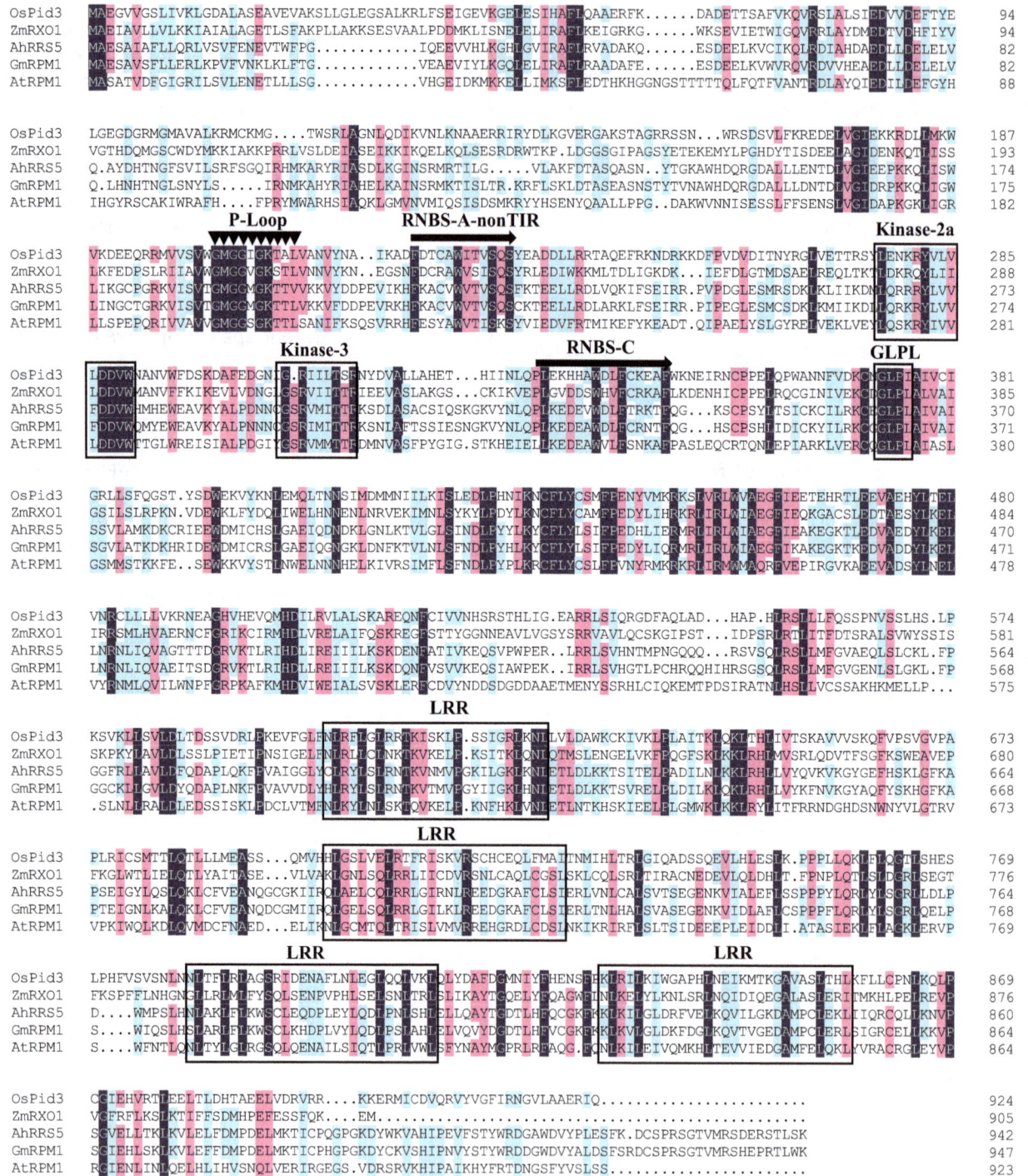

Figure 1 Conserved domain comparison between the deduced amino acid sequence of AhRRS5 and other resistance proteins. Sequences were aligned using the ClustalW2 program. Gaps have been introduced to optimize the alignment. Identical or conserved amino acids are shaded in dark and light, respectively. The sources of the proteins and GenBank accession numbers are as follows: OsPid3, blast resistance protein (ACN62383.1) from *Oryza sativa Indica* Group; AtRPM1 (AGC12570.1) from *Arabidopsis thaliana*; GmRPM1 (XP_006587620.1) from *Glycine max*; and ZmRXO1 disease resistance protein (AAX31149.1) from *Zea mays*.

a common ancestor R gene and belong to NBS-LRR-type resistance genes (Figure 2).

AhRRS5 functions in the nucleus

Sequence analysis indicated that the predicted AhRRS5 protein was localized in the nucleus (Data S1) (http://nls-mapper.iab.-keio.ac.jp/cgi-bin/NLS_Mapper_form.cgi). To confirm this

indication and the site of function, we generated an AhRRS5-green fluorescent protein (GFP) fusion driven by the constitutive CaMV35S promoter (Figure 3a). With 35S::GFP as a negative control, the *AhRRS5::GFP* fusion gene was transformed into *Agrobacterium* strain GV3101, which was further infiltrated into *N. benthamiana* leaves. Typical results indicated the exclusive localization of AhRRS5-GFP in the nucleus, whereas GFP alone

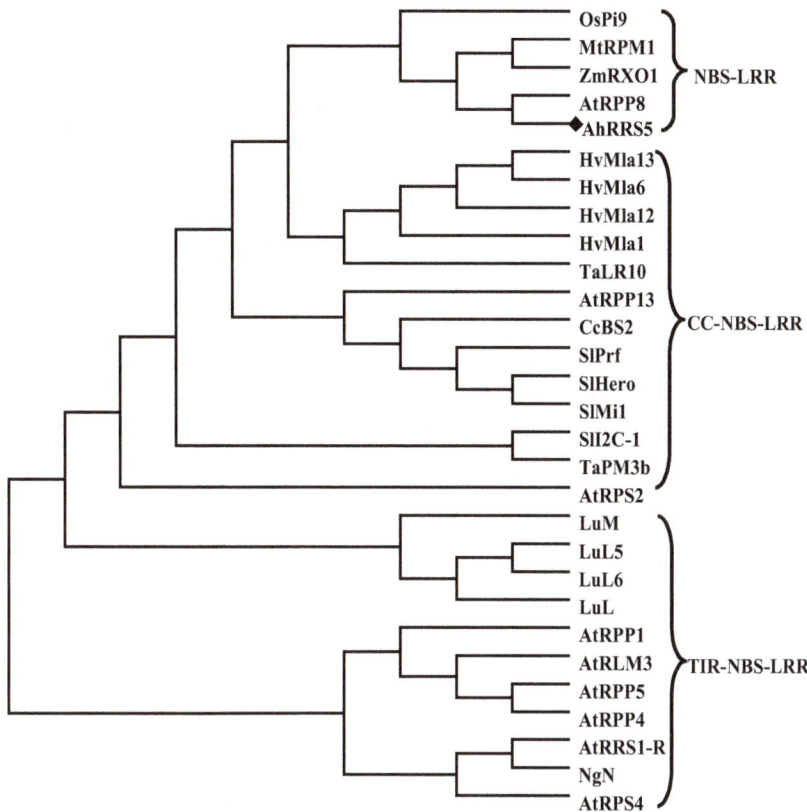

Figure 2 Phylogenetic tree was constructed using AhRRS5 and known different types of NBS-LRR resistant proteins. AhRRS5 is shown by a red rhombus. Alignments were performed in ClustalW2, and phylogenetic tree was constructed by the neighbour-joining algorithm of MEGA 5.01. Bootstrap values (1000 replicates) are shown in percentages at the branch nodes.

occurred in multiple subcellular compartments, including the cytoplasm and the nucleus (Figure 3b). The results indicate that AhRRS5 is localized and functions in the nucleus.

AhRRS5 showed varied expression patterns among tissues

In the microarray with a high density of unigenes, four unigenes including *AhRRS5* were found with a sequence identity of more than 97%. These unigenes apparently belong to the same *AhRRS5* gene family. Nonamplified double strain cDNA was used

for microarray hybridization to evaluate the transcript levels of the unigenes. All four members showed a synchronized expression pattern among tissues or organs. They showed tissue-specific expression manners; in particular, they were expressed the highest in the roots, then in the testa, pericarps and stem, but were weakly expressed in other tissues (Figure 4a). Embryos displayed the least expression levels of these genes. In addition, the transcripts of these genes obviously increased with pericarp development (Figure 4b) but remained almost constant with trace amounts during embryo development (Figure 4c).

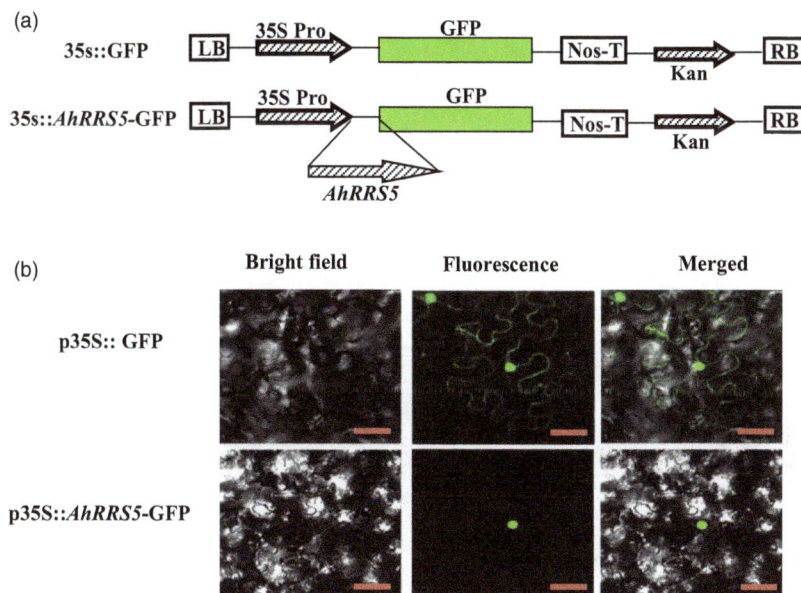

Figure 3 Subcellular localization of *AhRRS5*. (a) Schematic of p*35S::GFP* and p*35S:: AhRRS5-GFP* constructs used for the subcellular localization of *AhRRS5* by agroinfiltration into *N. benthamiana* cells. (b) AhRRS5-GFP localized in the nucleus of *N. benthamiana* cells, GFP alone localized throughout the whole cells. Bright field (left), fluorescence (middle) and merged images (right) were obtained at 48 h by using Leica confocal microscopy after agroinfiltration. Bars = 50 μm.

Therefore, *AhRRS5* may be involved in the resistant response and in plant development to some extent.

AhRRS5 showed a wide response to biotic and abiotic stresses

Response of AhRRS5 to exogenous hormones

The transcript level of *AhRRS5* was determined in the medium-resistant variety Minhua 6 at the eight-leaf stage after exogenous treatment with SA, ABA, ET and MeJA to identify the possible involvement of *AhRRS5* in signalling pathways relating to the phytohormones (Figure 5). Compared with the control plants, *AhRRS5* transcripts increased between 3 and 24 h with two peaks after SA treatment. The highest transcript level (6.7-fold up-regulation) was observed at 12 h post-treatment (hpt) (Figure 5a). *AhRRS5* transcription also increased with a single peak of 5.1-fold up-regulation at 3 hpt after ABA treatment (Figure 5b). In response to ET, *AhRRS5* expression was enhanced from 3 hpt to 12 hpt, and the highest transcript level (10-fold) was obtained at 12 hpt (Figure 5c). The application of 100 mM MeJA also elevated *AhRRS5* expression with two peaks, and the highest level was achieved at 6 hpt (Figure 5d).

Highly susceptible and resistant varieties Xinhuixiaoli and Yueyou 92, respectively, were used to clarify the relationship between *AhRRS5* and the hormones (Figure 6). Although *AhRRS5* showed a similar expression in response to these hormones in Minhua 6, this gene demonstrated distinct expression characteristics between the two varieties. *AhRRS5* was more significantly up-regulated after SA and ABA treatments in Xinhuixiaoli than in

Yueyou 92 (Figure 6a,b); however, this gene increased less after ET and JA treatments (Figure 6c,d). In particular, the application of ET down-regulated AhRRS5 in Xinhuixiaoli but up-regulated it in Yueyou 92 (Figure 6d). This result indicates that the regulation of *AhRRS5* differs between resistant and susceptible varieties in peanut.

Responses of *AhRRS5* transcripts to abiotic stresses

The responses of *AhRRS5* including three other orthologous NBS-LRR genes to low temperature (4 °C) and drought were studied by microarray hybridization using the cDNA of mixed double strains at different time points (Materials and methods) in eight-leaf Minhua 6. *AhRRS5* and three other NBS-LRR genes remained constant in response to low temperature but were up-regulated by nearly 8- to 10-fold in response to drought (Data S4). To clarify whether *AhRRS5* is involved in the response to abiotic stresses, the relative transcripts of *AhRRS5* were also examined in eight-leaf Minhua 6 seedlings under low temperature and drought treatments through quantitative real-time PCR analysis (Figure 5e,f). The transcript level of *AhRRS5* decreased and then increased in response to low temperature and drought. In specific, under low temperature, the transcript level of *AhRRS5* decreased by two- to three-fold at 3 and 6 hpt and then increased between 24 and 48 hpt, with the highest level (2.5-fold) at 48 hpt (Figure 5e). Compared with the control, the transcript level of *AhRRS5* was down-regulated by two-fold at 1 day post-treatment (dpt) but was up-regulated from 2 dpt to 4 dpt with a 3.3-fold induction at 4 dpt under drought (Figure 5f), thereby confirming the microarray results.

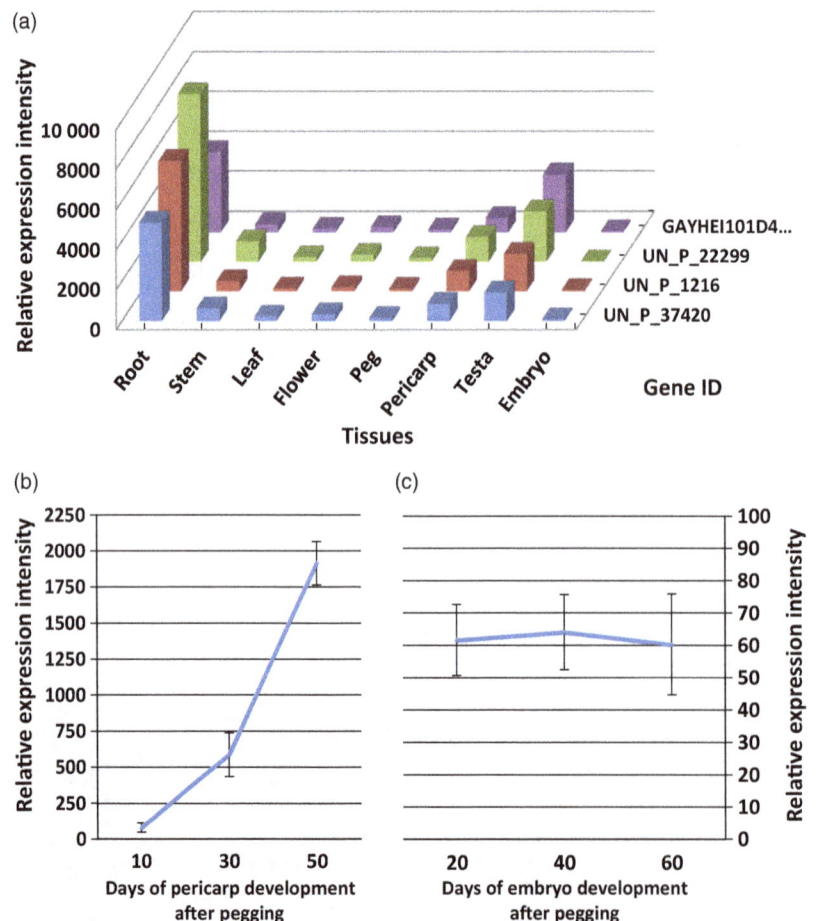

(a)

(b)   (c)

Figure 4 In silico identification of the expression characteristics of four members of the *AhRRS5* gene family. (a) The *AhRRS5* family showed tissue-specific expression in peanut, the highest level was in the root, followed by the testa and pericarp. Weak expression was found in the other tissues. (b) *AhRRS5* genes increased expression with pericarp development. (c) AhRRS5 had the least expression levels with developing embryos. UN_P_37420, UN_P_1216, UN_P_22299 and GAYHEI101D4L7C_pchu_p are *AhRRS5* and the three other members of the same family.

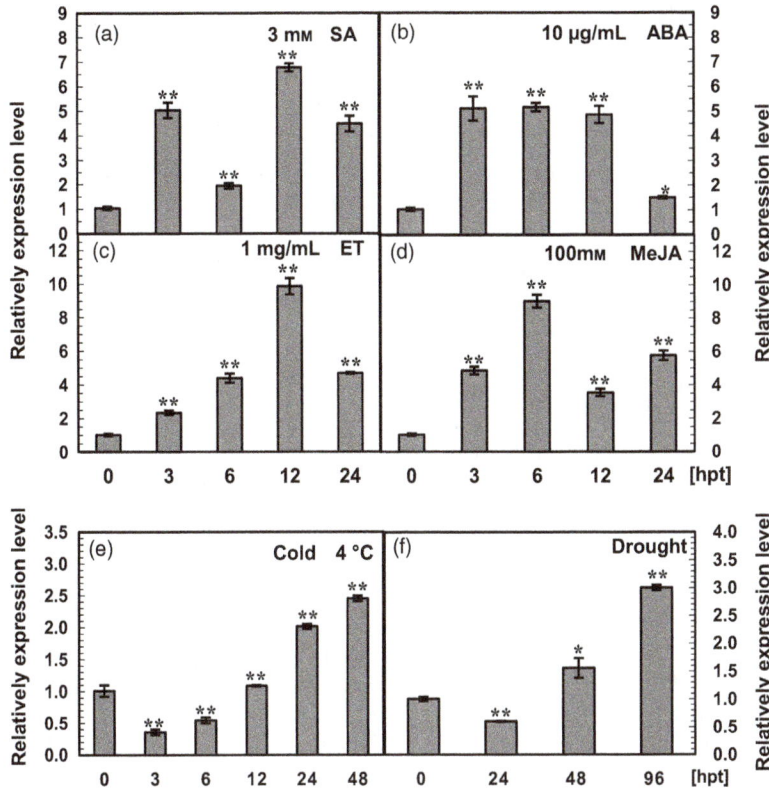

Figure 5 qRT-PCR analysis of AhRRS5 transcripts in peanut cultivar Minhua 6 under abiotic treatments. (a–d) AhRRS5 relative expression level in peanut leaves at different time points after treatment with salicylic acid (SA, 3 mM), abscisic acid (ABA, 10 μg/mL), ethylene (ET, 1 mg/mL) and methyl jasmonate (MeJA, 100 mM). (e and f) AhRRS5 expression was determined at various hour intervals after treatment with low temperature (4 °C) and drought in peanut plants at the eight-leaf stage. The relative expression level of AhRRS5 in peanut plants at various time points was compared with the mock control, which was set to 1. The asterisks indicate a significant difference (SNK test, *P < 0.05 or **P < 0.01). Error bars indicate the standard error.

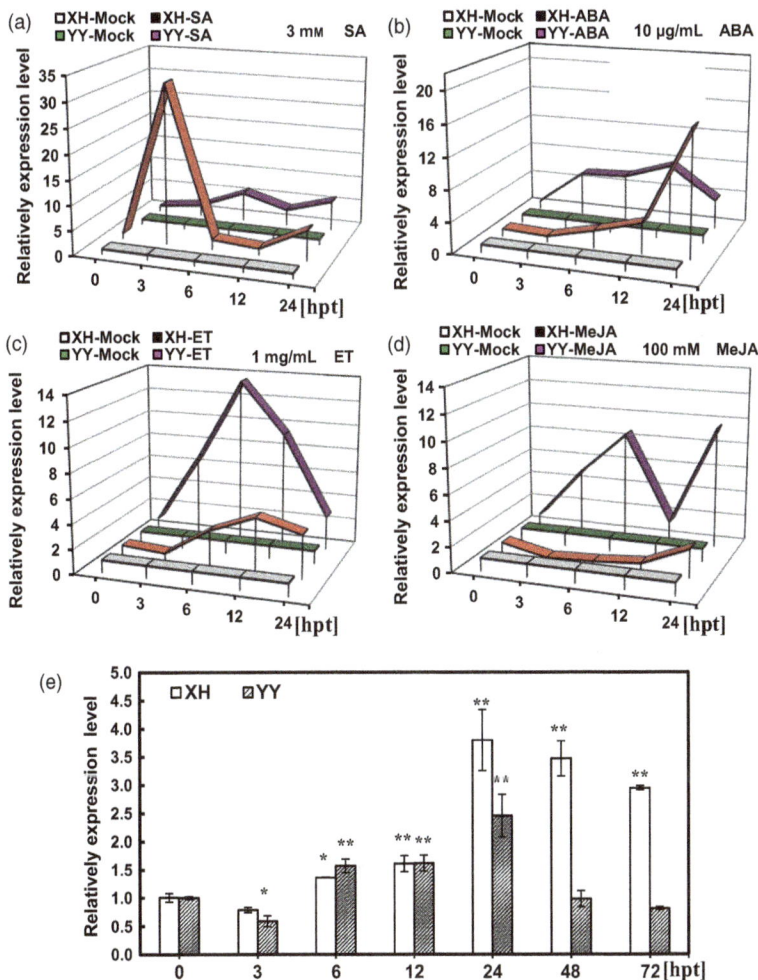

Figure 6 Comparative expression characteristics of AhRRS5 between resistant and susceptible varieties under hormones and R. solanacearum treatments. (a) AhRRS5 showed two expression peaks in response to SA within 24 h, and it up-regulated over 32-fold in susceptible variety at 3 HPT, much greater than in resistant one. (b) AhRRS5 increased expression under ABA treatment with one peak; it up-regulated later in the susceptible but >16-fold at 24 h. (c) AhRRS5 up-regulated with one peak within 24 h with nearly 16-fold at 6 h in the resistant variety. (d) AhRRS5 responded differently between resistant and susceptible peanut with MeJA treatment, down-regulated in the susceptible peanut and up-regulated in the resistant ones with two peaks of over eightfold increase. (e) AhRRS5 was up-regulated higher in susceptible variety especially after 24 hpt with inoculation of R. solanacearum. XH-Mock: susceptible variety Xinhuixiaoli without treatment; YY-Mock: resistant variety Yueyou 92 without treatment. XH-SA, XH-ABA, XH-ET and XH-MeJA: susceptible variety treated with SA, ABA, ET and MeJA, respectively; YY-SA, YY-ABA, YY-ET and YY-MJA, resistant variety treated with SA, ABA, ET and MeJA, respectively. The relative expression level of AhRRS5 in peanut plants at various time course was compared with mock or control, which was set to 1. The asterisk indicate a significant difference (SNK test,*P-value <0.05 or **P-value <0.01), Error bars indicate the standard error.

Expression pattern of *AhRRS5* in the resistant/susceptible peanut cultivars after *R. solanacearum* challenge

AhRRS5 was characterized using resistant and susceptible peanut cultivars after inoculation with *R. solanacearum* by microarray hybridization and qRT-PCR. The four members of *AhRRS5* in the microarray exhibited similar pattern of transcription with *R. solanacearum* inoculation. These genes were up-regulated by nearly one-fold under inoculation with *R. solanacearum* in Yueyou 92, but a higher up-regulation was observed in Xinhuixiaoli (Data S4). In addition, the expression patterns of *AhRRS5* at different time courses after *R. solanacearum* inoculation were compared in the two varieties. *AhRRS5* transcripts were induced between 0 and 24 h in the leaves of Yueyou 92 and then returned to their ground state at 72 hpi in response to *R. solanacearum* strain challenge. The expression level of *AhRRS5* in Xinhuixiaoli was up-regulated from 6 hpi, showed a peak of 3.75-fold transcript level at 24 hpi, and remained high between 24 and 96 hpi (Figure 6e). This finding suggests that *AhRRS5* participates in the immunity of peanut to *R. solanacearum*.

Transient overexpression of *AhRRS5* in *N. benthamiana* leaves induces hypersensitive response

Successful pathogens can attenuate PTI by secreting effector molecules into the host plant cell. Some R proteins could recognize pathogen effector molecules and induce ETI with HR resulting in cell death at the infection site. This process is followed by a series of downstream defence responses. Overexpression vector harbouring p35S::AhRRS5 was generated and transformed into *Agrobacterium* GV3101 to verify whether AhRRS5 overexpression causes HR cell death. AhRRS5 was transiently expressed in *N. benthamiana* leaves through agroinfiltration. Then, AhRRS5 overexpression in *N. benthamiana* leaves induced an intensive HR mimicking cell death 48 h after infiltration. However, no visible HR cell death was found in those infiltrated with GV3101 harbouring empty vector p35S::00. Furthermore, electrolyte significantly leaked at 24 and 48 hpt after treatment, and darker trypan blue staining was observed after AhRRS5 overexpression for 24 hpt. This result suggests that AhRRS5 can trigger HR response in *N. benthamiana* leaves (Figure 7a,b). In addition,

large amounts of H_2O_2 accumulation were found in the *N. benthamiana* leaves after AhRRS5 overexpression by DAB staining (Figure 7b). These results demonstrate that the transient overexpression of *AhRRS5* in tobacco leaves induces HR and H_2O_2 generation as a defence response to stresses.

Overexpression of *AhRRS5* in tobacco enhances resistance to *R. solanacearum*

The involvement of *AhRRS5* *R. solanacearum* resistance was evaluated by transforming CB-1, a conventional tobacco cv., medium-susceptible to bacterial wilt mediated by *Agrobacterium*, with *AhRRS5* driven by two copies of the CaMV35S promoter in the pBI121 binary vector. Transgenic T_0 and T_1 tobacco plants were generated to examine the role of *AhRRS5* in tobacco–*R. solanacearum* interaction (Figure 8a). Three T_2 transgenic homozygous lines were screened by inoculation and identified for their resistance to *R. solanacearum* (Figure 8b). The line *AhRRS5-OE-3* line which showed the greatest *AhRRS5* relative transcript levels and resistant to *R. solanacearum* (not shown) of all the tested lines, was selected for the detailed disease resistance assays. No apparent phenotypic differences between the wild-type and transgenic plants were observed. A highly virulent strain of *R. solanacearum* was used to inoculate individuals of *AhRRS5-OE-3* T_2 lines and wild-type plants. Vein injection was then used for *R. solanacearum* inoculation. All tested transgenic lines exhibited enhanced disease resistance. Evident wilting symptoms were detected in the leaves of wild-type plants at 7 dpi, whereas only faint wilting symptoms were exhibited by *AhRRS5-OE-3* lines (Figure 8b,d). Extremely severe wilting symptoms were developed in the wild-type plants at 20 dpi but not in the *AhRRS5-OE-3* transgenic lines. Wilting and contagion symptoms were evident on the stems of the infected wild-type tobacco at 7 and 20 dpi, but no significant symptoms were found in the transgenic lines (Figure 8e). Further evaluation of *AhRRS5* was performed in the Honghuadajinyuan cultivar, which is hypersusceptible to *R. solanacearum*. Five transgenic T_2 homozygous lines were inoculated compared with the wild type. All lines showed increased but distinct levels of resistance to *R. solanacearum* (Tables 1 and S2). Line 3 showed the highest resistance with a low infection index and a death rate of (7.08%) at 21 dpi, but the

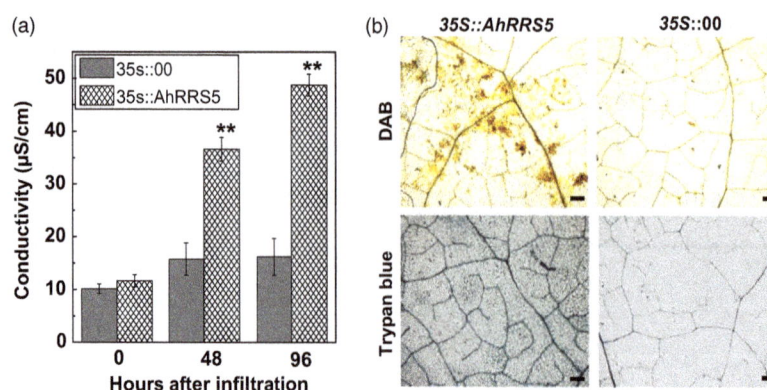

Figure 7 Effect of transient expression of *AhRRS5* in *Nicotiana benthamiana* on immunity induction. (a) Electrolyte leakage of *N. benthamiana* leaves were infiltrated with *Agrobacterium* strain GV3101 containing *35S::AhRRS5* and *35S::00*. (b) Trypan blue and DAB staining of cell death and H_2O_2 generation in *N. benthamiana* leaves 48 h after AhRRS5–*Agrobacterium* infiltration. Bars = 0.1 mm. Error bars indicate the standard error, Alphabet indicates statistically significant differences between wild-type and *35S::AhRRS5* tobacco by Student–Newman–Keuls test, *$P < 0.05$ or **$P < 0.01$), Error bars indicate the standard error.

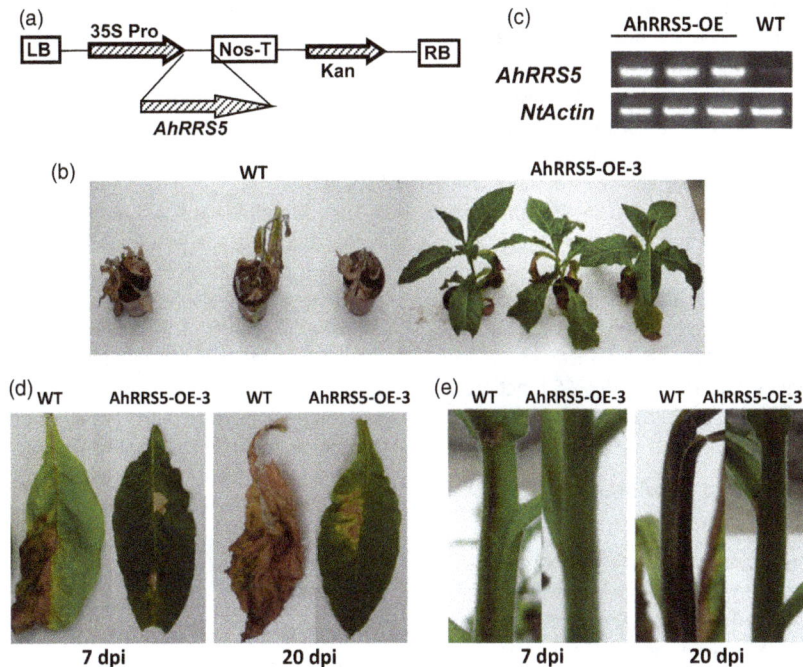

Figure 8 Overexpression of *AhRRS5* enhanced resistance to *Ralstonia solanacearum* in transgenic tobacco. (a) Schematic of the pBI121-*AhRRS5* construct. LB and RB, left and right borders of the T-DNA; 2 × 35SPro, two cauliflower mosaic virus 35S promoters; Nos-T, NOS terminator; Kan^r, kanamycin resistance. (b) The third leaves of 8-week-old wild-type tobacco and AhRRS5-OE-3 transgenic plants were inoculated with 10 μL of suspension of 10^8 cfu per millilitre of high-virulence *R. solanacearum* strain. The photograph was obtained 20 days postinoculation (dpi). (c) RT-PCR analysis of *AhRRS5* expression in transgenic and wild-type tobacco plants; expression level of Ntactin was visualized as endogenous control. (d) Disease symptoms of detached leaves of wild-type and AhRRS5-OE-3 transgenic plants after inoculation with *R. solanacearum*. Transgenic leaves showed immune resistance or high-resistance phenotype. Photos were obtained at 7 and 20 dpi. (e) Different phenotypes of the stem were observed between wild-type and transgenic AhRRS5-OE-3 plants after inoculation with *R. solanacearum*. Transgenic plant stem showed no or much week infections. Photos were taken at 7 and 20 dpi.

mock line showed serious wilting with 93.58% index and 81.08% death of plants, respectively, at 21 dpi. These results indicate that *AhRRS5* overexpression greatly enhances disease resistance against *R. solanacearum* in tobacco.

To further confirm the role of *AhRRS5* in disease resistance and elucidate its possible molecular mode of action, transcriptional responses of known defence genes to overexpression of *AhRRS5* in noninoculated tobacco plants were investigated by qPCR (Data S5). We examined transcript levels of the HR-associated genes *NtHIN1*, *NtHSR201*, *NtHSR203* and *NtHSR515* (Sohn *et al.*, 2007), SA-responsive genes *NtPR1a/c*, *NtPR3*, *NtPR4* and *NtNPR1* (Brogue *et al.*, 1991; Ward *et al.*, 1991), JA-responsive *NtPR1b* and *NtPR2* (Sohn *et al.*, 2007) and ET-associated genes such as *NtEFE26* and *NtACS6* (Chen *et al.*, 2003). Each of the tested tobacco genes was shown previously to be up-regulated in response to pathogen infection (Chen *et al.*, 2003; Rizhsky *et al.*, 2002; Sohn *et al.*, 2007). We found transcript levels of HR-associated genes, such as *NtHIN1*, *NtHSR201* and *NtHSR515* to be increased by 3.3-fold, 2.8-fold and 3.3-fold in the *AhRRS5-OE-3* line compared to wild type plants, respectively. Transcript levels of the SA-responsive *NtPR1a/c*, *NtPR3*, *NtPR4* and *NtNPR1* genes were increased in *AhRRS5-OE-3* plants by 11.9-fold, 3.0-fold, 2.0-fold and 3.0-fold, respectively, while those of the JA-responsive *NtPR2* and *NtPR1b* genes were 2.5-fold and 4.0-fold higher in *AhRRS5-OE-3* plants. These results show that *AhRRS5* overexpression enhances stress-related gene expression compared to the wild-type tobacco.

Table 1 Disease indexes and death ratios of different OE lines and the wild type after inoculation with *Ralstonia solanacearum*

OE lines	7 dpi		21 dpi	
	Disease index (%)	Death ratio (%)	Disease index (%)	Death ratio (%)
OE-2	22.90	2.80	45.79	34.58
OE-3	12.83	0.00	20.35	7.08
OE-4	26.51	4.82	64.46	56.63
OE-5	37.39	10.62	72.35	60.18
OE-8	19.92	6.50	31.10	14.63
Wild type	73.65	22.97	93.58	81.08

dpi, days postinoculation.

Up-regulation of marker genes in response to *R. solanacearum* infection

HR-responsive genes, namely *NtHIN1*, *NtHSR201* and *NtHSR515*, were significantly up-regulated in the transgenic plants ($P < 0.01$ or $P < 0.05$) but down-regulated in wild-type CB-1 to different extents at 48 hpi with *R. solanacearum* (Figure 9). By contrast, *NtHSR203* did not respond to the strain infection either in the transgenic or control plants (Figure 9a). The expression levels of *NtPR1a/c* and *NtPR3*, which are SA-responsive pathogenesis-

related (PR) genes, increased in the AhRRS5-OE-1 plants by 1,
453.0- and 14.5-fold, respectively, which are much higher than
those in CB-1. In addition, the NtRP4 gene was down-regulated
by 2.5-fold (Figure 9b). JA-responsive NtPR2 was up-regulated in
CB-1 but down-regulated in the transgenic plants in response
to the strain, whereas NtPR1b was 14.3-fold higher in the
AhRRS5-OE-3 plants than in CB-1 (Figure 9c). The transcript levels
of ET-responsive genes NtEFE26 and NtACS6 in the transgenic
plants were also significantly increased at 48 h after infection but
not in the wild-type plants (Figure 9d). Several pathogen-induced
HR- and defence-associated genes were enhanced by AhRRS5
overexpression, but few were reduced or remained unchanged,
which are consistent with the resistance enhancement in the
transgenic lines. These findings indicate that AhRRS5 functions in
the resistance of transgenic tobacco through a wide series of
signalling pathways.

NDR1 and NPR1 genes were up-regulated by R. solanacearum infection

Non-race-specific disease resistance 1 (NDR1) and nonexpressor
of pathogenesis-related gene 1 (NPR1) genes were involved in the
R gene resistance signalling pathway. In silico identification of
three NDR1-like and two NPR1-like gene expressions were
performed between AhRRS5-OE-3 transgenic plants and wild-
type plants, as well as hyper-resistant and hypersusceptible
varieties Yanyan 97 and Honghuadajinyuan after inoculation
with R. solanacearum, respectively (Figure 10). Two NPR1-like
genes were slightly up-regulated by 6%–23% in the AhRRS5-OE-
3 lines after inoculation but were down-regulated by 15%–21%
in the wild-type plants after inoculation (Figure 10a). The NPR1
gene, TC79797, considerably increased or decreased in response
to the pathogen, consistent with the resistant and susceptible

varieties after inoculation (Figure 10b). Furthermore, the results
of real-time PCR revealed that the transcript level of NPR1
increased by 14.5-fold in transgenic lines of AhRRS5-OE-3 as
compared with wild-type plants after inoculation with
R. solanacearum, much higher than the increase of transcripts
in inoculated wild-type over corresponding mock plants
(Figure 10c).

The transcript levels of three NDR1-like genes slightly increased
in the AhRRS5 transgenic lines but significantly decreased in the
wild type after inoculation; this result indicates that AhRRS5 can
maintain a high level of expression for the NDR1 gene (Figure 10a). However, three NDR1-like genes were considerably
down-regulated in both resistant and susceptible varieties after
inoculation (Figure 10b). The results indicate that both NDR1- and
NPR1-like genes in tobacco are involved in AhRRS5 resistance in
transgenic tobacco, but only NPR1 genes are required for the
hyper-resistant tobacco variety Yanyan 97. AhRRS5 might also be
involved in the R gene signalling for resistance against microbial
infection.

Discussion

AhRRS5 is a novel peanut NBS-LRR resistance protein localized in the nucleus

NBS-LRR genes are a class of resistance genes that function in
pathogen recognition and defence response signal transduction
(Ameline-Torregrosa et al., 2008; Gao et al., 2010). More than
70 disease resistance genes cloned from higher plants by map-
based methods belong to NBS-LRR domain genes resistant to
bacterial, fungal and viral diseases, as well as some environmental
stresses (Liu et al., 2007). AhRRS5 was isolated from peanut
using microarray analysis and could be up-regulated by

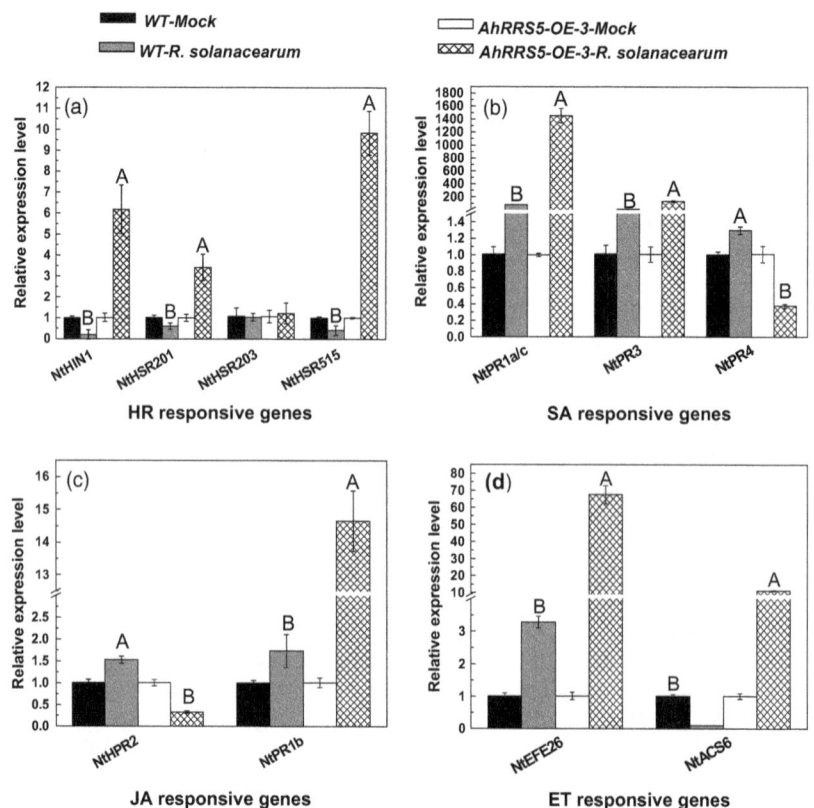

Figure 9 Transcript levels of tobacco defence-
related marker genes in wild-type CB-1 and
AhRRS5-OE-3 transgenic tobacco line 48 h after
inoculation with R. solanacearum. The transcript
levels of NtHIN1, NtHSR201, NtHSR203,
NtHSR515, NtPR1a/c, NtPR3, NtPR4, NtNPR1,
NtPR2, NtPR1b, NtEFE26 and NtACS6 were
determined by quantitative real-time PCR. Relative
transcript levels were normalized using the
transcripts of NtEF1α. The transcript levels of
nontreated wild-type or AhRRS5-OE-3 tobacco
plants were used as the control and assigned
value of 1. Alphabet indicates statistically
significant differences between wild-type and
AhRRS5-OE-3 tobacco plants by Student–
Newman–Keuls test (lowercase difference
indicates $P < 0.05$; uppercase difference indicates
$P < 0.01$).

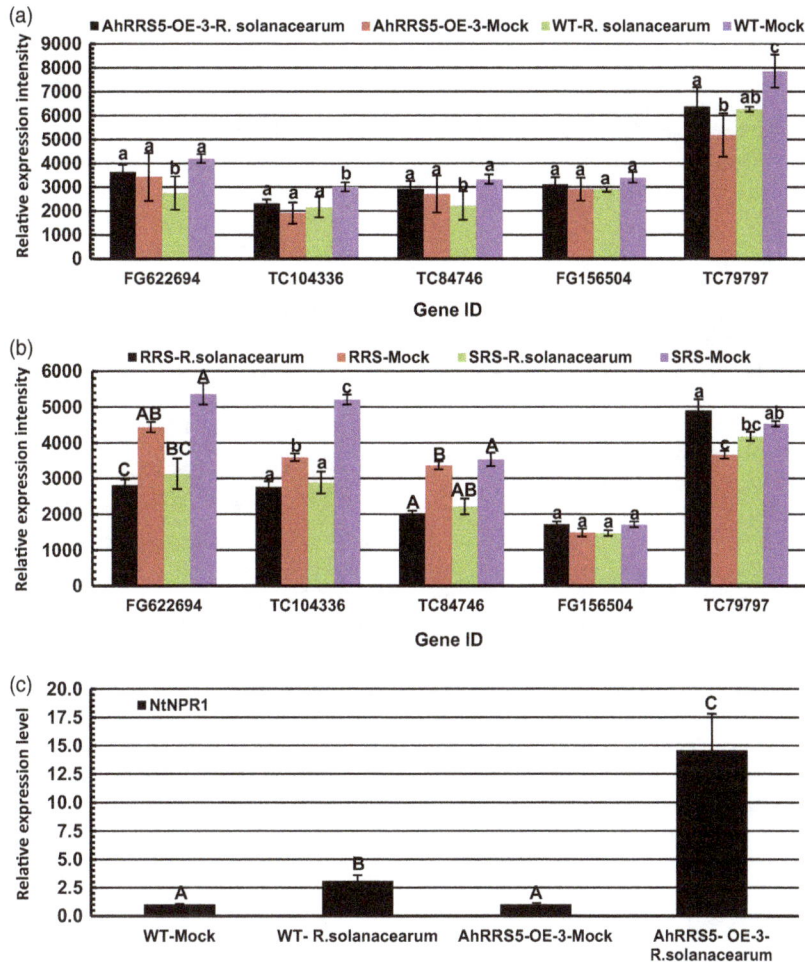

Figure 10 In silico and qPCR analysis of *NDR1-* and *NPR1-*like gene expression upon inoculation with *R. solanacearum*. (a and b) Microarray data. (a) Expression of three *NDR1-*like and two *NPR1-*like genes. *AhRRS5-OE-3-R. solanacearum* indicates tobacco CB-1 cultivar transformed with *AhRRS5* with inoculation; *AhRRS5-OE-3-*Mock, transgenic CB-1 without inoculation;
WT-*R. solanacearum*, CB-1 with inoculation; WT-Mock, CB-1 without inoculation. (b) Down-regulation of three *NDR1-*like genes in varieties after inoculation. RRS-*R. solanacearum* indicates hyper-resistant tobacco variety Yanyan 97 under inoculation; RRS-Mock, hyper-resistant variety Yanyan 97 without inoculation. SRS *R. solanacearum*, hypersusceptible variety Honghuadajinyuan with inoculation; SRS-Mock, hypersusceptible variety Honghuadajinyuan without inoculation. *FG622694*, *TC104336* and *TC84746* are *NDR1-*like genes; *FG156504* and *TC79797* are *NPR1/NIM1-*like genes, respectively. (c) Transcript level of *NtNPR1* gene in tobacco plants with or without inoculation with *R. solanacearum* through qRT-PCR analysis. WT-Mock and WT-*R. solanacearum*, *AhRRS5-OE-3-*Mock and *AhRRS5-OE-3-R. solanacearum* indicate wild-type tobacco without or with inoculation with pathogen, AhRRS5-OE-3 transgenic tobacco without or with inoculation with pathogen, respectively. Alphabets mark statistically significant differences between wild-type and transgenic tobacco plants, by Student–Newman–Keuls test (lowercase differences indicate *P*-value <0.05; uppercase differences indicate *P*-value <0.01).

R. solanacearum inoculation. The AhRRS5 protein has a typical NB-ARC domain containing P-loop, kinase-2, kinase-3a and GLPL and other conservative modules similar to *Arabidopsis RPM1*, RXO1 protein of maize, Pid3 of rice and so on (Figure 1; Leister *et al.*,1996; Zhao *et al.*, 2004; Chen *et al.*, 2011). Four normal LRR motifs, which may participate in the peanut pathogen interaction or defence responses against the pathogen, were found in AhRRS5 (Takken and Joosten, 2000). The revealed amino acid sequence of AhRRS5 most closely resembles those of R genes of known functions, such as RXO1 from *Z. mays* resistant to *Xanthomonas oryzae* pv. *Oryzicola* (Zhao *et al.*, 2004, 2005), RPM1 from *A. thaliana* resistant to *P. syringae* (Leister *et al.*, 1996) and Pid3 from rice resistant to *M. oryzae* (Chen *et al.*, 2011) (Data S2). Phylogenetic analysis with 29 R genes of known

functions showed that AhRRS5 could be classified into non-TIR-NBS-LRR type and NBS-LRR subclass of resistance genes.

Subcellular localization visualized by the AhRRS5::GFP fusion protein in *N. benthamiana* leave cells showed that the AhRRS5:: GFP fusion protein appeared solely in the nucleus and was associated with its nuclear localization signal GKFKKLKILGLDRF at positions 816–829 (Figure 3b; Data S1). This result agrees with the subcellular localization features of most NBS-LRR disease resistance genes (Meyers *et al.*, 2003). The first identified resistance gene to bacterial wilt is *RRS1-R* (a TIR-NBS-LRR gene) in *Arabidopsis*, which is mainly cytoplasm-localized but nuclear-localized only depending on the presence of effector PopP2 from *R. solanacearum* (Deslandes *et al.*, 2003). The RRS1-R protein contains TIR-, NBS- and LRR-conserved domains, aside from a

WRKY motif, which activates transcription in plants (Eulgem and Somssich, 2007). Another NBS-LRR resistance gene, *RPS4*, to *R. solanacearum* in *Arabidopsis* is also localized in both the nucleus and cytoplasm (Wirthmueller *et al.*, 2007). However, RPM1 activated in the plasma membrane functions independent of the nucleus (Boyes *et al.*, 1998; Gao *et al.*, 2011). Therefore, AhRRS5 possibly functions mainly in the nucleus.

AhRRS5 is widely involved in defence responses to biotic/abiotic stresses

AhRRS5 transcripts were up-regulated in both resistant and susceptible varieties challenged with *R. solanacearum* and highly up-regulated in the susceptible variety at 24 hpi (Figure 6e). These results indicate that *AhRRS5* participates in the defence response to the pathogen. *AhRRS5* was up-regulated in response to all exogenous hormones applied, namely SA, ABA, ET and JA, in the leaves, although this gene was specifically expressed in the peanut root, testa and pericarp, and weakly in other organs, such as the leaf (Figure 4a). These phytohormones are well-known signalling molecules involved in controlling the defence gene expression against biotic and abiotic stresses (Divi *et al.*, 2010; Ton *et al.*, 2009). SA is usually associated with R gene-mediated disease resistance, and SA-deficient mutants often compromise R gene-mediated resistance (Yang *et al.*, 2013). Exogenous application of SA induces PR genes and enhances resistance to a broad range of pathogens (Bari and Jones, 2009). *Arabidopsis* RRS1-R-mediated resistance to *R. solanacearum* is partially dependent on SA and NDR1 (Deslandes *et al.*, 2002). *Arabidopsis* RCY1 gene, which encodes a CC-NBS-LRR protein for resistance to the yellow strain of cucumber mosaic virus, requires SA and ET signalling (Takahashi *et al.*, 2002). ET regulates various growth and developmental processes and is also involved in responses to stresses, such as salt, drought, cold, flooding and infection caused by microbes and insects (Yoo *et al.*, 2009). ET could modulate disease resistance (Broekaert *et al.*, 2006; Van Loon *et al.*, 2006). MeJA regulate defence to herbivores and necrotrophic pathogens (Browse, 2009). SA and JA/ET defence pathways are usually antagonistic, but synergistic interactions have also been reported in defence response to pathogens (Beckers and Spoel, 2006; Mur *et al.*, 2006; Nahar *et al.*, 2012; Vos *et al.*, 2015), which is also consistent with the results on *AhRRS5* responding to phytohormones such as SA, JA and ET. Rice ET, JA and SA biosynthetic pathways are prerequisites for defence against *Hirschmanniella oryzae*, and ABA participates in the antagonistic interaction to SA/JA/ET-dependent basal defence to the pathogen (Nahar *et al.*, 2012). We found *AhRRS5* was up-regulated in response to all of the four hormones including ABA. ABA functions in abiotic stress tolerance, antagonizes the SA signalling pathway in higher plants and enhances disease susceptibility (Bari and Jones, 2009; Jiang *et al.*, 2010; Nahar *et al.*, 2012). However, ABA plays a positive role in papilla-mediated defence against *Leptosphaeria maculans* in *Arabidopsis* (Ton *et al.*, 2009). Exogenous application of ABA strengthens rice basal resistance against the brown spot caused by *Cochliobolus miyabeanus* (De Vleesschauwer *et al.*, 2012). The role of ABA in defence depends on the type of pathogens, timing of the defence response and plant tissues (Ton *et al.*, 2009). In general, hormone balance plays a vital role in fine tuning appropriate defence responses to the recognized pathogen.

In the present study, the results of qRT-PCR and microarray analysis showed that *AhRRS5* was up-regulated by SA, ABA, ET and JA and was enhanced differently in the response to

R. solanacearum in three resistant varieties. Concentration curves showed that *AhRRS5* was up-regulated with two optimal peaks in response to SA and JA, but with a single peak to ABA and ET in Minhua 6 (Figure 5a–d). Similar patterns were also found in Xinhuixiaoli and Yueyou 92, although *AhRRS5* was down-regulated in Xinhuixiaoli 24 h after JA treatment (Figure 6a–d). These results indicate that *AhRRS5* may involve in the crosstalk between these phytohormones against pathogen infection, such as *R. solanacearum*. *AhRRS5* also showed an altered response to low temperature and drought (Figure 5e,f), indicating its association with biotic/abiotic stresses. Our data suggest that peanut AhRRS5 plays a role in the defence response to bacterial wilt via the synergistic interaction of diverse signalling pathways. Therefore, *AhRRS5* in response to *R. solanacearum* may adopt a distant mechanism in comparison with other pathogen-associated genes.

AhRRS5 confers resistance to bacterial wilt in heterozygous tobacco transformant

The resistance genes against *R. solanacearum* have not been cloned and characterized except for model plant *Arabidopsis* (Deslandes *et al.*, 2002; Godiard *et al.*, 2003). *AtRRS1-R*, genetically identified as recessive, confers dominant resistance to *R. solanacearum* GMI1000 in transgenic *Arabidopsis*. This gene presents a novel R gene structure combining domains of a TIR-NBS-LRR protein and a WRKY motif (Deslandes *et al.*, 2002). Deslandes *et al.* (2003) showed that RRS1 can recognize the pathogen by directly interacting with effector PopP2 and depends on PopP2 to colocalize at the nucleus for pathogen defence. The *Arabidopsis* LRR-RLK gene ERECTA, located in the QTL QRS1, shows resistance to *R. solanacearum* and also affects the development of aerial organs (Godiard *et al.*, 2003). The NB-LRR gene RPS4 from *Arabidopsis* ecotype Ws-0 functions as a dual resistance gene system with RRS1 to prevent three distinct pathogens, namely *R. solanacearum*, *Pst-avrRps4* and *Colletotrichum higginsianum* (Narusaka *et al.*, 2009). *RPS4* was suggested to function downstream of, or together with, *RRS1-Ws* in the signalling pathway resistant to *R. solanacearum*.

AhRRS5 induced by *R. solanacearum* challenge is a non-TIR-NBS-LRR gene different from *RRS1-R* in *Arabidopsis*. Overexpression transgenic tobacco constitutively expressing *AhRRS5* showed enhanced disease resistance to bacterial wilt. In specific, *AhRRS5* overexpression in transgenic CB-1, a medium-susceptible cultivar, showed strong resistance to the pathogen infection (Figure 8). The hypersusceptible cultivar Honghuadajinyuan overexpressing *AhRRS5* also increased the resistance to *R. solanacearum* infection, although different transgenic lines demonstrated distinct levels of resistance in response to the pathogen (Table 1). Lines OE-3 and OE-8 showed much higher resistance or immune response to bacterial wilt than other lines, which may have resulted from the effect of insertion locations of the gene in chromosomes. The transient overexpression of *AhRRS5* in *N. benthamiana* showed that it can induce hypersensitive response causing cell death and also produce H_2O_2 in HR (Figure 7). These results indicate that *AhRRS5* may participate in resistance against *R. solanacearum* involving ROS signalling. Therefore, *AhRRS5* is a novel NBS-LRR resistance gene cloned from peanut, which confers resistance to the *R. solanacearum*.

AhRRS5 resistance is involved in multidefence signalling pathways

A complex network of different signalling transductions exists in plant–pathogen interactions, and different signalling pathways are associated with the transcription of some marker genes in their mediated disease resistance reaction. Many marker genes, such as NtHIN1, HSR201 and HSR515, are activated in HR signalling (Sohn et al., 2007). SA-mediated defence responses could activate system-acquired resistance (SAR) and are accompanied with the expression of several PR genes, such as PR1a/c, PR3, PR4 and PR5 (Dong, 1998; Glazebrook, 2005). PR genes PR2 and PR1b are activated and expressed in ET-mediated defence response, whereas EFE26 and ACS6 are activated in JA-mediated defence response (Koornneef and Pieterse, 2008; Kunkel and Brooks, 2002; Thomma et al., 1998). Changes in the expression levels of these markers directly indicate the involvement of plant defence responses and signal transduction pathways (Chen et al., 2003; Rizhsky et al., 2002; Sohn et al., 2007). We examined the transcripts of these marker genes in AhRRS5 overexpression tobacco lines by qPCR. Results showed that AhRRS5 overexpression up-regulated not only the transcript levels of NtHIN1, NtHSR201 and NtHSR515 in HR signalling but also those of SA-regulated genes (PR1a/c, PR3) in the T_2 tobacco plants inoculated with virulent R. solanacearum (Figure 9a,b). The transcript levels of JA-regulated PR1b and ET-responsive NtEFE26 and NtACS6 were also greatly enhanced (Figure 9c,d). The results conform to the data in peanut, in which AhRRS5 was up-regulated by the exogenous applications of SA, ET, JA and ABA. The RRS1-R-mediated bacterial wilt resistance in Arabidopsis involves ABA participation, and the effect of ABA is greater than that of SA (Deslandes et al., 2003; Hernández-Blanco et al., 2007). These results are relatively similar to AhRRS5 response to R. solanacearum, indicating that these hormone signals perform synergistically against the pathogen. The overexpression of AhRRS5 conferring increased resistance to bacterial wilt in tobacco was achieved by the increase the gene expression in defence signal transduction pathways.

AhRRS5 resistance requires the involvement of NDR1 and NPR1

AhRRS5 overexpression up-regulated NDR1 transcripts in response to R. solanacearum challenge, concurring with the report that RRS1-R in Arabidopsis is SA-dependent and requires the downstream gene NDR1 for its resistance to bacterial wilt (cf., Chen et al., 2003). However, NDR1 was significantly down-regulated in the nontransgenic resistance variety Yanyan 97 in response to the pathogen. This finding indicates that other resistance mechanisms exist in response to bacterial wilt. NDR1 primarily mediates signalling derived from the CC-NB-LRR type of R proteins, whereas EDS1 involves those from the TIR-NB-LRR class of R proteins (Aarts et al., 1998; Wang et al., 2014). These results are apparently contradictory to the events of AhRRS5 and AtRRS1-R (Deslandes et al., 2002; Lahaye, 2002). NDR1 involves R protein-mediated resistance to many pathogens (Day et al., 2006; Lu et al., 2013; Repetti et al., 2004). Soya bean GmNDR1a and GmNDR1b bind pathogen effectors and regulate resistance signalling (Selote et al., 2014). Arabidopsis resistance signalling pathways to P. syringae 2 and P. syringae pv. maculicola 1 exhibit different mechanisms of activation in terms of effector action, but both require NDR1 participation (Kim, 2006). Thus, AhRRS5 is associated with NDR1 for its mediated resistance to bacterial wilt.

NPR1 is a key regulator of SAR and is essential for the SA signal transduction to activate PR gene expression (Pieterse and Van Loon, 2004; Sandhu et al., 2009). We examined NPR1 transcription by employing microarray analysis and found that the transgenic plants overexpressing AhRRS5 up-regulated the two NPR1 transcripts after inoculation with R. solanacearum but down-regulated them after pathogen challenge in wild-type plants (Figure 10a,b). These results were confirmed in the resistant and susceptible varieties, indicating that NPR1 plays an important role in pathogen resistance. We further found that the AhRRS5-OE-3 line significantly up-regulated the transcript level of NPR1 by 14.5-fold in response to the R. solanacearum challenge (Figure 10c). The PR marker genes of SA signalling in the transgenic plants of AhRRS5 were then up-regulated (Figure 9b). NPR1-mediated signalling resisting viral and bacterial pathogens and repressing NPR1 transcript would increase the susceptibility of plants to pathogens (Li et al., 2012; Xiao and Chye, 2011). Thus, our results suggest that AhRRS5 participates in pathogen resistance by employing the NPR1-mediated SA signalling and the R gene pathway associated with NDR1.

Experimental procedures

Plant materials and growth conditions

Peanut cultivars (Arachis hypogaea cv. Minhua 6, cv. Yueyou 92 and cv. Xinhuixiaoli, as medium-resistant, hyper-resistant and hyper-susceptible variants to R. solanacearum, respectively) were provided by the Oil Crop Institute in Fujian Agriculture and Forestry University. Seeds were sown in sterile sands in plastic pots. Seedlings of transgenic lines and wild-type tobacco (Nicotiana tabacum cv. CB-1, cv. Yanyan 97 and cv. Honghuadajinyuan, with medium susceptibility, hyper-resistance and hyper-susceptibility to R. solanacearum, respectively) were provided by Fujian Tobacco Agricultural Research Institute. N. benthamiana is available in this laboratory. T_2 seeds of transgenic tobacco lines were surface-sterilized with 75% alcohol for 20 sec, 10% H_2O_2 for 10 min, washed five times with sterile water and finally placed on MS medium supplemented with 75 mg/L kanamycin for 2–3 weeks. The survivals were then transferred into a soil mix containing peat moss/perlite (2/1, v/v) in a plastic tray and grown in a greenhouse for another 2–3 weeks. Transgenic and wild-type tobacco plants of the same size were transferred into a soil mix containing peat moss/general soil (2/1, v/v) in plastic pots for another 3–4 weeks. Peanut and tobacco plants were grown in the greenhouse at 26 °C and 70% relative humidity under a 16 h/8 h light/dark cycle.

Pathogens and inoculation

Virulent strains Rs-P.362200 and FJ1003 strain of R. solanacearum were from peanut and tobacco, respectively. The pathogen strains were streaked on TTC agar medium (0.5 g/L 2,3,5-triphenyltetrazolium chloride, 5 g/L peptone, 0.1 g/L casein hydrolysate, 2 g/L D-glucose and 15 g/L agar) (Kelman, 1954) and then incubated at 28 °C for 48 h. Virulent colonies were harvested with sterile water (with 0.02% Tween-20), and the inoculum was prepared by adjusting the concentration of bacterial cells to an optical density of 0.5 at 600 nm wavelength (NanoDrop 2000c; Thermo Fisher Scientific, San Jose, CA, USA), corresponding to approximately 10^8 cfu/mL.

Then, 4-week-old peanut seedlings of Yueyou 92 and Xinhuixiaoli were inoculated at the third and fourth leaves from the upperpart by leaflet cutting (perpendicular to the midrib of leaflet, 2/3 deep cut to the midrib), and four leaflets were inoculated per plant. Control plants were inoculated with distilled water containing 0.02% Tween-20. Two uncut leaflets of the treated leaves were harvested at the indicated time points for future analysis.

Tobacco was inoculated by infiltrating 10 μL of *R. solanacearum* suspension with 10^8 cfu/mL concentration into the third leaves from the upperpart using a syringe with a needle, and then, the fourth leaves were harvested at the indicated time points for future analysis. The typical symptoms of bacterial wilt were monitored daily in five disease severity ratings from 0 to 4, where 0 = no symptoms, 1 = 1/4 inoculated leaves wilted, 2 = 1/4–1/2 inoculated leaves wilted, 3 = 1/2–3/4 inoculated leaves wilted and 4 = whole plant wilted, plant death. Disease index (DI) and death ratio (DR) were calculated using the following formula: DI (%) = [∑ (ni × vi) ÷ (V × N)] × 100, DR (%) = (ni ÷ N) × 100, where ni = number of plants with the respective disease rating; vi = the disease rating; V = the highest disease rating; and N = the total number of observed plants.

Application of plant hormones and abiotic/biotic stresses

One-month-old peanut seedlings (Minhua 6) were sprayed with 3 mM SA, 10 μg/mL ABA, 1 mg/mL ET and 100 mM MeJA in distilled water (H_2O). Control seedlings were sprayed with distilled water (H_2O). The leaves of the treated seedlings were harvested at indicated time points, frozen in liquid nitrogen and then stored at −80 °C until used. Yueyou 92 and Xinhuixiaoli were used in another trial. Seven-leaf peanut Minhua 6 plants were treated at 4 °C and 25 °C. Leaves were harvested at indicated time points after treatments. Minhua 6 plants at the seven-leaf stage were treated by stopping and normal watering for drought stress. Leaves were harvested at different time points, frozen in liquid nitrogen and then stored at −80 °C until use. Three biological replicates were set for all stress treatments.

Full-length cDNA cloning

The candidate gene was screened through microarray analysis with approximately 100 000 unigene probes on the basis of the available fragment sequence. The 5′- and 3′-end cDNA sequences were cloned by RACE using the SMARTᴛᴍ RACE cloning kit (Clontech, Palo Alto, CA) in accordance with the manufacturer's instructions with minor revisions. Total RNA was extracted from the leaves of resistant peanut cultivar to *R. solanacearum* by the CTAB method. RACE-F and 3′ PCR adaptor primers were joined on both ends of the cDNA. Then, 5′ RACE was generated by PCR using the primary primer set of RACE-F primer and PRRS_1EW9-R, followed by the reaction system: 94 °C for 5 min; 35 cycles of 30 s at 95 °C, 30 s at 60 °C and 1 min 30 s at 72 °C; and 72 °C for 10 min. Similarly, 3′ RACE was generated by the set of PRRS_1EW9_F and the 3′ PCR primer with the following PCR programme: 94 °C for 5 min; 5 cycles of 30 s at 95 °C and 2 min at 72 °C; and 30 cycles of 30 s at 95 °C, 60 °C, 30 s, 2 min at 72 °C; and 72 °C for 10 min. The RACE products were cloned and sequenced. After assembly, full-length cDNA and DNA sequences of *AhRRS5* were cloned from the reverse transcription products and genomic DNA by using the set of AhRRS5- FL-F and AhRRS5-FL-R. All primers used in this study are listed in Table S1.

Sequence analysis and phylogenetic tree construction

AhRRS5 sequence similarity analysis was performed using BLASTN and BLASTX (http://www.ncbi.nlm.nih.gov/BLAST). Four known functional resistant proteins with close similarities were obtained from the BLASTX results. Multiple sequence alignments were performed with ClustalW2 (Data S1). A phylogenetic tree was generated using 29 resistant proteins of known function by using MEGA 5.10 (Data S3).

Subcellular localization

The full-length *AhRRS5* ORF without the termination codon was amplified by high-fidelity PCR polymerase with gene-specific primers AhRRS5-BamH1-F and AhRRS5-Asc1-R harbouring BamHI and AscI sites, respectively. The PCR products were inserted into the vector pBI-GFP between *BamHI* and *AscI* and formed a construct with the p35S::AhRRS5-GFP fusion gene. With pBI-GFP containing 35S::GFP as a control, p35S::AhRRS5-GFP and p35S::GFP were transformed into *Agrobacterium tumefaciens* strain GV3101. which was cultured in induction medium (10 mM ethanesulfonic acid, pH 5.7, 10 mM $MgCl_2$ and 200 mM acetosyringone), harvested and diluted to OD_{600} = 0.8, and then injected into *Nicotiana benthamiana* leaves using a syringe without a needle. Forty-eight hours after agroinfiltration, GFP fluorescence was imaged in a fluorescence microscope, with an excitation wavelength of 488 nm and a 505–530 nm bandpass emission filter. GFP florescence was imaged using laser confocal florescence microscopy (Leica TCS SP8, Solms, Germany).

Vector construction and transient expression

The complete ORF of *AhRRS5* was amplified by high-fidelity PCR polymerase with AhRRS5-OE-F and AhRRS5-OE-R primers harbouring *BamHI* and *AscI* sites, respectively. The PCR products were cloned into the modified vector pBI121-GUSA between *BamHI* and *AscI* sites to replace the GUSA gene. The obtained vector containing *AhRRS5* driven by the 2 × CaMV35S promoter was named *p35S::AhRRS5*. The *p35S::AhRRS5* vector was transferred into *Agrobacterium tumefaciens* strains GV3101 and EHA105.

Agrobacterium tumefaciens strain GV3101 harbouring the *p35S::AhRRS5* vector was cultured to OD_{600} = 1.0 in induction medium (10 mM ethanesulfonic acid, pH 5.7, 10 mM $MgCl_2$ and 200 mM acetosyringone) and diluted to OD600 = 0.8. Th diluted culture was injected into *Nicotiana benthamiana* leaves using a syringe without a needle. For the DAB and trypan blue staining, the tobacco (*N. benthamiana*) leaf was infiltrated of *AhRRS5* in a small syringe with 1.0 cm diameter, the volume was about 100 μL. For the electrolyte leakage analysis, the second leaf was infiltrated with about 1 mL agrobacterium until spread to the whole leaf. The infiltrated leaves were harvested at the indicated time points for future analysis. Three biological replicates were set for the experiment.

Tobacco transformation

N. tabacum cv. CB-1, cv. Honghuadajinyuan were used as the host, and p35S::AhRRS5 fusion gene was transformed by the leaf-disc method mediated by EHA105 to generate transgenic plants (Rizhsky *et al.*, 2002). The initial transgenic T_0 and T_1 offspring were selected by kanamycin and confirmed by RT-PCR to verify transgene integration. The T_2 transgenic homozygous lines were obtained and used in this study.

Quantitative real-time RT-PCR

Total RNA was extracted from peanut, transgenic tobacco and wild-type seedlings through CTAB extraction (Chen *et al.*, 2015). Reverse transcription was performed with PrimeScriptᴛᴍ RTase (TaKaRa, Dalian, China) in accordance with the manufacturer's instructions. Real-time PCR for the relative expression level of target genes was performed with specific primers (see Table S1 for gene-specific primers) essentially provided for the Master cyclerepealplex (Eppendorf, Hamburg, Germany) and SYBR

Premix Ex Taq II (Perfect Real Time; TaKaRa, Dalian China). Each reaction mix (20 μL) contained 10 μL of SYBR Premix ExTaq (2×), 0.2 μL of PCR forward/reverse gene-specific primers (10 μM) and diluted cDNA (2 μL). Three experimental replicates were performed for each gene using different cDNAs synthesized from three biological replicates. The PCR programme was as follows: 95 °C for 5 min; 40 cycles of 5 s at 95 °C, 30 s at 60 °C and 30 s at 72 °C; and 95 °C for 15 s, 60 °C for 1 min, 95 °C for 15 s and 60 °C for 15 s. The specificity of amplification was confirmed by melting curve analysis after 40 cycles. The relative expression level of the target gene was calculated using the comparative CT method ($2^{-\Delta\Delta CT}$ method) (Schmittgen and Livak, 2008) by normalizing the PCR threshold cycle number (Ct value) of the target gene with that of the reference gene. The Ct value was calculated as follows: $\Delta\Delta Ct = (CT_{gene} - CT_{actin})_{treat} - (CT_{gene} - CT_{actin})_{control}$. Ahactin was used as an internal reference to detect the relative transcript level of AhRRS5 under different treatments in peanut. Tobacco NtEF1α was used as an internal reference to detect the relative transcript levels of related defence genes after treatment with R. solanacearum between the wild-type and transgenic tobacco plants.

Histochemical analysis and ion conductivity determination

Transient expression development was assessed 48 h after the transient overexpression of AhRRS5 in tobacco leaves by staining the infected plants with 3, 3′-diaminobenzidine (DAB; Sigma, St. Louis, MO) and lactophenol–ethanol–trypan blue. The infected tobacco leaves were incubated in 1 mg/mL DAB solution overnight at room temperature, boiled for 5 min in a solution of 3:1:1 ethanol/lactic acid/glycerol and then placed in absolute ethanol before observation to measure H_2O_2 level. Cell death was detected by boiling the inoculated leaves in trypan blue staining solution (10 mL of lactic acid, 10 mL of glycerol, 10 g of phenol, 30 mL of absolute ethanol and 10 mg of trypan blue, dissolved in 10 mL of ddH$_2$O) for 2 min. The leaves were left at room temperature overnight, transferred into chloral hydrate solution (2.5 g of chloral hydrate dissolved in 1 mL of distilled water) and then boiled for 20 min to destain. The leaves were observed under a light microscope.

Ion conductivity was measured as previously described with minor modifications (Hwang and Hwang, 2011). Six round leaf discs (11 mm in diameter) per agroinfiltrated leave were cut, washed in ddH$_2$O and then incubated in 20 mL of ddH$_2$O with evacuation for 10 min at room temperature. Electrolyte leakage was measured using MettlerToledo 326.

Microarray analysis

In silico analysis of AhRRS5 gene expression pattern in peanut, microarray designing, hybridization, washing, and scanning and data analysis were performed as described by Chen et al. (2015). The gene expression intensity of all hybridizations was analysed, and expression levels were estimated among different tissues and under diverse stress conditions. The expression data of genes were normalized using quantile normalization (Bolstad et al., 2003) and generated using the Robust Multichip Average algorithm (Irizarry et al., 2003a,b). Three replicates were performed for all experiments.

Tobacco microarray analysis was performed using the leaves of the hyper-resistant tobacco variety Yanyan 97, hypersusceptible tobacco variety Honghuadajinyuan, T$_2$ generation transgenic tobacco of AhRRS5-OE-3, and wild-type tobacco after R. solanacearum inoculation. Microarray designing, hybridization,

washing, and scanning and data analysis were conducted as previously described (Zhang et al., 2016). Gene expression data were analysed as follows.

Acknowledgement

This work was supported by the International Cooperation Program (2008DFA31450) and the National 863 Program (2013AA102602) of the Ministry of Science and Technology of P.R. China.

References

Aarts, N., Metz, M., Holub, E., Staskawicz, B.J., Daniels, M.J. and Parker, J.E. (1998) Different requirements for EDS1 and NDR1 by disease resistance genes define at least two R gene-mediated signaling pathways in Arabidopsis. Proc. Natl Acad. Sci. 95, 10306–10311.

Ameline-Torregrosa, C., Wang, B.-B., O'Bleness, M.S., Deshpande, S., Zhu, H., Roe, B., Young, N.D. et al. (2008) Identification and characterization of nucleotide-binding site-leucine-rich repeat genes in the model plant Medicago truncatula. Plant Physiol. 146, 5–21.

Baker, B., Zambryski, P., Staskawicz, B. and Dinesh-Kumar, S.P. (1997) Signaling in plant-microbe interactions. Science, 276, 726–733.

Bari, R. and Jones, J.D.G. (2009) Role of plant hormones in plant defence responses. Plant Mol. Biol. 69, 473–488.

Beckers, G.J.M. and Spoel, S.H. (2006) Fine-tuning plant defence signalling: salicylate versus jasmonate. Plant Biol. 8, 1–10.

Ben, C., Debellé, F., Berges, H., Bellec, A., Jardinaud, M.-F., Anson, P., Huguet, T. et al. (2013) MtQRRS1, an R-locus required for Medicago truncatula quantitative resistance to Ralstonia solanacearum. New Phytol. 199, 758–772.

Bertioli, D.J., Leal-Bertioli, S.C.M., Lion, M.B., Santos, V.L., Pappas, G., Cannon, S.B. and Guimarães, P.M. (2003) A large scale analysis of resistance gene homologues in Arachis. Mol. Genet. Genomics, 270, 34–45.

Bhatnagar-Mathur, P., Sunkara, S., Bhatnagar-Panwar, M., Waliyar, F. and Sharma, K.K. (2015) Biotechnological advances for combating Aspergillus flavus and aflatoxin contamination in crops. Plant Sci. 234, 119–132.

Bolstad, B.M., Irizarry, R.A., Åstrand, M. and Speed, T.P. (2003) A comparison of normalization methods for high density oligonucleotide array data based on variance and bias. Bioinformatics, 19, 185–193.

Bonardi, V., Tang, S., Stallmann, A., Roberts, M., Cherkis, K. and Dangl, J.L. (2011) Expanded functions for a family of plant intracellular immune receptors beyond specific recognition of pathogen effectors. Proc. Natl Acad. Sci. 108, 16463–16468.

Boyes, D.C., Nam, J. and Dangl, J.L. (1998) The Arabidopsis thaliana RPM1 disease resistance gene product is a peripheral plasma membrane protein that is degraded coincident with the hypersensitive response. Proc. Natl Acad. Sci. 95, 15849–15854.

Broekaert, W.F., Delauré, S.L., De Bolle, M.F.C. and Cammue, B.P.A. (2006) The role of ethylene in host-pathogen interactions. Annu. Rev. Phytopathol. 44, 393–416.

Brogue, K., Chet, I., Holliday, M., Cressman, R., Biddle, P., Knowlton, S., Mauvais, C.J. et al. (1991) Transgenic plants with enhanced resistance to the fungal pathogen Rhizoctonia solani. Science, 254, 1194–1197.

Browse, J. (2009) Jasmonate passes muster: a receptor and targets for the defense hormone. Annu. Rev. Plant Biol. 60, 183–205.

Carmeille, A., Caranta, C., Dintinger, J., Prior, P., Luisetti, J. and Besse, P. (2006) Identification of QTLs for Ralstonia solanacearum race 3-phylotype II resistance in tomato. Theor. Appl. Genet. 113, 110–121.

Cesari, S., Thilliez, G., Ribot, C., Chalvon, V., Michel, C., Jauneau, A., Rivas, S. et al. (2013) The rice resistance protein pair RGA4/RGA5 recognizes the Magnaporthe oryzae effectors AVR-Pia and AVR1-CO39 by direct binding. Plant Cell, 25, 1463–1481.

Chen, N., Goodwin, P.H. and Hsiang, T. (2003) The role of ethylene during the infection of Nicotiana tabacum by Colletotrichum destructivum. J. Exp. Bot. 54, 2449–2456.

Chen, J., Shi, Y., Liu, W., Chai, R., Fu, Y., Zhuang, J. and Wu, J. (2011) A Pid3 allele from rice cultivar Gumei2 confers resistance to *Magnaporthe oryzae*. *J. Genet. Genomics*, **38**, 209–216.

Chen, H., Zhang, C., Deng, Y., Zhou, S., Zheng, Y., Ma, S., Tang, R. *et al.* (2016) Identification of low Ca2 + stress-induced embryo apoptosis response genes in *Arachis hypogaea* by SSH-associated library lift (SSHaLL). *Plant Biotechnol. J.* **14**, 682–698.

Danesh, D., Aarons, S., McGill, G.E. and Young, N.D. (1994) Genetic dissection of oligogenic resistance to bacterial wilt in tomato. *Mol. Plant Microbe Interact.* **7**, 464–471.

Dangl, J.L., Dietrich, R.A. and Richberg, M.H. (1996) Death don't have no mercy: cell death programs in plant-microbe interactions. *Plant Cell*, **8**, 1793.

Day, B., Dahlbeck, D. and Staskawicz, B.J. (2006) NDR1 interaction with *RIN4* mediates the differential activation of multiple disease resistance pathways in *Arabidopsis*. *Plant Cell*, **18**, 2782–2791.

De Vleesschauwer, D., Van Buyten, E., Satoh, K., Balidion, J., Mauleon, R., Choi, I.-R., Vera-Cruz, C. *et al.* (2012) Brassinosteroids antagonize gibberellin-and salicylate-mediated root immunity in rice. *Plant Physiol.* **158**, 1833–1846.

Deslandes, L., Olivier, J., Theulières, F., Hirsch, J., Feng, D.X., Bittner-Eddy, P., Beynon, J. *et al.* (2002) Resistance to *Ralstonia solanacearum* in *Arabidopsis thaliana* is conferred by the recessive *RRS1-R* gene, a member of a novel family of resistance genes. *Proc. Natl Acad. Sci.* **99**, 2404–2409.

Deslandes, L., Olivier, J., Peeters, N., Feng, D.X., Khounlotham, M., Boucher, C., Somssich, I. *et al.* (2003) Physical interaction between RRS1-R, a protein conferring resistance to bacterial wilt, and PopP2, a type III effector targeted to the plant nucleus. *Proc. Natl Acad. Sci. USA*, **100**, 8024–8029.

Divi, U.K., Rahman, T. and Krishna, P. (2010) Brassinosteroid-mediated stress tolerance in *Arabidopsis* shows interactions with abscisic acid, ethylene and salicylic acid pathways. *BMC Plant Biol.* **10**, 151.

Dong, X. (1998) SA, JA, ethylene, and disease resistance in plants. *Curr. Opin. Plant Biol.* **1**, 316–323.

Eulgem, T. and Somssich, I.E. (2007) Networks of WRKY transcription factors in defense signaling. *Curr. Opin. Plant Biol.* **10**, 366–371.

Flor, H.H. (1971) Current status of the gene-for-gene concept. *Annu. Rev. Phytopathol.* **9**, 275–296.

Gabriëls, S.H.E.J., Vossen, J.H., Ekengren, S.K., van Ooijen, G., Abd-El-Haliem, A.M., Berg, G., Rainey, D.Y. *et al.* (2007) An NB-LRR protein required for HR signalling mediated by both extra-and intracellular resistance proteins. *Plant J.* **50**, 14–28.

Gao, Y., Xu, Z., Jiao, F., Yu, H., Xiao, B., Li, Y. and Lu, X. (2010) Cloning, structural features, and expression analysis of resistance gene analogs in tobacco. *Mol. Biol. Rep.* **37**, 345–354.

Gao, Z., Chung, E.-H., Eitas, T.K. and Dangl, J.L. (2011) Plant intracellular innate immune receptor Resistance to *Pseudomonas syringae* pv. maculicola 1 (RPM1) is activated at, and functions on, the plasma membrane. *Proc. Natl Acad. Sci.* **108**, 7619–7624.

Glazebrook, J. (2005) Contrasting mechanisms of defense against biotrophic and necrotrophic pathogens. *Annu. Rev. Phytopathol.* **43**, 205–227.

Godiard, L., Sauviac, L., Torii, K.U., Grenon, O., Mangin, B., Grimsley, N.H. and Marco, Y. (2003) ERECTA, an LRR receptor-like kinase protein controlling development pleiotropically affects resistance to bacterial wilt. *Plant J.* **36**, 353–365.

Greenberg, J.T. (1997) Programmed cell death in plant-pathogen interactions. *Annu. Rev. Plant Biol.* **48**, 525–545.

Gururani, M.A., Venkatesh, J., Upadhyaya, C.P., Nookaraju, A., Pandey, S.K. and Park, S.W. (2012) Plant disease resistance genes: current status and future directions. *Physiol. Mol. Plant Pathol.* **78**, 51–65.

Henry, E., Yadeta, K.A. and Coaker, G. (2013) Recognition of bacterial plant pathogens: local, systemic and transgenerational immunity. *New Phytol.* **199**, 908–915.

Hernández-Blanco, C., Feng, D.X., Hu, J., Sánchez-Vallet, A., Deslandes, L., Llorente, F., Berrocal-Lobo, M. *et al.* (2007) Impairment of cellulose synthases required for *Arabidopsis* secondary cell wall formation enhances disease resistance. *Plant Cell Online*, **19**, 890–903.

Hulbert, S.H., Webb, C.A., Smith, S.M. and Sun, Q. (2001) Resistance gene complexes: evolution and utilization. *Annu. Rev. Phytopathol.* **39**, 285–312.

Hwang, I.S. and Hwang, B.K. (2011) The pepper mannose-binding lectin gene *CaMBL1* is required to regulate cell death and defense responses to microbial pathogens. *Plant Physiol.* **155**, 447–463.

Irizarry, R.A., Bolstad, B.M., Collin, F., Cope, L.M., Hobbs, B. and Speed, T.P. (2003a) Summaries of Affymetrix GeneChip probe level data. *Nucleic Acids Res.* **31**, e15.

Irizarry, R.A., Hobbs, B., Collin, F., Beazer-Barclay, Y.D., Antonellis, K.J., Scherf, U., Speed, T.P. *et al.* (2003b) Exploration, normalization, and summaries of high density oligonucleotide array probe level data. *Biostatistics*, **4**, 249–264.

Jiang, C.-J., Shimono, M., Sugano, S., Kojima, M., Yazawa, K., Yoshida, R., Inoue, H. *et al.* (2010) Abscisic acid interacts antagonistically with salicylic acid signaling pathway in rice-*Magnaporthe grisea* interaction. *Mol. Plant Microbe Interact.* **23**, 791–798.

Jones, J.D.G. and Dangl, J.L. (2006) The plant immune system. *Nature*, **444**, 323–329.

Keen, N.T. (1990) Gene-for-gene complementarity in plant-pathogen interactions. *Annu. Rev. Genet.* **24**, 447–463.

Kelman, A. (1954) The relationship of pathogenicity of *Pseudomonas solanacearum* to colony appearance in a tetrazolium medium. *Phytopathology*, **44**, 693–695.

Keneni, G., Bekele, E., Imtiaz, M. and Dagne, K. (2012) Genetic vulnerability of modern crop cultivars: causes, mechanism and remedies. *Int. J. Plant Res.* **2**, 69–79.

Kim, M.G. (2006) *The Molecular Battle between Virulence Weapons of Pseudomonas syringae* and Integrated Defense Responses of *Arabidopsis thaliana*. Ohio: The Ohio State University.

Koornneef, A. and Pieterse, C.M.J. (2008) Cross talk in defense signaling. *Plant Physiol.* **146**, 839–844.

Kunkel, B.N. and Brooks, D.M. (2002) Cross talk between signaling pathways in pathogen defense. *Curr. Opin. Plant Biol.* **5**, 325–331.

Lahaye, T. (2002) The *Arabidopsis RRS1-R* disease resistance gene-uncovering the plant's nucleus as the new battlefield of plant defense? *Trends Plant Sci.* **7**, 425–427.

Lamb, C.J., Lawton, M.A., Dron, M. and Dixon, R.A. (1989) Signals and transduction mechanisms for activation of plant defenses against microbial attack. *Cell*, **56**, 215–224.

Lebeau, A., Gouy, M., Daunay, M.C., Wicker, E., Chiroleu, F., Prior, P., Frary, A. *et al.* (2013) Genetic mapping of a major dominant gene for resistance to *Ralstonia solanacearum* in eggplant. *Theor. Appl. Genet.* **126**, 143–158.

Leister, R.T., Ausubel, F.M. and Katagiri, F. (1996) Molecular recognition of pathogen attack occurs inside of plant cells in plant disease resistance specified by the *Arabidopsis* genes *RPS2* and *RPM1*. *Proc. Natl Acad. Sci.* **93**, 15497–15502.

Li, B., Gao, R., Cui, R., Lü, B., Li, X., Zhao, Y., You, Z. *et al.* (2012) Tobacco TTG2 suppresses resistance to pathogens by sequestering NPR1 from the nucleus. *J. Cell Sci.* **125**, 4913–4922.

Liu, J., Liu, X., Dai, L. and Wang, G. (2007) Recent progress in elucidating the structure, function and evolution of disease resistance genes in plants. *J. Genet. Genomics*, **34**, 765–776.

Lu, H., Zhang, C., Albrecht, U., Shimizu, R., Wang, G. and Bowman, K.D. (2013) Overexpression of a citrus NDR1 ortholog increases disease resistance in *Arabidopsis*. *Front. Plant Sci.* **4**, 10–3389.

Mangin, B., Thoquet, P., Olivier, J. and Grimsley, N.H. (1999) Temporal and multiple quantitative trait loci analyses of resistance to bacterial wilt in tomato permit the resolution of linked loci. *Genetics*, **151**, 1165–1172.

McDowell, J.M. and Woffenden, B.J. (2003) Plant disease resistance genes: recent insights and potential applications. *Trends Biotechnol.* **21**, 178–183.

McHale, L., Tan, X., Koehl, P. and Michelmore, R.W. (2006) Plant NBS-LRR proteins: adaptable guards. *Genome Biol.* **7**, 212.

Meyers, B.C., Kozik, A., Griego, A., Kuang, H. and Michelmore, R.W. (2003) Genome-wide analysis of NBS-LRR–encoding genes in Arabidopsis. *Plant Cell Online*, **15**, 809–834.

Meyers, B.C., Kaushik, S. and Nandety, R.S. (2005) Evolving disease resistance genes. *Curr. Opin. Plant Biol.* **8**, 129–134.

Mindrinos, M., Katagiri, F., Yu, G.-L. and Ausubel, F.M. (1994) The *A. thaliana* disease resistance gene *RPS2* encodes a protein containing a nucleotide-binding site and leucine-rich repeats. *Cell*, **78**, 1089–1099.

Mur, L.A.J., Kenton, P., Atzorn, R., Miersch, O. and Wasternack, C. (2006) The outcomes of concentration-specific interactions between salicylate and jasmonate signaling include synergy, antagonism, and oxidative stress leading to cell death. *Plant Physiol.* **140**, 249–262.

Nahar, K., Kyndt, T., Nzogela, Y.B. and Gheysen, G. (2012) Abscisic acid interacts antagonistically with classical defense pathways in rice–migratory nematode interaction. *New Phytol.* **196**, 901–913.

Narusaka, M., Shirasu, K., Noutoshi, Y., Kubo, Y., Shiraishi, T., Iwabuchi, M. and Narusaka, Y. (2009) RRS1 and RPS4 provide a dual resistance-gene system against fungal and bacterial pathogens. *Plant J.* **60**, 218–226.

Peart, J.R., Mestre, P., Lu, R., Malcuit, I. and Baulcombe, D.C. (2005) NRG1, a CC-NB-LRR protein, together with N, a TIR-NB-LRR protein, mediates resistance against tobacco mosaic virus. *Curr. Biol.* **15**, 968–973.

Pieterse, C.M.J. and Van Loon, L.C. (2004) NPR1: the spider in the web of induced resistance signaling pathways. *Curr. Opin. Plant Biol.* **7**, 456–464.

Qian, Y., Wang, X., Wang, D., Zhang, L., Zu, C., Gao, Z., Zhang, H. *et al.* (2013) The detection of QTLs controlling bacterial wilt resistance in tobacco (*N. tabacum* L.). *Euphytica*, **192**, 259–266.

Reddy, P.P. (2016) Viral diseases and their management. *Sustain. Crop Prot. Prot. Cultiv.* 161–176.

Repetti, P.P., Day, B., Dahlbeck, D., Mehlert, A. and Staskawicz, B.J. (2004) Overexpression of the plasma membrane-localized NDR1 protein results in enhanced bacterial disease resistance in *Arabidopsis thaliana*. *Plant J.* **40**, 225–237.

Rizhsky, L., Liang, H. and Mittler, R. (2002) The combined effect of drought stress and heat shock on gene expression in tobacco. *Plant Physiol.* **130**, 1143–1151.

Ryals, J.A., Neuenschwander, U.H., Willits, M.G., Molina, A., Steiner, H.-Y. and Hunt, M.D. (1996) Systemic acquired resistance. *Plant Cell*, **8**, 1809.

Sandhu, D., Tasma, I.M., Frasch, R. and Bhattacharyya, M.K. (2009) Systemic acquired resistance in soybean is regulated by two proteins, orthologous to *Arabidopsis* NPR1. *BMC Plant Biol.* **9**, 105.

Schmittgen, T.D. and Livak, K.J. (2008) Analyzing real-time PCR data by the comparative CT method. *Nat. Protoc.* **3**, 1101–1108.

Selote, D., Shine, M.B., Robin, G.P. and Kachroo, A. (2014) Soybean NDR1-like proteins bind pathogen effectors and regulate resistance signaling. *New Phytol.* **202**, 485–498.

Sinapidou, E., Williams, K., Nott, L., Bahkt, S., Tör, M., Crute, I., Bittner-Eddy, P. *et al.* (2004) Two TIR: NB: LRR genes are required to specify resistance to Peronospora parasitica isolate Cala2 in *Arabidopsis*. *Plant J.* **38**, 898–909.

Sohn, S.-I., Kim, Y.-H., Kim, B.-R., Lee, S.-Y., Lim, C.K., Hur, J.H. and Lee, J.-Y. (2007) Transgenic tobacco expressing the hrpN (EP) gene from *Erwinia pyrifoliae* triggers defense responses against botrytis cinerea. *Mol. Cells*, **24**, 232.

Sohn, K.H., Segonzac, C., Rallapalli, G., Sarris, P.F., Woo, J.Y., Williams, S.J., Newman, T.E. *et al.* (2014) The nuclear immune receptor RPS4 is required for RRS1 SLH1-dependent constitutive defense activation in *Arabidopsis thaliana*. *PLoS Genet.* **10**, e1004655.

Somers, D.J., Palomino, C., Satovic, Z., Cubero, J.I. and Torres, A.M. (2006) Identification and characterization of NBS-LRR class resistance gene analogs in faba bean (*Vicia faba* L.) and chickpea (*Cicer arietinum* L.). *Genome*, **49**, 1227–1237.

Sunkara, S., Bhatnagar-Mathur, P. and Sharma, K.K. (2014) Transgenic interventions in peanut crop improvement: progress and prospects. *Genet. Genomics Breed. Peanuts*, **178**, 179–216.

Takahashi, H., Miller, J., Nozaki, Y., Takeda, M., Shah, J., Hase, S., Ikegami, M. *et al.* (2002) *RCY1*, an *Arabidopsis thaliana* RPP8/HRT family resistance gene, conferring resistance to cucumber mosaic virus requires salicylic acid, ethylene and a novel signal transduction mechanism. *Plant J.* **32**, 655–667.

Takken, F.L.W. and Joosten, M.H.A.J. (2000) Plant resistance genes: their structure, function and evolution. *Eur. J. Plant Pathol.* **106**, 699–713.

Thomma, B.P.H.J., Eggermont, K., Penninckx, I.A.M.A., Mauch-Mani, B., Vogelsang, R., Cammue, B.P.A. and Broekaert, W.F. (1998) Separate jasmonate-dependent and salicylate-dependent defense-response pathways in *Arabidopsis* are essential for resistance to distinct microbial pathogens. *Proc. Natl Acad. Sci.* **95**, 15107–15111.

Thomma, B.P.H.J., Nürnberger, T. and Joosten, M.H.A.J. (2011) Of PAMPs and effectors: the blurred PTI-ETI dichotomy. *Plant Cell*, **23**, 4–15.

Thoquet, P., Olivier, J., Sperisen, C., Rogowsky, P., Laterrot, H. and Grimsley, N. (1996) Quantitative trait loci determining resistance to bacterial wilt in tomato cultivar Hawaii7996. *Mol. Plant Microbe Interact.* **9**, 826–836.

Ton, J., Flors, V. and Mauch-Mani, B. (2009) The multifaceted role of ABA in disease resistance. *Trends Plant Sci.* **14**, 310–317.

Van Loon, L.C., Rep, M. and Pieterse, C.M.J. (2006) Significance of inducible defense-related proteins in infected plants. *Annu. Rev. Phytopathol.* **44**, 135–162.

Van Ooijen, G., Mayr, G., Kasiem, M.M.A., Albrecht, M., Cornelissen, B.J.C. and Takken, F.L.W. (2008) Structure–function analysis of the NB-ARC domain of plant disease resistance proteins. *J. Exp. Bot.* **59**, 1383–1397.

Vos, I.A., Moritz, L., Pieterse, C.M.J. and Van Wees, S.C.M. (2015) Impact of hormonal crosstalk on plant resistance and fitness under multi-attacker conditions. *Front. Plant Sci.* **6**, 639.

Wan, H., Yuan, W., Ye, Q., Wang, R., Ruan, M., Li, Z., Zhou, G. *et al.* (2012) Analysis of TIR- and non-TIR-NBS-LRR disease resistance gene analogous in pepper, characterization, genetic variation, functional divergence and expression patterns. *BMC Genom.* **13**, 502.

Wang, J.-F., Olivier, J., Thoquet, P., Mangin, B., Sauviac, L. and Grimsley, N.H. (2000) Resistance of tomato line Hawaii7996 to *Ralstonia solanacearum* Pss4 in Taiwan is controlled mainly by a major strain-specific locus. *Mol. Plant Microbe Interact.* **13**, 6–13.

Wang, J., Shine, M.B., Gao, Q.-M., Navarre, D., Jiang, W., Liu, C., Chen, Q. *et al.* (2014) Enhanced disease susceptibility1 mediates pathogen resistance and virulence function of a bacterial effector in soybean. *Plant Physiol.* **165**, 1269–1284.

Ward, E.R., Payne, G.B., Moyer, M.B., Williams, S.C., Dincher, S.S., Sharkey, K.C., Beck, J.J. *et al.* (1991) Differential regulation of beta-1,3-glucanase messenger RNAs in response to pathogen infection. *Plant Physiol.* **96**, 390–397.

Wicker, E., Grassart, L., Coranson-Beaudu, R., Mian, D., Guilbaud, C., Fegan, M. and Prior, P. (2007) *Ralstonia solanacearum* strains from Martinique (French West Indies) exhibiting a new pathogenic potential. *Appl. Environ. Microbiol.* **73**, 6790–6801.

Wirthmueller, L., Zhang, Y., Jones, J.D.G. and Parker, J.E. (2007) Nuclear accumulation of the *Arabidopsis* immune receptor RPS4 is necessary for triggering EDS1-dependent defense. *Curr. Biol.* **17**, 2023–2029.

Xiao, S. and Chye, M.-L. (2011) Overexpression of Arabidopsis ACBP3 enhances NPR1-dependent plant resistance to *Pseudomonas syringe* pv tomato DC3000. *Plant Physiol.* **156**, 2069–2081.

Yang, D.-L., Yang, Y. and He, Z. (2013) Roles of plant hormones and their interplay in rice immunity. *Mol. Plant*, **6**, 675–685.

Yoo, S.-D., Cho, Y. and Sheen, J. (2009) Emerging connections in the ethylene signaling network. *Trends Plant Sci.* **14**, 270–279.

Yu, S.L., Wang, C.T., Yang, Q.L., Zhang, D.X., Zhang, X.Y., Cao, Y.L., Liang, X.Q. *et al.* (2011) *Peanut Genetics and Breeding in China*, pp. 565. Shanghai: Shanghai Scientific and Technology Press.

Zhang, J. and Zhou, J.M. (2010) Plant immunity triggered by microbial molecular signatures. *Mol. Plant*, **3**, 783–793.

Zhang, C., Pan, S., Chen, H., Cai, T.C., Zhuang, C.H., Deng, Y., Zhuang, Y.H. *et al.* (2016) Characterization of *NtREL1*, a novel root-specific gene from tobacco, and upstream promoter activity analysis in homologous and heterologous hosts. *Plant Cell Rep.* **35**, 757–769.

Zhao, B., Ardales, E.Y., Raymundo, A., Bai, J., Trick, H.N., Leach, J.E. and Hulbert, S.H. (2004) The avrRxo1 gene from the rice pathogen *Xanthomonas oryzae* pv. oryzicola confers a nonhost defense reaction on maize with resistance gene Rxo1. *Mol. Plant Microbe Interact.* **17**, 771–779.

Zhao, B., Lin, X., Poland, J., Trick, H., Leach, J. and Hulbert, S. (2005) A maize resistance gene functions against bacterial streak disease in rice. *Proc. Natl Acad. Sci.* **102**, 15383–15388.

Zheng, Y., Li, C., Liu, Y., Yan, C., Zhang, T., Zhuang, W. and Shan, S. (2012) Cloning and characterization of a NBS-LRR resistance gene from peanut. *J. Agric. Sci.* **4**, 243–252.

Zvereva, A.S. and Pooggin, M.M. (2012) Silencing and innate immunity in plant defense against viral and non-viral pathogens. *Viruses*, **4**, 2578–2597.

16

Targeted mutation of Δ12 and Δ15 desaturase genes in hemp produce major alterations in seed fatty acid composition including a high oleic hemp oil

Monika Bielecka[1,†,§], Filip Kaminski[1,§], Ian Adams[1,‡], Helen Poulson[1], Raymond Sloan[2], Yi Li[1], Tony R. Larson[1], Thilo Winzer[1] and Ian A. Graham[1,*]

[1]Centre for Novel Agricultural Products, Department of Biology, University of York, York, UK
[2]Biorenewables Development Centre, The Biocentre, York Science Park, Heslington, York, UK

*Correspondence
 email
ian.graham@york.ac.uk
†Present address: Department of Biology and Pharmaceutical Botany, Faculty of Pharmacy, Medical University of Wroclaw, Borowska 211, 50-556 Wroclaw, Poland.
‡Present address: Crop and Food Security Programme, Food and Environment Research Agency, York, UK
§These authors contributed equally to this work.

Keywords: Cannabis sativa, TILLING, reverse genetics, metabolic engineering, genome mining, transcriptomics.

Summary

We used expressed sequence tag library and whole genome sequence mining to identify a suite of putative desaturase genes representing the four main activities required for production of polyunsaturated fatty acids in hemp seed oil. Phylogenetic-based classification and developing seed transcriptome analysis informed selection for further analysis of one of seven Δ12 desaturases and one of three Δ15 desaturases that we designate CSFAD2A and CSFAD3A, respectively. Heterologous expression of corresponding cDNAs in *Saccharomyces cerevisiae* showed CSFAD2A to have Δx+3 activity, while CSFAD3A activity was exclusively at the Δ15 position. TILLING of an ethyl methane sulphonate mutagenized population identified multiple alleles including non-sense mutations in both genes and fatty acid composition of seed oil confirmed these to be the major Δ12 and Δ15 desaturases in developing hemp seed. Following four backcrosses and sibling crosses to achieve homozygosity, csfad2a-1 was grown in the field and found to produce a 70 molar per cent high oleic acid (18:1^{Δ9}) oil at yields similar to wild type. Cold-pressed high oleic oil produced fewer volatiles and had a sevenfold increase in shelf life compared to wild type. Two low abundance octadecadienoic acids, 18:2^{Δ6,9} and 18:2^{Δ9,15}, were identified in the high oleic oil, and their presence suggests remaining endogenous desaturase activities utilize the increased levels of oleic acid as substrate. Consistent with this, CSFAD3A produces 18:2^{Δ9,15} from endogenous 18:1^{Δ9} when expressed in *S. cerevisiae*. This work lays the foundation for the development of additional novel oil varieties in this multipurpose low input crop.

Introduction

The seeds of *Cannabis sativa* L. (hemp, marijuana) have been an important source of oil and protein in human nutrition dating back to Neolithic times in ancient China (Li, 1974). *Cannabis sativa* has an annual life cycle and is mostly dioecious with male and female flowers borne on separate individuals. Selective breeding has produced marijuana strains accumulating high levels of psychoactive cannabinoids in the female flowers and hemp cultivars typically having low levels of cannabinoids but good fibre and/or seed oil traits. The draft genome of the marijuana drug strain Purple Kush and comparison of its female flower transcriptome with that of the hemp cultivar Finola demonstrated much higher expression levels of genes involved in cannabinoid production in the drug strain (van Bakel et al., 2011). Furthermore, single nucleotide variant analysis revealed a distinct separation between the hemp and marijuana strains.

Finola (breeder code FIN-314) was developed in Finland as an oilseed hemp variety (Callaway and Laakkonen, 1996). It is dioecious and suitable for seed harvest by conventional agricultural machinery and can yield over 2000 kg/ha seed under good conditions (Callaway, 2004). Hemp has modest agrochemical requirements, is an excellent break crop and is suited to warm-to-temperate growing conditions (Callaway, 2004). At over 80% in polyunsaturated fatty acids (PUFAs), hemp seed oil rivals most of the commonly used vegetable oils. At 56% linoleic acid (LA,

18:2^{Δ9,12}) and 22% *alpha*-linolenic acid (ALA, 18:3^{Δ9,12,15}), hemp oil is a rich source of these essential fatty acids. In addition, hemp oil also contains *gamma*-linolenic acid (GLA, 18:3^{Δ6,9,12}) and stearidonic acid (SDA, 18:4^{Δ6,9,12,15}), which occur at about 4% and 2%, respectively, in Finola (Callaway, 2004). Here, we have used the recognized chemical nomenclature for fatty acids, which indicate the position of double bonds relative to the carboxyl group. Another frequently used terminology references the methyl end of the fatty acid with, for example, ALA (18:3^{Δ9,12,15}) and SDA (18:4^{Δ6,9,12,15}) being referred to as omega-3 fatty acids, whereas LA (18:2^{Δ9,12}) and GLA (18:3^{Δ6,9,12}) are referred to as omega-6 fatty acids.

Two multifunctional classes of desaturases have been found in plants, one soluble and the other membrane bound (Shanklin and Cahoon, 1998). In plants, C16- and C18-fatty acids are synthesized in the stroma of plastids and, following desaturation of 18:0 to 18:1 by a soluble Δ9 stearoyl-ACP desaturase, contribute to the assembly of complex membrane lipids (Ohlrogge and Browse, 1995). Further desaturation of fatty acids in membrane lipids of the chloroplast and endoplasmic reticulum (ER) is carried out by the membrane-bound desaturases, a number of which have been designated FAD2 TO FAD8 based on work in Arabidopsis (Ohlrogge and Browse, 1995). In the current study, we were particularly interested in the FAD2 and FAD3 enzymes because these are responsible for the Δ12 desaturation of oleic acid (18:1^{Δ9}) to LA (18:2^{Δ9,12}) and the Δ15 desaturation of LA to ALA

$(18:3^{\Delta 9,12,15})$, respectively. Production of GLA $(18:3^{\Delta 6,9,12})$ from LA and SDA $(18:4^{\Delta 6,9,12,15})$ from ALA requires the action of a Δ6 desaturase which we also expect to find expressed in developing hemp seeds. Our principle objective was to perform a first characterization of the membrane-bound desaturases that determine fatty acid composition in hemp seed oil using a combination of heterologous expression and reverse genetics. The publication of the Purple Kush and Finola genomes and transcriptome sequences during the course of our work (van Bakel et al., 2011) also allowed us to perform a comprehensive in-silico analysis of relevant fatty acid desaturases. A valuable outcome of the reverse genetic approach is that seed oil fatty acid composition is altered with the potential to improve the crop for industry. We were interested in developing a high oleic acid hemp seed oil because similar developments in crops such as soybean have opened up new markets due to increased oxidative stability (Kinney, 1998), which is a particular problem with hemp oil given the high levels of PUFAs.

Results

Genome mining reveals multiple copies of soluble and membrane-bound desaturases in *Cannabis sativa*

We isolated mRNA from the upturned (U) stage of embryo development of the Finola variety, because this represents a stage of significant storage oil deposition in the form of triacylglycerol (TAG) in dicotyledonous oilseeds (Baud et al., 2002), and used this for cDNA library construction. We generated 1893 expressed sequence tags (ESTs) from the U stage cDNA library by conventional Sanger sequencing and a BLASTX similarity search revealed 11 ESTs with homology to desaturase genes. Two of the resulting unigenes contained an incomplete open reading frame (ORF) giving a predicted amino acid sequence with homology to the Δ12 desaturases. Two homologous full-length cDNA sequences were obtained by RACE PCR, and the corresponding genes were named *CSFAD2A* and *CSFAD2B* (Table S1). We also cloned a *FAD3* desaturase fragment by PCR amplification using degenerate primers (Lee et al., 1998) on the U stage cDNA. RACE PCR produced a 1188-bp full-length cDNA sequence that we name *CSFAD3A* (Table S1).

We used *CSFAD2A*, *CSFAD2B*, *CSFAD3A* and other previously characterized plant membrane-bound Δ12 (FAD2), Δ15 (FAD3) and Δ6/Δ8 sphingo-lipid, as well as the soluble Δ9 stearoyl-ACP desaturases as queries to retrieve additional membrane-bound and soluble desaturase sequences from the genome sequences of Purple Kush (canSat3) and Finola (Finola1) (van Bakel et al., 2011). This resulted in the identification of putative sequences for seven FAD2 (designated *CSFAD2A* to *CSFAD2G*), three FAD3 (designated *CSFAD3A*, *CSFAD3B* and *CSFAD3C*), two genes with homology to both Δ8-sphingo-lipid desaturases and Δ6 fatty acid desaturases (designated *CSD8* and *CSD6*) and five Δ9 stearoyl-ACP desaturases (designated *CSSACPD-A* to *CSSACPD-E*) in the more complete genome of the Purple Kush variety. For all but *CSFAD2F* and *CSFAD3C* orthologous sequences were also identified in the Finola genome (Table S1), which probably reflects the draft nature of this genome.

All deduced *C. sativa* amino acid sequences from both varieties were aligned with desaturase sequences from other plant species using ClustalX (Figures S1–S4). Phylogenetic trees were calculated from distance matrices by the neighbour-joining algorithm using desaturase sequences retrieved from the Purple Kush genome (Figure 1). All three subclasses of membrane-

bound desaturases from *C. sativa* contain three histidine cluster motifs involved in binding the di-iron complex. CSD8 and CSD6 are homologous, but CSD8 shows greatest similarity with a number of functionally characterized Δ8 sphingo-lipid desaturases, while CSD6 is most similar to functionally characterized Δ6 fatty acid desaturases (Figures 1 and S3). In addition to the three histidine boxes, both genes contain a conserved HPGG motif within a cyt b5-like domain, the histidine residue of which is essential for enzyme activity (Sayanova et al., 1999b). SACPD family members contain the EXXH motif involved in binding the di-iron complex together with additional glutamine residues involved in coordinating the di-iron complex, which are typical for this class of desaturases (Shanklin and Cahoon, 1998). Alignment of genomic and cDNA sequences revealed that all members of the *CSFAD2* and *CSD6* subclasses of microsomal desaturases consist of a single exon, while the gene arrangement of all members of the *CSFAD3* subclass contain eight exons and seven introns within the gene. This arrangement was conserved across the members of this subclass (summarized in Table S1). For the plastidial stearoyl-ACP desaturases, two (*CSSACPD-A* and *CSSACPD-B*) contain three exons and the remaining three contain two exons consistent with the phylogenetic arrangement (Table S1, Figure 1).

Deep sequencing of the developing seed transcriptome identifies candidate desaturases involved in modifying fatty acid composition of seed oil

Expressed sequence tag libraries were prepared by deep sequencing cDNA prepared from RNA isolated from torpedo (T), U and filled-not-desiccated (FND) stages of Finola embryo development as depicted in Figure 2a. Raw reads were mapped to the open reading frames of 17 putative desaturase genes as detailed in Table S1. Three of the five plastidial stearoyl-ACP desaturases are expressed, with *CSSACPD-C* transcripts being the most abundant; three of the seven *CSFAD2* genes are expressed, with *CSFAD2A* being the highest; all three of the *CSFAD3* genes are expressed, but of these, only *CSFAD3A* increases during embryo development, with *CSFAD3B* and *CSFAD3C* present at very low levels. *CSD8* and *CSD6* show similar low levels of expression up until the U stage with transcripts of both genes being absent at the later FND stage (Figure 2a). Based on homology and expression analysis, lead candidates for each of the desaturation steps shown in Figure 2b can be identified as *SACPD-C*, *CSFAD2A*, *CSFAD3A* and *CSD6*. We focused our efforts on functionally characterizing *CSFAD2A* and *CSFAD3A*.

Characterization of the *Cannabis sativa* microsomal desaturase *CSFAD2A*

Quantitative RT-PCR analysis confirmed high-level expression of *CSFAD2A* during embryo development, peaking at the FND stage where it was more than 1000 times higher than in young leaves (Figure 3a). A similar pattern of expression but at much lower levels was observed for the *CSFAD2B* gene with the difference in expression between leaves and embryo much less pronounced, being about 20 times higher at the FND stage (Figure 3a). To confirm the functional identity of *CSFAD2A*, we cloned the corresponding ORF into the expression vector *pESC-TRP* containing the galactose-inducible GAL1 promoter and heterologously expressed this in *Saccharomyces cerevisiae*. This yeast has been used successfully for functional expression of several plant microsomal desaturases, because it acts as a

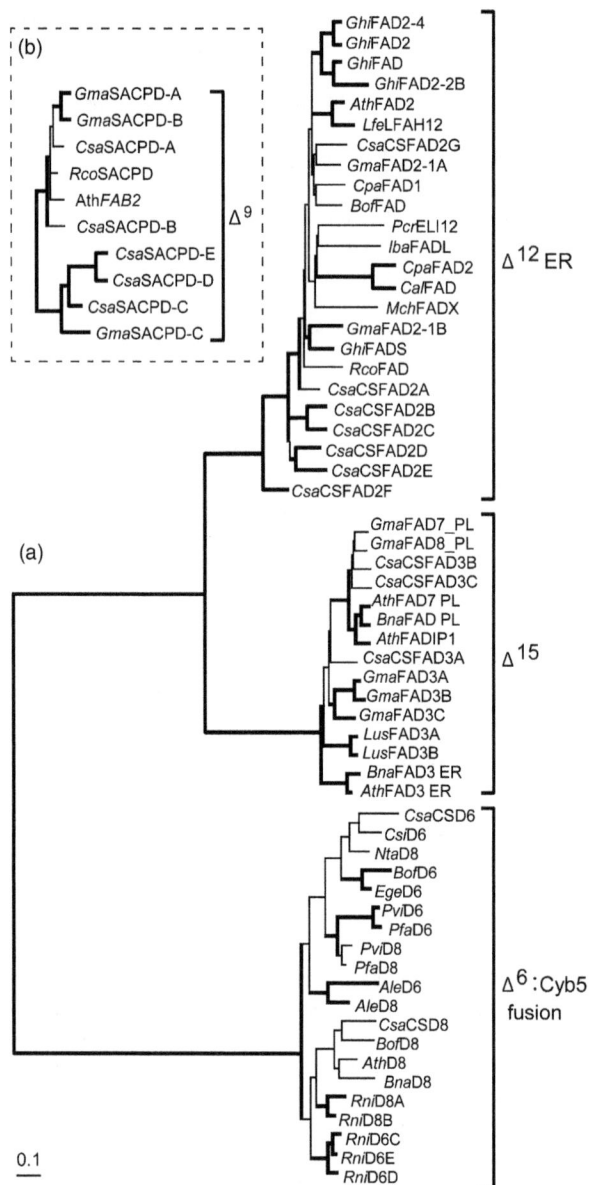

Figure 1 Evolutionary relationship of *Cannabis sativa* putative desaturases (retrieved from the Purple Kush genome) with known desaturases from selected species. (a) Genes encoding membrane-bound Δ6, Δ8 sphingo-lipid, Δ12 and Δ15 desaturases. (b) Soluble Δ9 Stearoyl-ACP desaturases. Species-specific identifiers: *Anemone lendsquerelli* (*Ale*), *Arabidopsis thaliana* (*Ath*), *Brassica napus* (*Bna*), *Borago officinalis* (*Bof*), *Crepis alpine* (*Cal*), *Crepis palaestina* (*Cpa*), *Cannabis sativa* (*Csa*), *Camellia sinensis* (*Csi*), *Echium gentianoides* (*Ege*), *Gossypium hirsutum* (*Ghi*), *Glycine max* (*Gma*), *Impatiens balsamina* (*Iba*), *Lesquerella fendleri* (*Lfe*), *Linum usitatissimum* (*Lus*), *Momordica charantia* (*Mch*), *Nicotiana tabacum* (*Nta*), *Petroselinum crispum* (*Pcr*), *Primula farinosa* (*Pfa*), *Primula vialli* (*Pvi*), *Ricinus communis* (*Rco*), *Ribes nigrum* (*Rni*). All branches are drawn to scale as indicated by the scale bar (0.1 substitutions/site). Strongly supported nodes with above 70% bootstrap values are highlighted with thickened lines.

convenient host with a simple fatty acid profile due to the presence of only a Δ9 desaturase producing palmitoleate and oleate, and the appropriate redox chain in a suitable membrane (Reed *et al.*, 2000). Fatty acid analysis of transformed yeast cells

revealed the presence of two new fatty acids that were not present in either wild-type yeast or the empty vector control (Figure 3b, Table 1). Gas chromatography (GC) analysis of fatty acid methyl esters (FAMEs) demonstrated that the major novel peak is linoleic acid. As shown in Table 1, 72% of the endogenous oleic acid ($18:1^{\Delta 9}$) appears to have served as substrate for CSFAD2A and been converted into linoleic acid ($18:2^{\Delta 9,12}$), confirming CSFAD2A to have Δ12 desaturase activity. We transesterified the FAME fraction to 3-pyridylcarbinol esters and used GCMS to identify the second novel peak as ($16:2^{\Delta 9,12}$; Figure S5). We conclude that CSFAD2A can also use palmitoleic acid ($16:1^{\Delta 9}$) as substrate, with a conversion efficiency to $16:2^{\Delta 9,12}$ of 43% (Figure 3b, Table 1). Feeding eicosenoic acid ($20:1^{\Delta 11}$) to *CSFAD2A*-transformed yeast cultures resulted in 62% conversion to $20:2^{\Delta 11,14}$ (Figure S6) demonstrating that the enzyme can accept 16–20 C fatty acids and that the specificity is most accurately described as Δx+3 (Schwartzbeck *et al.*, 2001).

Identification and characterization of two *CSFAD2A* desaturase mutants

To establish the *in vivo* role of *CSFAD2A*, we screened an ethyl methane sulphonate (EMS) mutagenized M2 outcrossed population of Finola using the TILLING method (Till *et al.*, 2006). We identified an allelic series of mutations among which *csfad2a-1* carries a stop codon at amino acid position 167. We performed two rounds of backcrossing of *csfad2a-1* to Finola and obtained homozygous *csfad2a-1* individuals (BC_2F_1) by crossing heterozygous male and female BC_2 siblings. *csfad2a-1* homozygotes displayed a dramatic increase in oleic acid content to 77 molar% in seed oil (Figure 3c, Table S2). In parallel, the levels of LA and ALA were strongly decreased compared to the fatty acid profile of the segregating wild-type seed oil from the same population, suggesting that this decrease was at the expense of the increase in oleic acid (Figure 3c, Table S2). Two novel fatty acids appeared in *csfad2a-1* at five and two molar per cent (Table S2). GC retention times indicated these to be 18:2 fatty acids and GCMS following derivatization to 3-pyridylcarbinol esters revealed these to be $18:2^{\Delta 6,9}$ and $18:2^{\Delta 9,15}$, respectively (Table S2). These may arise through the action of other desaturases on the high percentage oleic acid present in the developing embryos of *csfad2a-1*. The dramatic fatty acid level changes observed in *csfad2a-1* seed confirmed that the predicted truncated CSFAD2A protein is nonfunctional. Interestingly, no major changes in seed fatty acid profile were observed if the mutation was present in the heterozygous state, indicating that only one copy of this highly expressed *CSFAD2A* gene is sufficient to maintain the near wild-type level of fatty acids in hemp seed.

We also identified a second allele, *csfad2a-2*, which carries two point mutations giving rise to a proline to leucine transition at positions 218 and 375 of the predicted amino acid sequence of CSFAD2A. Homozygous *csfad2a-2* (BC_1F_1) seed accumulate nearly 70 molar per cent of oleic acid, low levels of $18:2^{\Delta 6,9}$ and $18:2^{\Delta 9,15}$ and decreased levels of LA and ALA compared to heterozygous and segregating wild-type seeds from the same population (Figure 3d, Table S2). This seed oil phenotype is very similar to that of *csfad2a-1* (Figure 3c) and is consistent with one or both of the P to L transitions disrupting protein function. This is expected given the importance of proline amino acids in determining protein structure. Interestingly, the levels of oleic acid, linoleic acid and α-linolenic acid remained unchanged in

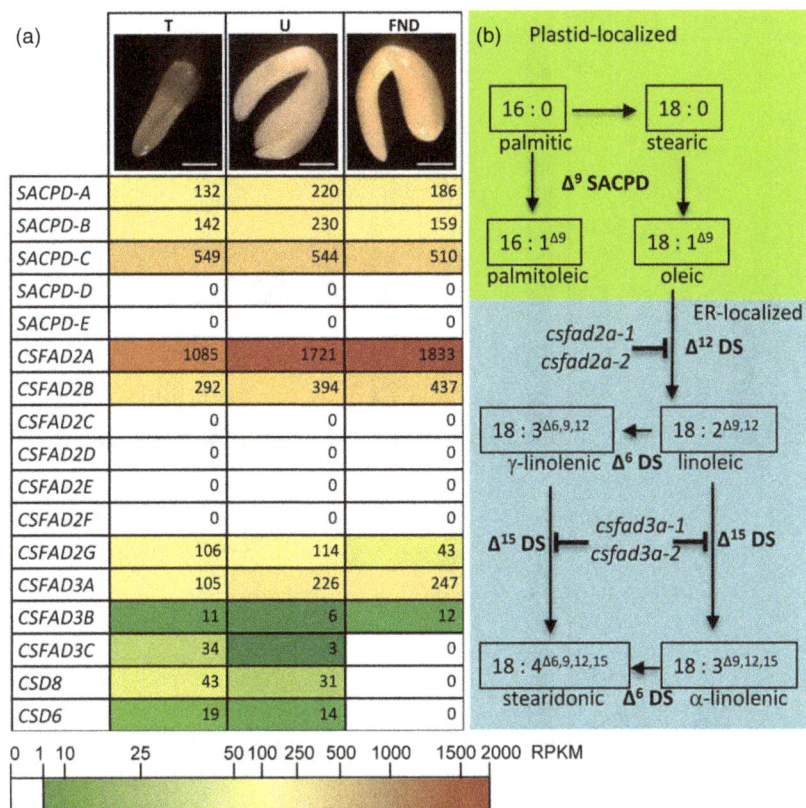

Figure 2 Expression of putative desaturase genes in developing embryos of the hemp cultivar Finola and metabolic context. (a) Embryos representative of each developmental stage used for RNA isolation are shown (Scale bar = 1 mm). Gene expression is depicted in a heat map format with RPKM values included. (b) Schematic presentation of the biosynthetic pathway giving rise to the major fatty acids in hemp seed oil. SACPD—stearoyl-ACP desaturase, DS—desaturase. Enzymatic steps are shown in bold and those steps compromised by mutation in specific *CSFAD2* and *CSFAD3* genes as detailed in Figures 3 and 4 are indicated.

leaf tissues of both *csfad2a-1* and *csfad2a-2* compared to wild-type plants (Table S3), which is consistent with the gene expression data showing *CSFAD2A* to be largely seed specific (Figure 3a).

Characterization of the *Cannabis sativa* microsomal desaturase *CSFAD3A*

Quantitative RT-PCR confirmed expression of *CSFAD3A* in both leaves and embryos and showed it to be induced during seed development peaking at the FND stage where it is about 14 times higher than levels in young leaves (Figure 4a). Heterologous expression of *CSFAD3A* in *S. cerevisiae* followed by fatty acid feeding resulted in desaturation of linoleic ($18:2^{\Delta9,12}$) to α-linolenic acid ($18:3^{\Delta9,12,15}$) and γ-linolenic acid ($18:3^{\Delta6,9,12}$) to stearidonic acid ($18:4^{\Delta6,9,12,15}$) at a conversion efficiency of 56% and 23%, respectively (Figure 4b, Table 1). The yeast *CSFAD3A* transformants also exhibited low level activity with endogenous $16:1^{\Delta9}$ and $18:1^{\Delta9}$ resulting in what we identified as $16:2^{\Delta9,15}$ and $18:2^{\Delta9,15}$, respectively (Table 1, Figures S5 and S7). *CSFAD3A* transformants did not show any activity on exogenously supplied $20:1^{\Delta11}$ after 28 h incubation. Together these results confirm that CSFAD3A acts as a Δ15 desaturase when expressed in *S. cerevisiae*.

Identification of mutations in *CSFAD3* confirms Δ15 desaturase activity

We screened our EMS-mutagenized hemp population and identified an allelic series of mutations in *CSFAD3* including one, designated *csfad3a-1*, that results in a stop codon being introduced at codon position 255. We performed three rounds of backcrossing to Finola and obtained homozygous *csfad3a-1* (BC$_3$F$_1$) seeds by crossing BC$_3$ siblings. Seed oil of the homozygous *csfad3a-1* contained near zero and zero levels of ALA and

SDA, respectively, an elevation of LA from 55 to 75 molar per cent and no significant effect on GLA compared to the segregating wild type and heterozygotes in the M5 generation (Figure 4c, Table S2). A similar seed oil phenotype was seen in BC$_2$F$_1$ material (Table S2). These dramatic changes in the homozygous *csfad3a-1* seed oil profile confirmed that CSFAD3A acts as a Δ15 desaturase *in vivo* as well as in a heterologous host. Interestingly, when the mutation is in the heterozygous state, an intermediate phenotype is displayed in the seed oil with just half the levels of ALA and SDA compared to wild type. A second mutant, *csfad3a-2*, carried a point mutation resulting in conversion of proline to leucine at amino acid position 190 and this resulted in a similar seed oil phenotype to *csfad3a-1* (Figure 4d, Table S2). In contrast to seed oil, the production of ALA in the leaf tissue of both *csfad3a-1* and *csfad3a-2* is decreased by only 6% and 7%, respectively, compared to wild type (Table S3). This suggests the expression of other genes encoding Δ15 desaturase enzymes in leaf tissue, with *CSFAD3B* and *CSFAD3C* being obvious candidates.

High oleic hemp oil product performance

We selected *csfad2a-1* for further analysis, extended the back-crossing to generate BC$_4$ material and bulked up *csfad2a-1* seed by crossing homozygous mutant siblings. This material, now referred to as 'High Oleic Hemp' was grown in a single block field trial in Yorkshire, UK, during the 2011 growing season. Overall plant growth habit, flowering time and seed yield per plant were similar to the Finola wild type. Seed was cold-pressed giving a percentage oil of approximately 36% in the wild type and *csfad2a-1* material (Figure 5a). Fatty acid composition analysis confirmed the high oleic status of cold-pressed field grown *csfad2a-1* material, on a par with a commercial high oleic rapeseed material (Figure 5b). Rancimat determination of oxida-

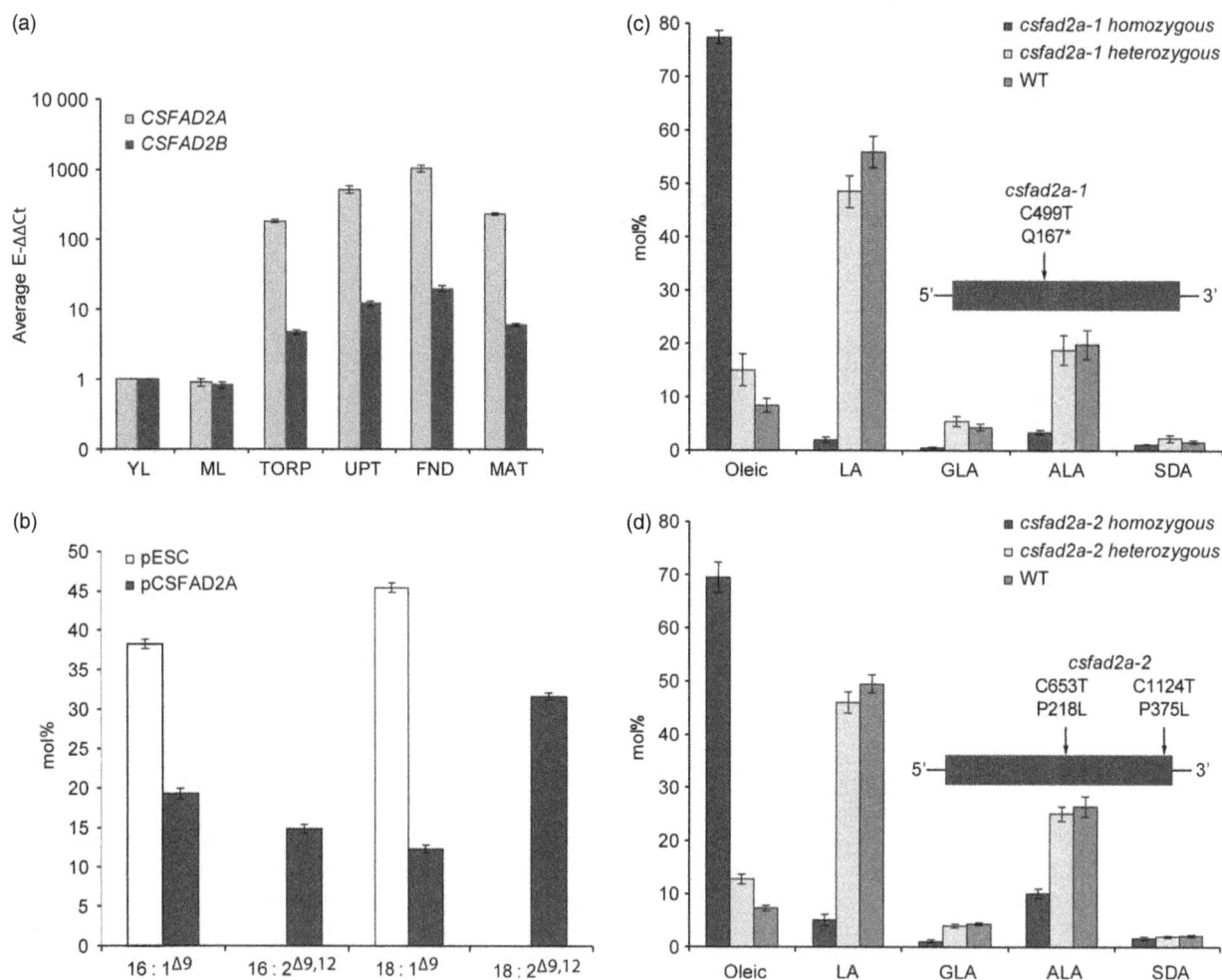

Figure 3 Characterization of *CSFAD2A* gene function. (a) Expression of *CSFAD2A* and *CSFAD2B* in developing embryo and mature leaf tissue compared to levels in young hemp leaves. Mean values represent the average of three biological replicas each consisting of three technical replicates. Abbreviations: YL: young leaves; ML: mature leaves; TORP: torpedo stage of embryo; UPT: upturned stage of embryo; FND: filled-not-desiccated stage of embryo; MAT: mature seed embryo. (b) Fatty acid composition of *Saccharomyces cerevisiae* transformed with either *CSFAD2A* cDNA or an empty vector (pESC-TRP) control. Each value is the mean ± SD from three independent experiments. Fatty acid composition of seed oil from (c) homozygous *csfad2a-1* (BC$_2$F$_1$) and (d) homozygous *csfad2a-2* (BC$_1$F$_1$) plants compared to respective segregating heterozygous and wild-type plants from the same generation as detailed in Table S2. Each value is the mean ± SD from 8 to 28 seeds from same line and generation.

tion stability of the pressed oil is an industry standard methodology that allows shelf life to be determined by extrapolation of oxidation at elevated temperatures. We found that high oleic hemp *csfad2a-1* oil had an increased shelf life from 1.5 to 10 months at 20 °C, 4.1 to 28.6 months at 4 °C and 5.3 to 37.1 months at 0 °C (Figure 5c). Shelf life of high oleic rapeseed oil is also longer than standard rapeseed oil (Figure 5c), but shelf life of the high oleic hemp oil exceeds that calculated for high oleic rapeseed oil despite them having equivalent amounts of oleic and polyunsaturated fatty acids (Figure 5b, c). Plant seeds contain antioxidants such as tocopherols, which are thought to play a role in preventing oxidation of polyunsaturated fatty acids. We measured levels of tocopherols in high oleic hemp oil and found these to be significantly higher than that present in Finola hemp oil (Figure 5d), and also significantly higher than in both standard rapeseed oil and high oleic rapeseed oil (Figure S8). Consistent with the increased stability of high oleic hemp and rapeseed oils, we found that both produced decreased levels of volatile aldehydes as determined by head space analysis (Figure S9).

Not surprisingly, the high oleic hemp TAG composition consisted mainly of triolein, which was completely absent from Finola hemp oil (Figure S10).

Discussion

Mutagenesis has been used to increase the amount of genetic variation available for selective breeding since the 1940s but as induced mutations are mostly recessive, resulting phenotypes are only observed when the mutant allele is in the homozygous state. In dioecious species such as hemp, obtaining homozygous mutations is a particular challenge, hence the TILLING (Till *et al.*, 2006) method, which allows identification of mutations in the heterozygous state, is particularly valuable. The advent of next generation sequencing technologies permits cost-effective, routine identification of genes from species not previously described at the molecular level and the present work demonstrates that comprehensive target gene selection can give rise to predictable breeding outcomes even in such species.

Table 1 Fatty acid composition of yeast transformants and pESC-TRP empty vector controls fed with fatty acid substrates. Percentage conversion was calculated as product/(substrate + product) * 100 at the assay endpoint. Each value is the mean ± SD from three independent experiments

Substrate	Substrate endpoint mol% total fatty acids	Product	Product endpoint mol% total fatty acids	% conversion
pESC-TRP				
$16:1^{\Delta 9}\ast$	38.3 ± 0.6	–	–	–
$18:1^{\Delta 9}\ast$	45.4 ± 0.6	–	–	–
$18:2^{\Delta 9,12}$	10.9 ± 0.3	–	–	–
$18:3^{\Delta 6,9,12}$	14.1 ± 0.5	–	–	–
$20:1^{\Delta 11}$	0.8 ± 0.05	–	–	–
pCSFAD2A				
$16:1^{\Delta 9}\ast$	19.3 ± 0.7	$16:2^{\Delta 9,12}$	14.8 ± 0.6	43 ± 1.4
$18:1^{\Delta 9}\ast$	12.3 ± 0.5	$18:2^{\Delta 9,12}$	31.7 ± 0.4	72 ± 0.8
$20:1^{\Delta 11}$	0.3 ± 0.01	$20:1^{\Delta 11,14}$	0.5 ± 0.01	62 ± 0.3
pCSFAD3A				
$16:1^{\Delta 9}\ast$	38.2 ± 0.3	$16:2^{\Delta 9,15}$	1.6 ± 0.04	3.9 ± 0.1
$18:1^{\Delta 9}\ast$	38.0 ± 0.8	$18:2^{\Delta 9,15}$	3.6 ± 0.1	8.7 ± 0.4
$18:2^{\Delta 9,12}$	4.2 ± 0.2	$18:3^{\Delta 9,12,15}$	5.4 ± 0.2	56.3 ± 0.5
$18:3^{\Delta 6,9,12}$	10.7 ± 0.8	$18:4^{\Delta 6,9,12,15}$	3.2 ± 0.2	23.1 ± 0.4
$20:1^{\Delta 11}$	0.59 ± 0.6	–	–	–

*Endogenous substrate; no fatty acid added to medium.

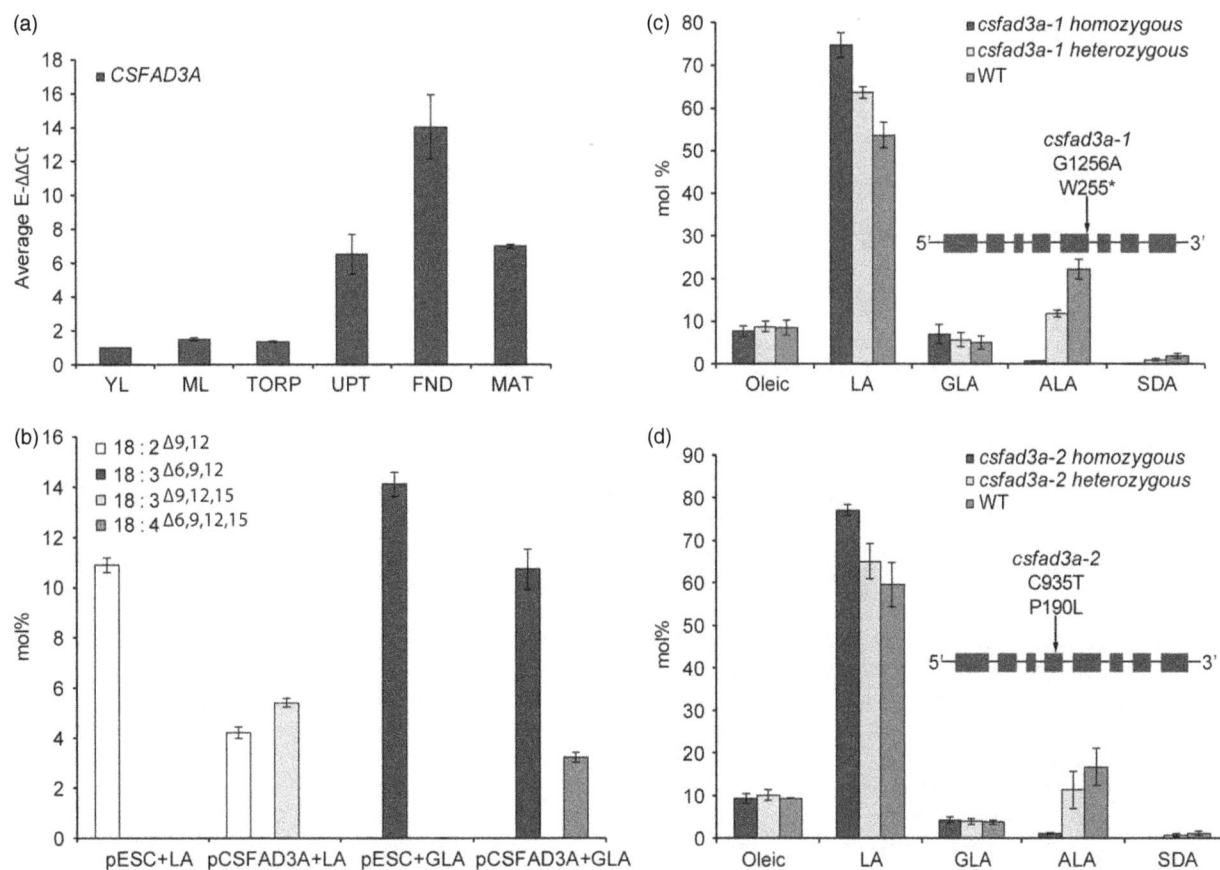

Figure 4 Characterization of *CSFAD3A* gene function. (a) Expression of *CSFAD3A* in developing embryo and mature leaf tissue compared to levels in young hemp leaves. Mean values represent the average of three biological replicates each consisting of three technical replicates. Abbreviations: YL: young leaves; ML: mature leaves; TORP: torpedo stage of embryo; UPT: upturned stage of embryo; FND: filled-not-desiccated stage of embryo; MAT: mature seed embryo. (b) Fatty acid composition of *Saccharomyces cerevisiae* transformed with either *CSFAD3A* cDNA or an empty vector (pESC-TRP) control. Each value is the mean ± SD from three independent experiments. Fatty acid composition of seed oil from (c) homozygous *csfad3a-1* (BC₃F₁) and (d) homozygous *csfad3a-2* (BC₂F₁) plants compared to respective segregating heterozygous and wild-type plants from the same generation as detailed in Table S2. Each value is the mean ± SD from 4 to 20 seeds from same line and generation.

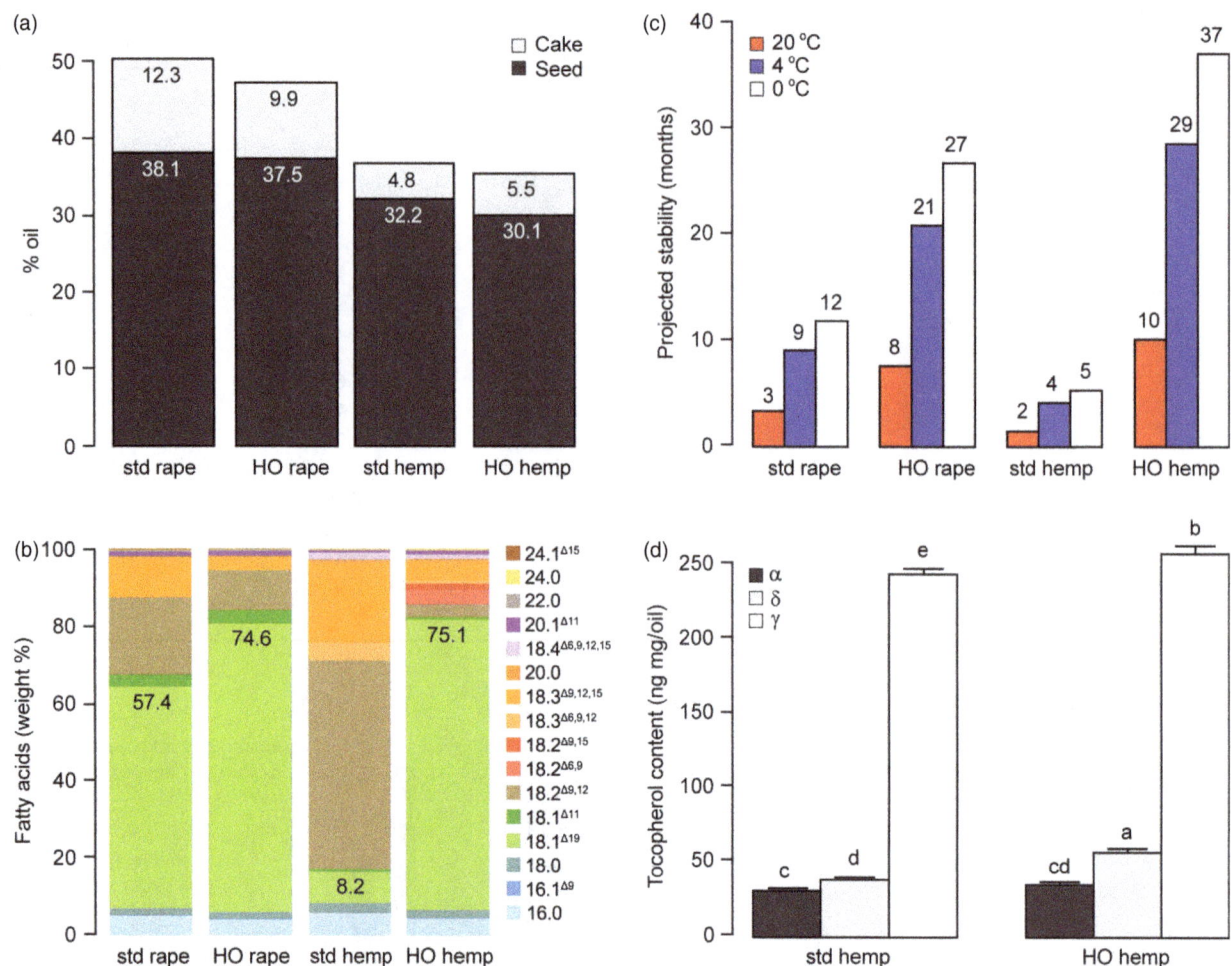

Figure 5 Cold-pressed oil analyses from standard (std) or high oleic (HO) hempseed and rapeseed. Small batches of seed harvested from field plots (~150 g) were cold-pressed and analysed for total oil content in the cake and seed (a), relative distribution of fatty acids (b) and rancimat-assayed stability at three different temperatures (c). Tocopherol assays are shown for hemp seed only (d). All data are representative assay values taken from the second or third pressed oil batches after the press had been preconditioned with appropriate seed and reached uniform operating temperatures. For tocopherol analyses (d), values are means ± 1 standard error from five analyses from the same oil batches with letters above bars indicating significantly different groups (ANOVA and Tukey's HSD; $P < 0.05$).

Heterologous expression in *S. cerevisiae* revealed that CSFAD2A desaturates $16:1^{\Delta 9}$ to $16:2^{\Delta 9,12}$ and $18:1^{\Delta 9}$ to $18:2^{\Delta 9,12}$ at 43% and 72% efficiency, respectively (Figure 3b, Table 1). However, the fact that the heterologously expressed protein also desaturates $20:1^{\Delta 11}$ eicosenoate to $20:2^{\Delta 11,14}$ demonstrates that this desaturase does not have strict $\Delta 12$ regiospecificity. Different integral membrane desaturases have evolved at least three distinct counting mechanisms for positioning double bonds as previously summarized (Shanklin and Cahoon, 1998). In addition to enzymes that count from either the carboxyl end or the methyl end of the molecule, there are also examples of enzymes that appear to follow the $\Delta x+3$ rule whereby, rather than measuring from the carboxyl or methyl ends, the desaturase references the existing double bond at position 'x' and places the new double bond (methylene interrupted) at position $x+3$ towards the methyl group (Hitz *et al.*, 1994; Schwartzbeck *et al.*, 2001). For example, the FAD2 enzyme from developing seeds of peanut (*Arachis hypogaea* L.) which, in addition to using palmitoleate and oleate as substrates also converts $19:1^{\Delta 10}$ to $19:2^{\Delta 10,13}$, leading the authors to conclude that it follows the $\Delta x+3$ rule. This rule also best

describes the consistent placement of the second double bond in the three monounsaturated substrates used by CSFAD2A.

Heterologous expression of CSFAD3A demonstrates that it introduces a double bond at the $\Delta 15$ position of $18:2^{\Delta 9,12}$ producing $18:3^{\Delta 9,12,15}$ and $18:3^{\Delta 6,9,12}$ producing $18:4^{\Delta 6,9,12,15}$ at conversion efficiencies of 56.3 and 23.1%, respectively (Table 1). This enzyme also produces $18:2^{\Delta 9,15}$ and $16:2^{\Delta 9,15}$ from $18:1^{\Delta 9}$ and $16:1^{\Delta 9}$ but at lower conversion efficiencies of 3.9 and 8.7%. Unlike CSFAD2A, CSFAD3A does not use $20:1^{\Delta 11}$ as substrate. Thus, while efficient introduction of a $\Delta 15$ double bond by CSFAD3A requires a substrate with $\Delta 9$ and $\Delta 12$ double bonds consistent with previous reports (Shanklin and Cahoon, 1998), it can also use both 16C and 18C substrates carrying only a $\Delta 9$ double bond but at lower efficiency. We therefore conclude that CSFAD3 exhibits $\Delta 15$ regioselectivity. The ability to use oleic acid as substrate, albeit at lower conversion efficiency than linoleic acid or γ-linolenic acid, provides an explanation for the appearance of the $18:2^{\Delta 9,15}$ at two molar per cent in the high oleic seed oil of *csfad2a-1* (Table S2). These results are consistent with the observation that overexpression of a FAD3 in seeds of Arabidopsis leads to production of a polymethylene-interrupted dienoic fatty acid (Puttick *et al.*, 2009). A similar explanation involving an endogenous Δ6 activity could account for the presence of small amounts of 18:2^{Δ6,9} in *csfad2a* mutants.

The seed oil phenotypes that we observe for both *csfad2a* and *csfad3a* are largely consistent with the activity data obtained from heterologous expression. Production of high levels of oleic acid in both *csfad2a-1* and *csfad2a-2* confirm that this gene is responsible for the major Δ12 activity in developing hemp seeds as depicted in Figure 2b. Small amounts of LA, ALA, GLA and SDA totalling seven molar per cent in the *csfad2a-1* non-sense mutant (Figure 3c, Table S2) suggest a low level of Δ12 desaturase activity remains in developing seeds of this mutant. This residual activity could be encoded by *CSFAD2B*, which is also expressed in developing seeds but at much lower levels than *CSFAD2A* (Figure 3a). In contrast, leaf fatty acid composition is unaffected in *csfad2a-1* and *csfad2a-2* (Table S3), consistent with the seed specific expression of the gene (Figure 3a).

The decrease of ALA (18:2^{Δ9,12,15}) from approximately 20 molar per cent in wild type to <1% in both alleles of the *csfad3a* mutant together with the near elimination of SDA (18:4^{Δ6,9,12,15}; Figure 4c, d, Table S2) confirms that *CSFAD3A* encodes the major Δ15 activity in developing hemp seeds. Whether or not an endogenous Δ6 desaturase activity contributes to the relatively low levels of SDA in wild-type seeds as depicted in Figure 2b cannot be ascertained because the *csfad3a* mutant has severely decreased levels of the ALA (18:3^{Δ9,12,15}) substrate. The increase in LA (18:2^{Δ9,12}) from 55 to 75 molar per cent in seed oil of *csfad3a* mutants accounts for the decrease in ALA. However, this increase in LA has no effect on GLA (18:2^{Δ6,9,12}) levels, suggesting that the Δ6 desaturation of LA to GLA is limited by Δ6 enzyme rather than substrate availability. Consistent with this, the transcript levels of both the *CSD8* and *CSD6* are very low from the T to U stages of embryo development and are not detected at the FND stage (Figure 2a). Up-regulation of *CSD6* may therefore be the best strategy to increase amounts of GLA as has been demonstrated by genetic engineering of tobacco (Reddy and Thomas, 1996; Sayanova *et al.*, 1999a), *Brassica juncea* (Hong *et al.*, 2002), *Brassica napus* (Liu *et al.*, 2001) and evening primrose (de Gyves *et al.*, 2004).

Initial field trial results from backcrossed *csfad2a-1* material confirm high-level production of oleic acid in the field (Figure 5a). Rancimat assays showed an approximately sevenfold increase in high oleic hemp oil stability at three different temperatures compared to Finola oil (Figure 5c). High oleic rapeseed oil showed approximately twofold increase in stability compared to the control at three temperatures and overall, high oleic hemp was more stable despite the two oils having a similar percentage of oleic acid (Figure 5b). The major cause of increased oil stability is likely due to the decrease in relatively unstable polyunsaturated fatty acids and increase in the relatively stable oleic acid. In addition, it is tempting to speculate that the higher overall levels of tocopherols in high oleic hemp compared to high oleic rapeseed oil could account for the increase in stability of the former (Figure S8). Interestingly, there is a small but significant increase in levels of tocopherols in high oleic hemp oil compared to Finola oil (Figures 5d and S8), which could be due to either increased synthesis or decreased degradation of these antioxidants in high oleic oil seeds.

The availability of the complete suite of desaturase genes together with the ability to generate allelic series of mutations in candidate genes by TILLING makes Finola hemp an excellent choice for further investigation of seed oil metabolism as well as an important target for crop improvement.

Experimental procedures

cDNA library construction and EST preparation from developing seeds of *Cannabis sativa*

U stage tissue was ground to a fine powder in liquid nitrogen and RNA extracted using the RNAeasy kit (Qiagen, Hilden, Germany). cDNA was synthesized with the Creator™ SMART™ cDNA Library Construction Kit (Clontech, Mountain View, CA) and cloned into the pDNR-LIB vector (Clontech). 1893 ESTs were generated by Single-pass Sanger sequencing and a Blast similarity search identified two unigene sequences with homology to FAD2 desaturases. Random amplification of cDNA ends (RACE) was performed to obtain full-length sequences using primers for CSFAD2A: 5'-AAAATGGGAGCCGGTGGCCGAAT-3' and 5'-GGGCGGAATTGCTTTCTTGATTTCGC-3'; RACE primers for CSFAD2B: 5'-GCAGACGATATGACCGTTTCGCTTCTCA-3' and 5'-GCGAGTTGGTACAACACGAATGTGGTGA-3'.

To generate the FAD3 gene from hemp cDNA, the following degenerate primers were used 5'-ACNCAYCAYCARAAYCAYGG-3' and 5'-CAYTGYTTNCCNCKRTACCA-3'. To obtain full-length CSFAD3, the following RACE primers were used: 5'-CACGGCCATGTTGAGAATGACGAG-3' and 5'-GGACAAACAGACAAGCAAAGCAGCCA-3'.

Genome mining, sequence alignment and phylogenetic analysis

Membrane-bound and soluble desaturase sequences were retrieved by local tBLASTn searches of the genome sequences from *C. sativa* varieties Purple Kush (canSat3) and Finola (Finola1), both of which were downloaded from the Cannabis Genome Browser (http://genome.ccbr.utoronto.ca/downloads.html; van Bakel *et al.*, 2011), using plant membrane-bound Δ6, Δ8 sphingo-lipid, Δ12 (FAD2) and Δ15 (FAD3), as well as the soluble Δ9 stearoyl-ACP desaturases as queries. Initial gene models of these desaturases were predicted with the FGENESH software (Salamov and Solovyev, 2000), readjusted by multiple sequence alignment using ClustalX (Thompson *et al.*, 1994) and further clarified by comparing their genomic coding sequences with corresponding EST sequences where possible. Undetermined residues were denoted as X for the corresponding N base calls in the nucleotide sequences based on the most homologous sequence in the multiple sequence alignment. Assembled nucleotide and predicted amino acid sequences for all identified desaturases are provided in Table S1 and are available from the Cannabis Genome Browser website: http://genome.ccbr.utoronto.ca/. The full-length and cDNA sequences of the functionally characterized CSFAD2A and CSFAD3A have been deposited in GenBank (Accession numbers KF679783 and KF679784, respectively).

All desaturase sequences from other plant species were retrieved from GenBank (http://www.ncbi.nlm.nih.gov/), these include Δ6 desaturases and Δ8 sphingo-lipid desaturases (Song *et al.*, 2010), Δ12 and Δ15 desaturases (Andreu *et al.*, 2010; Bilyeu *et al.*, 2003; Li *et al.*, 2008; Sperling and Heinz, 2003; Vrinten *et al.*, 2005; Zhang *et al.*, 2009), and the soluble Δ9 stearoyl-ACP desaturases (Zhang *et al.*, 2008). Sequences have been assigned a three-letter species-specific identifier as follows: *Anemone lendsquerelli* (Ale), *Arabidopsis thaliana* (Ath), *Brassica napus* (Bna), *Borago officinalis* (Bof), *Crepis alpine* (Cal), *Crepis palaestina* (Cpa), *Cannabis sativa* (Csa), *Camellia sinensis* (Csi), *Echium gentianoides* (Ege), *Gossypium hirsutum* (Ghi), *Glycine max* (Gma), *Impatiens balsamina* (Iba), *Lesquerella fendleri* (Lfe), *Linum usitatissimum* (Lus), *Momordica charantia* (Mch), *Nicotiana tabacum* (Nta), *Petroselinum crispum* (Pcr), *Primula farinosa* (Pfa), *Primula vialli* (Pvi), *Ricinus communis* (Rco), *Ribes nigrum* (Rni).

GenBank accession numbers for the protein sequences are as follows: Δ6, Δ8 sphingo-lipid desaturases: AleD6/8 (AAQ10731/AAQ10732), AthD8 (NP_191717), BnaD8 (CAA11857), BofD6/8 (AAC49700/AAG43277), CsiD6 (AAO13090), EgeD6 (AAL23580), NtaD8 (ABO31111), PfaD6/8 (AAP23034/AAP23033), PviD6/8 (AAP23036/AAP23035), RniD8A/B (ADA60228/ADA60229), and RniD6C/D/E (ADA60230/ADA60231/ADA60232); Δ12 desaturases: AthFAD2 (AAA32782), BofFAD (AAC31698), CalFAD (CAA76158), CpaFAD1 (CAA76157), CpaFAD2 (CAA76156), GhiFAD2-4 (AAQ16653), GhiFAD2 (AAL37484), GhiFAD (CAA71199), GhiFAD2-2B (ABY71269), GhiFAD5 (CAA65744), GmaFAD2-1A/B (AAB00860/ABF84062), IbaFADL (AAF05915), LfeLFAH12 (AAC32755), MchFADX (AAF05916), PcrELI12 (AAB80697), RcoFAD (AAC49010); Δ15 desaturases: AthFAD3_ER (P48623), AthFAD7_PL (BAA03106), AthFADIP1_PL (BAA04504), BnaFAD_PL (AAA61774), BnaFAD3_ER (P48624), GmaFAD3A/B/C (AAO24263/AAO24264/AAO24265), GmaFAD7/8 (ACF19424/ACU17817), LusFAD3A/3B (ABA02172/ABA02173); Δ9 stearoyl-ACP desaturases: AthFAB2 (AEC10310), GmaSACPD-A/B/C, (AAX86050/AAX86049/ABM45911), RcoSACPD (AAA74692).

All deduced amino acid sequences from *C. sativa* were aligned with desaturases of other plant species using ClustalX. The alignments were reconciled and further adjusted by eye to minimize insertion/deletion events, and only the most conserved alignment regions were used in the subsequent phylogenetic analyses. Distance analyses used the Protdist program with a Jones-Taylor-Thornton substitution matrix of the Phylip 3.6b package (Felsenstein, 1989). Phylogenetic trees were calculated from the distance matrices by the neighbour-joining algorithm. Bootstrap analyses consisted of 1000 replicates using the same protocol. Groups with above 70% bootstrap value were considered as strongly supported.

Deep sequencing the developing hemp seed transcriptome

Pyrosequencing was carried out on three cDNA libraries prepared from dissected embryos at the T, U and FND stages at the GenePool genomics facility at the University of Edinburgh on the 454 GS-FLX sequencing platform (Roche Diagnostics, Branford, CT). Raw sequence analysis, contiguous sequence assembly and annotation were performed as described (Graham *et al.*, 2010). Abundance of membrane-bound and soluble desaturase transcripts were analysed *in silico* by determining read counts in the three EST libraries. The raw reads were mapped to the reference sequence, which consisted of the open reading frames of the 17 desaturase genes (included in Table S1) with BWA mapping software (Li and Durbin, 2009). The raw read counts were retrieved from the resulting output file for each gene in the libraries, and the counts were then normalized to a reads per kilobase per million reads (RPKM) value as an approximation of gene expression.

Quantitative real-time PCR

Total RNA from leaves of 2-week-old and 4-week-old hemp plants was extracted with TRI Reagent Solution (Ambion®; Life Technologies, Carlsbad, CA) and treated with Turbo DNA-*free*

(Ambion®). Single-strand cDNA was synthesized from DNase-treated RNA using SuperScript II (Invitrogen™; Life Technologies, Carlsbad, CA) reverse transcriptase with oligo(dT)$_{16-18}$ primer (Invitrogen) and diluted to 50 ng/μL. Quantitative real-time PCR was performed using an ABI Prism 7300 detection system (Applied Biosystems®; Life Technologies, Carlsbad, CA) and SYBR Green PCR Master mix (Applied Biosystems®) to monitor dsDNA synthesis. Gene-specific primers: 5'CTCGGACATAGGGATTTTCATTG3' and 5'CAACCCAACCTAACCCTTTGG3' for *CSFAD2A*; 5'CGATGTGGGGGTTTTCATCA-3' and 5'-AACCCAACTCAACCCTCTTGCT-3' for *CSFAD2B*; 5'TCAAATCCCACACTACCATCTTGT3' and 5'TTTCTAGGCTCCCTGTAATACTTTCC3' for *CSFAD3*. Amplification plots were analysed with an R_n threshold of 0.2 to obtain C_T (threshold cycle) values. Each transcript was normalized to the hemp *actin-2* gene amplified with primers: 5'GGGTCACACTGTGCCAATCTAC3' and 5'CCCAGCAAGGTCAAGACGAA3' and compared among samples.

Establishment and screening of an EMS-mutagenized population

Hemp seed of the Finola variety were purchased from the Finola company (http://www.Finola.com), Finland, and grown in controlled glasshouse facilities at the University of York. The seed was treated with 300 mM EMS for 5 h before sowing onto soil-based John Innes Compost No. 2. Mutagenized M1 female plants were outcrossed with male wild-type Finola plants to produce a heterozygous M2 screening population. Five nanogram per microlitre DNA from individual plants was pooled fourfold for screening. A 1140-bp fragment of *CSFAD2A* was amplified in a two-step PCR amplification. The first step was carried out with unlabelled primers (5'CCCATTGCTTTAAACGCTCTCTAATTCGCT3' (left) and 5'CACCCCTAACCACATTAAGCCATACCCCAT3' (right)) on 12.5 ng pooled gDNA in 10 μL volumes. Labelling of the amplified gene fragment with infrared dyes occurred during the second PCR step, where a mixture of labelled and unlabelled primers was used for further amplification and simultaneous labelling using appropriately diluted product from the first PCR step as template (left primer labelled with IRDye 700, right primer labelled with IRDye 800 (MWG, Ebersberg, Germany), left primer labelled:unlabelled ratio = 3:2; right primer labelled:unlabelled ratio = 4:1).

A 1500-bp fragment of *CSFAD3A* was also amplified in a two-step PCR reaction using nonlabelled gene-specific primers: 5'CGCCATTCCTAAGCATTGTT3' (left) and 5'ATAGTGGTCCTGGCTGATGC3' (right) in the first step. As for the Δ12 desaturase fragment, labelling with infrared dyes occurred during the second PCR step but using 5'M13-tailed primers: 5'TGTAAAACGACGGCCAGTGGGCTGCTCAAGGAACCATGTTCT3' (left) and 5'AGGAAACAGCTATGACCATCCTTGGTAGCTTCCACAAGATGG3' (right) mixed with M13 primers labelled with IRDye 700 and IRDye 800. The ratios of labelled to unlabelled primers were as for the *CSFAD2A* fragment. Heteroduplex formation, CEL I nuclease digestion and analysis on the LI-COR 4300 DNA sequencer platform were carried out as described by Till *et al.* (2006).

Cloning and expression of *Cannabis sativa CSFAD2A* and *CSFAD3A* in *Saccharomyces cerevisiae*

The open reading frames of *CSFAD2A* and *CSFAD3A* were amplified by PCR using Phusion Hot Start DNA polymerase (Finnzymes) and the following pairs of specific primers: 5'ATAGGATCCAAA**ATG**GGAGCCGGT3' (left) and 5'GCCTCGAGC**CTA**AAACTTGTTTTTGTACC3' (right) for *CSFAD2A* and

5'GGG<u>GAATTC</u>ATA**ATG**ACAGAATCACATGC3' (left) and 5'TA<u>G CGGCCGC</u>ATA**CTA**CATTTGCTTGGC3' (right) for *CSFAD3*. Both open reading frames were cloned into the pESC-TRP (Stratagene) and transformed into *S. cerevisiae* strain G175 (Gietz and Woods, 2002). Cultures were grown at 28 °C in the presence of 2% (w/v) raffinose and 1% (w/v) Tergitol NP-40 (Sigma-Aldrich, St. Louis, MO). Expression of the transgene was induced when OD_{600} reached 0.2–0.3 by supplementing galactose to 2% (w/v) and appropriate fatty acids were added to a final concentration of 50 μM. Incubation was carried out at 25 °C for four generations (28 h).

Fatty acid and oil analysis

Fatty acid methyl esters (FAMEs) were prepared by direct transmethylation of single seeds or ~10 mg oil samples (Browse *et al.*, 1986). FAME content was determined by gas chromatography with flame ionization detection (GC-FID; GC Trace Ultra, Thermoquest Separation Products, Manchester, UK). A 1-μL aliquot of FAMEs in hexane was injected into a 3-mm internal diameter FocusLiner containing glass wool (SGE, Milton Keynes, UK) at 230 °C in programmed flow mode with H_2 as carrier gas. The H_2 flow program was as follows: initial hold 0.3 mL/min for 0.1 min, then ramped at 5 mL/min^2 to 0.5 mL/min for the remainder of the run. The split ratio was maintained at 1 : 250, and a gas saver slow of 20 mL/min was initiated at 1.5 min into the run. Separation was achieved using a narrow-bore cyanopropyl polysilphenylene-siloxane capillary column (SGE BPX70; 10 m length × 0.1 mm internal diameter × 0.2 μm film thickness). FAMEs were separated using the following temperature program: initial hold 150 °C 0.1 min, then ramped at 16 °C/min to 220 °C, followed by cool down to initial conditions at 120 °C/min. The FID was run at 300 °C with air, H_2 and make-up N_2 gases flowing at 350, 35 and 30 mL/min, respectively. The signal was collected and peaks detected and integrated using ChromQuest 4.2 software (Thermo Electron Corporation, Manchester, UK). FAMEs were identified and quantified relative to the Supelco 37 component FAME mix (Sigma-Aldrich, Gillingham, UK).

Extracts containing FAMEs that did not coelute with standards or whose identity was unclear were concentrated and further derivatized to their 3-pyridylcarbinol esters (Dubois *et al.*, 2006), chromatographed using a longer, thicker-film BPX70 column using He as carrier gas with an extended thermal gradient, and 70 eV electron impact mass spectra generated using a Leco Pegasus IV mass spectrometer running ChromaTof 4.5 software (Leco, Stockport, UK). Under these conditions, retention time order was preserved as per the GC-FID analyses. Mass spectra were interpreted to localize dienoic double bond positions as described by Christie *et al.* (1987).

Phenotyping for fatty acid content was carried out on single cotyledons dissected from 2-days-old seedlings germinated on moist filter paper. The surviving seedlings were transferred to soil, grown and genotyped, and then, selected individuals were used for subsequent crosses.

Oil pressing was carried out using a small capacity Komet screw press (Model CA 59 G; IBG Monforts, Mönchengladbach, Germany), with a 6-mm press nozzle die and a screw speed of 20 r.p.m. Running temperature was checked with a digital thermometer inserted into the restriction die, with screw-press barrel temperature not exceeding 60 °C. After each sample, all press devices were cleaned and dried.

The oxidative stability of the pressed oils was determined using a Metrohm Rancimat model 743, according to AOCS Official Method Cd 12b-92. Briefly, the induction times (n = 4) for portions of oil (3.0 g) were determined at 100, 110 and 120 °C and 20 L/h air throughput. Projected shelf life stability was calculated by extrapolation of the relationship between the measured induction time and the temperature (Metrohm Application Bulletin No. 141/3e).

Tocopherol analysis

Tocopherols were measured in 100 mg/mL dilutions of pressed oils in methyl tertiary butyl ether. Aliquots (10 μL) were separated by a nonaqueous reverse phase HPLC method used for neutral lipid separation (Burgal *et al.*, 2008) and quantified by their UV absorbance at 297 nm against calibration curves of authentic standards (Sigma-Aldrich, Gillingham, UK).

Acknowledgements

Financial support for this work came from the UK Technology Strategy Board, reference TP/3/BIO/6/1/17190, EU 7th Framework Programme award, project number 311849, 'MultiHemp' and to the Centre for Novel Agricultural Products from the Garfield Weston Foundation. We thank Martin Farrow, ADM Erith, UK, for providing the Hi-Oleic rapeseed, the University of York, Department of Biology horticulture staff for expert plant care and Judith Mitchell for administrative support.

References

Andreu, V., Lagunas, B., Collados, R., Picorel, R. and Alfonso, M. (2010) The GmFAD7 gene family from soybean: identification of novel genes and tissue-specific conformations of the FAD7 enzyme involved in desaturase activity. *J. Exp. Bot.* **61**, 3371–3384.

van Bakel, H., Stout, J.M., Cote, A.G., Tallon, C.M., Sharpe, A.G., Hughes, T.R. and Page, J.E. (2011) The draft genome and transcriptome of *Cannabis sativa*. *Genome Biol.* **12**, R102.

Baud, S., Boutin, J.P., Miquel, M., Lepiniec, L. and Rochat, C. (2002) An integrated overview of seed development in *Arabidopsis thaliana* ecotype WS. *Plant Physiol. Biochem.* **40**, 151–160.

Bilyeu, K.D., Palavalli, L., Sleper, D.A. and Beuselinck, P.R. (2003) Three microsomal omega-3-fatty acid desaturase genes contribute to soybean linolenic acid levels. *Crop Sci.* **43**, 1833–1838.

Browse, J., McCourt, P.J. and Somerville, C.R. (1986) Fatty acid composition of leaf lipids determined after combined digestion and fatty acid methyl ester formation from fresh tissue. *Anal. Biochem.* **152**, 141–145.

Burgal, J., Shockey, J., Lu, C., Dyer, J., Larson, T., Graham, I. and Browse, J. (2008) Metabolic engineering of hydroxy fatty acid production in plants: RcDGAT2 drives dramatic increases in ricinoleate levels in seed oil. *Plant Biotechnol. J.* **6**, 819–831.

Callaway, J. (2004) Hempseed as a nutritional resource: an overview. *Euphytica*, **140**, 65–72.

Callaway, J.C. and Laakkonen, T.T. (1996) Cultivation of *Cannabis* oil seed varieties in Finland. *J. Int. Hemp Assoc.* **3**, 32–34.

Christie, W.W., Brechany, E.Y. and Holman, R.T. (1987) Mass spectra of the picolinyl esters of isomeric mono- and dienoic fatty acids. *Lipids*, **22**, 224–228.

Dubois, N., Barthomeuf, C. and Bergé, J.-P. (2006) Convenient preparation of picolinyl derivatives from fatty acid esters. *Eur. J. Lipid Sci. Technol.* **108**, 28–32.

Felsenstein, J. (1989) PHYLIP – Phylogeny inference package (Version 3.2). *Cladistics*, **5**, 164–166.

Gietz, R.D. and Woods, R.A. (2002) Transformation of yeast by lithium acetate/single-stranded carrier DNA/polyethylene glycol method. In *Methods in*

Enzymology, Vol. 350 (Christine, G. and Gerald, R.F., eds), pp. 87–96. Salt Lake City: Academic Press.

Graham, I.A., Besser, K., Blumer, S., Branigan, C.A., Czechowski, T., Elias, L., Guterman, I., Harvey, D., Isaac, P.G., Khan, A.M., Larson, T.R., Li, Y., Pawson, T., Penfield, T., Rae, A.M., Rathbone, D.A., Reid, S., Ross, J., Smallwood, M.F., Segura, V., Townsend, T., Vyas, D., Winzer, T. and Bowles, D. (2010) The genetic map of *Artemisia annua* L. identifies loci affecting yield of the antimalarial drug artemisinin. *Science*, **327**, 328–331.

de Gyves, E.M., Sparks, C.A., Sayanova, O., Lazzeri, P., Napier, J.A. and Jones, H.D. (2004) Genetic manipulation of γ-linolenic acid (GLA) synthesis in a commercial variety of evening primrose (*Oenothera* sp.). *Plant Biotechnol. J.* **2**, 351–357.

Hitz, W.D., Carlson, T.J., Booth, J.R.J., Kinney, A.J., Stecca, K.L. and Yadav, N.S. (1994) Cloning of a higher-plant plastid ω-6 fatty acid desaturase cDNA and its expression in a cyanobacterium. *Plant Physiol.* **105**, 635–641.

Hong, H., Datla, N., Reed, D.W., Covello, P.S., MacKenzie, S.L. and Qiu, X. (2002) High-level production of γ-linolenic acid in *Brassica juncea* using a Δ6 desaturase from *Pythium irregulare*. *Plant Physiol.* **129**, 354–362.

Kinney, A.J. (1998) Production of specialised oils for industry. In *Plant Lipid Biosynthesis: Fundamentals and Agricultural Applications*(Harwood, J.L., ed), pp. 273–286. Cambridge, UK: Cambridge University Press.

Lee, M., Lenman, M., Banaś, A., Bafor, M., Singh, S., Schweizer, M., Nilsson, R., Liljenberg, C., Dahlqvist, A., Gummeson, P.-O., Sjödahl, S., Green, A. and Stymne, S. (1998) Identification of non-heme diiron proteins that catalyze triple bond and epoxy group formation. *Science*, **280**, 915–918.

Li, H.-L. (1974) An archaeological and historical account of cannabis in China. *Econ. Bot.* **28**, 437–448.

Li, H. and Durbin, R. (2009) Fast and accurate short read alignment with Burrows-Wheeler transform. *Bioinformatics*, **25**, 1754–1760.

Li, L., Wang, X., Gai, J. and Yu, D. (2008) Isolation and characterization of a seed-specific isoform of microsomal omega-6 fatty acid desaturase gene (FAD2-1B) from soybean. *DNA Seq.* **19**, 28–36.

Liu, J.-W., Huang, Y.-S., DeMichele, S., Bergana, M., Bobik, E., Hastilow, C., Chuang, L.-T., Mukerji, P. and Knutzon, D. (2001) Evaluation of the seed oils from a canola plant genetically transformed to produce high level of γ-linolenic acid. In *γ-Linolenic Acid: Recent Advances in Biotechnology and Clinical Applications*(Huang, Y.-S. and Ziboh, A., eds), pp. 61–71. Champaign, IL: AOCS Press.

Ohlrogge, J. and Browse, J. (1995) Lipid biosynthesis. *Plant Cell*, **7**, 957–970.

Puttick, D., Dauk, M., Lozinsky, S. and Smith, M.A. (2009) Overexpression of a FAD3 desaturase increases synthesis of a polymethylene-interrupted dienoic fatty acid in seeds of *Arabidopsis thaliana* L. *Lipids*, **44**, 753–757.

Reddy, A.S. and Thomas, T.L. (1996) Expression of a cyanobacterial Δ6-desaturase gene results in γ-linolenic acid production in transgenic plants. *Nat. Biotechnol.* **14**, 639–642.

Reed, D.W., Schäfer, U.A. and Covello, P.S. (2000) Characterization of the *Brassica napus* extraplastidial linoleate desaturase by expression in *Saccharomyces cerevisiae*. *Plant Physiol.* **122**, 715–720.

Salamov, A.A. and Solovyev, V.V. (2000) *Ab initio* gene finding in Drosophila genomic DNA. *Genome Res.* **10**, 516–522.

Sayanova, O., Davies, G.M., Smith, M.A., Griffiths, G., Stobart, A.K., Shewry, P.R. and Napier, J.A. (1999a) Accumulation of Δ6-unsaturated fatty acids in transgenic tobacco plants expressing a Δ6-desaturase from *Borago officinalis*. *J. Exp. Bot.* **50**, 1647–1652.

Sayanova, O., Shewry, P.R. and Napier, J.A. (1999b) Histidine-41 of the cytochrome *b*5 domain of the borage Δ6 fatty acid desaturase is essential for enzyme activity. *Plant Physiol.* **121**, 641–646.

Schwartzbeck, J.L., Jung, S., Abbott, A.G., Mosley, E., Lewis, S., Pries, G.L. and Powell, G.L. (2001) Endoplasmic oleoyl-PC desaturase references the second double bond. *Phytochemistry*, **57**, 643–652.

Shanklin, J. and Cahoon, E.B. (1998) Desaturation and related modifications of fatty acids. *Annu Rev Plant Physiol Plant Mol Biol.* **49**, 611–641.

Song, L.Y., Lu, W.X., Hu, J., Zhang, Y., Yin, W.B., Chen, Y.H., Hao, S.T., Wang, B.L., Wang, R.R. and Hu, Z.M. (2010) Identification and functional analysis of the genes encoding Δ6-desaturase from *Ribes nigrum*. *J. Exp. Bot.* **61**, 1827–1838.

Sperling, P. and Heinz, E. (2003) Plant sphingolipids: structural diversity, biosynthesis, first genes and functions. *Biochim. Biophys. Acta*, **1632**, 1–15.

Thompson, J.D., Higgins, D.G. and Gibson, T.J. (1994) CLUSTAL W: improving the sensitivity of progressive multiple sequence alignment through sequence weighting, position-specific gap penalties and weight matrix choice. *Nucleic Acids Res.* **22**, 4673–4680.

Till, B.J., Zerr, T., Comai, L. and Henikoff, S. (2006) A protocol for TILLING and Ecotilling in plants and animals. *Nat. Protoc.* **1**, 2465–2477.

Vrinten, P., Hu, Z., Munchinsky, M.-A., Rowland, G. and Qiu, X. (2005) Two FAD3 desaturase genes control the level of linolenic acid in flax seed. *Plant Physiol.* **139**, 79–87.

Zhang, P., Burton, J.W., Upchurch, R.G., Whittle, E., Shanklin, J. and Dewey, R.E. (2008) Mutations in a Δ9–Stearoyl-ACP-desaturase gene are associated with enhanced stearic acid levels in soybean seeds. *Crop Sci.* **48**, 2305–2313.

Zhang, D., Pirtle, I.L., Park, S.J., Nampaisansuk, M., Neogi, P., Wanjie, S.W., Pirtle, R.M. and Chapman, K.D. (2009) Identification and expression of a new delta-12 fatty acid desaturase (FAD2-4) gene in upland cotton and its functional expression in yeast and *Arabidopsis thaliana* plants. *Plant Physiol. Biochem.* **47**, 462–471.

17

Targeted silencing of *BjMYB28* transcription factor gene directs development of low glucosinolate lines in oilseed *Brassica juncea*

Rehna Augustine[1], Arundhati Mukhopadhyay[2] and Naveen C. Bisht[1,]*

[1]*National Institute of Plant Genome Research (NIPGR), New Delhi, India*
[2]*Department of Genetics, Center for Genetic Manipulation of Crop Plants (CGMCP), University of Delhi South Campus, New Delhi, India*

*Correspondence
email ncbisht@nipgr.ac.in
Accession numbers: JQ666166, JQ666167,
JQ666168, JQ666169, JX316031,
JX316032.

Keywords: *Brassica juncea*,
glucosinolates, MYB28, RNAi
suppression, secondary metabolites,
transgenic plants.

Summary

Brassica juncea (Indian mustard), a globally important oilseed crop, contains relatively high amount of seed glucosinolates ranging from 80 to 120 µmol/g dry weight (DW). One of the major breeding objectives in oilseed *Brassicas* is to improve the seed-meal quality through the development of low-seed-glucosinolate lines (<30 µmol/g DW), as high amounts of certain seed glucosinolates are known to be anti-nutritional and reduce the meal palatability. Here, we report the development of transgenic *B. juncea* lines having seed glucosinolates as low as 11.26 µmol/g DW, through RNAi-based targeted suppression of *BjMYB28*, a R2R3-MYB transcription factor family gene involved in aliphatic glucosinolate biosynthesis. Targeted silencing of *BjMYB28* homologs provided significant reduction in the anti-nutritional aliphatic glucosinolates fractions, without altering the desirable nonaliphatic glucosinolate pool, both in leaves and seeds of transgenic plants. Molecular characterization of single-copy, low glucosinolate homozygous lines confirmed significant down-regulation of *BjMYB28* homologs *vis-à-vis* enhanced accumulation of *BjMYB28*-specific siRNA pool. Consequently, these low glucosinolate lines also showed significant suppression of genes involved in aliphatic glucosinolate biosynthesis. The low glucosinolate trait was stable in subsequent generations of the transgenic lines with no visible off-target effects on plant growth and development. Various seed quality parameters including fatty acid composition, oil content, protein content and seed weight of the low glucosinolate lines also remained unaltered, when tested under containment conditions in the field. Our results indicate that targeted silencing of a key glucosinolate transcriptional regulator MYB28 has huge potential for reducing the glucosinolates content and improving the seed-meal quality of oilseed *Brassica* crops.

Introduction

Glucosinolates are a group of amino acid-derived plant secondary metabolites distinctive to the order Capparales. This order includes the family Brassicaceae, containing agriculturally important vegetables, condiments and oilseed crops and the model species *Arabidopsis thaliana*. Glucosinolates are broadly classified into three major groups, namely aliphatic, indole and aromatic glucosinolates, based on their precursor amino acid. The biosynthesis of both aliphatic and indole glucosinolates can be divided into three steps: (i) side-chain elongation of precursor amino acid, (ii) core-glucosinolate structure formation and (iii) modification of side group (Halkier and Gershenzon, 2006; Wittstock and Halkier, 2002). Together with side-chain elongation of the R-group, side-chain modifications generate a plethora of diverse glucosinolate compounds, with more than 200 structures identified till date (Clarke, 2010).

In recent years, glucosinolate research has gained increasing importance as these metabolites have both useful and harmful effects. Besides imparting distinct flavour to the cruciferous vegetables, certain glucosinolates and their degradation products have been reported to act as anti-carcinogenic and chemo-protective

agents in mammalian systems (Fahey *et al.*, 1997; Hayes *et al.*, 2008). Both aromatic and indole glucosinolates are known to play important roles in plant defence against pathogens and herbivores (Clay *et al.*, 2009; Hopkins *et al.*, 2009). However, on the negative side, some of the aliphatic glucosinolates and their degradation products in seed meal are associated with anti-nutritional properties (Cartea and Velasco, 2008; Fahey *et al.*, 2001; Mawson *et al.*, 1993). For example, 2-hydroxy-3-butenyl glucosinolate (progoitrin) degrades to goitrogenic products while other glucosinolate such as 3-butenyl and 4-pentenyl produce isothiocyanates, which ultimately reduce the meal palatability (Campos de Quiros and Mithen, 1996) and food intake of cattle, swine and poultry.

Over the last few decades, one of the major breeding objectives in oilseed *Brassica* crops (*B. rapa*, *B. napus*, and *B. juncea*) has been the improvement of its seed-meal quality through the development of lines having total glucosinolate <30 µmol/g DW in seed-meal (Potts *et al.*, 1999). Canola quality lines (having low glucosinolate and low erucic acid in seed) have been successfully developed in *B. napus* and *B. rapa*. The development of these varieties with major improvements in agronomic, oil and meal quality has greatly influenced the rapid increase of canola consumption over the last two decades mainly in the North American continent and the European Union. In contrast, the *B. juncea* gene pools contain significantly higher amounts of seed aliphatic glucosinolates, ranging from 80 to 120 µmol/g DW (Pradhan *et al.*, 1993; Sodhi *et al.*, 2002). The presence of such a high quantity of seed aliphatic glucosinolates limits the value of this crop in the international market both as edible oil and animal feed. However, *B. juncea* has several potential advantages over *B. napus* (canola rapeseed) as a global oilseed crop of semi-arid regions, viz., more vigorous seedling growth, less pod shattering problem, greater tolerance to heat and drought and enhanced resistance to the blackleg fungus, *Leptosphaeria maculans* (Burton *et al.*, 1999; Potts *et al.*, 1999).

Till date, no productive and agronomically viable low glucosinolate line has been reported in *B. juncea* (Indian mustard). Glucosinolate trait in *B. juncea* follows a complex inheritance pattern being controlled by multiple quantitative trait loci, QTL (Bisht *et al.*, 2009; Lionneton *et al.*, 2004; Mahmood *et al.*, 2003; Ramchiary *et al.*, 2007; Sodhi *et al.*, 2002). Even though a canola quality line is available in exotic gene pool of *B. juncea* (Burton *et al.*, 1999; Potts *et al.*, 1999), the introgression of low glucosinolate alleles into well adapted Indian *B. juncea* cultivars through conventional breeding has been largely impeding. Some of the inherent problems associated with breeding low glucosinolate QTLs were the existence of epistatic QTL, and context-dependent interactions of the loci (Bisht *et al.*, 2009; Ramchiary *et al.*, 2007). In addition, a significant amount of linkage drag from donor genome was also reported because of which negative linkage between QTL alleles of low glucosinolate and yield-related QTL in few linkage groups was observed (Ramchiary *et al.*, 2007). Thus, recombination-based transfer inevitably requires fine mapping of each QTL and screening a large number of progeny in every back-cross generation, which makes the approach time-consuming, labour-intensive and more importantly germplasm dependent.

Alternatively, transgenic technologies could be exploited to engineer low glucosinolate lines in oilseed *Brassica* crops. Our understanding of genes involved in the complex glucosinolate biosynthesis pathway has been possible from accumulating studies in the closest model species *A. thaliana* (Grubb and Abel, 2006; Halkier and Gershenzon, 2006; Sønderby *et al.*, 2010b). Briefly, till date, >20 biosynthesis pathway genes and three transcriptional regulator genes have been reported to control the aliphatic glucosinolate biosynthesis in *A. thaliana* (Figure S1, Gigolashvili *et al.*, 2007, 2009; Hirai *et al.*, 2007; Sønderby *et al.*, 2007, 2010a). However, because of the inherent polyploidy associated with the *Brassica* crops, multiplicity of glucosinolate biosynthesis and regulatory genes are quite expected. Recently, multiple homologs of glucosinolate candidate genes were observed in *B. rapa* genome (Wang *et al.*, 2011; Zang *et al.*, 2009). It is likely that allotetraploid *B. juncea* and *B. napus* might have even greater inventory of glucosinolate candidate genes, possibly resulting in more complex regulatory and metabolic networks in these oilseed crops. In view of the above, metabolic engineering of low glucosinolate content in *B. juncea* is a major challenge for plant biotechnology and requires a robust genetic engineering strategy.

Here, we describe the development of low glucosinolate *B. juncea* transgenic lines, having seed glucosinolate content <30 µmol/g DW, through RNAi-based targeted suppression of a key glucosinolate transcriptional regulator, *BjMYB28*. The development of low glucosinolate *B. juncea* lines will significantly improve its seed-meal quality and raise its international trade potential both as oilseed and feed crop.

Results and discussion

BjMYB28-3 was differentially expressed in glucosinolate contrasting lines of *B. juncea*

In order to study the differential regulation of glucosinolate biosynthesis in *B. juncea* cultivars, we selected a well adapted high glucosinolate Indian line (Varuna) and the canola quality exotic line (Heera) and measured their leaf and seed glucosinolate contents and profiles. In general, the Indian line showed a very high amount of total glucosinolates both in leaves and seeds compared to the canola quality line (Table S1). More than 95% of the total glucosinolate in *B. juncea* was found to be aliphatic glucosinolates. The two *B. juncea* lines were highly contrasting for aliphatic glucosinolate pools both in leaves and seeds; however, the nonaliphatic glucosinolate pools were found to be almost similar. For example, the total aliphatic seed glucosinolate content of *B. juncea* Indian line was found to be 105.12 ± 5.8 µmol/g DW, where as the canola quality line showed a total aliphatic seed glucosinolate content of 12.46 ± 0.9 µmol/g DW (Table S1).

Recent reports in *Arabidopsis* showed that levels of aliphatic glucosinolate are transcriptionally regulated by at least three members, namely AtMYB28, AtMYB29 and AtMYB76 of subgroup-12 of R2R3-MYB transcription factor family genes (Gigolashvili *et al.*, 2007, 2009; Hirai *et al.*, 2007; Sønderby *et al.*, 2007, 2010a). These studies concluded that AtMYB28 is the principal regulator of aliphatic glucosinolate biosynthesis and affects the production of both short- and long-chain aliphatic glucosinolates, whereas AtMYB29 and AtMYB76 (encoded by tandemly duplicated genes in *A. thaliana*) are additional independent control elements affecting biosynthesis of only short-chain aliphatic glucosinolates.

We therefore investigated the inventory of these key regulators for aliphatic glucosinolates in allopolyploid *B. juncea*. Four full-length homologs of the *Arabidopsis AtMYB28* (At5 g61420) were isolated from *B. juncea* high glucosinolate line Varuna, namely *BjMYB28-1*, *BjMYB28-2*, *BjMYB28-3* and *BjMYB28-4* (Augustine *et al.*, communicated; accession nos. JQ666166, JQ666167, JQ666168, JQ666169). In contrast, only two full-length *BjMYB29* homologs viz., *BjMYB29-1* and *BjMYB29-2* (accession nos., JX316031, JX316032) were isolated from *B. juncea* (Figure S2). We could not identify any *MYB76*-specific orthologs from *B. juncea* genome, which was also found to be absent in *B. rapa* genome (Wang *et al.*, 2011; Zang *et al.*, 2009).

The expression profile of *BjMYB28* and *BjMYB29* homologs across different tissue types was determined in two *B. juncea* lines contrasting for glucosinolate contents. Differential expression of *BjMYB28* homologs was observed across the developing stages of glucosinolate contrasting lines, wherein the fold difference was found to be highest in the developing silique (Figure 1a–d). Interestingly, expression of one of the homolog, namely *BjMYB28-3*, was observed exclusively in the high glucosinolate line, whereas the other three *BjMYB28* homologs showed expression in both the *B. juncea* lines, even though the levels of transcripts were variable. The *BjMYB28-3* transcript was undetectable in the low glucosinolate canola quality line, despite using primers from different regions of the same gene (Figures 1c and S3). The expression of two *BjMYB29* homologs was found to be significantly lower compared to that of *BjMYB28* homologs

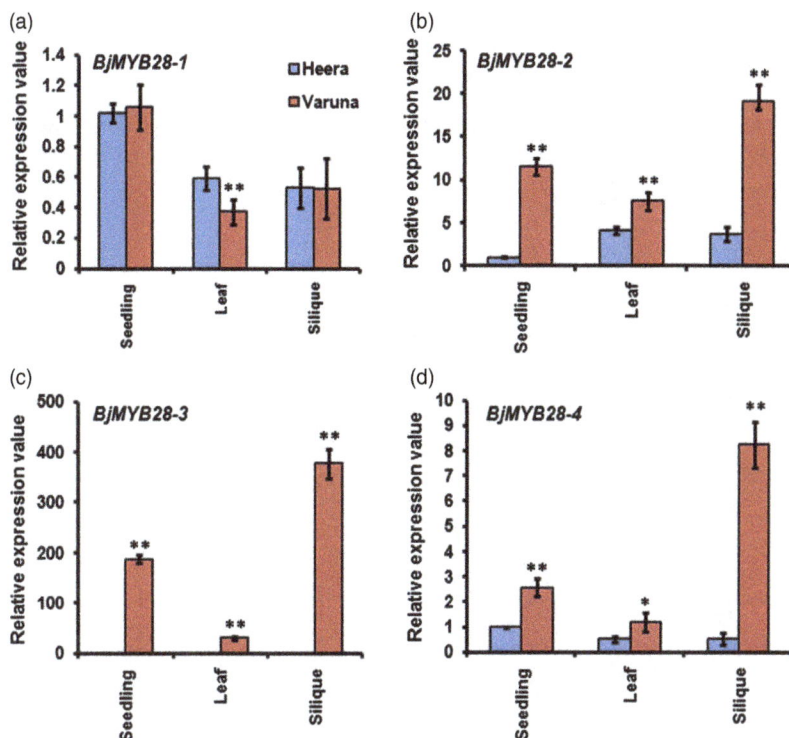

Figure 1 Expression analysis of *BjMYB28* homologs between *Brassica juncea* lines contrasting for total glucosinolate content. qRT-PCR analysis of (a) *BjMYB28-1*, (b) *BjMYB28-2*, (c) *BjMYB28-3* and (d) *BjMYB28-4* homologs. Seedling stage of *B. juncea* line, Heera was used as a reference calibrator (set as 1). The qRT-PCR experiments were conducted with three independent sets of plants and values represent mean ± standard error. Significant difference in expression profile of *BjMYB28* homologs of high glucosinolate line Varuna in comparison with the low glucosinolate line Heera at different stages are indicated by asterisks (*$P < 0.05$, **$P < 0.01$ in Fishers LSD test determined by ANOVA).

across the developing stages of *B. juncea* (Figure S4). Our observation was in accordance with that of *Arabidopsis*, wherein *AtMYB29* showed lower expression level compared to the *AtMYB28* across all the developmental stages (Gigolashvili *et al.*, 2009).

The differential expression of *BjMYB28-3* across glucosinolate contrasting lines of *B. juncea* could be an important cue suggesting its involvement towards controlling glucosinolate variation. In order to investigate the differential regulation of *BjMYB28-3* expression, we performed Southern blot analysis on glucosinolate contrasting *B. juncea* cultivars using probe from 5′ upstream region specific to *BjMYB28-3* homolog. The low glucosinolate cultivar of *B. juncea* did not show any hybridization signal (Figure S5). Very recently, genomic deletions underlying the two QTLs for seed glucosinolate contents in *B. napus* were identified using associative transcriptomics approach (Harper *et al.*, 2012). The deleted regions of *B. napus* were found to contain orthologs of *AtMYB28*. Our data also suggest that there could be genomic deletion of *BjMYB28-3* homolog in the low glucosinolate cultivar, Heera. The Polish spring rape variety 'Bronowski' is regarded as the sole donor source of low glucosinolate trait to the present day 'canola' quality cultivars of *Brassica* species (Bisht *et al.*, 2009; Potts *et al.*, 1999). With the advent of genomics tools, genetic dissection of glucosinolate QTLs across *Brassica* species can be undertaken to identify the historically important 'Bronowski gene(s)' controlling the low glucosinolate trait in *Brassica* crops.

Additionally, mapping and association of pathway genes with seed-glucosinolate QTLs in *B. juncea* and *B. napus* also suggested that the *MYB28* orthologs contribute prominently towards

controlling variability in glucosinolate contents across these complex allopolyploid genomes (Bisht *et al.*, 2009; Feng *et al.*, 2012; Ramchiary *et al.*, 2007). For example, in *B. juncea*, two of six QTLs identified for seed-glucosinolate trait harbour *BjMYB28* homologs (Bisht *et al.*, 2009; Ramchiary *et al.*, 2007). Further, one of the A-genome-specific QTLs identified in *B. juncea* was orthologous to that identified in *B. napus* genome (Feng *et al.*, 2012). Thus, in accordance with our data, it seems that *MYB28* orthologs could be important genetic determinants controlling the variation of seed-glucosinolate content across *Brassica* species.

Development of low glucosinolate transgenic lines in *B. juncea*

Our initial investigations led us to a tentative conclusion that *BjMYB28-3* could be a potential candidate for the development of low glucosinolate lines in *B. juncea*. An intron spliced hairpin RNAi (ihpRNAi) construct (Figure 2a) was therefore developed, wherein a 342 bp region downstream to the conserved R2R3 repeats of *BjMYB28-3* homolog was used, hereafter referred to as BjMYB28(RNAi) construct. The *BjMYB28-3* target sequence used to develop the said construct showed 77.0–87.4% sequence identity to other three *BjMYB28* homologs and shared lower levels of sequence identity with the two *BjMYB29* homologs (56.4–61.5%) (Figure S6). Because *MYB28* is a member of large MYB transcription factor family genes, engineering with a strong, constitutive promoter could provide off-shoot targets and alter other biological processes which may be deleterious to the plant. Therefore, to ensure controlled and specific silencing across the plant developmental stages and cell-types, the said cassette was

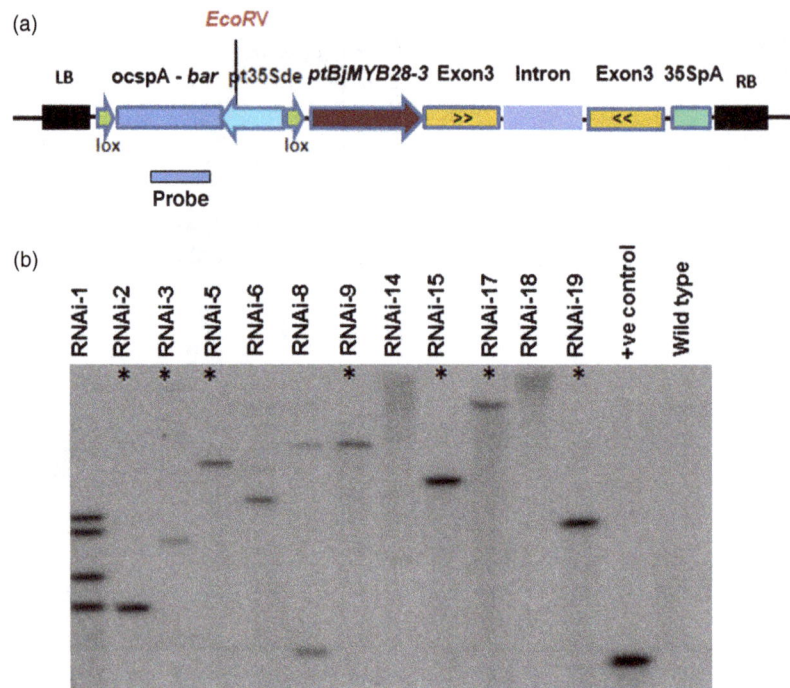

Figure 2 Development and characterization of BjMYB28(RNAi) transgenic lines. (a) Map of the T-DNA construct of BjMYB28(RNAi) transformation vector. A 342 bp region of the *BjMYB28-3* gene was cloned in sense and antisense orientations, with gene-specific second intron (409 bp) as spacer, under the control of native promoter. (b) Southern blot analysis of BjMYB28(RNAi) transgenic lines. A 390 bp fragment of *bar* gene was used as the probe. Asterisks (*) represent transgenic lines with single-copy integration of T-DNA.

driven by the native promoter of *BjMYB28-3* gene, the homolog showing differential expression profile between the contrasting glucosinolate lines of *B. juncea* (Figure 1c). The transformation constructs were introduced into the well-adapted high seed glucosinolate Indian line of *B. juncea* (cv. Varuna; total seed glucosinolate content ca. 107.52 μmol/g DW) through *Agrobacterium*-mediated genetic transformation. A total of 36 and 15 primary transgenic (T0) lines were generated using the BjMYB28 (RNAi) and vector control constructs, respectively. Southern blot analysis of the T0 transgenic events showed stable integration of T-DNA cassette in variable copies (Figure 2b). The transgenic lines were maintained by selfing in each generation under the containment field conditions and were analyzed for glucosinolate phenotype.

Transgenic lines developed were initially subjected to near infrared reflectance spectroscopy (NIRS) to estimate total glucosinolate contents in T1 seeds. The total seed glucosinolate content in the BjMYB28(RNAi) transgenic lines ranged from 15.21 to 124.76 μmol/g DW, suggesting variable degree of silencing efficiency in these lines (Table S2). We classified the transgenic lines into four different classes based on their seed glucosinolate content (Figure 3a) viz., class I (<30 μmol/g DW; the internationally acceptable limit), class II (31–60 μmol/g DW), class III (61–90 μmol/g DW) and class IV (>90 μmol/g DW). Ten (30.6%) transgenic lines generated using the BjMYB28(RNAi) construct showed seed glucosinolate content <30 μmol/g DW (hereafter, referred as low glucosinolate transgenic lines). Further, a significant reduction in total seed glucosinolate content (down to 86% of the wild-type level) suggested a very high silencing efficiency of the BjMYB28(RNAi) construct. All transgenic lines developed using the vector control construct showed seed glucosinolate content similar to that of transformation host,

Varuna (107.52 ± 5.90 μmol/g DW). Transgene-induced RNAi has been shown to be an effective mechanism for silencing un-desirous genes/traits in plants; however, a great deal of variation in the silencing efficacy and silencing-induced phenotype among transgenic events has been reported in crop plants (Ali et al., 2010; Mansoor et al., 2006).

The BjMYB28(RNAi) lines showed reduced accumulation of aliphatic glucosinolates without compromising the nonaliphatic glucosinolate pool

In order to determine the effect of BjMYB28(RNAi) construct on suppression of total seed glucosinolates, we analyzed representative transgenic lines for individual glucosinolate components. HPLC analysis was carried out to estimate the glucosinolate profile in seeds of nine representative single-copy transgenic lines, which showed >50% reduction in the total glucosinolate content (i.e. class I and class II) compared to the wild-type plants in initial NIRS analysis. HPLC estimation of component glucosinolates in T1 seeds of these nine transgenic lines showed significant reduction in total aliphatic glucosinolate levels (Table 1). Gluconapin (3-butenyl), the predominant aliphatic glucosinolate which is also the precursor of progoitrin (a major anti-nutritional glucosinolate) was reduced to as low as 9.51 ± 1.06 μmol/g DW in the transgenic lines (wild-type level: 100.03 ± 7.77 μmol/g DW). Similarly, sinigrin (2-propenyl), the second major seed glucosinolate (wild-type levels: 14.66 ± 1.56 μmol/g DW) was also reduced to 1.16 ± .07 μmol/g DW in the transgenic lines. In addition to these major anti-nutritional glucosinolates, other aliphatic glucosinolate fractions were also reduced significantly (down to ca. 84%) in the BjMYB28(RNAi) lines compared to that of wild-type (Varuna) and vector control lines. However, the transgenic lines showed only marginal changes in the total indole

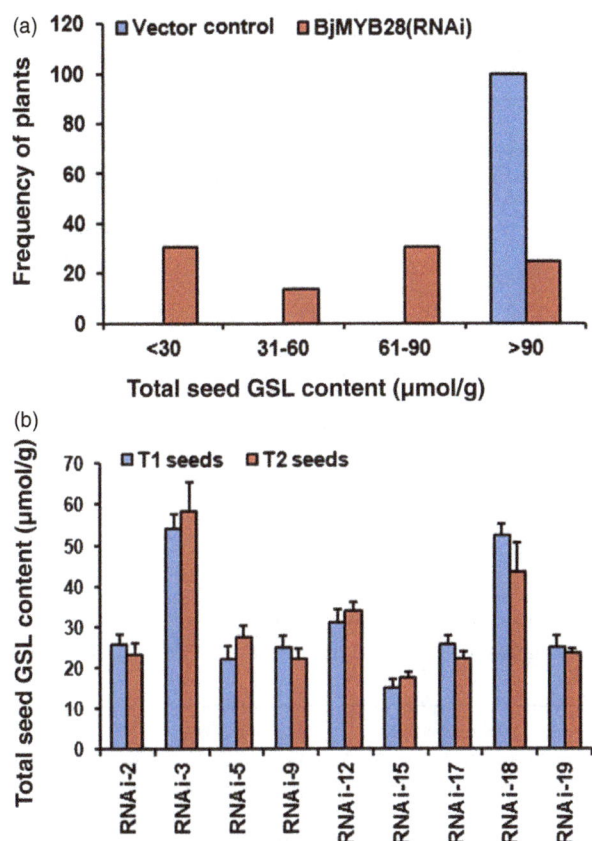

Figure 3 Average seed-glucosinolate estimation of BjMYB28(RNAi) transgenic lines. (a) Frequency (%) of T0 plants within different classes based on total seed glucosinolates content in T1 seeds. The T0 transgenic plants obtained using BjMYB28(RNAi) and vector control constructs were categorized under four different classes viz., <30, 31–60, 61–90, and >90 μmol/g DW. (b) Comparison of average total glucosinolate observed in T1 and T2 seeds of single-copy transgenic events having total seed glucosinolate <60 μmol/g DW. T2 seeds of five independent T1 plants were analyzed for the total glucosinolate content using NIRS, in replicates.

glucosinolates content (Table 1), thereby reflecting highly specific suppression of seed aliphatic glucosinolates by the BjMYB28 (RNAi) construct.

In recent years, both glucosinolate content and profiles in oilseed B. napus has been engineered by silencing Brassica homologs of aliphatic glucosinolate pathway genes including MAM3 and GSL-ALK (Liu et al., 2011, 2012). RNAi-based silencing of MAM3 (Methylthio-alkyl-malate synthase), a biosynthetic pathway gene involved in the chain elongation step of aliphatic glucosinolate biosynthesis, was used to reduce the aliphatic glucosinolate contents in B. napus seeds (Liu et al., 2011). Although total seed glucosinolate contents in MAM3 (RNAi) transgenic plants were found to be reduced in B. napus canola (low glucosinolate) line; silencing of MAM3 in high glucosinolate background of B. napus could elicit only marginal decrease in total glucosinolate contents. The MAM3(RNAi) transgenic lines showed induced production of 2-propenyl glucosinolate while reducing the level of both C4 (2-hydroxy-3-butenyl and 3-butenyl) and C5 (5-methylsulfinylpentyl and 4-pentenyl) glucosinolates. In contrast, our results clearly demonstrated that the targeted silencing of a key transcriptional regulator, BjMYB28 in a high glucosinolate background of oilseed

B. juncea could provide significant and stable suppression of total seed glucosinolate content (including C3, C4 and C5 glucosinolates) in the transgenic lines, without affecting the overall profile of aliphatic glucosinolate components.

One of the greatest applicability of transgene-induced RNAi is towards multigene families and polyploids, as it is not straightforward to create loss-of-function phenotype (knock-out) of multiple genes and their homologs by conventional breeding approaches, particularly, if members of the gene family are tightly linked (Lawrence and Pikaard, 2003). Thus, for agronomical traits like glucosinolates which are quantitative, sufficient degree of suppression of a target gene (as well as its multiple homologs) may be required for stable propagation of low glucosinolate trait in subsequent generations.

The low glucosinolate trait was stable in subsequent generations

To check the stability of BjMYB28 silencing, nine BjMYB28(RNAi) single-copy transgenic lines showing >50% reduction in total seed glucosinolate were analyzed in subsequent generations under containment field conditions. The T2 seeds derived from independent Basta resistant T1 progeny of each transgenic event showed glucosinolate levels comparable to or lower than that of T1 seeds (Figure 3b; Table S3), thereby indicating stable integration and performance of the BjMYB28(RNAi) silencing cassette in the following generation. The homozygous T3 seeds showed even lower levels of total seed glucosinolate contents compared to the T1 seeds (Table 1). For example, the transgenic line, RNAi-9 with seed glucosinolate content of 22.90 ± 2.50 μmol/g DW in T1 seeds had average seed glucosinolate content of 13.78 ± 0.62 μmol/g DW in homozygous T3 seeds (Table 2). This in all possibility could be attributed to the dosage-mediated silencing in the homozygous state. The single-copy homozygous T3 lines showed total aliphatic glucosinolate content ranging from 9.64 to 18.19 μmol/g DW. The nonaliphatic glucosinolate profile in the low glucosinolate lines remained almost similar to that of wild-type and vector control lines.

Glucosinolates are known to be synthesized mainly in the vegetative organs such as leaves and silique walls, and then transported actively to seeds through phloem transporters (Nour-Eldin et al., 2012), although synthesis in immature seeds has also been proposed (Chen et al., 2001; Toroser et al., 1995). In order to investigate the effect of BjMYB28(RNAi) suppression in vegetative tissues, we estimated the glucosinolate content and profile in leaves of homozygous T2 transgenic lines having seed glucosinolate level <30 μmol/g DW. The six transgenic lines showed total leaf glucosinolates ranging from 5.92 to 11.43 μmol/g DW. A drastic reduction in total aliphatic glucosinolates to 80–90% in the young leaves of transgenic lines was observed (Figure 4a). The indole glucosinolates pool in leaves was found to be almost unaltered compared to the control lines (Table S4). Thus, use of native BjMYB28 promoter can ensure sufficient suppression of glucosinolate biosynthesis in leaves as well as other vegetative tissues, which in turn reduces the total seed glucosinolate accumulation in a significant way.

Low glucosinolate transgenic lines showed increased siRNA accumulations and reduced levels of BjMYB28 transcripts

One of the major aims of our study was to develop the low glucosinolate lines in B. juncea which could meet international standard as canola quality oilseed (Potts et al., 1999). We

Table 1 Average glucosinolate contents and profile (in μmol/g DW) in T1 seeds of BjMYB28(RNAi) transgenic lines. The glucosinolate profiles of single-copy transgenic lines with total glucosinolates <60 μmol/g DW were estimated using HPLC. Value represent mean ± SD of three independent biological replicates

| Plant code | Aliphatic glucosinolates | | | Nonaliphatic glucosinolates | Total glucosinolates |
	Sinigrin (2-propenyl)	Gluconapin (3-butenyl)	Glucobrassica-napin (4-pentenyl)		
Control	14.66 ± 1.56	100.03 ± 7.77	1.39 ± 0.12	1.95 ± 0.26	118.05 ± 6.67
RNAi-2	1.58 ± 0.41*	15.97 ± 0.70*	0.47 ± 0.12*	1.72 ± 0.43	19.83 ± 0.97*
RNAi-3	6.90 ± 2.90*	47.56 ± 3.68*	0.93 ± 0.26*	1.81 ± 0.44	57.20 ± 3.55*
RNAi-5	1.98 ± 0.44*	16.40 ± 1.01*	0.42 ± 0.12*	1.80 ± 0.34	20.74 ± 1.06*
RNAi-9	2.70 ± 1.14*	17.78 ± 1.06*	0.55 ± 0.26*	1.71 ± 0.36	22.90 ± 2.50*
RNAi-12	2.68 ± 0.50*	22.94 ± 2.40*	0.60 ± 0.18*	1.98 ± 0.40	28.21 ± 2.41*
RNAi-15	1.16 ± 0.07*	9.51 ± 1.06*	0.23 ± 0.14*	1.46 ± 0.32	12.58 ± 1.20*
RNAi-17	1.83 ± 0.66*	15.54 ± 0.89*	0.42 ± 0.22*	1.67 ± 0.42	19.48 ± 0.42*
RNAi-18	5.47 ± 2.40*	43.43 ± 5.32*	0.83 ± 0.25*	1.78 ± 0.45	51.67 ± 4.42*
RNAi-19	2.17 ± 0.41*	21.90 ± 4.83*	0.63 ± 0.32*	1.96 ± 0.46	26.67 ± 5.05*

*Significantly different ($P < 0.01$) in comparison with wild-type control (Varuna) plants.

Table 2 Average glucosinolate contents and profile (in μmol/g DW) in T3 homozygous seeds of BjMYB28(RNAi) transgenic lines. Only single-copy low glucosinolate lines (<30 μmol/g DW) were analyzed in advance generations. Value represent mean ± SD of three independent biological replicates

Transgenic event	Sinigrin (2-propenyl)	Gluconapin (3-butenyl)	Glucobrassica-napin (4-pentenyl)	Total seed aliphatic glucosinolates	Non aliphatic glucosinolates	Total seed glucosinolates
Control	13.45 ± 1.61	90.88 ± 6.11	2.39 ± 0.31	106.69 ± 7.57	1.74 ± 0.45	108.43 ± 8.00
RNAi-2	1.88 ± 0.14*	13.09 ± 1.01*	0.59 ± 0.13*	15.57 ± 1.23*	1.98 ± 0.10	17.55 ± 1.22*
RNAi-5	2.18 ± 0.28*	14.59 ± 2.04*	0.68 ± 0.09*	17.46 ± 2.26*	1.81 ± 0.11	19.27 ± 2.15*
RNAi-9	1.48 ± 0.09*	10.19 ± 0.21*	0.39 ± 0.05*	12.06 ± 0.33*	1.72 ± 0.32	13.78 ± 0.62*
RNAi-15	1.27 ± 0.28*	8.01 ± 1.57*	0.36 ± 0.12*	9.64 ± 1.84*	1.62 ± 0.55	11.26 ± 2.04*
RNAi-17	1.55 ± 0.06*	11.98 ± 0.95*	0.52 ± 0.04*	14.06 ± 0.49*	1.77 ± 0.60	15.84 ± 0.78*
RNAi-19	2.81 ± 0.41*	14.96 ± 1.63*	0.41 ± 0.08*	18.19 ± 0.99*	1.78 ± 0.30	19.97 ± 2.26*

*Significantly different ($P < 0.01$) in comparison with wild-type control (Varuna) plants.

therefore carried out detailed molecular characterization on six of the nine BjMYB28(RNAi) single-copy transgenic lines showing strong reduction in total seed glucosinolate content, that is, <30 μmol/g DW.

To check whether the reduced glucosinolate pool in low glucosinolate transgenic lines was due to *BjMYB28*-specific siRNA accumulation, small RNA Northern blot was performed in seedling stage of T2 homozygous plants. *BjMYB28*-specific siRNAs were detected in the low glucosinolate transgenic lines (Figure 4b). The decrease in glucosinolate accumulation was found to be associated with increased levels of siRNA accumulation in the homozygous transgenic lines. The siRNA accumulation was not observed either in the wild-type or in the vector control plants. Thus, our data clearly demonstrated RNAi-based suppression of *BjMYB28* transcripts in low glucosinolate transgenic lines.

We further determined the target specificity of BjMYB28(RNAi) construct towards regulating the *BjMYB28* homologs at post-transcriptional level. Steady-state transcript accumulation, using qRT-PCR analysis, was measured in seedlings, the stage at which all four *BjMYB28* homologs showed optimal expression (Figure 1). There was a significant 4–10-fold reduction in *BjMYB28-3* transcript, in the transgenic seedlings compared to the control plants (Figure 4c). Expression of other three *BjMYB28* homologs was also affected in a limited manner across the low glucosinolate transgenic lines, with *BjMYB28-1* being the second-

most down-regulated transcript. The differential suppression of *BjMYB28* homologs in transgenic lines could be possibly attributed to the presence of variable sequence identity among these homologs at the RNAi target region (Figure S6). A significant reduction in *BjMYB28-3* transcript was also observed in the siliques of low glucosinolate transgenic lines (Figure S7). The steady-state level of *BjMYB29* transcripts however remained almost unaltered. Thus, use of the native *BjMYB28-3* promoter ensured controlled and specific knock-down of *BjMYB28* homologs *vis-à-vis* reduced accumulation of aliphatic glucosinolates across *B. juncea* developmental stages.

Suppression of *BjMYB28* homologs resulted in significant down-regulation of aliphatic glucosinolate biosynthetic genes in transgenic lines

If *BjMYB28* is a major regulator of the aliphatic glucosinolate biosynthesis pathway, the expression of the genes involved in aliphatic glucosinolate biosynthesis pathway in low glucosinolate lines should be consequently down-regulated. To investigate this, steady-state mRNA levels of the downstream genes were assessed in the seedling stage of single-copy, homozygous T2 low glucosinolate lines. Our analysis revealed significant down-regulation of all the selected genes of aliphatic glucosinolate biosynthetic pathway in *B. juncea*, including those involved in side-chain elongation (*GSL-ELONG*), core structure biosynthesis

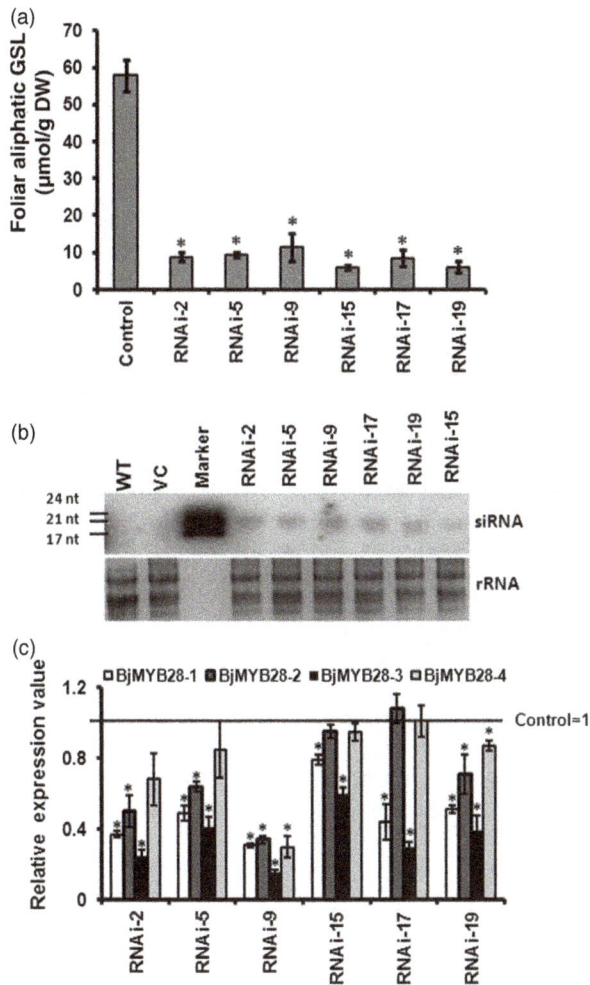

Figure 4 Average leaf glucosinolate analysis and molecular characterization of low glucosinolate BjMYB28(RNAi) lines. Single-copy homozygous T2 lines with total seed glucosinolate <30 µmol/g DW were analyzed. (a) Average leaf aliphatic glucosinolate content in low glucosinolate transgenic lines. The mean ± SD of aliphatic glucosinolate content of three independent biological measurements were determined using HPLC. (b) BjMYB28-specific siRNA accumulation in low glucosinolate lines. Upper panel shows Northern blot analysis of small RNA using BjMYB28-3 target region. Equal loading of RNA samples is shown in the lower panel. (c) Reduction in steady-state levels of four BjMYB28 transcripts using qRT-PCR analysis. Wild-type (Varuna) and a vector line were used as control for qRT-PCR (set as 1) as well as siRNA accumulation experiments. Asterisks (*) indicate significant difference at $P < 0.01$ (Fishers LSD test determined by ANOVA).

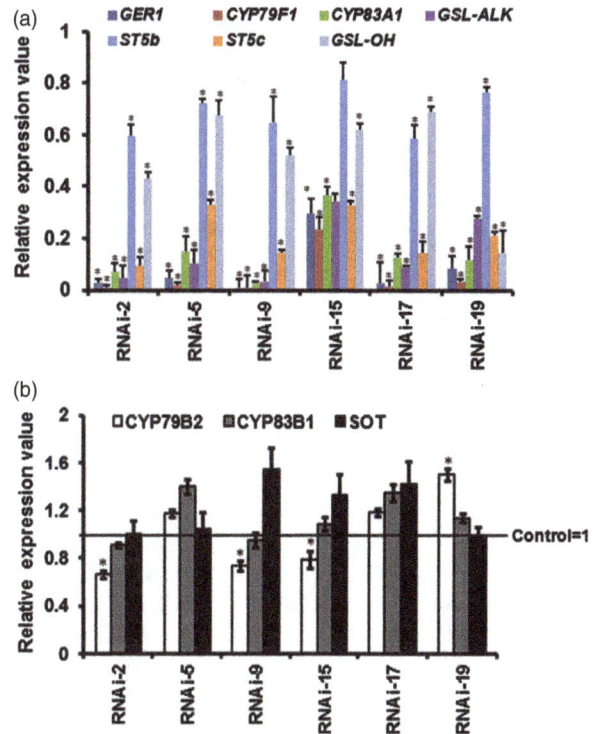

Figure 5 Gene expression analysis of glucosinolate candidate genes in the seedling stage of single-copy homozygous T2 low glucosinolate BjMYB28(RNAi) lines. Expression profiles of genes involved in (a) aliphatic glucosinolate biosynthesis pathway and (b) indole glucosinolate biosynthesis pathway. qRT-PCR experiments were conducted thrice with at least two technical replicates each. Wild-type (Varuna) and a vector line were used as control for qRT-PCR analysis (set as 1). Asterisks (*) indicate significant difference at $P < 0.01$ (Fishers LSD test determined by ANOVA).

(CYP79F1, CYP83A1) and secondary modifications of aliphatic glucosinolate (GSL-ALK, GSL-OH) (Figure 5a). When expression levels of indole glucosinolate biosynthetic pathway genes like CYP79B2, CYP83B1 and SOT16 were measured in the low glucosinolate transgenic lines, no significant change in the transcripts levels was observed for CYP83B1 and SOT16 (Figure 5b). The low glucosinolate lines, RNAi-2, RNAi-9 and RNAi-15 showed reduction in the transcript level of CYP79B2; however, the indole glucosinolate pools in these lines remained almost unaltered. This could be possibly because of the redundancy of CYP79B2 gene homologs in polyploid B. juncea. Thus, our results clearly indicated that BjMYB28 specifically regulate the

aliphatic glucosinolate biosynthetic genes in B. juncea but not the indole glucosinolate biosynthetic genes, which was in complete accordance with the glucosinolate profiles observed in the transgenic lines.

Low glucosinolate transgenic lines had unaltered levels of seed quality and yield parameters

MYB28 belongs to a gigantic R2R3 transcription factor multigene family, the members of which are known to regulate a large array of biological processes and functions in plants (Dubos et al., 2010). For biotechnological applications, alteration of these transcription regulators needs to be performed in a very controlled and precise manner so as to have no or minimum detrimental effects to the plants. Recently, ectopic over-expression of aliphatic glucosinolate regulators in Arabidopsis showed retardation of plant growth, impaired gravitropic response and fertility in transgenic events (Gigolashvili et al., 2007, 2009). Thus, for an important oilseed crop like B. juncea, the effect of transgene-mediated RNAi suppression of BjMYB28 in low glucosinolate lines needs to be tested stringently for various growth and agronomical parameters. The low glucosinolate transgenic lines developed in this study showed proper seed germination, growth phenotype, male- and female-fertility and seed setting in subsequent generations, thereby reflecting no visible off-target effects on plant growth and development.

B. juncea is an important oilseed crop of the Indian subcontinent, yielding 37%–40% seed oil. In recent years, demand for *Brassica* oil for the production of biofuels has also increased to a greater extent (Hill *et al.*, 2006). Hence, both seed quality and yield parameters of the low glucosinolate transgenic lines developed in this study also need to be considered. The six low glucosinolate transgenic lines homozygous for the T-DNA cassette were grown for two successive generations under the containment field conditions and the open-pollinated T2 and T3 seeds were analyzed for various oil-quality parameters using NIRS. The results showed no significant difference in the total oil content among the wild-type control plants and the six low glucosinolate transgenic lines developed (Figure 6a). After expelling oil, the oilcake is left out which is rich in proteins and therefore serves as an animal feed. The total protein content of the low glucosinolate transgenic lines and the control plants were also comparable, ranging from 25 to 27% (Figure 6b). Fatty acid compositions of the transgenic lines were also analyzed. The levels of three major fatty acids, namely oleic acid (C18:1), linolenic acid (C18:3) and erucic acid (C22:1) were found to be unchanged in the low glucosinolate transgenic lines compared to the control plants (Figure 6c). One of the major yield parameters namely 100 seed weight of the transgenic lines was also found to be unaltered compared to Varuna, the transformation host and the national check cultivar of *B. juncea* (Figure 6d). Detailed assessment of various seed quality and yield traits of the low glucosinolate *B. juncea* transgenic lines needs to be performed at multilocation open field trials in future.

Future prospects of low glucosinolate transgenic *B. juncea*

One of the major concern of development of low glucosinolate lines in *Brassica* crops is its susceptibility to pest and diseases as glucosinolates forms integral part of plants immune system. There are various reports from *Brassica* species suggesting mixed response of pests and pathogens towards glucosinolate contents and profile (Bodnaryk, 1997; Giamoustaris and Mithen, 1997; Williams, 1989). The glucosinolate–herbivore interactions are known to be very complex, wherein specific glucosinolate fraction may act as attractant or repellent for specialist or generalist herbivores (Hopkins *et al.*, 2009). In general, indole glucosinolates are known to be critical determinants for pest and disease resistance (Kim and Jander, 2007), whereas aliphatic glucosinolates play major role as deterrents for herbivore attack as well as for the overall fitness of the plant (Lankau and Kliebenstein, 2009).

In this study, we found that indole glucosinolate content of the low glucosinolate *B. juncea* transgenic lines remained almost unaltered compared to that of the wild-type control (Table S4). When these lines were grown under the containment field conditions for three successive growing seasons, we did not observe any altered susceptibility to pests and diseases compared to the *B. juncea* control (cv. Varuna). However, detailed assessment of the vulnerability of low glucosinolate *B. juncea* lines developed in the current study to various pests and pathogens will be undertaken in future.

Consumer acceptability is another major concern related to genetically modified crops. The hpRNAi cassette used to develop the low glucosinolate *B. juncea* lines in the current study contains only *cis*-DNA sequences (including the promoter, target exon sequence and the intronic spacer of the native *BjMYB28-3* homolog) derived from the host *B. juncea*. Besides, the *bar* gene cassette (conferring resistance to herbicide glufosinate) was cloned within the *loxP* sites, which in turn will facilitate excision of the selectable marker from the homozygous low glucosinolate stock in the subsequent generations by using the bacterial *Cre-lox* system of recombination (Arumugam *et al.*, 2007). Thus taking into account all these facts, we presume that the low glucosinolate *B. juncea* transgenic lines will broaden the scope and acceptability of this oilseed crop in international market.

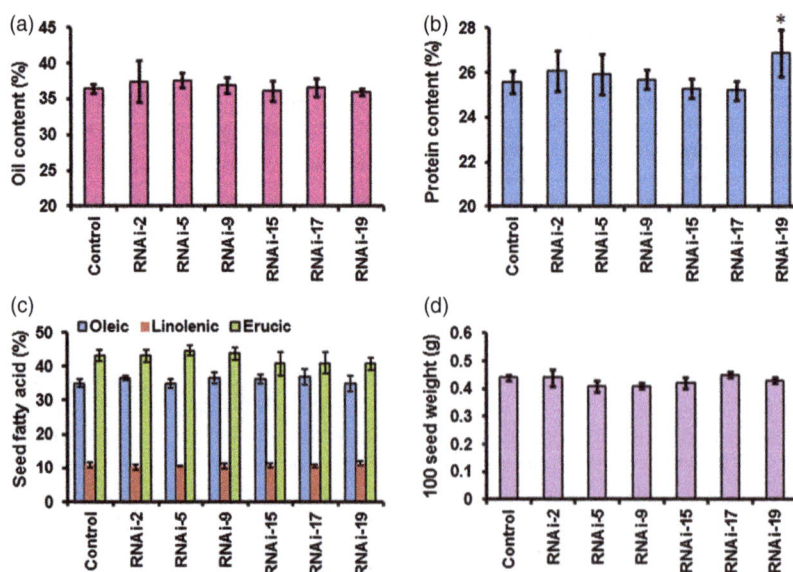

Figure 6 Characterization of low glucosinolate transgenic *B. juncea* lines for various seed quality parameters. Six single-copy, low glucosinolate BjMYB28 (RNAi) lines were analyzed for (a) Total oil content; (b) Protein content; (c) Fatty acid profiles (oleic, linolenic and erucic acid); and (d) Hundred (100) seed weight. The NIRS data of T3 homozygous seeds obtained from at least 12 independent T2 progeny are represented along with SD. Wild-type (Varuna) and vector control lines were used as control. Asterisks (*) indicate significant difference, $P < 0.05$ (one-way ANOVA).

Conclusion

Quantitative traits like glucosinolates are known to be controlled by highly complex gene expression networks comprising multiple biosynthetic pathway genes and their transcriptional regulators (Hirai et al., 2007). Further, in polyploid crops like B. juncea and B. napus, because of the existence of structural and expression divergence among multiple homologs, the manipulation of such quantitative trait is quite challenging either through conventional breeding or transgenic approaches.

In this study, we identified BjMYB28 as major transcription factor gene controlling the aliphatic glucosinolates accumulation in B. juncea. We also showed that using transgene-based RNAi-suppression strategy against BjMYB28, a significant suppression of aliphatic glucosinolate contents could be achieved, without compromising the nonaliphatic glucosinolates pool. We successfully developed stable low glucosinolate transgenic lines in B. juncea having total seed glucosinolate levels as low as 11.26 μmol/g DW (ca. 89% reduction compared to the wild-type level). The low glucosinolate transgenic lines performed well for growth and various oil-quality parameters under the containment field conditions. The work provides a significant advancement over the previously adopted breeding approaches which are encumbered by linkage-drags, and necessitate introgression of multiple loci to achieve the low glucosinolate trait. The study will contribute immensely towards developing low aliphatic glucosinolate lines across related Brassica species thus raising their oil- and seed-meal quality in the global market.

Experimental procedures

Plant materials and growth condition

Brassica juncea, high glucosinolate Indian line (Varuna) and canola quality east-European line (Heera) were grown in a growth chamber (Conviron) at 10 h light/14 h dark cycle, with temperature of 22 °C/15 °C and 70% relative humidity, respectively. Different developmental stages were collected, immediately frozen in liquid nitrogen and stored at −80 °C.

Generation of BjMYB28 hairpin RNAi transformation construct

A modified binary vector (pPZP200lox) containing the 'lox' tandem repeats was used (Arumugam et al., 2007). A PCR amplified 35Sde-bar-ocspA fragment conferring resistance to the herbicide phosphinothricin was cloned between the lox repeats at EcoRV site, and used as vector control construct. A 1054 bp fragment of BjMYB28-3 promoter was cloned directionally within the SmaI/SpeI sites of the above mentioned vector to create the vector pPZP200lox:35Sde-bar-ocspA::ptBjMYB28. A 409 bp fragment of the second intron of BjMYB28-3 gene was amplified using specific primers and cloned into the SmaI and SacI site of vector pRT100 (Topfer et al., 1987) to create the vector pRT100: int2. To this vector, a 342 bp fragment from third exon (encompassing 721–1062 bp sequence from ATG start codon) of BjMYB28-3 was amplified and cloned in both sense and antisense orientations. The Exon(s)-int2-Exon(as)-35SpA cassette thus created was excised and finally cloned directionally at XhoI/PstI sites of binary vector pPZP200lox:35Sde-bar-ocspA:: ptBjMYB28 to construct the final binary vector BjMYB28(RNAi). The transformation vectors were mobilized into A. tumefaciens strain GV3101 using freeze thaw method (Nishiguchi et al., 1981). All DNA manipulations were performed using standard protocols. Primers used in the current study are provided in Table S5.

B. juncea genetic transformation

Genetic transformation of B. juncea was performed using the protocol described earlier (Jagannath et al., 2003) with minor modifications. Briefly, seeds were cleaned with detergent, followed by treating with 70% ethanol. Surface sterilization was carried out with 0.05% HgCl₂ for 10 min. Seeds were then inoculated on to basal MS media and grown in culture room set at 10 h light (day)/14 h dark (night) cycle at a constant temperature of 23 ± 1 °C. Hypocotyls of 5 day old seedlings were cut into 0.5–1.0 cm pieces, and used as explants. Explants were cultured in MS liquid media supplemented with 1 mg/ml each of NAA and BAP (N1B1) for 24 h. Agrobacterium strain harbouring desired construct was grown in YEB medium supplemented with proper antibiotics (Rif₁₀Gent₂₀Spec₅₀). Bacterial cells at OD₆₀₀ of 0.5–0.6 were harvested and re-suspended in N1B1 medium to a final OD₆₀₀ of 0.3. Explants were incubated with the bacterial suspension for 30 min and cocultivated at 23 °C, 110 r.p.m. for 16–24 h. After cocultivation, the explants were washed with liquid N1B1 supplemented with Augmentin (200 mg/L) to remove excess Agrobacterium cells. The explants were plated onto shoot induction media containing N1B1, Augmentin (200 mg/L), AgNO₃ (20 μM) and 10 mg/L of Basta (Agrevo, active ingredient phosphinothricin). Regenerated shoots obtained within 25–35 days were transferred onto rooting media supplemented with 2 mg/L of IBA along with Augmentin (200 mg/L) and 10 mg/L Basta. The well-rooted transformants were transferred directly on to soil during the growing season. Transgenic plants were grown in a containment net-house in the field according to the guidelines of Department of Biotechnology, Government of India.

Southern blotting of transgenic plants

Genomic DNA from (approximately 1 month old) B. juncea transgenic lines as well as the wild-type Varuna (−ve control) was isolated using CTAB method. About 10 μg of genomic DNA was digested with EcoRV. After proper treatments (depurination, denaturation and neutralization), the DNA fragments were transferred onto the nylon membrane (Amersham Hybond XL, GE Healthcare, Buckinghamshire, UK) and UV cross-linked. Prehybridization and hybridization steps were carried out at 42 °C in Denhardts buffer for 16–18 h. Probe (ca. 390 bp fragment of bar gene) was labelled with dCTP [³²Pα] by incubating at 37 °C for 1 h using Amersham Megaprime DNA Labelling Systems (GE Healthcare). After hybridization and washing, the blot was kept in a cassette, exposed to X-ray film in dark for 48–72 h and developed in an automatic film processor (Hyperprocessor, GE Healthcare).

Gene expression analysis using qRT-PCR

Total RNA from plant samples was isolated using Spectrum Total RNA Isolation Kit (Sigma-Aldrich, St. Louis, MO). Approximately, two microgram of total RNA was reverse transcribed using high capacity first strand cDNA synthesis kit (Applied Biosystems, Foster City, CA) according to manufacturer's instructions. The relative expression of glucosinolate candidate genes was analyzed by real-time qRT-PCR in ABI 7900HT Fast Real-time PCR machine (ABI) using SYBR green protocol. BjACTIN2 gene was used as endogenous control (Chandna et al., 2012). Data were analyzed in three independent sets of plants with three technical replicates each. Statistical analyses were conducted using one-way ANOVA following Fishers LSD test of significance.

Specific primers for *BjMYB28* homologs were designed based on the nucleotide sequence alignment; preferably from the 3′ region of the gene. Nucleotide sequence of genes involved in both aliphatic and indole glucosinolate biosynthetic pathways was adapted from the recently published A-genome sequences of *B. rapa* (Wang *et al.*, 2011; Zang *et al.*, 2009). Various primers used for qRT-PCR analysis are tabulated in Table S5.

Detection of small RNA by Northern blotting

Total RNA was extracted with TRIZol reagent (Invitrogen, Life Technologies Corp., Carlsbad, CA) and RNA integrity was checked in Agilent-2100 Bioanalyzer and approximately, 20 μg of total RNA was used for blotting. RNA molecules were fractionated on 17% polyacrylamide gel containing 7 M urea. Miniprotean cell unit (Bio-Rad, Hercules, CA) was used to resolve the fragments in the gel. The transfer of RNA to nylon membrane was set up overnight at cold room with 0.5X TBE. After transfer, the membrane was soaked in 2X SSC and UV cross-linked and proceeded for hybridization. Prehybridization was carried out for 6 h followed by hybridization for 12–16 h at 50 °C. Probes specific to the *BjMYB28-3* region selected for RNAi design was radiolabelled with dCTP [^{32}Pα] using Amersham Megaprime DNA Labelling Systems (GE Healthcare) according to the manufacturer's instructions. Probes to marker (New England Biolabs, Ipswich, MA) were labelled with dATP [^{32}Pγ] using polynucleotide kinase. The probes were added to the blot simultaneously. Blots were exposed to phosphor screen (GE Healthcare) for 12 h and developed in multi-image scanner (Typhoon 9210, Amersham Biosciences, Buckinghamshire, UK).

Estimation of total glucosinolates, protein, oil content and fatty acid composition

Total glucosinolates, oil, protein contents and fatty acid composition of field grown seeds (T2 and T3 homozygous seeds) were determined using near infrared reflectance spectroscopy (NIRS-5000, FOSS, Denmark) in duplicates as per manufacturer's instructions. Seeds of wild-type Varuna and the vector control plants were used as control. The 100 seed weight of open-pollinated T3 seeds of at least 10 independent T2 progeny was calculated, in duplicates. Statistical analyses were conducted using one-way ANOVA following Fishers LSD test of significance.

Glucosinolate analysis using HPLC

The transgenic events were analyzed for component seed/leaf glucosinolate profiles using HPLC as per the protocols described earlier (Kraling *et al.*, 1990) with minor modifications. Samples were analyzed in Prominence UFLC 20A (Shimadzu, Japan) machine with reverse phase C18 column (Shimadzu LC column-XR-ODS; 100 mm × 3.0 mm with 5 μm internal diameter). Benzyl glucosinolate, glucotropaeolin (Applichem, Darmstadt, Germany) was used as the internal standard. A gradient of water (solvent A) and acetonitrile (solvent B) was used with a flow rate of 1 ml/min at an oven temperature of 35 °C. Elution was achieved with a gradient of 1–19% of solvent B over a period of 10 min. Glucosinolate components were detected at 229 nm and peaks were identified with reference to the retention time of already published chromatograms (Brown *et al.*,

2003). Concentrations of individual glucosinolates were calculated in micromoles per gram dry weight (μmol/g DW) relative to the area of the internal standard peak applying their relative response factors. At least three independent biological replicates were analyzed and statistical test was conducted using one-way ANOVA following Fishers LSD test of significance.

Data deposition

The sequences reported in this paper are deposited in the National Center for Biotechnology Information GeneBank database with accession nos. JQ666166 (*BjMYB28-1*), JQ666167 (*BjMYB28-2*), JQ666168 (*BjMYB28-3*), JQ666169 (*BjMYB28-4*), JX316031 (*BjMYB29-1*) and JX316032 (*BjMYB29-2*).

Acknowledgements

The work was supported by Department of Biotechnology, India (project schemes: BT/PR271/AGR/36/687/2011 and Rapid Grant for Young Investigators) and the core-grant provided by NIPGR, India to NCB. RA was funded with Junior Research Fellowship from Council of Scientific and Industrial Research, India. We are grateful to Central Instrumentation Facility at NIPGR. Critical suggestions from Prof. Deepak Pental, Prof. Akshay K. Pradhan, Prof. Roger Beachy and Dr. Swarup K. Parida are highly acknowledged. Two anonymous reviewers are also acknowledged.

References

Ali, N., Datta, S.K. and Datta, K. (2010) RNA interference in designing transgenic crops. *GM Crops*, **1**, 207–213.

Arumugam, N., Gupta, V., Jagannath, A., Mukhopadhyay, A., Pradhan, A.K., Burma, P.K. and Pental, D. (2007) A passage through *in vitro* culture leads to efficient production of true marker-free transgenic plants in *Brassica juncea* using the Cre-loxP system. *Trans. Res.* **16**, 703–712.

Bisht, N.C., Gupta, V., Ramchiary, N., Sodhi, Y.S., Mukhopadhyay, A., Arumugam, N., Pental, D. and Pradhan, A.K. (2009) Fine mapping of loci involved with glucosinolate biosynthesis in oilseed mustard (*Brassica juncea*) using genomic information from allied species. *Theor. Appl. Genet.* **118**, 413–421.

Bodnaryk, R.P. (1997) Will low-glucosinolate lines of the mustards *Brassica juncea* and *Sinapis alba* be vulnerable to insect pests? *Can. J. Plant Sci.* **77**, 283–287.

Brown, P.D., Tokuhisa, J.G., Reichelt, M. and Gershenzon, J. (2003) Variation of glucosinolate accumulation among different organs and developmental stages of *Arabidopsis thaliana*. *Phytochemistry*, **62**, 471–481.

Burton, W.A., Pymer, S.J., Salisbury, P.A., Kirk, J.T.O. and Oram, R.N. (1999) *Performance of Australian canola quality Indian mustard breeding lines*. Proceedings 10th Int. Rapeseed Congress, Canberra, Australia, 26–29 September (CD-ROM).

Campos de Quiros, H.C. and Mithen, R. (1996) Molecular markers for low-glucosinolate alleles in oilseed rape (*Brassica napus* L.). *Mol. Breed.* **2**, 277–281.

Cartea, M.E. and Velasco, P. (2008) Glucosinolates in Brassica foods: bioavailability in food and significance for human health. *Phytochem. Rev.* **7**, 213–229.

Chandna, R., Augustine, R. and Bisht, N.C. (2012) Evaluation of candidate reference genes for gene expression normalization in *Brassica juncea* using real time quantitative RT-PCR. *PLoS ONE*, **7**, e36918.

Chen, S., Petersen, B.L., Olsen, C.E., Schulz, A. and Halkier, B.A. (2001) Long-distance phloem transport of glucosinolates in Arabidopsis. *Plant Physiol.* **127**, 194–201.

Clarke, D.B. (2010) Glucosinolates, structures and analysis in food. *Anal. Methods*, **2**, 310–325.

Clay, N.K., Adio, A.M., Carine, C., Jander, G. and Ausubel, F.M. (2009) Glucosinolate metabolites required for an Arabidopsis innate immune response. *Science*, **323**, 95–101.

Dubos, C., Stracke, R., Grotewold, E., Weisshaar, B., Martin, C. and Lepiniec, L. (2010) MYB transcription factors in *Arabidopsis*. *Trends Plant Sci.* **15**, 573–581.

Fahey, J.W., Zhang, Y. and Talalay, P. (1997) Broccoli sprouts: an exceptionally rich source of inducers of enzymes that protect against chemical carcinogens. *Proc. Natl Acad. Sci. USA*, **94**, 10367–10372.

Fahey, J.W., Zalcmann, A.T. and Talalay, P. (2001) The chemical diversity and distribution of glucosinolates and isothiocyanates among plants. *Phytochemistry*, **56**, 5–51.

Feng, J., Long, Y., Shi, L., Shi, J., Barker, G. and Meng, J. (2012) Characterization of metabolite quantitative trait loci and metabolic networks that control glucosinolate concentration in the seeds and leaves of *Brassica napus*. *New Phytol.* **193**, 96–108.

Giamoustaris, A. and Mithen, R. (1997) Glucosinolates and disease resistance in oilseed rape (*Brassica napus* spp *oleifera*). *Plant. Pathol.* **46**, 271–275.

Gigolashvili, T., Yatusevich, R., Berger, B., Muller, C. and Flugge, U.I. (2007) The R2R3-MYB transcription factor HAG1/MYB28 is a regulator of methionine-derived glucosinolate biosynthesis in *Arabidopsis thaliana*. *Plant J.* **51**, 247–261.

Gigolashvili, T., Berger, B. and Flugge, U.I. (2009) Specific and coordinated control of indole and aliphatic glucosinolate biosynthesis by R2R3-MYB transcription factors in *Arabidopsis thaliana*. *Phytochem. Rev.* **8**, 3–13.

Grubb, C.D. and Abel, S. (2006) Glucosinolate metabolism and its control. *Trends Plant Sci.* **11**, 89–100.

Halkier, B.A. and Gershenzon, J. (2006) Biology and biochemistry of glucosinolates. *Annu. Rev. Plant Biol.* **57**, 303–333.

Harper, A.L., Trick, M., Higgins, J., Fraser, F., Clissold, L., Wells, R., Hattori, C., Werner, P. and Bancroft, I. (2012) Associative transcriptomics of traits in the polyploid crop species *Brassica napus*. *Nat. Biotechnol.* **30**, 798–802.

Hayes, J.D., Kelleher, M.O. and Eggleston, I.M. (2008) The cancer chemopreventive action of phytochemicals derived from glucosinolates. *Eur. J. Nutr.* **47**, 73–88.

Hill, J., Nelson, E., Tilman, D., Polasky, S. and Tiffany, D. (2006) Environmental, economic, and energetic costs and benefits of biodiesel and ethanol biofuels. *Proc. Natl Acad. Sci. USA*, **103**, 11206–11210.

Hirai, M.Y., Sugiyama, K., Sawada, Y., Tohge, T., Obayashi, T., Suzuk, A., Araki, R., Sakurai, N., Suzuki, H., Aoki, K., Goda, H., Nishizawa, O.I., Shibata, D. and Saito, K. (2007) Omics-based identification of Arabidopsis Myb transcription factors regulating aliphatic glucosinolate biosynthesis. *Proc. Natl Acad. Sci. USA*, **104**, 6478–6483.

Hopkins, R.J., van Dam, N.M. and van Loon, J.J.A. (2009) Role of glucosinolates in insect plant relationships and multitrophic interactions. *Annu. Rev. Entomol.* **54**, 57–83.

Jagannath, A., Bandyopadhyay, P., Mehra, S., Arumugam, N., Burma, P.K. and Pental, D. (2003) Agrobacterium-mediated genetic transformation of *Brassica juncea*. In *Plant Genetic Engineering Volume 2: Improvement of Food Crops* (Jaiwal, P.K. and Singh, R.P., eds), pp. 349–360. Houstan, TX: USA: Sci Tech Publishing LLC.

Kim, J.H. and Jander, G. (2007) *Myzus persicae* (green peach aphid) feeding on Arabidopsis induces the formation of a deterrent indole glucosinolate. *Plant J.* **49**, 1008–1019.

Kraling, K., Robbelen, G., Thies, W., Herrmann, M. and Ahmadi, M.R. (1990) Variation in seed glucosinolate in lines of *Brassica napus*. *Plant Breed.* **105**, 33–39.

Lankau, R.A. and Kliebenstein, D.J. (2009) Competition, herbivory and genetics interact to determine the accumulation and fitness consequences of a defense metabolite. *J. Ecol.* **97**, 78–88.

Lawrence, R.J. and Pikaard, C.S. (2003) Transgene-induced RNA interference: a strategy for overcoming gene redundancy in polyploids to generate loss-of-function mutations. *Plant J.* **36**, 114–121.

Lionneton, E., Aubert, G., Ochatt, S. and Merah, O. (2004) Genetic analysis of agronomic and quality traits in mustard (*Brassica juncea*). *Theor. Appl. Genet.* **109**, 792–799.

Liu, Z., Hammerlindl, J., Keller, W., McVetty, P.B.E., Daayf, F., Quiros, C.F. and Li, G. (2011) MAM gene silencing leads to the induction of C3 and reduction

of C4 and C5 side-chain aliphatic glucosinolates in *Brassica napus*. *Mol. Breed.* **27**, 467–478.

Liu, Z., Hirani, A.H., McVetty, P.B.E., Daayf, F., Quiros, C.F. and Li, G. (2012) Reducing progoitrin and enriching glucoraphanin in *Brassica napus* seeds through silencing of the *glucosinolate-ALK* gene family. *Plant Mol. Biol.* **79**, 179–189.

Mahmood, T., Ekuere, U., Yeh, F., Good, A.G. and Stringam, G.R. (2003) Molecular mapping of seed aliphatic glucosinolates in *Brassica juncea*. *Genome*, **46**, 753–760.

Mansoor, S., Amin, I., Hussain, M., Zafar, Y. and Briddon, R.W. (2006) Engineering novel traits in plants through RNA interference. *Trends Plant Sci.* **11**, 559–565.

Mawson, R., Heaney, R.K., Zdunczyk, Z. and Kozlowska, H. (1993) Rapeseed meal-glucosinolates and their antinutritional effects. Part II. Flavour and palatability. *Mol. Nutr. Food Res.* **37**, 336–344.

Nishiguchi, R., Takanami, M. and Oka, A. (1981) Characterization and sequence determination of the replicatory region in the hairy-root-inducing plasmid pRiA4b. *Mol. Gen. Genet.* **206**, 1–8.

Nour-Eldin, H.H., Andersen, T.G., Burow, M., Madsen, S.R., Jørgensen, M.E., Olsen, C.E., Dreyer, I., Hedrich, R., Geiger, D. and Halkier, B.A. (2012) NRT/PTR transporters are essential for translocation of glucosinolate defence compounds to seeds. *Nature*, **488**, 531–534.

Potts, D., Rakow, G. and Males, D.R. (1999) *Canola quality Brassica juncea*, a new oilseed crop for the Canadian prairies. Proceedings 10th Int. Rapeseed Congress, Canberra, Australia, 26–29 September (CD-ROM).

Pradhan, A.K., Sodhi, Y.S., Mukhopadhyay, A. and Pental, D. (1993) Heterosis breeding in Indian mustard (*Brassica juncea* L. Czern & Coss): analysis of component characters contributing to heterosis for yield. *Euphytica*, **69**, 219–229.

Ramchiary, N., Bisht, N.C., Gupta, V., Mukhopadhyay, A., Arumugam, N., Sodhi, Y.S., Pental, D. and Pradhan, A.K. (2007) QTL analysis reveals context-dependent loci for seed glucosinolate trait in the oilseed *Brassica juncea*: importance of recurrent selection backcross (RSB) scheme for the identification of 'true' QTL. *Theor. Appl. Genet.* **116**, 77–85.

Sodhi, Y.S., Mukhopadhyay, A., Arumugam, N., Verma, J.K., Gupta, V., Pental, D. and Pradhan, A.K. (2002) Genetic analysis of total glucosinolate in crosses involving a high glucosinolate Indian variety and a low glucosinolate line of *Brassica juncea*. *Plant Breed.* **121**, 508–511.

Sønderby, I.E., Hansen, B.G., Bjarnholt, N., Ticconi, C., Halkier, B.A. and Kliebenstein, D.J. (2007) A systems biology approach identifies a R2R3MYB gene subfamily with distinct and overlapping functions in regulation of aliphatic glucosinolate. *PLoS ONE*, **2**, e1322.

Sønderby, I.E., Burow, M., Rowe, H.C., Kliebenstein, D.J. and Halkier, B.A. (2010a) A Complex interplay of three R2R3 MYB transcription factors determines the profile of aliphatic glucosinolate in *Arabidopsis*. *Plant Physiol.* **153**, 348–363.

Sønderby, I.E., Geu-Flores, F. and Halkier, B.A. (2010b) Biosynthesis of glucosinolates-gene discovery and beyond. *Trends Plant Sci.* **15**, 283–290.

Topfer, R., Matzei, V., Gronenborn, B., Schell, J. and Steinbiss, H.-H. (1987) A set of plant expression vectors for transcriptional and translational fusions. *Nucleic Acids Res.* **15**, 5890.

Toroser, D., Griffiths, H., Wood, C. and Thomas, D.R. (1995) Biosynthesis and partitioning of individual glucosinolates between pod walls and seeds and evidence for the occurrence of PAPS: desulphoglucosinolate sulphotransferase in seeds of oilseed rape (*Brassica napus* L.). *J. Exp. Bot.* **46**, 1753–1760.

Wang, H., Wu, J., Sun, S., Liu, B., Cheng, F., Sun, R. and Wang, X. (2011) Glucosinolate biosynthetic genes in *Brassica rapa*. *Gene*, **487**, 135–142.

Williams, I.H. (1989) Pest incidence on single low and double low oilseed rape lines. *Asp. Appl. Biol.* **23**, 277–286.

Wittstock, U. and Halkier, B.A. (2002) Glucosinolate research in the Arabidopsis era. *Trends Plant Sci.* **6**, 263–270.

Zang, Y.X., Kim, H.U., Kim, J.A., Lim, M.H., Jin, M., Lee, S.C., Kwon, S.J., Lee, S.I., Hong, J.K., Park, T.H., Mun, J.H., Seol, Y.J., Hong, S.B. and Park, B.S. (2009) Genome-wide identification of glucosinolate synthesis genes in *Brassica rapa*. *FEBS J.* **276**, 3559–3574.

18

MicroRNA156 as a promising tool for alfalfa improvement

Banyar Aung[1,2], Margaret Y. Gruber[3], Lisa Amyot[1], Khaled Omari[1], Annick Bertrand[4] and Abdelali Hannoufa[1,2],*

[1]*Agriculture and Agri-Food Canada, London, ON, Canada*
[2]*Biology Department, Western University, London, ON, Canada*
[3]*Agriculture and Agri-Food Canada, Saskatoon, SK, Canada*
[4]*Agriculture and Agri-Food Canada, St. Foy, QC, Canada*

Correspondence

email Abdelali.Hannoufa@agr.gc.ca

Keywords: miR156, alfalfa, forage yield, flowering, shoot branching, nodulation.

Summary

A precursor of miR156 (MsmiR156d) was cloned and overexpressed in alfalfa (*Medicago sativa* L.) as a means to enhance alfalfa biomass yield. Of the five predicted *SPL* genes encoded by the alfalfa genome, three (*SPL6, SPL12* and *SPL13*) contain miR156 cleavage sites and their expression was down-regulated in transgenic alfalfa plants overexpressing miR156. These transgenic plants had reduced internode length and stem thickness, enhanced shoot branching, increased trichome density, a delay in flowering time and elevated biomass production. Minor effects on sugar, starch, lignin and cellulose contents were also observed. Moreover, transgenic alfalfa plants had increased root length, while nodulation was maintained. The multitude of traits affected by miR156 may be due to the network of genes regulated by the three target *SPL*s. Our results show that the miR156/*SPL* system has strong potential as a tool to substantially improve quality and yield traits in alfalfa.

Introduction

Medicago sativa (alfalfa) is the world's leading forage crop and a low-input bioenergy crop. The vigorous and deep rooting systems of alfalfa are ideal for preventing soil erosion and for reducing contamination in surface and ground water (Putnam *et al.*, 2001). As a legume crop, alfalfa is able to fix a large amount of atmospheric nitrogen (approximately 135–605 kg/ha year) (Putnam *et al.*, 2001), and thus has the potential to play an important role in the production of biomass for bioenergy. Its ability to fix atmospheric N_2 allows alfalfa to grow in a wide range of soil types and also provides N_2 to subsequent crops when alfalfa is included in a crop rotation scheme.

Alfalfa improvement efforts have led to the release of cultivars with very modest improvements in yield and disease resistance (Barnes *et al.*, 1977). However, classical breeding efforts towards larger improvements of alfalfa biomass yield and control of flowering are still ineffective. Forage yields of varieties released in the 1990s were no higher than initial/second harvests of varieties released in the 1950s (Volenec *et al.*, 2002), and substantial improvements in disease resistance and winter hardiness were also not observed (Lamb *et al.*, 2006; Volenec *et al.*, 2002). Alfalfa is an obligate outcross-pollinating tetraploid crop; hence, seeds from the same plant are heterozygous genotypic variants (Putnam *et al.*, 2001), and its polyploidy nature further adds to its genomic diversity (Lesins and Lesins, 1979) and complicates the use of classical breeding to improve alfalfa.

The use of modern biotechnology tools is challenging in alfalfa (Volenec *et al.*, 2002) as gene sequence information is relatively rare in alfalfa databases. Thus, identification of genes that play important roles in the control of plant development and an understanding of their regulation are keys which may lead to a more substantial improvement of alfalfa.

MicroRNAs are relatively new tools in the arsenal of plant molecular breeders (Jiao *et al.*, 2010; Macovei *et al.*, 2012;

Sunkar *et al.*, 2012; Wang *et al.*, 2012; Zhou and Luo, 2013). These 16–24 nucleotides (nt) regulators of post-transcriptional gene expression in eukaryotes (Bartel, 2004; Jones-Rhoades, 2012; Sun, 2012; Voinnet, 2009) have been shown to affect important plant traits, ranging from biotic and abiotic stress to yield (Zhou and Luo, 2013).

MicroRNA156 (miR156) regulates members of the *SQUAMOSA PROMOTER BINDING PROTEIN-LIKE* (*SPL*) gene family which are transcription factors that, in turn, affect the expression of downstream genes and result in the regulation of a large plant growth and development network (Cardon *et al.*, 1999). Thus, a number of traits, including emergence of vegetative leaves, shoot branching, floral transition, fertility, biomass production and grain yield, are modulated (Sun, 2012; Voinnet, 2009; Zhou and Luo, 2013).

In Arabidopsis, emergence of vegetative leaves is regulated by miR156-targeted *SPL13* because transgenic plants overexpressing *SPL13* mutated in the region complementary to miR156 sequence show a delay in the production of vegetative leaves (Martin *et al.*, 2010a,b). Arabidopsis loss-of-function mutants *spl9* and *spl15* display enhanced shoot branching, shortened plastochron and abnormal inflorescence phenotypes (Schwarz *et al.*, 2008). Moreover, Shikata *et al.* (2009) showed that production of lateral organs and shoot maturation in the reproductive stage are regulated by miR156-targeted *SPL2, SPL10* and *SPL11*. Likewise, Wu and Poethig (2006) demonstrated that overexpression of Arabidopsis miR156 reduces the transcript levels of *SPL3, SPL4* and *SPL5*, prolongs the vegetative stage and delays flowering time.

Effects of the miR156/*SPL* system on gene regulation are also found in other plant species, including *Oryza sativa* (Xie *et al.*, 2006), *Brassica napus* (Wei *et al.*, 2010), *Panicum virgatum* (Fu *et al.*, 2012) and *Solanum tuberosum* ssp. *andigena* (Bhogale *et al.*, 2014). While modest expression of *O. sativa* miR156 (OsmiR156) enhances apical dominance and biomass production in switchgrass, stunted growth and reduced biomass are also observed in plants expressing a high level of OsmiR156 (Fu et al., 2012). In addition, overexpression of maize Corngrass1 (Cg1) miR156 completely inhibits flowering in switchgrass (Chuck et al., 2011). MiR156 also impairs N_2 fixation through regulation of miR172-regulated network of downstream genes in soya bean (Yan et al., 2013).

In this work, we hypothesized that overexpression of miR156 would impact forage yield and quality traits in alfalfa. To prove this, we first isolated a precursor of miR156 from alfalfa (MsmiR156d) and generated transgenic alfalfa overexpressing MsmiR156. Next, we identified the SPL gene targets of miR156. Finally, we investigated the effects of miR156 overexpression (miR156-OE) on important traits and phenotypes of this forage crop.

Results

Potential miR156 precursors in *M. sativa* genome

A search of the literature showed that 15 members of the miR156 family (miR156a to miR156o) are deposited in the miRBase database. These miR156 precursors were identified from different plant species, and many of them are found in a single plant species, including the model legume Medicago truncatula. Our in silico analysis, however, revealed no homologues to miR156 precursors in M. sativa public databases. We next identified potential miR156 precursors in-house from unpublished next generation alfalfa RNA sequencing reads (Austin et al., unpublished), but this yielded only short miR156-like sequences (approximately 60 bp), and hairpin structures could not be predicted in these short sequences (data not shown).

To work around the lack of a M. sativa sequence, we isolated a miR156 precursor from alfalfa by PCR using primers designed based on M. truncatula sequence (Accession: CU019603). Sequencing and BLAST search results showed that the isolated sequence was very similar to miR156d precursor in M. truncatula. In addition, the sequence contained mature miR156 sequence and produced the secondary structure (hairpin) as in typical miR156 precursors (Figure S1). Based on these results, we confirmed that the isolated sequence was a miR156 precursor and we designated it as MsmiR156d.

Prediction and validation of MsmiR156 targets in alfalfa

No homologues of SPL genes could be found in publicly available alfalfa sequence databases. A search for SPL homologues in M. truncatula databases revealed 10 SPL gene accessions. However, only five of them (Accessions: XM_003614178, XM_003601719, XM_003602747, XM_003625188 and XM_003593569) contained sequences that had complimentary regions to the mature miR156 sequence, and thus, these five genes were predicted to be putative targets of miR156. Sequences homologous to these five genes were then amplified from alfalfa using primers designed from the M. truncatula sequences. Of the five SPL genes, only three (SPL6, SPL12 and SPL13) contained miR156 cleavage sites, based on 5'-RACE analysis in alfalfa. Among the 20 SPL6 clones that were sequenced, 15 contained only one cleavage site after the 11th nucleotide in the binding site of miR156 (Figure 1a). In contrast, two cleavage sites were found in SPL12 and SPL13 transcripts, respectively. For SPL12, 2 clones had cleavage occurring after the 10th nucleotide and 11 clones after the 11th nucleotide of the miR156 binding site (Figure 1b). In SPL13, however, cleavages

were observed after the 9th nucleotide (9 clones) and 10th nucleotide (6 clones) of the miR156 sequences (Figure 1c).

Based on 5' RACE results, the other two predicted SPLs showed no sequence similarity to other SPL genes even though the primers for PCR were designed based on M. truncatula gene-specific sequences containing miR156 cleaved sites. To further validate these 5' RACE results, an additional 10 clones from the 5' RLM-RACE PCR products of each of the two SPL genes were sequenced, but none revealed miR156 cleavage sites, confirming the earlier data. Hence, these two genes were excluded from further analysis.

Expression profiles of *MsmiR156* and target *SPL* genes in transgenic alfalfa

To validate the importance of SPL6, SPL12 and SPL13 as miR156 targets in alfalfa, the transcript levels of miR156 as well as those of putative target SPL genes were determined in transgenic alfalfa using qRT-PCR and small RNA gel blot. As the abundance of MsmiR156 increased eightfold to 650-fold (compared to control), the transcript levels of the three target SPL genes were reduced in two-month-old transgenic alfalfa (Figure 2a–e). However, miR156 dose-dependent SPL transcript levels were not found in all transgenic alfalfa. For instance, relative SPL6 transcripts were equally reduced in almost all MsmiR156-OE plants except for the A16 genotype (Figure 2c). Regardless of the high abundance of miR156, the relative transcript levels of SPL12 and SPL13 in A8 and A11 genotypes were higher than those in A16 and A8a genotypes, which contained less abundant miR156 transcripts

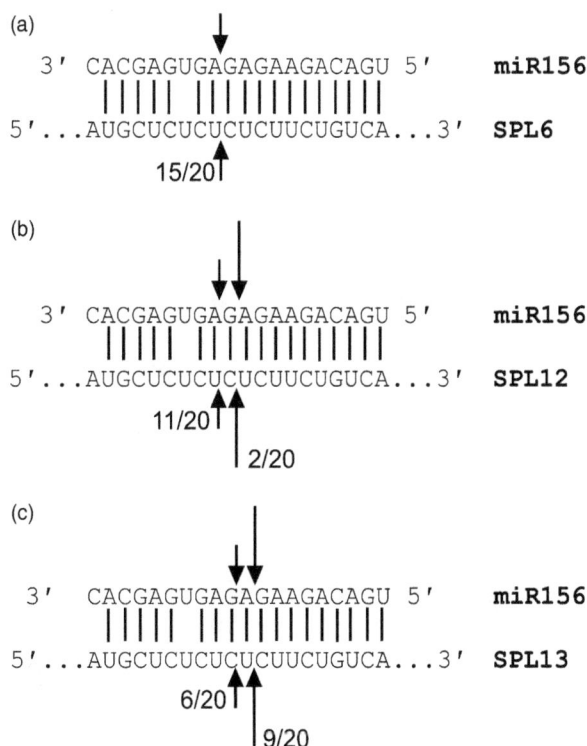

Figure 1 Validation of the miR156 cleavage sites in SPL transcripts in transgenic alfalfa. (a) Cleaved sites in SPL6, (b) SPL12 and (c) SPL13. Complementary miR156 sequences and a fragment of target SPL sequences are shown. Arrows indicate the 5' termini of cleaved target mRNA. Denominators refer to the number of clones sequenced whereas the nominators represent the number of clones cleaved at a particular site.

(Figure 2d,e). Six months after propagation, when most plants were flowering, *miR156* transcript levels were still elevated relative to the control plants (Figure 2f,g), but *SPL* expression was not reduced (Figure 2h–j). At this point, it is unclear whether miR156 regulates the target genes using other mechanisms, or alternatively, high expression of *SPL* genes in the reproductive

stage represses the effects of miR156 via feedback mechanisms, as was found in Arabidopsis (Wei *et al.*, 2012).

In addition to *SPL* genes, we also checked the expression of homologues of two flowering-related genes, *AP2* and *AP3*, and found that *AP3* of the floral meristem identity pathway (Kim *et al.*, 2013) was significantly reduced in MsmiR156-OE alfalfa at the 6-month stage (Figure 2k). Unlike *AP3*, expression of *AP2*, which functions as both a transcriptional activator and repressor to regulate flowering (Yant *et al.*, 2010), was not significantly affected in MsmiR156-OE plants (Figure 2l).

Apart from leaves and shoots, miR156 also functions in other parts of plant organs (Voinnet, 2009; Yan *et al.*, 2013). To understand miR156/*SPL*-gene regulatory networks in nodulated alfalfa roots, we analysed the expression of miR156 and target genes in the roots of one-month-old transgenic plants and found that MsmiR156-OE plants also contained as abundant transcripts for miR156 as in shoots (Figure 3a). However, no significant effect of miR156 was found on the transcript levels of *SPL6*, and the regulation of *SPL13* was variable in the nodulated roots (Figure 3b,c). In contrast, overexpression of miR156 significantly reduced the transcript of *SPL12* in roots (Figure 3d). Expression analysis also revealed that transcript levels of *AP3*, *AP2* and nodule-specific *leghemoglobin* genes were decreased by miR156 in at least one of the plants (Figure 3e–g).

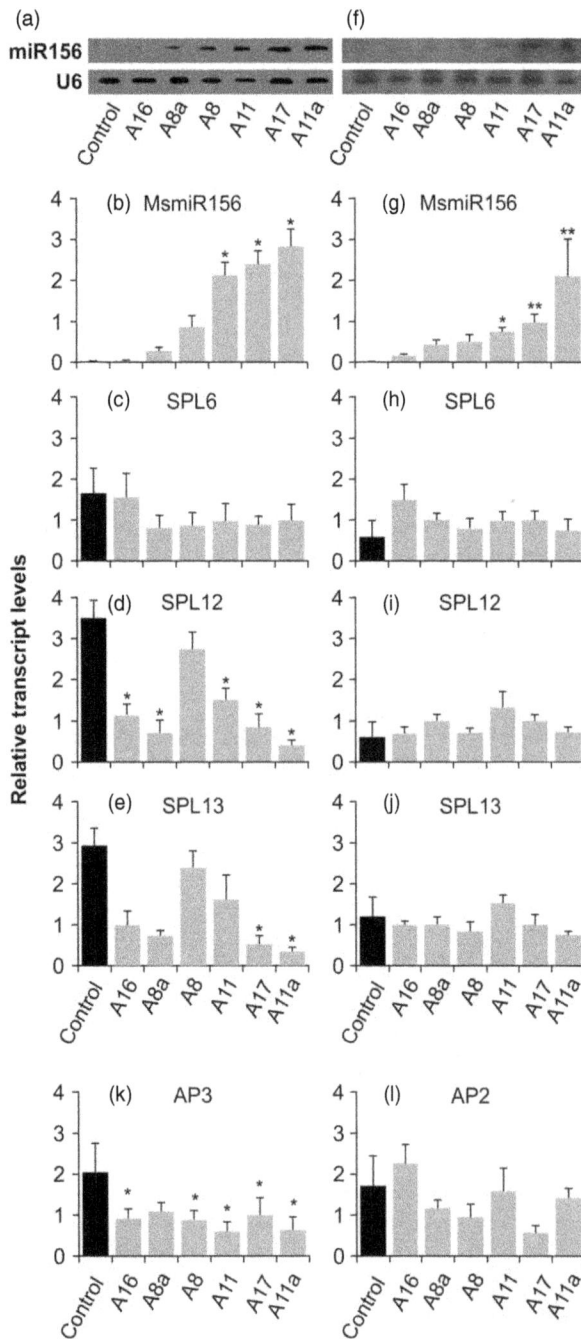

Figure 2 Expression profiles of *MsmiR156* and target *SPL* genes in alfalfa shoots. (a–e) two-month-old alfalfa, (f–l) six-month-old alfalfa. Abundance of mature miR156 in (a) two-month-old and (f) six-month-old alfalfa. Transcript levels of (b, g) *miR156*, (c, h) *SPL6*, (d, i) *SPL12*, (e, j) *SPL13*, (k) *AP3* and (l) *AP2*. Asterisks indicate means ± SE ($n = 3$) are significantly different from empty vector control at $P < 0.05$ (*) or $P < 0.01$ (**) using one-way ANOVA and a Dunnett's test.

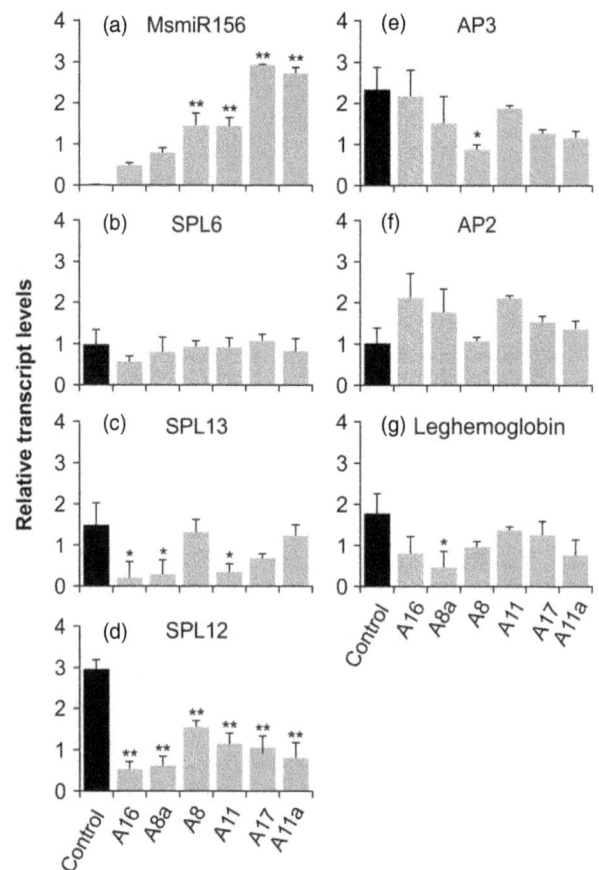

Figure 3 Expression profiles of *MsmiR156* and target *SPL* genes in nodulated roots of one-month-old alfalfa roots. Transcript levels of (a) *miR156*, (b) *SPL6*, (c) *SPL13*, (d) *SPL12*, (e) *AP3* (f) *AP2* and (g) *Leghemoglobin*. Asterisks indicate means ± SE ($n = 3$) are significantly different from empty vector control plants at $P < 0.05$ (*) or $P < 0.01$ (**) using one-way ANOVA and a Dunnett's test.

Morphological characterization of transgenic alfalfa

To investigate the effects of miR156 overexpression on plant growth and development, transgenic plants were propagated by vegetative cuttings. Enhanced shoot branching was observed on 40-day-old alfalfa but the effect of miR156 on this trait was more obvious as the plants continued to grow (Figure 4a,b). At 2 months after propagation, the number of branches (main and lateral branches) rose 1.5-fold to 6-fold in the transgenic plants overexpressing miR156 (Table 1, Figure S2). As the plants aged, however, increases of only 1.3-fold to 2.3-fold in the number of lateral branches were observed and 1.6- to 3.9-fold increases in the number of main branches were measured (Table 1, Figure S3). MsmiR156-dose dependency also occurred for plant height, internode length, number of nodes on the stems and stem thickness (Table 1, Figure 5). Plant height in low miR156-enhanced plants did not differ significantly from the control plants, but for high expressing plants (1600- to 3400-fold greater than the control plant), plant height was reduced. Overexpression of miR156 also substantially reduced stem thickness and increased trichome density in leaves of all transgenic plants (Figures 5b,c and S4, respectively). In some cases, longer hairs developed compared to the control plants (Figure S4).

Overexpression of MsmiR156 also affected flowering times in alfalfa. Control alfalfa plants started flowering 135 days after vegetative propagation, whereas flowering was delayed for 3–5 days in genotypes A16, A8a and A8 concurrent with a 250-fold to 840-fold increase in the transcript level of miR156 precursor. Moreover, a 1200-fold increase in the abundance of miR156 in genotype A11 delayed flowering for 45 days, and no sign of flowering was found in genotypes A17 and A11a (with a 1600- and 3400-fold increase in miR156, respectively) at the 6-month harvest date (Table 1). These phenotypic changes reflect the changes in the expression of miR156, as well as of target *SPLs* and the flowering genes, *AP2* and *AP3*.

Figure 4 Representative effects of miR156 enhancement on shoot branching in alfalfa genotype A11a. (a) Forty days after vegetative propagation. (b) Six-month-old transgenic alfalfa. Remaining six-month-old genotypes and those of two-month-old transgenic plants can be found in supplementary Figure S3 and S4.

Effects of MsmiR156 on forage yield and quality

To investigate the effect of miR156 on forage yield, plant material was harvested by cutting all the above-ground tissue down to 1 cm above the soil level at 2 months and 6 months after root initiation, followed by drying in an oven (65 °C) for 7 days and measuring dry weight. At the two-month stage, biomass was significantly increased in most miR156-OE alfalfa genotypes except A16 (Figure 6a) and at 6 months, biomass production of most miR156-enhanced plants was still at least 10% higher than that of control plants (Figure 6b).

Total soluble sugars (sucrose, glucose, pinitol and fructose) were not significantly different among the transgenic alfalfa even though the abundance of miR156 in MsmiR156-OE plants was 250- to 3400-fold higher than that in control plants (Tables 2 and S2). Similarly, starch contents were not significantly different regardless of the levels of miR156 expression. Surprisingly, a 3400-fold increase in miR156 abundance in genotype A11a reduced the structural sugar released from cell wall by approximately 10% following enzymatic saccharification in spite of a higher accumulation of total acid-digestible cellulose (glucose) in the miR156-enhanced alfalfa (Table 2). Effect of miR156 overexpression on pectin content was variable. Although pectin levels were decreased by 6–9% in genotypes A8a and A11, increases of up to 5% were found in other transgenic alfalfa genotypes (Table 2). Moreover, lignin analysis on cell wall residues showed that miR156 caused an approximately 10% reduction in lignin content in two of the miR156 overexpressing genotypes (A17 and A11a) (Table 2), but there was no significant effect on lignin composition in terms of the ratio of syringyl-like (S) to guaiacyl-like (G) lignin structures (Table 2).

Effect of miR156 on root length and nodulation

To investigate the effect of miR156 on root length and nodulation, rooted transgenic plants (2 weeks after cutting) were inoculated with *Sinorhizobium meliloti* (Sm1021) for 2 weeks. Overexpression of miR156 increased nodulation in A11 and A17, but had no effect on nodule numbers in other genotypes (Figure 7a–c). Root length was also increased in A11a and A11 genotypes (Figure 7d,e). Moreover, longer root lengths were also found in other transgenic genotypes (not included in this manuscript) compared to control plants.

Discussion

MiR156 has been shown to affect biomass production and grain yield in plants (Chuck *et al.*, 2011; Fu *et al.*, 2012; Jiao *et al.*, 2010; Wang *et al.*, 2012). We thus anticipated that miR156 could be used as a tool to improve important yield-related traits in alfalfa. *In silico* analysis in our study revealed neither miR156 precursors nor homologues of *SPL* nor regulators of floral development in alfalfa databases. Next Generation sequence reads of alfalfa contained only short miR156-like sequences (approximately 60 bp) without the unique secondary structure which distinguishes miRNA. Consequently, we relied on sequence similarity between *M. truncatula* and alfalfa to isolate a precursor (MsmiR156d) from the latter species, and we used it to conduct functional characterization studies in alfalfa. 5′-RACE showed only three homologues *(SPL6, SPL12* and *SPL13)* to be targets of miR156 in alfalfa. MiR156 regulates 10 members of *SPL* in *A. thaliana* and 11 in *O. sativa* (Xie *et al.*, 2006; Xing *et al.*,

Table 1 Effects of miR156 on phenotypes of alfalfa

Plant genotype	Number of main branches	Number of lateral branches	Plant height (cm)	Internode length (mm)	Number of nodes	Flowering (Days after propagation)
Two months after propagation						
Control	2 ± 0.25	4 ± 1.08	32 ± 1.41	38 ± 5.07	24 ± 3.11	NF
A16	3 ± 0.65	7 ± 1.75	28 ± 1.11	25 ± 5.65*	25 ± 4.49	NF
A8a	3 ± 0.48	7 ± 1.25	28 ± 2.84	23 ± 2.4*	25 ± 5.15	NF
A8	4 ± 0.29	8 ± 0.5	31 ± 2.46	30 ± 5.52	34 ± 4.17	NF
A11	6 ± 0.75	19 ± 1.94*	30 ± 0.63	28 ± 1.03	52 ± 6.51*	NF
A11a	10 ± 1.68*	19 ± 1.49*	26 ± 0.82*	20 ± 2.58*	61 ± 2.69*	NF
A17	11 ± 1.62*	25 ± 1.89*	23 ± 1.25*	20 ± 0.71*	52 ± 3.61*	NF
Six months after propagation						
Control	23 ± 0.71	31 ± 4.15	56.75 ± 2.77	5.75 ± 0.25	14 ± 0.48	135
A16	23 ± 4.5	40 ± 1.87	55.75 ± 3.77	5.63 ± 0.43	17 ± 0.65	138
A8a	36 ± 5.22*	44 ± 7.95	63.25 ± 5.76	4.67 ± 0.33*	16 ± 0.5	139
A8	46 ± 3.03**	51 ± 2.92**	55 ± 5.96	4.88 ± 0.35	15 ± 0.48	140
A11	54 ± 10.69**	65 ± 10.2**	48.75 ± 5.54	3 ± 0.17**	20 ± 1.55**	180
A17	88 ± 9.01**	70 ± 7.07**	34 ± 4.3**	3 ± 0.22**	17 ± 1.94	>180
A11a	88 ± 10.31**	61 ± 12.2**	39.5 ± 2.96**	2.71 ± 0.32**	18 ± 1.29*	>180

NF, not flowering. Phenotypes were measured in two-month and six-month-old transgenic alfalfa. Values are mean ± SD, (n = 4). Asterisks indicate means (±SD) are significantly different from empty vector control plants at *P < 0.05 or **P < 0.01 (n = 4) using one-way ANOVA and a Dunnett's test.

Figure 5 Effects of miR156 on internode length and stem thickness of alfalfa. (a) Representative internode length. (b) Stem thickness phenotypes. (c) Quantitative measurement of stem thickness. Asterisks indicate means ± SD (n = 3) are significantly different from empty vector control plants at P < 0.05 (*) or < 0.01 (**)using one-way ANOVA and a Dunnett's test. All phenotypes were observed in six-month-old plants.

2010). With its tetraploid nature, it is highly probable that more members of miR156-targeted *SPL* or other genes will be found in alfalfa once the genome is fully sequenced.

Improvement of biomass yield and quality with miR156

For more than a century, researchers have been trying to control alfalfa flowering and improve alfalfa forage yield, but progress on these agronomic traits has been very limited. Ludwick (2000) reported that the alfalfa cultivar Mesa-Sirsa could yield 24 ton/acre provided that a combination of 230.42 kg N_2 fertilizer and 3.84 metres total water per acre are supplied. However, such investment would radically reduce the profit margin available to alfalfa growers, and the yield would still not come sufficiently close to high-yielding grasses like Miscanthus, which can recycle nitrogen back into roots and produce well in warmer climates. In fact, evaluation of the performance of the alfalfa cultivars released over the last five decades showed no substantial improvement in alfalfa forage yield (Lamb et al., 2006).

Here, we highlight the world's first demonstration that miR156 overexpression enhances the vegetative state and improves forage yield, while inhibiting floral transition in alfalfa. An increase in miR156 abundance of up to 840-fold in A16, A8a and A8 delayed flowering by 3–5 days, whereas a 1200-fold and higher increase in A11, A17 and A11a delayed flowering time by more than 45 days. Transcription of the flowering time regulatory gene *AP3* was down-regulated in these transgenic plants. Biomass was increased in most of our alfalfa genotypes, especially at the vegetative stage. While biomass of transgenic alfalfa was still high at 6 months, it was less significant than at the two-month stage.

Delayed flowering is highly advantageous in alfalfa for several reasons. First, it allows for greater flexibility in harvest time and higher forage yield. Second, it reduces the risk of cross-pollination between transgenic and/or nontransgenic plants, which is one of the most contentious issues for farmers (and environmentalists) and is an impediment to the widespread adoption of transgenic alfalfa. Third, forage quality will be improved with delayed flowering because undesirable lignin will be maintained at a lower level characteristic of the vegetative phase (Hatfield et al., 1994).

Schwarz et al. (2008) showed that overexpression of miR156 promoted the production of trichome density in Arabidopsis, but this trait was not affected when Arabidopsis miR156 was expressed in heterologous *B. napus* (Wei et al., 2010). Enhanced trichome density and trichome lengthening was also found in our transgenic alfalfa overexpressing MsmiR156. This provides an additional advantage to alfalfa as trichomes can offer a physical barrier against insect pests, which are a major problem for alfalfa cultivation (Summers, 1998).

In our study, we found higher levels of cellulose and lower levels of lignin in two MsmiR156 overexpressing genotypes, A17

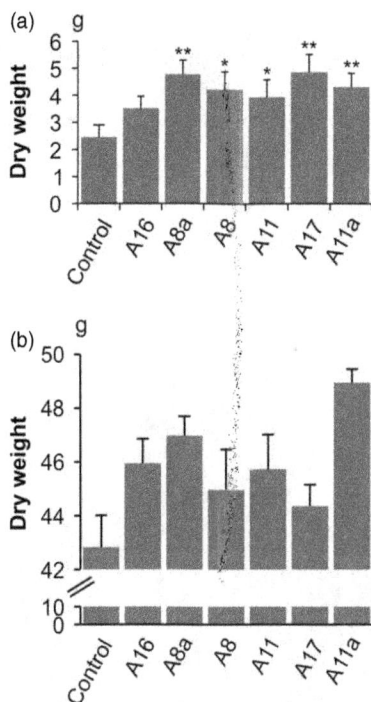

Figure 6 Effect of miR156 on forage yield of alfalfa. Biomass of above-ground tissues for (a) two-month-old and (b) six-month-old alfalfa genotypes. Asterisks indicate means ± SD (n = 4) are significantly different from empty vector control plants at P < 0.05 (*) or P < 0.01 (**) using one-way ANOVA and a Dunnett's test.

and A11a, where flowering is delayed by more than 90 days. Curiously, cellulose appears more resistant to enzyme saccharification than to acid-digestion in these plants. Lignin composition analysis showed that the ratio of syringyl to guaiacyl moieties (S/G), which influences sugar release in plants (Davison et al., 2005; Studer et al., 2011), is not significantly different among the different transgenic plants. We also observed that overexpression of miR156 improved cell wall pectin, which is considered an alternative to starch in high-energy feeds for dairy producers (Samac et al., 2004). Together, these studies show the value of miR156 to alfalfa plant breeding and that manipulation of flowering time (and potentially much larger improvements in forage yield and quality) using the miR156/*SPL* system is now within reach.

Auxin signalling plays an important role during root development (Overvoorde et al., 2010) and is regulated by a number of plant miRNAs (Rubio-Somoza and Weigel, 2011). For instance, networks of miR164 and target *NAM/ATAF/CUC 1* (*NAC1*) genes regulate lateral root initiation, while miR167 and its target auxin response factor 8 (ARF8) (Guo et al., 2005; Mallory et al., 2004) and miR393 and its target auxin signalling F-Box 3 (AFB3) affect the emergence and elongation of lateral roots (Vidal et al., 2010). MiRNAs have also been reported to influence root symbiosis with nitrogen fixing rhizobia. While some miRNAs promote root symbiosis, others interfere with this process. MiRNAs that regulate plant defence response genes enhance the infection of rhizobia and microrrhizal symbiosis and increase nodulation in roots (Devers et al., 2011; Jagadeeswaran et al., 2009; Li et al., 2010). Li et al. (2010) showed that regulation of the genes encoding GSK3-like protein kinase (a regulator of plant immunity) and the TIR-NBS-LRR encoding disease-resistant protein by miR482 increase the number of nodules in soya bean. In addition, Yan et al. (2013) recently demonstrated that overexpression of miR172 promoted nodulation, but increased expression of miR156 repressed the expression of miR172 which, in turn, controlled the expression of leghemoglobin through regulation of the AP2 transcription factor and reduced the number of nodules in transgenic soya bean.

In our study, we found that overexpression of miR156 increased root length, and this was coupled with the maintenance of nodulation in some transgenic alfalfa genotypes. Increased nodulation in A11 and A17 appears to be due to

Table 2 Analysis of forage quality of transgenic alfalfa

Plant genotype	Soluble sugar (mg/g DW)*	Structural sugar (mg/g CW)[†]	Starch (mg/g DW)	Pectin (%)	Lignin (% w/w DW)	Lignin composition (S/G)[‡]	Cellulose (% w/w DW)
Control	24.1 ± 4.3	253.5 ± 15.8	4 ± 1.2	14.8 ± 0.2	12.9 ± 0.1	0.62 ± 0.1	35.5 ± 0.1
A16	24.4 ± 3.7	254.5 ± 15.8	6 ± 1.6	15.1 ± 0.2	12.8 ± 0.2	0.60 ± 0.3	36.0 ± 0.3
A8a	20.8 ± 4.4	222.3 ± 13.3	4 ± 1.3	14.2 ± 0.2	12.7 ± 0.1	0.63 ± 0.1	33.6 ± 0.3
A8	25.2 ± 5.2	225.9 ± 19.1	5 ± 1.5	15.0 ± 0.1	12.7 ± 0.2	0.62 ± 0.1	33.5 ± 0.2
A11	21.1 ± 6.4	246.7 ± 9.3	5 ± 1.9	13.9 ± 0.1	12.9 ± 0.1	0.60 ± 0.2	32.6 ± 0.3
A17	21.4 ± 4.1	281.6 ± 15.0	5 ± 1.3	15.3 ± 0.2	12.0 ± 0.3*	0.65 ± 0.1	37.2 ± 0.4*
A11a	25.7 ± 7.9	217 ± 2.06*	5 ± 0.3	15.2 ± 0.2	11.7 ± 0.1*	0.63 ± 0.1	38.3 ± 0.4*

*Soluble sugars are averages of sum of sucrose, glucose, fructose and pinitol.

[†]Structural sugars include averages of sum of glucose, xylose, galactose, arabinose and mannose that were released from purified cell walls after 48 h of incubation with enzymes.

[‡]S: syringyl-like lignin structures/G: guaiacyl-like lignin structures.

Asterisks indicate means (±SD) are significantly different from empty vector control plants at P < 0.05 (*) (n = 4) using one-way ANOVA and a Dunnett's test.

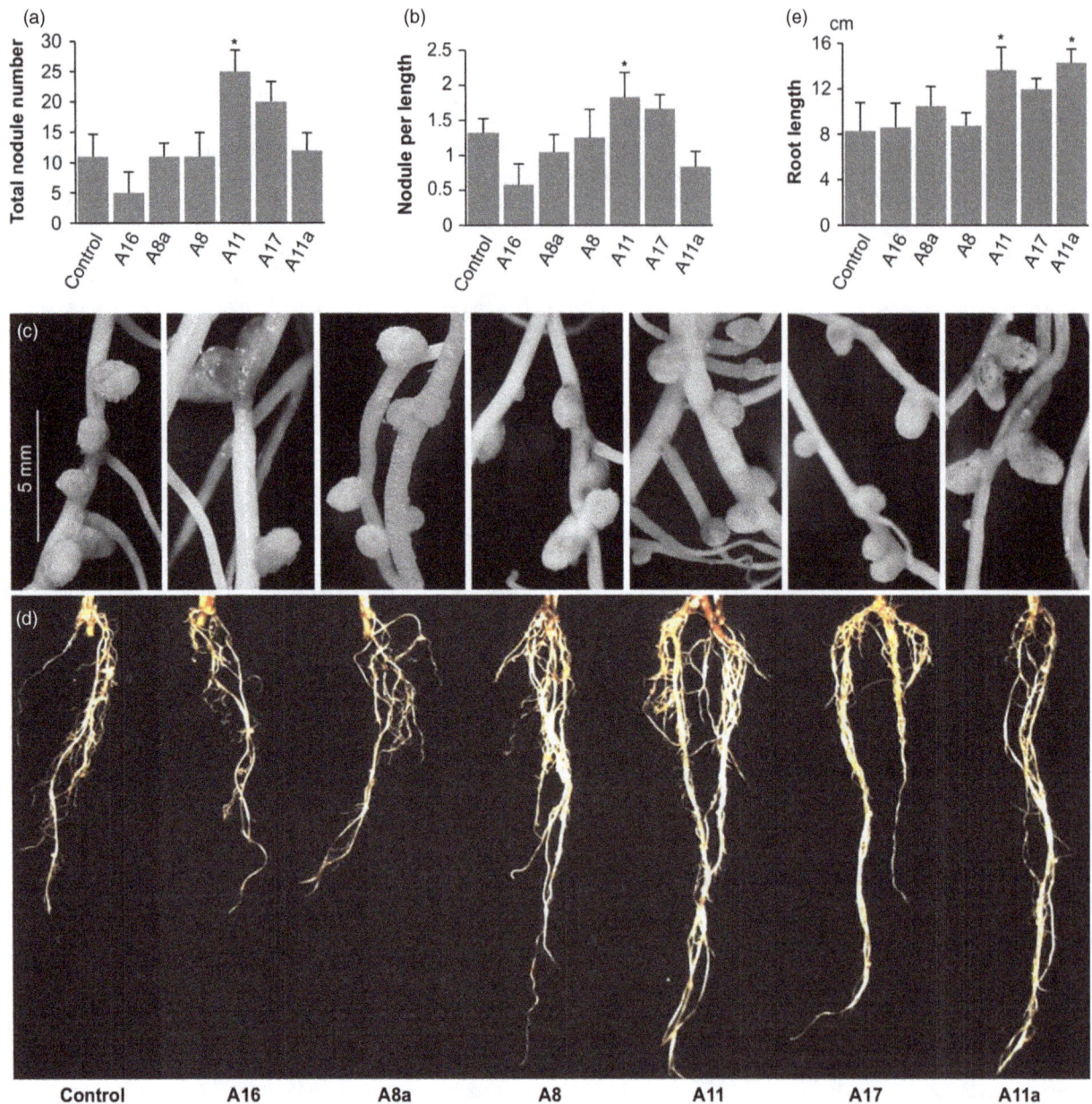

Figure 7 Effects of miR156 on root architecture and nodulation in alfalfa. (a) Total number of nodules in each transgenic alfalfa genotype. (b) Nodule number per root length. (c) Phenotypes of nodules. (d) Root phenotypes. (e) Quantitative measurement of root lengths. Asterisks indicate means $\pm$ SD ($n = 8$) are significantly different from empty vector control plants at $P < 0.05$ (*) using one-way ANOVA and a Dunnett's test.

enhanced nodulating efficiency as manifested by higher numbers of nodules per cm root length in these two genotypes. These transgenic plants should now be field tested to determine the impact of this improved trait in a range of soil types and the ability to tolerate drought, salinity and to efficiently use water, as roots in these plants are now able to penetrate deeper at least into loose soil. It is also important to explore the *SPL* genes and their downstream targets that are affected by miR156 to determine their individual capacity to affect traits, such as nodule function, that can impact on biomass yield.

SPL genes in alfalfa

Our study showed that *SPL6*, *SPL12* and *SPL13* are down-regulated by miR156-enhancement in alfalfa. Functions have

been described for miR156-regulated *SPL* genes in a range of species now, including *A. thaliana*, *G. max*, *O. sativa*, *Ricinus communis*, *Manihot escunlenta*, *Phaseolus vulgaris*, *Populus trichocarpa*, *Solanum lycopersicum*, *Pinus taeda*, *Physcomitrella patens* and *Vitisv inifera* (Sun, 2012). However, miR156 regulates different members of the *SPL* gene family in different plant species. For example, in Arabidopsis miR156-targeted *SPL6* is a novel positive regulator of nucleotide binding-leucine rich repeat (NB-LRR) receptor-mediated plant innate immunity (Padmanabhan *et al.*, 2013). *SPL13* is a negative regulator of emergence of vegetative leaves at the cotyledon stage in Arabidopsis (Martin *et al.*, 2010a,b). In addition to *SPL6* and *SPL13*, we showed that *SPL12* is a miR156 target in alfalfa, even though it was not predicted to be a target in Arabidopsis (Xing *et al.*, 2010). With the new tools of next generation sequencing available at this time, the transcriptome of these miR156-enhanced alfalfa plants can now be examined for other genes that are affected in the miR156 network of downstream genes. An anthocyanin bHLH-MYB-WD40 complex has already been shown to be a target of SPL9 in M. trunculata (Gou et al., 2011).

Conclusion

MiR156 regulates three SPL genes in alfalfa, and these SPL genes are expected to regulate their own set of cognate downstream genes, although we recognize that there may be some overlap in their function. Overexpression of miR156 in alfalfa must regulate a large network of genes in alfalfa, because numerous positive traits are affected, such as improved forage yield, delayed flowering, increased root length and increased trichome density. Some negative traits, such as reduced stem diameter, reduced leghemoglobin transcripts and excessive flowering delay are also observed, but it should be possible to select desirable traits from among the multitude of transgenic genotypes we have produced. Functional analysis of these SPLs and downstream genes will also enable the development of gene manipulation strategies and molecular markers to improve alfalfa traits by manipulating expression of only one or two genes to affect only positive traits while avoiding negative ones.

Experimental procedures

Plant materials and growth conditions

Wild type (WT) alfalfa (Medicago sativa) clone N4.4.2 (Badhan et al., 2014) was obtained from Dr. Daniel Brown (Agriculture and Agri-Food Canada) and was used in transformation assays and all other aspects of this work. WT and transgenic alfalfa plants were maintained in a greenhouse kept at 21–23 °C, 16 h light per day (halogen lights were applied after 18:00 h), light intensity of 380–450 W/m^2 (approximately 500 W/m^2 at high noontime) and a relative humidity (RH) of 70%.

In silico search for miR156 homologues in M. sativa databases

An in silico search was conducted using publicly available sequences and guidelines from Meyers et al. (2008). Briefly, miR156 precursor sequences from miRBase (http://www.mirbase.org/) were used to conduct a BLAST search against M. sativa and M. truncatula sequences in the NCBI database. Identified sequences that were similar to the precursors (expectation, e-value: 1e-34) were used to predict RNA secondary structure using miRTour (Milev et al., 2011).

Nucleic acid extraction and polymerase chain reaction

Genomic DNA was collected from alfalfa shoots using CTAB buffer [2% CTAB, 1.4 M NaCl, 20 mM ETDA, 100 mM Tris, pH 8.0]. To extract DNA, samples were ground in 1.5-mL tubes using disposable pestles. To the homogenized samples, 0.5 mL CTAB buffer and 0.5 mL chloroform were added and centrifuged at $900 \times g$ for 5 min. The DNA was precipitated with isopropanol and subsequently washed with 70% ethanol. The DNA pellet was then re-suspended in Milli-Q water and used directly for PCR.

Total RNA was isolated from shoots of alfalfa using Trizol reagent (Invitrogen, Burlington, ON, Canada). The cDNA was synthesized using SuperScript™III First-Strand Synthesis System (Invitrogen) as described in the instruction manual. PCR was

performed using either Phusion® High-Fidelity DNA Polymerase (New England Biolabs, Whitby, ON, Canada) as described in the instruction manuals.

Primers

Gene sequences of miR156, SPLs, AP2, AP3 and Leghemoglobin genes were retrieved from the sequence database of M. truncatula, the model legume species closely related to alfalfa. Primers used in this study were initially designed based on M. truncatula sequences (Table S1). Using these primers, the corresponding gene sequences were amplified from M. sativa cDNA by PCR, and then gene-specific primers (Table S1) were designed using primer BLAST (http://www.ncbi.nlm.nih.gov/tools/primer-blast/) and IDT (UNFOLD) website (https://www.idtdna.com/UNA Fold?) to ensure the primers were neither self-complementary nor produced secondary structures.

Isolation of alfalfa miR156 precursor

A precursor of miR156 (approximately 300 bp) was amplified by PCR from M. sativa cDNA template using two primers; Mt2139F11-156bF and Mt2139F11-156bR primers (Table S1) that were designed based on a miR156 precursor identified in a M. truncatula sequence database (Accession: CU019603). The PCR product was cloned, sequenced and compared to sequences of counterparts in other plant species using BLASTx.

Generation of vector construct and plant transformation

To generate an overexpression construct, the alfalfa miR156 (MsmiR15) precursor was cloned into pENTR/D-TOPO entry vector (Invitrogen). The cloned MsmiR156 was then subcloned into the pBINPLUS vector downstream of a CaMV35S (35S) promoter between the BamHI and SacI restriction sites (insert replaced the GUS gene in the vector). Expression clones were checked by colony PCR using a 35S promoter primer (forward) and gene-specific reverse primer (Table S1) and confirmed by sequencing. The expression clones were then transferred to Agrobacterium tumefaciens (LBA4404) competent cells using the freeze-thaw method. Finally, the A. tumefaciens strain harbouring MsmiR156 was used to generate MsmiR156 overexpression (MsmiR156-OE) alfalfa, whereas the A. tumefaciens strain harbouring empty vector (pBINPLUS) was used as an empty vector control.

Plant transformation was performed as described by Tian et al., 2002. The presence of the transgene in miR156-OE alfalfa genotypes was confirmed by PCR using 35S promoter primer (forward) and a gene-specific reverse primer (Mt2139F11-156bR) (Table S1). Transgenic plants transformed with an empty vector construct (pBINPLUS) were identified using 35S promoter and vector-specific (NOS-R2) primers (Table S1).

Detection of miR156 cleavage targets

The cleavage sites in alfalfa SPL genes were detected using 5′ rapid amplification of cDNA end (5′-RACE) as described by Llave et al., 2002. The experiment was conducted using FirstChoice® RLM-RACE Kit (Ambion, Burlington, ON, Canada) according to the manufacturer's instructions with slight modifications. PCR products with expected sizes (approximately 250 bp) from Inner 5′ RLM-RACE PCR were purified using gel purification kit (Qiagen, Toronto, ON, Canada) and cloned into a pJET1.2/blunt cloning vector (Fermentas, Ottawa, ON, Canada). Positive colonies were screened by digesting with HindIII and XhoI restriction enzymes. Finally, at least 20 positive clones for each SPL gene were subjected to sequencing using a pJET1.2/blunt sequencing primer.

Propagation of alfalfa by stem cuttings

Prior to vegetative propagation by rooted stem cuttings, alfalfa plants were cut back twice and allowed to grow for 2 months to synchronize growth and ensure all plant materials were at the same developmental stage. For each alfalfa genotype, at least four rooted cuttings (biological replicates) were used in experimental protocols. For each replicate, about 3–4 stem sections containing 3 nodes each were cut and inserted into moistened growing media (Pro-Mix®, Mycorrhizae™, Premier Horticulture Inc., Woodstock, ON, Canada) in a 5 plastic pot. The pots were covered with propagation domes (Ontario grower's supply, London, ON, Canada) and kept in the greenhouse for 3 weeks to allow rooting from the cut-stem. The plants were then thinned and only one rooted stem cutting was maintained in each pot and advanced for further characterization.

Morphological characterization of alfalfa

Transgenic plants were characterized at 2 months and 6 months growth after propagation. The phenotypes included in the characterization were number of main branches, number of lateral branches, plant height, internode length, number of nodes, stem thickness, trichome density, flowering time and dry weight. The branches directly emerging from the soil were considered as main branches while those that were on the main branches were counted as lateral branches. The longest stem in each biological replicate was used to measure plant height and the length of internode. In addition, the number of nodes in each replicate was also counted on the longest stem. To represent stem thickness in each replicate, the base of the stem was cut and the thickness (diameter) was measured under a dissecting microscope using 10-mm magnification. The appearance of trichomes in the leaves was determined under stereo microscope using 1-mm magnification. Phenotypes of transgenic plants were photographed using a digital camera (Canon, EOS 300D).

Cell wall residue preparation

Cell wall residue (CWR) was isolated from alfalfa as described by Brinkmann et al. (2002) using phosphate buffer and a Triton X-100 approach.

Determination of lignin content and lignin composition (S/G ratio)

Lignin content was analysed based on the thioglycolate-alkaline hydrolysis assay of Brinkmann et al. (2002) using 10 mg alfalfa CWR. Quantification was carried out using a UV/VIS spectrophotometer (Helios™ Zeta, Thermo Scientific, Madison, WI, USA) at 280 nm. A calibration curve was developed using commercial lignin (Sigma-Aldrich, Oakville, ON, Canada). The r^2 of the calibration curve was 0.9943 and the recovery of quality control (QC) samples was 95–110%. The relative standard deviation (RSD) was ≤2.0%.

Lignin composition (S/G ratio) was determined using the method of Foster et al. (2010). The analysis was performed using gas chromatography–mass spectrometry (GC-MS, a combination of GC 7890A and MS 5975C from Agilent Technologies).

Determination of total acid-releasable cellulosic glucose using anthrone

Cellulose content of CWR was measured as described by Updegraff (1969) using 2 mg alfalfa stem. A UV/VIS spectrophotometer (Helios™ Zeta, Thermo Scientific, Madison, WI, USA) was used for quantification purposes at 280 nm. A calibration curve was developed using cellulose (Sigma-Aldrich). The r^2 of the calibration curve was 0.9979, and the recovery of quality control (QC) samples was 93–106%. The relative standard deviation (RSD) was ≤2.1%.

Analysis of nonstructural carbohydrates

Removal of sugar and starch from alfalfa biomass was conducted based on the method of Theander et al. (1995). Approximately 200 mg of ground stem material was incubated in 7 mL of deionized H_2O at 100 °C for 90 min. Tubes were subsequently centrifuged for 10 min at 1500 × g and 500 µL of the supernatant was quantified for soluble carbohydrates (sucrose, glucose, pinitol and fructose) using HPLC and a Model 2410 refractive index detector (Waters, Milford, MA). Carbohydrates were separated on an Aminex HPX-87P column (300 mm × 7.8 mm × 9 µm) preceded by a Carbo-P precolumn (Bio-Rad, Mississauga, ON, Canada) and eluted isocratically at 80 °C, at a flow rate of 0.5 mL/min with deionized H_2O.

Tubes containing the remaining supernatant and pellet were subsequently vortexed, and starch was hydrolysed by adding 3 mL of digestion buffer (200 mM sodium acetate, pH 4.5) containing amyloglucosidase (15 U/mL; Sigma-Aldrich). Tubes were incubated 60 min at 55 °C and then centrifuged 10 min at 1500 × g. The supernatant was collected for quantification of glucose by HPLC, as described above. Starch was quantified by subtracting the amount of soluble glucose from the total amount of glucose measured following digestion with amyloglucosidase. Nonstructural carbohydrates (NSC) were estimated by the sum of soluble carbohydrates and starch. Pellets remaining after extraction of soluble carbohydrates and starch were washed three times with methanol at 60 °C and air dried to obtain cell wall (CW) preparations.

Analysis of pectin content

Pectin content was determined according to Filisetti-Cozzi and Carpita (1991) using 1.0 mg alfalfa CWR. Measurements were carried out using a UV/VIS spectrophotometer (Helios™ Zeta, Thermo Scientific, Madison, WI, USA) at 525 nm. A calibration curve was developed using apple pectin (BDH Chemicals). The r^2 of the calibration curve was 0.9984, and the recovery of quality control (QC) samples was 87–103%. The relative standard deviation (RSD) was ≤5.0%.

Analysis of structural carbohydrate composition using enzyme saccharification

Structural carbohydrates from CW cellulose and hemi-cellulose (i.e. glucose, xylose, galactose, arabinose, mannose) were quantified by HPLC following enzymatic saccharification as described by Selig et al., 2008; Duceppe et al., 2012.

Nodulation test

To minimize contamination, soil media (vermiculite mixed with sand) were autoclaved for 1.5 h at 120 °C and equipment used in the experiment were surface sterilized using 1% sodium hypochlorite. To test for nodulation, Sinorhizobium meliloti Sm1021 was cultured on TY medium containing tryptone, yeast extract, 10 mM $CaCl_2$ and agar for 2 days at 28 °C. A single colony was then inoculated in liquid TY medium and incubated at 28 °C to a cell density of 10^8 cells/mL. The cell culture was then pelleted and re-suspended in sterilized distilled water. One mL of the bacterial suspension was used to inoculate each pot (5") containing rooted alfalfa stems (2 week after propagation), and the pot was covered with propagation domes and kept in the greenhouse. As a negative control, 1 mL sterilized distilled water was used to inoculate rooted stems. Root lengths were measured, and the total numbers of nodules were counted at 2 weeks after inoculation with the *S. meliloti*. Nodule phenotypes were photographed under stereo microscope using 5-mm magnification.

Detection of mature miR156 abundance

The abundance of mature miR156 was determined using small RNA gel blots as described in Wang *et al.*, 2014. Briefly, a miR156 probe was synthesized using a mirVana™miRNA probe construction kit (Ambion) and labelled with Digoxigenin(DIG)-11-dUTP (Roche). Total RNA (15 µg) was separated on a 15% polyacrylamide gel and transferred to a nylon membrane. After incubating in Ultrahyb® hybridization buffer (Ambion), the membrane was hybridized with the probe for 19 h at 45 °C. The Roche DIG system was used for blocking, secondary antibody (Anti-Digoxygenin-AP) incubation and detection (CDP-Star) according to the manufacturer's instructions.

Quantitative real-time RT-PCR (qRT-PCR) analysis

The transcript levels of *miR156*, *SPLs*, *AP2*, *AP3* and *leghemoglobin* genes were analysed by quantitative real-time RT-PCR (qRT-PCR) using a CFX96 Touch™ Real-Time PCR Detection System (Bio-Rad) and the guidelines for the minimum information for publication of quantitative real-time PCR experiments (Bustin *et al.*, 2009; Taylor, 2010).

Total RNA was isolated from shoots, roots and nodules using PowerPlant®RNA Isolation Kit (MO BIO Laboratory, Mississauga, ON, Canada) as described in the instruction manual. The reverse transcription reaction was performed using 900 ng total RNA and qScript™ cDNA SuperMix (Quanta Bioscience, Mississauga, ON, Canada) following the manufacturer's instruction. The cDNA was diluted with sterilized distilled water (1 : 3), a total volume of 10 µL containing 0.2 µM of each forward and reverse primer, 1× PerfeCta SYBR Green FastMix (Quanta Biosciences, Canada), and 2 µL cDNA was used in each qRT-PCR. For each alfalfa plant genotype, at least three rooted cuttings (biological replicates) and three technical (assay) replicates were tested. The PCR was performed in two steps: 95 °C for 3 min followed by 44 cycles at 95 °C for 10 s and 58 °C for 30 s. Two reference genes (acetyl CoA carboxylase 1 and acetyl CoA carboxylase 2; Alexander *et al.*, 2007) were used to normalize the transcript levels in qRT-PCR. Finally, transcript levels of the respective genes were analysed using relative quantification by the comparative Ct ($2^{-\Delta\Delta CT}$) method (Livak and Schmittgen, 2001).

Statistical analysis

Minitab 17 was used to conduct one-way ANOVA. For each alfalfa plant genotype, at least four biological replicates (Four rooted stems per transgenic genotype) were used for morphological characterization and three technical (assay) replicates for molecular characterization. The differences of the means were tested using one-way ANOVA and a Dunnett's statistical test.

Acknowledgements

This project was funded by grants from the National Science and Engineering Research Council to AH and from Agriculture and Agri-Food Canada to MYG and AH. We thank Chenlong Li for help with small RNA blot and M. Udvari (Noble Foundation) for a kind gift of S. melilti.

References

Alexander, T.W., Reuter, T. and McAllister, T.A. (2007) Qualitative and quantitative polymerase chain reaction assays for an alfalfa (*Medicago sativa*)-specific reference gene to use in monitoring transgenic cultivars. *J. Agric. Food Chem.* **55**, 2918–2922.

Badhan, A., Jin, L., Wang, Y., Han, S., Kowalczys, K., Brown, D.C.W., Ayala, C.J., Latoszek-Green, M., Miki, B., Tsang, A. and McAllister, T. (2014) Expression of a fungal ferulic acid esterase in alfalfa modifies cell wall digestibility. *Biotechnol. Biofuels*, **12**, 663–673.

Barnes, D.K., Bingham, E.T., Murphy, R.P., Hunt, O.J., Beard, D.F., Skrdla, W.H. and Teuber, L.R. (1977) Alfalfa germplasm in the United States: genetic vulnerability, use, improvement, and maintenance. *Tech. Bull.*, **1571**, 1–21, USDA, Washington DC.

Bartel, D.P. (2004) MicroRNAs: genomics, biogenesis, mechanism, and function. *Cell*, **116**, 281–297.

Bhogale, S., Mahajan, A.S., Natarajan, B., Rajabhoj, M., Thulasiram, H.V. and Banerjee, A.K. (2014) MicroRNA156: a potential graft-transmissible microRNA that modulates plant architecture and tuberization in *Solanum tuberosum* ssp. andigena. *Plant Physiol.* **164**, 1011–1027.

Brinkmann, K., Blaschke, L. and Polle, A. (2002) Comparison of different methods for lignin determination as a basis for calibration of near-infrared reflectance spectroscopy and implications of lignoproteins. *J. Chem. Ecol.* **28**, 2483–2501.

Bustin, S.A., Benes, V., Garson, J.A., Hellemans, J., Huggett, J., Kubista, M., Mueller, R., Nolan, T., Pfaffl, M.W., Shipley, G.L., Vandesompele, J. and Wittwer, C.T. (2009) The MIQE Guidelines: minimum information for publication of quantitative Real-Time PCR experiments. *Clin. Chem.* **4**, 1–12.

Cardon, G.H., Hohmann, S., Klein, J., Nettesheim, K., Saedler, H. and Huijser, P. (1999) Molecular characterisation of the Arabidopsis SBP-box genes. *Gene*, **237**, 91–104.

Chuck, G.M., Tobias, C., Sun, T., Kraemer, F., Li, C., Dibble, D., Arora, R., Bragg, J.N., Vogel, J.P. and Singh, S. (2011) Overexpression of the maize *Corngrass 1* microRNA prevents flowering: improves digestibility, and increases starch content of switchgrass. *Proc. Natl Acad. Sci. USA*, **108**, 17550–17555.

Davison, B.H., Drescher, S.R., Tuskan, G.A., Davis, M.F. and Nghiem, N.P. (2005) Variation of S/G ratio and lignin content in a *Populus* family influences the release of xylose by dilute acid hydrolysis. *Appl. Biochem. Biotechnol.* **427**, 129–132.

Devers, E.A., Branscheid, A., May, P. and Krajinski, F. (2011) Stars and symbiosis: microRNA- and microRNA-mediated transcript cleavage involved in arbuscular mycorrhizal symbiosis. *Plant Physiol.* **156**, 1990–2010.

Duceppe, M.-O., Bertrand, A., Pattathil, S., Miller, J., Castonguay, Y., Hahn, M.G., Michaud, R. and Dubé, M.-P. (2012) Assessment of genetic variability of cell wall degradability for the selection of alfalfa with improved saccharification efficiency. *Bioenergy Res.* **4**, 1–11.

Filisetti-Cozzi, T.M.C.C. and Carpita, N.C. (1991) Measurement of uranic scids without interference from neutral dugars. *Anal. Biochem.* **197**, 157–162.

Foster, C.E., Martin, T.M. and Pauly, M. (2010) Comprehensive compositional analysis of plant cell walls (lignocellulosic biomass) Part I: lignin. *J. Vis. Exp.* **37**, 1–4.

Fu, C., Sunkar, R., Zhou, C., Shen, H., Zhang, J.Y., Matts, J., Wolf, J., Mann, D.G., Stewart, C.N. Jr, Tang, Y. and Wang, Z.Y. (2012) Overexpression of miR156 in switchgrass (*Panicum virgatum* L.) results in various morphological alterations and leads to improved biomass production. *Plant Biotechnol. J.* **10**, 443–452.

Gou, J.Y., Felippes, F.F., Liu, C.J., Weigel, D. and Wang, J.W. (2011) Negative regulation of anthocyanin biosynthesis in Arabidopsis by a miR156-targeted SPL transcription factor. *Plant Cell*, **23**, 1512–1522.

Guo, H.S., Xie, Q., Fei, J.F. and Chua, N.H. (2005) MicroRNA directs mRNA cleavage of the transcription factor NAC1 to downregulate auxin signals for Arabidopsis lateral root development. *Plant Cell*, **17**, 1376–1386.

Hatfield, R.D., Jung, H.-J.G., Ralph, J., Buxton, D.R. and Weimer, P.J. (1994) A comparison of the insoluble residues produced by the klason lignin and acid detergent lignin procedures. *J. Sci. Food Agric.* **65**, 51–58.

Jagadeeswaran, G., Zheng, Y., Li, Y.F., Shukla, L.I., Matts, J., Hoyt, P., Macmil, S.L., Wiley, G.B., Roe, B.A., Zhang, W. and Sunkar, R. (2009) Cloning and characterization of small RNAs from *Medicago truncatula* reveals four novel legume-specific microRNA families. *New Phytol.*, **184**, 85–98.

Jiao, Y., Wang, Y., Xue, D., Wang, J., Yan, M., Liu, G., Dong, G., Zeng, D., Lu, Z. and Zhu, X. (2010) Regulation of OsSPL14 by osmiR156 defines ideal plant architecture in rice. *Nat. Genet.* **42**, 541–544.

Jones-Rhoades, M.W. (2012) Conservation and divergence in plant microRNAs. *Plant Mol. Biol.* **80**, 3–16.

Kim, M.Y., Kang, Y.J., Lee, T. and Lee, S.-H. (2013) Divergence of flowering-related genes in three legume species. *Plant Genome*, **6**, 1–12.

Lamb, J.F.S., Sheaffer, C.C., Rhodes, L.H., Mark Sulc, R., Underander, D.J. and Brummer, E.C. (2006) Five decades of alfalfa cultivar improvement: impact on forage yield, persistence, and nutritive value. *Crop Sci.*, **46**, 902–909.

Lesins, K.A. and Lesins, I. (1979) *Genus Medicago* (Leguminosae): A taxogenetic study. Springer, Netherlands, ISBN-**10**, 94009-96365.

Li, H., Deng, Y., Wu, T., Subramanian, S. and Yu, O. (2010) Misexpression of miR482, miR1512, and miR1515 increases soybean nodulation. *Plant Physiol.* **153**, 1759–1770.

Livak, K.J. and Schmittgen, T.D. (2001) Analysis of relative gene expression data using real-time quantitative PCR and the $2^{-\Delta\Delta CT}$ method. *Methods*, **25**, 402–408.

Llave, C., Xie, Z., Kasschau, K.D. and Carrington, J.C. (2002) Cleavage of scarecrow-like mRNA targets directed by a class of Arabidopsis miRNA. *Science*, **297**, 2053–2056.

Ludwick, A.E. (2000) High yield alfalfa: 24 tons irrigated..12 tons non-irrigated. *Better Crops*, **84**, 18–19.

Macovei, A., Gill, S.S. and Tuteja, N. (2012) microRNAs as promising tools for improving stress tolerance in rice. *Plant Signal Behav.* **7**, 1296–1301.

Mallory, A.C., Dugas, D.V., Bartel, D.P. and Bartel, B. (2004) MicroRNA regulation of NAC-domain targets is required for proper formation and separation of adjacent embryonic, vegetative, and floral organs. *Curr. Biol.* **14**, 1035–1046.

Martin, R.C., Asahina, M., Liu, P.P., Kristof, J.R., Coppersmith, J.L. and Pluskota, W.E. (2010a) The regulation of post-germinative transition from the cotyledon- to vegetative-leaf stages by microRNA-targeted SQUAMOSA PROMOTER-BINDING PROTEINLIKE13 in Arabidopsis. *Seed Sci. Res.* **20**, 89–96.

Martin, R.C., Asahina, M., Liu, P.P., Kristof, J.R., Coppersmith, J.L., Pluskota, W.E., Bassel, G.W., Goloviznina, N.A., Nguyen, T.T., Martinez-Andujar, C., Kumar, M.B.A., Pupel, P. and Nonogaki, H. (2010b) The microRNA156 and microRNA172 gene regulation cascades at post-germinative stages in Arabidopsis. *Seed Sci. Res.* **20**, 79–87.

Meyers, B.C., Axtell, M.J., Bartel, B., Bartel, D.P., Baulcombe, D., Bowman, J.L., Cao, X., Carrington, J.C., Chen, X. and Green, P.J. (2008) Criteria for annotation of plant microRNAs. *Plant Cell*, **20**, 3186–3190.

Milev, I., Yahubyan, G., Minkov, I. and Baev, V. (2011) miRTour: plant miRNA and target prediction tool. *Bioinformation*, **6**, 248–249.

Overvoorde, P., Fukaki, H. and Beeckman, T. (2010) Auxin control of root development. *Cold Spring Harb. Perspect. Biol.* **2**, a001537.

Padmanabhan, M.S., Ma, S., Burch-Smith, T.M., Czymmek, K., Huijser, P. and Dinesh-Kumar, S.P. (2013) Novel positive regulatory role for the *SPL6* transcription factor in the N TIR-NB-LRR receptor-mediated plant innate immunity. *PLoS Pathog.* **9**, e1003235.

Putnam, D., Russelle, M., Orloff, S., Kuhn, J., Fitzhugh, L., Godfrey, L., Kiess, A. and Long, R. (2001) Alfalfa, wildlife and the environment. In: *The Importance and Benefit of Alfalfa in the 21st Century*, pp. 1–24, California Alfalfa and Forage Association.

Rubio-Somoza, I. and Weigel, D. (2011) MicroRNA networks and developmental plasticity in plants. *Trends Plant Sci.* **864**, 1–7.

Samac, D.A., Litterer, L., Temple, G., Jung, H.-J.G. and Somers, D.A. (2004) Expression of UDP-Glucose dehydrogenase reduces cell-wall polysaccharide concentration and increases xylose content in alfalfa stems. *Appl. Biochem. Biotechnol.* **1167**, 113–116.

Schwarz, S., Grande, A.V., Bujdoso, N., Saedler, H. and Huijser, P. (2008) The microRNA regulated SBP-box genes *SPL9* and *SPL15* control shoot maturation in Arabidopsis. *Plant Mol. Biol.* **67**, 183–195.

Selig, M., Weiss, N. and Ji, Y. (2008) Enzymatic saccharification of lignocellulosic biomass. In: *Laboratory Analytical Procedure* (NREL, ed.), pp. 1–5. Golden: Midwest Research Institute.

Shikata, M., Koyama, T., Mitsuda, N. and Ohme-Takagi, M. (2009) Arabidopsis SBP-box genes *SPL10*, *SPL11* and *SPL2* control morphological change in association with shoot maturation in the reproductive phase. *Plant Cell Physiol.* **50**, 2133–2145.

Studer, M.H., DeMartini, J.D., Davis, M.F., Sykes, R.W., Davison, B., Keller, M., Tuskan, G.A. and Wyman, C.E. (2011) Lignin content in natural *Populus* variants affects sugar release. *Proc. Natl Acad. Sci. USA*, **10**, 1–6.

Summers, C.G. (1998) Integrated pest management in forage alfalfa. *Integrated Pest Manag. Rev.* **3**, 127–154.

Sun, G. (2012) MicroRNAs and their diverse functions in plants. *Plant Mol. Biol.* **80**, 17–36.

Sunkar, R., Li, Y.-F. and Jagadeeswaran, G. (2012) Functions of microRNAs in plant stress responses. *Cell*, **17**, 360–1385.

Taylor, S. (2010) A practical approach to RT-qPCR-publishing data that conform to the MIQE guidelines. *Methods*, **50**, S1–S5.

Theander, O., Aman, P., Westerlund, E., Andersson, R. and Petersson, D. (1995) Total dietary fiber determined as neutral sugar residues, and Klason lignin (The Uppsala Method): collaborative study. *J. AOAC Int.* **78**, 1030–1044.

Tian, L., Wang, H., Wu, K., Latoszek-Green, M., Hu, M., Miki, B. and Brown, D.C.W. (2002) Efficient recovery of transgenic plants through organogenesis and embryogenesis using a cryptic promoter to drive marker gene expression. *Plant Cell Rep.* **20**, 1181–1187.

Updegraff, D.M. (1969) Semimicro determination of cellulose in biological materials. *Anal. Biochem.* **3**, 420–424.

Vidal, E.A., Araus, V., Lu, C., Parry, G., Green, P.J., Coruzzi, G.M. and Guiterrez, R.A. (2010) Nitrate-responsive miR393/AFB3 regulatory module controls root system architecture in *Arabidopsis thaliana*. *Proc. Natl Acad. Sci. USA*, **107**, 4477–4482.

Voinnet, O. (2009) Origin, biogenesis, and activity of plant microRNAs. *Cell*, **136**, 669–687.

Volenec, J.J., Cunningham, S.M., Haagenson, D.M., Berg, W.K., Joern, B.C. and Wiersma, D.W. (2002) Physiological genetics of alfalfa improvement: past failures, future prospects. *Field. Crop. Res.* **75**, 97–110.

Wang, S., Wu, K., Yuan, Q., Liu, X., Liu, Z., Lin, X., Zeng, R., Zhu, H., Dong, G. and Qian, Q. (2012) Control of grain size, shape and quality by ssSPL16 in rice. *Nat. Genet.* **44**, 950–954.

Wang, Y., Wang, Z., Amyot, L., Tian, L., Xu, Z., Gruber, M.Y. and Hannoufa, A. (2014) Ectopic expression of miR156 represses nodulation and causes morphological and developmental changes in Lotus japonicus. *Mol. Genet. Genomics*, DOI 10.1007/s00438-014-0931-4.

Wei, S., Yu, B., Gruber, M.Y., Khachatourians, G.G., Hegedus, D.D. and Hannoufa, A. (2010) Enhanced seed carotenoid levels and branching in transgenic *Brassica napus* expressing the Arabidopsis miR156b gene. *J. Agric. Food Chem.* **58**, 9572–9578.

Wei, S., Gruber, M.Y., Yu, B., Gao, M.-J., Khachatourians, G.G., Hegedus, D.D., Parkin, I.A. and Hannoufa, A. (2012) Arabidopsis mutant *sk156* reveals complex regulation of *SPL15* in a miR156-controlled gene network. *BMC Plant Biol.* **18**, 169.

Wu, G. and Poethig, R.S. (2006) Temporal regulation of shoot development in *Arabidopsis thaliana* by miR156 and its target *SPL3*. *Development*, **133**, 3539–3547.

Xie, K., Wu, C. and Xiong, L. (2006) Genomic organization, differential expression, and interaction of SQUAMOSA PROMOTER-BINDING-LIKE transcription factors and microRNA156 in rice. *Plant Physiol.* **142**, 280–293.

Xing, S.S.M., Hohmann, S., Berndtgen, R. and Huijser, P. (2010) miR156-targeted and nontargeted SBP-Box transcription factors act in concert to secure male fertility in Arabidopsis. *Plant Cell*, **22**, 3935–3950.

Yan, Z., Hossain, M.S., Wang, J., Valdés-López, O., Liang, Y., Libault, M., Qiu, L. and Stacey, G. (2013) miR172 regulates soybean nodulation. *Mol. Plant Microbe Interact.* **12**, 1371–1377.

PERMISSIONS

LIST OF CONTRIBUTORS

William M. Ainley, Lakshmi Sastry-Dent, Mary E. Welte, Michael G. Murray, David R. Corbin, Rebecca R. Miles, Nicole L. Arnold, Tonya L. Strange, Matthew A. Simpson, Zehui Cao, Carley Carroll, Katherine S. Pawelczak, Ryan Blue, Kim West, Lynn M. Rowland, Douglas Perkins, Pon Samuel, Cristie M. Dewes, Liu Shen, Shreedharan Sriram, Steven L. Evans, Steven R. Webb and Joseph F. Petolino
Dow Agrosciences Llc, Indianapolis, IN, USA

Bryan Zeitler, Rainier Amora, Edward J. Rebar, Lei Zhang, Phillip D.Gregory and Fyodor D. Urnov
Sangamo BioSciences, Inc., Richmond, CA, USA

Hatice Boke, Esma Ozhuner, Mine Turktas and Turgay Unver
Department of Biology, Faculty of Science, Cankiri Karatekin University, Cankiri, Turkey

Iskender Parmaksiz
Department of Molecular Biology and Genetics, Faculty of Science, Gaziosmanpasa University, Tokat, Turkey

Sebahattin Ozcan
Department of Field Crops, Faculty of Agriculture, Ankara University, Ankara, Turkey

Ozgur Cakir
Department of Molecular Biology and Genetics, Faculty of Science, Istanbul University, Istanbul, Turkey
Department of Biology, East Carolina University, Greenville, NC, USA

Bilgin Candar-Cakir
Program of Molecular Biology and Genetics, Institute of Science, Istanbul University, Istanbul, Turkey
Department of Biology, East Carolina University, Greenville, NC, USA

Baohong Zhang
Department of Biology, East Carolina University, Greenville, NC, USA

Zhulong Chan, Patrick J. Bigelow, Wayne Loescher and Rebecca Grumet
Plant Breeding, Genetics and Biotechnology Program and Department of Horticulture, Plant and Soil Sciences Building, Michigan State University, East Lansing MI, USA

Hui-Ting Chan and Yi-Yin Do
Department of Horticulture and Landscape Architecture, National Taiwan University, Taiwan, Republic of China

Pung-Ling Huang
Department of Horticulture and Landscape Architecture, National Taiwan University, Taiwan, Republic of China
Graduate Institute of Biotechnology, Chinese Culture University, Taiwan, Republic of China

Min-Yuan Chia, Victor Fei Pang and Chian-Ren Jeng
Graduate Institute of Veterinary Medicine, National Taiwan University, Taiwan, Republic of China

Guang Chen, Huimin Feng, Qingdi Hu, Hongye Qu, Aiqun Chen, Ling Yu and Guohua Xu
State Key Laboratory of Crop Genetics and Germplasm Enhancement, MOA Key Laboratory of Plant Nutrition and Fertilization in Lower-Middle Reaches of the Yangtze River, Nanjing Agricultural University, Nanjing, China

Xiaoli Guo, Carola M. De La Torre, John Smeda and Melissa G. Mitchum
Division of Plant Sciences and Bond Life Sciences Center, University of Missouri, Columbia, MO, USA

Demosthenis Chronis
Robert W. Holley Center for Agriculture and Health, US Department of Agriculture, Agricultural Research Service, Ithaca, NY, USA

Xiaohong Wang
Robert W. Holley Center for Agriculture and Health, US Department of Agriculture, Agricultural Research Service, Ithaca, NY, USA
Department of Plant Pathology and Plant-Microbe Biology, Cornell University, Ithaca, NY, USA

Ling Li and Eve Syrkin Wurtele
Department of Genetics, Development and Cell Biology, Iowa State University, Ames, IA, USA

Qing Liu, Anna El Tahchy, Madeline Mitchell, Zhongyi Li, Pushkar Shrestha, Thomas Vanhercke, Jean-Philippe Ral, Ming-Bo Wang, Rosemary White, Philip Larkin, Surinder Singh and James Petrie
Commonwealth Scientific and Industrial Research Organisation Agriculture, Black Mountain, Act, Australia

Qigao Guo
Commonwealth Scientific and Industrial Research Organisation Agriculture, Black Mountain, Act, Australia
College of Horticulture and Landscape Architecture, Southwest University, Chongqing, China

Sehrish Akbar
Commonwealth Scientific and Industrial Research Organisation Agriculture, Black Mountain, Act, Australia
National University of Science and Technology (NUST) Islamabad, Islamabad, Pakistan

Yao Zhi
Commonwealth Scientific and Industrial Research Organisation Agriculture, Black Mountain, Act, Australia
State Key Laboratory of Agricultural Microbiology, Huazhong Agricultural University, Wuhan, China

Guolu Liang
College of Horticulture and Landscape Architecture, Southwest University, Chongqing, China

Dani Satyawan
Department of Plant Science and Research Institute of Agriculture and Life Sciences, Seoul National University, Seoul, Korea
Indonesian Center for Agricultural Biotechnology and Genetic Resources Research and Development, Bogor, Indonesia

Moon Young Kim and Suk-Ha Lee
Department of Plant Science and Research Institute of Agriculture and Life Sciences, Seoul National University, Seoul, Korea
Plant Genomics and Breeding Institute, Seoul National University, Seoul, Korea

Jinrui Shi, Huirong Gao, Hongyu Wang, H. Renee Lafitte, Rayeann L. Archibald, Meizhu Yang, Salim M. Hakimi, Hua Mo and Jeffrey E. Habben
DuPont Pioneer, Johnston, IA, USA

Yali Wang, Feijie Wu, Jinjuan Bai and Yuke He
National Key Laboratory of Plant Molecular Genetics, Shanghai Institute of Plant Physiology and Ecology, Shanghai Institutes for Biological Sciences, Chinese Academy of Sciences, Shanghai, China

Neele Wendler, Martin Mascher, Axel Himmelbach, Uwe Scholz and Nils Stein
Leibniz Institute of Plant Genetics and Crop Plant Research (IPK), Gatersleben, Germany

Christiane Nöh and Brigitte Ruge-Wehling
Julius Kühn-Institut (Jki), Institute for Breeding Research on Agricultural Crops, Groß Lüsewitz, Germany

Hong Zhai, Feibing Wang, Zengzhi Si, Jinxi Huo, Lei Xing, Yanyan An, Shaozhen He and Qingchang Liu
Beijing Key Laboratory of Crop Genetic Improvement/ Laboratory of Crop Heterosis and Utilization, Ministry of Education, China Agricultural University, Beijing, China

Chong Zhang, Hua Chen, Tiecheng Cai, Ye Deng and Weijian Zhuang
College of Plant Protection, Fujian Agriculture and Forestry University, Fuzhou, China
Fujian Key Laboratory of Crop Molecular and Cell Biology, Fujian Agriculture and Forestry University, Fuzhou, Fujian, China

Ruirong Zhuang, Ning Zhang and Yuanhuan Zeng
Fujian Key Laboratory of Crop Molecular and Cell Biology, Fujian Agriculture and Forestry University, Fuzhou, Fujian, China

Yixiong Zheng
Fujian Key Laboratory of Crop Molecular and Cell Biology, Fujian Agriculture and Forestry University, Fuzhou, Fujian, China
College of Agronomy, Zhongkai Agriculture and Engineering College, Guangzhou, Guangdong, China

Ronghua Tang
Cash Crops Research Institute, Guangxi Academy of Agricultural Sciences, Nanning, China

Ronglong Pan
Department of Life Science and Institute of Bioinformatics and Structural Biology, College of Life Science, National Tsing Hua University, Hsinchu, Taiwan

Monika Bielecka, Filip Kaminski, Ian Adams, Helen Poulson, Yi Li, Tony R. Larson, Thilo Winzer and Ian A. Graham
Centre for Novel Agricultural Products, Department of Biology, University of York, York, UK

Raymond Sloan
Biorenewables Development Centre, The Biocentre, York Science Park, Heslington, York, UK

Rehna Augustine and Naveen C. Bisht
National Institute of Plant Genome Research (NIPGR), New Delhi, India

Arundhati Mukhopadhyay
Department of Genetics, Center for Genetic Manipulation of Crop Plants (CGMCP), University of Delhi South Campus, New Delhi, India

Lisa Amyot and Khaled Omari
Agriculture and Agri-Food Canada, London, ON, Canada

Banyar Aung and Abdelali Hannoufa
Agriculture and Agri-Food Canada, London, ON, Canada

Biology Department, Western University, London, ON, Canada

Margaret Y. Gruber
Agriculture and Agri-Food Canada, Saskatoon, SK, Canada

Annick Bertrand
Agriculture and Agri-Food Canada, St. Foy, QC, Canada

Index

www.ingramcontent.com/pod-product-compliance
Lightning Source LLC
Chambersburg PA
CBHW080644200326
41458CB00013B/4728